D1615301

ANDERSONIAN LIBRARY
WITHDRAWN
FROM
LIBRARY
STOCK
UNIVERSITY OF STRATHCLYDE

McGraw-Hill Electronic Testing Handbook

Procedures and Techniques

John D. Lenk

McGraw-Hill, Inc.

New York San Francisco Washington, D.C. Auckland Bogotá
Caracas Lisbon London Madrid Mexico City Milan
Montreal New Delhi San Juan Singapore
Sydney Tokyo Toronto

Library of Congress Cataloging-in-Publication Data

Lenk, John D.
McGraw-Hill electronic testing handbook : procedures and techniques / by John D. Lenk.
p. cm.
Includes index.
ISBN 0-07-037602-6 (H)
1. Electronic instruments—Handbooks, manuals, etc. 2. Electronic apparatus and appliances—Testing—Handbooks, manuals, etc. I. Title. II. Title: Electronic testing handbook.
TK7878.4.L4514 1993
621.38'028'7—dc20 93-19595
CIP

1 2 3 4 5 6 7 8 9 0 DOH/DOH 9 9 8 7 6 5 4 3

ISBN 0-07-037602-6

The sponsoring editor for this book was Daniel A. Gonneau, the editor was Andrew Yoder, the designer was Jaclyn J. Boone, and the production supervisor was Katherine G. Brown. This book was set in ITC Century Light. It was composed by TAB Books.

Printed and bound by R. R. Donnelley & Sons Company.

For more information on other McGraw-Hill materials, in the United States please call 1-800-822-8158. In other countries, contact your local McGraw-Hill representative.

MH94

Dedication

Greetings from the Villa Buttercup!
To my wonderful wife Irene,
thank you for being by my side all these years!
To my lovely family, Karen, Tom, Brandon, and Justin.
And to our Lambie and Suzzie, be happy wherever you are!
To my special readers, may good fortune
find your doorways to good health and happy things.
Thank you for buying my books!
And a special thanks to Dan Gonneau, Stephen Fitzgerald,
and Robert McGraw of McGraw-Hill/TAB
for making me an international bestseller again!
This is book number 77.
Abundance!

Contents

Acknowledgments

Many professionals have contributed to this book. I gratefully acknowledge that the tremendous effort needed to make this book such a comprehensive work is impossible for one person and wish to thank all who contributed, both directly and indirectly.

I wish to give special thanks to the following: Joe Cagle and Rinaldo Swayne of Alpine/Luxman; Bob Carlson and Martin Plude of B&K-Precision Dynascan Corporation; Syd Coppersmith of Dallas Semiconductor; Kellie Garcia and Rosie Hinojosa of EXAR Corporation; Jeff Salter of GEC Plessey; Linda daCosta of Harris Semiconductor; Dick Harmon, Ross Snyder, Nancy Teater, Mike Arnold, Owen Brown, Glen Green, Mike Ryman, Michael Slater, Barry Bronson, and Joel Salsberg of Hewlett-Packard; Tom Roscoe, Dennis Yuoka, and Terrance Miller of Hitachi; Ron Denchfield of Linear Technology Corporation; David Fullagar of Maxim; Fred Swymer of Microsemi Corporation; Linda Capcara of Motorola Inc.; Andrew Jenkins of National Semiconductor Corporation; Antonio Ortiz of Optical Electronics Incorporated; Pat Wilson and Ray Krenzer of Philips Consumer Electronics; Thomas Lauterback of Quasar; Lorraine Jenkins of Raytheon Company Semiconductor Division; Ed Oxner and Robert Decker of Siliconix Incorporated; Donald Woolhouse of Sanyo; Theodore Zrebiec of Sony; Amy Sullivan of Texas Instruments; J. W. Phipps of Thomson Consumer Electronics (RCA); Alan Campbell of Unitrode Corporation; John Taylor and Matthew Mirapaul of Zenith.

I also wish to thank Joseph A. Labok of Los Angeles Valley College for help and encouragement throughout the years.

And a very special thanks to Daniel Gonneau, Editor-in-Chief, Steve Fitzgerald, Robert McGraw, Nancy Young, Bert Peterson, Barbara McCann, Kimberly Martin, Kriss Helman, Susan Wahlman, Jane Schmidt, Robert Ostrander, JoAnn Bishop, Andrew Yoder, Roland Phelps, Sandra Johnson-Bottomley, Wayne Smith, Charles Love, Peggy Lamb, Thomas Kowalczyk, Suzanne Babeuf, Judith Reiss,

Nancy Rosenblum, Kathy Green, Jodi Tyler, Jeanne Glassner, Stacey Spurlock and Tracy Baer of the McGraw-Hill/TAB organization for having that much confidence in the author. I recognize that all books are a team effort and am thankful that I am working with the First Team!

And to my wife Irene, my research analyst and agent, I wish to extend my thanks. Without her help, this book could not have been written.

Introduction

This is a dual-purpose book that can be put to immediate use by anyone working in electronics. As the title implies, the book fills two basic needs by describing both the operation and the applications for electronic test equipment.

All electronic technicians and engineers use some form of test equipment. Most technicians use many types of test equipment in their daily work. It is therefore essential that technicians know what test instruments are available, what the equipment does, and how the instruments operate. The first half of this book fills that specific need.

The second half of the book is devoted to step-by-step procedures for the testing of electronic devices, circuits, and components. The book not only tells how to perform the test, but describes what is being tested and why the test is required.

This combination of theory plus applications makes the book suitable as a reference text for student technicians, hobbyists, and experimenters, and as a guidebook for experienced working technicians and field-service engineers.

No single book can list all items of test equipment available in today's electronic field, much less describe such equipment. However, this book describes both the purpose and operating principles that are related to the most common types of test equipment. By concentrating on such test instruments, and on generalized circuits, you are led through the maze of information so often found in test-equipment catalogs. The book therefore serves as a guide or refresher for the experienced technician, and as a basic text for the student.

The most commonly used test instruments in both shop and lab are meters, scopes, generators, and counters. For that reason, the basic operating procedures for these instruments are included in applicable chapters. The operating procedures are typical, and can be used as a supplement to the procedures found in the instruction manual for the test instrument. The operating procedures included here are especially useful when no instruction manual is available.

Where practical throughout the second half of the book, three sets of procedures are given for the component/circuit tests.

The first set of procedures describes tests that can be performed with elementary test equipment, such as meters. These procedures are especially useful for the home experimenter and hobbyist.

The second set of procedures describes the same tests using more advanced equipment (such as scopes). These procedures are primarily for the advanced student and working technician.

The third set of procedures covers the same ground, but using even more specialized and sophisticated test equipment, such as curve tracers and special test sets. These last procedures, although presented in a simple, readily understandable fashion, are slanted for the lab technician and/or field-service engineer.

Although the emphasis is on practical tests, the book goes much further than the usual collection (or cookbook) of test procedures. For example, this book shows you what kind of scope displays or meter readouts to expect from both good and bad components and circuits.

In many cases, the book goes on to show what is probably wrong if the expected test results are not obtained. Thus, the procedures described in this book form the basis for troubleshooting, and are starting points for analysis of experimental and/or design circuits.

Chapter 1 is devoted to electronic meters, covering such common types as analog and digital, as well as the not-so-common differential meters. Descriptions of the internal circuits and external operating controls are provided for a cross section of meters (including older types that are still in common use). The chapter concludes with calibration procedures for various types of meters.

A special problem is found with shop meters. Most shops are not equipped with precision standards, and it is expensive and time-consuming to send meters out for calibration. To overcome this problem, the chapter includes procedures to permit the calibration and testing of shop meters against commonly available standards, and to make maximum use of such standards.

If the test results reveal that a particular meter is not up to standard, the manufacturer's service literature can then be consulted to find such information as location of calibration controls, test limits, and so on.

Chapters 2 through 6 provide coverage similar to that of chapter 1 for scopes, generators, counters, probes/transducers, and specialized test equipment.

Chapter 7 is devoted entirely to test procedures for bipolar (two-junction) transistors. The first sections of this chapter describe transistor characteristics and test procedures from the practical standpoint. The information in these sections permits you to test all important transistor characteristics (both in-circuit and out-of-circuit) using basic shop equipment. The section also helps you understand the basis for such tests. The remaining section of the chapter describes how the same tests, and additional tests, are performed using more sophisticated equipment.

Chapters 8 through 11 provide coverage similar to that of chapter 7 for field-effect transistors (FETs), unijunction transistors (UJTs), solid-state diodes, and thyristors (SCRs, triacs, sidacs, SUSs, etc.).

Chapter 12 is devoted entirely to test procedures for audio circuits. These procedures can be applied to complete audio equipment (such as a stereo system), or to specific circuits (such as the audio circuits of a radio transmitter or receiver).

The procedures can be applied to the circuits at any time during design or experimentation, and include a series of notes regarding the effects of changes in component values on test results. This information is summarized at the end of the chapter, and is of particular importance to hobbyists and experimenters.

Chapters 13 through 15 provide coverage similar to that of chapter 12 for power-supply circuits, RF circuits, and communications equipment.

Basic operating procedures

A thorough study of this handbook will make the reader familiar with the basic principles and operating procedures for all types of electronic test equipment. It is assumed that the reader will take the time to become equally familiar with the principles and operating controls for any particular test instrument being used. Such information is contained in the instruction manual for the particular equipment. It is absolutely essential that the operators become thoroughly familiar with their particular test instruments. No amount of textbook instruction makes the operator an expert; it requires actual practice.

It is strongly recommended that readers establish a routine operating procedure or sequence of operation for each item of test equipment in the shop or lab. This saves time and familiarizes the readers with the capabilities and limitations of the particular equipment, thus eliminating false conclusions based on unknown operating conditions.

The first step in placing a test instrument in operation is reading the instruction manual. Although most manuals are weak in applications data, they do describe how to read scales, how to connect test leads or probes, and the logical sequence for operating the controls. This is particularly important with new test equipment, where the operator is not familiar with the instrument's capabilities and limitations.

After the manual's instructions are digested, they can be compared with the procedures in this handbook. Remember that the procedures here are typical or general, and applicable regardless of the test to be performed or the type of instrument used. On the other hand, instruction-manual procedures apply to a specific instrument. If there is a conflict between the manual procedures and this handbook, follow the manual. Remember the old electronics rule; when all else fails, follow instructions.

Safety precautions

In addition to a routine operating procedure, certain precautions must be observed during operation of any electronic test equipment. Many of these precautions are the same for all types of test equipment; others are unique to particular test instruments, such as meters, scopes, signal generators, and so on. Some of the precautions are designed to prevent damage to the test equipment or the circuit under test; others are to prevent injury to the operator. Where applicable, special safety precautions are included throughout the various chapters of this handbook.

The following *general safety precautions* should be studied thoroughly and then compared to any specific precautions called for in the equipment instruction manuals and in the related chapters of this handbook.

1. Many test instruments are housed in metal cases. These cases are connected to the ground of the internal circuit. For proper operation, the ground terminal of the instrument should always be connected to the ground of the equipment under test. Make certain that the chassis of the equipment under test is not connected to either side of the ac line or to any potential above ground. If there is any doubt, connect the equipment under test to the power line through an isolation transformer.
2. Remember that there is always danger in testing electrical equipment that operates at hazardous voltages. Familiarize yourself with the equipment under test before working on it, and remember that high voltage might appear at unexpected points in defective equipment.
3. It is good practice to remove power before connecting test leads to high-voltage points (high-voltage probes are often provided with alligator clips). It is preferable to make all test connections with the power removed. If this is impractical, be especially careful to avoid accidental contact with equipment and objects that can produce a ground. Working with one hand in your pocket and standing on a properly insulated floor lessens the danger of shock.
4. Filter capacitors can store a charge large enough to be hazardous. Therefore, discharge filter capacitors (after power is turned off) before attaching the test leads.
5. Remember that leads with broken insulation offer the additional hazard of high voltages appearing at exposed points along the leads. Check test leads for frayed or broken insulation.
6. To lessen the danger of accidental shock, disconnect test leads immediately after the test is completed.
7. Remember that the risk of severe shock is only one of the possible hazards. Even a minor shock can place you in danger of more serious risks, such as a bad fall or contact with a source of higher voltage.
8. If practical, do not work on a high-voltage or otherwise hazardous circuit unless another person is available to assist in case of accident.
9. Even if you have had considerable experience with test equipment, always study the instruction manual of any instrument with which you are not familiar.
10. Use only shielded leads and probes. Never allow your fingers to slip down to the metal probe tip when the probe is in contact with a "hot" circuit.
11. Avoid operating test equipment in strong magnetic fields. Most high-quality instruments are shielded against magnetic interference. However, even digital readouts can be affected by strong magnetic fields.

12. Most test instruments have some maximum input voltage and/or current specified in the instruction manual. Do not exceed this maximum. Also, do not exceed the maximum line voltage or use a different power frequency on those instruments that operate from line power.
13. Avoid vibration and mechanical shock. Most electronic test equipment is delicate.
14. Do not attempt repair of electronic test equipment unless you are a qualified instrument technician. If you must adjust any internal controls, follow the instruction manual.
15. Study the circuit under test before making any test connections. Try to match the capabilities of the test instrument to the circuit under test. If the circuit has a range of measurements to be made (ac, dc, RF modulated signals, pulses, or complex waves), it might be necessary to use more than one instrument.

    For example, most meters will measure dc and low-frequency ac. If an unmodulated RF carrier is to be measured, use an RF probe. If the carrier to be measured is modulated with low-frequency signals, a demodulator probe must be used. If pulses, square waves, or complex waves (combinations of ac, dc, and pulses) are to be measured, a peak-to-peak reading meter provides some meaningful indications, but a scope is the best instrument.
16. There are two standard international operator warning symbols found on some test instruments. One symbol, a *triangle with an exclamation point at the center*, advises the operator to refer to the operating manual before using a particular terminal or control. The other symbol, a *zig-zag line simulating a lightning bolt*, warns the operator that there might be a dangerously high voltage at a particular location, or that there is a voltage limitation to be considered when using a terminal or control.

1
Electronic meters

It is almost impossible to get by in any phase of electronics without some form of meter. Both hobbyists and professional technicians find it necessary to check on circuits and components (to find what voltage is available, how much current is flowing, and so on). At one time, the simplest and most common instrument used to measure the three basic electrical values (voltage, current, and resistance) was the *voltohmmeter (VOM)*. Sometimes, the terms *multimeter* or *multitester* were used in place of VOM. Today, the *digital multimeter (DMM)* has generally replaced the VOM. Notice that the DMM is also known as the DVM or digital voltmeter, although must DVMs are capable of measuring current and resistance as well. There are dozens, if not hundreds of DMMs (and a few VOMs) available in all price ranges. As the price goes up, accuracy is increased, more scales of functions are added, and the scales are given greater range.

The first improvement on the VOM was the *vacuum-tube voltmeter (VTVM)*. Today, the VTVM has been replaced by the *transistorized* or *electronic meter*. The sensitivity of these instruments is much greater than that of the VOM because electronic meters contain an amplifier. Electronic meters have another advantage over the VOM in that the electronic-meter amplifier presents a high impedance to the circuit or component being measured. Thus, electronic meters draw little or no current from the circuit and have little effect on circuit operation.

Those electronic meters using the field-effect transistor or FET in their amplifiers present the highest impedance and draw the least current from the circuit because FETs have a very high impedance compared to that of other transistors. FET meters are thus used in very sensitive electronic circuits.

The VOM, VTVM, electronic meter, and FET meter are all *analog meters*. That is, the meters use rectifiers, amplifiers, and other circuits to generate a current that is proportional to the quantity being measured. In turn, this current drives a meter movement. The *digital meter* displays measurement in discrete numerals, rather than as a pointer deflection on a continuous scale, as is commonly used in analog instruments.

In addition to digital and analog meters, the *differential meter* is often used in laboratory applications. The differential meter operates by comparing an unknown voltage to a known voltage. Notice that the differential instrument can have either a pointer-type display (moving needle over a calibrated scale) or a digital readout.

It would be almost impossible and beyond the scope of this book to describe all of the circuits used in modern meters. Many of the circuits are special-purpose. Likewise, many basic circuits are used in various combinations. Rather than attempt to describe every known meter, the remainder of this chapter is devoted to typical meter circuits. The paragraphs that cover how a VOM is formed from a basic meter movement are especially useful to the student.

In the early days of electronics, it was common for the hobbyist/experimenter to build their own VOMs. Today, because of the reduced prices and the difficulty (for practical considerations) of making accurate meter scales, the homemade VOM is almost unknown. In a way, this trend is unfortunate because much can be learned building a VOM. As resistance is added to make a basic meter movement into a working ammeter and voltmeter, or as power and resistance are added to a basic movement for conversion to an ohmmeter, many of Ohm's and Kirchhoff's laws become practical values instead of dull theories.

1.1 Digital meters

To fully understand the operation of digital meters, it is necessary to have a full understanding of digital logic circuits. These include gates, amplifiers, switching elements, delay elements, binary counting systems, truth tables, registers, encoders, decoders, D/A converters, A/D converters, adders, scalers, counter/readouts, and so on. Full descriptions of these devices are contained in *Lenk's Digital Handbook* (McGraw-Hill, 1993). However, it is possible to have an adequate understanding of digital meters if you can follow simplified block diagrams, as presented here.

A knowledge of electronic counter/readouts is especially important. This is because a digital meter performs two functions: (1) conversion of voltage (or other quantity being measured) to time or frequency (usually in the form of pulses), and (2) conversion of the time or frequency data to a digital readout. In effect, a digital meter is a conversion circuit (voltage to time, etc.) plus an electronic counter/readout. Counters are covered further in chapter 4.

1.1.1 Basic digital meter

Both DMMs and DVMs display measurements as discrete numerals, rather than a pointer deflection on a continuous scale. Direct numerical readout reduces human error, eliminates parallax error, and increases reading speed. Automatic polarity and range-changing features on some digital meters reduce operator training, measurement error, and possible instrument damage through overload.

Figure 1-1 shows the operating controls and readout of a typical DMM. Notice the simplicity of controls. Once the power is turned on, the operator has only to se-

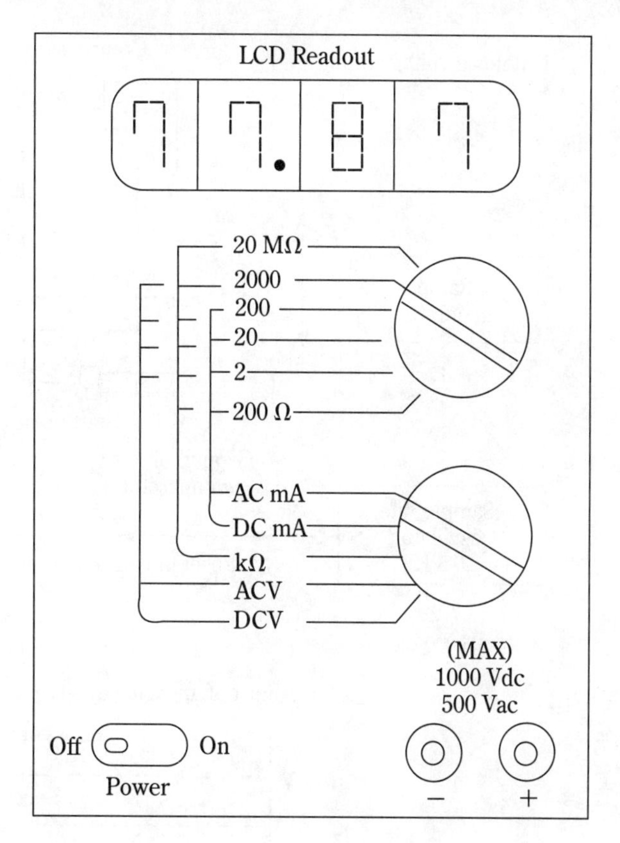

1-1 Operating controls and readout of a typical DMM.

lect the desired function and range before connecting the meter to the circuit. The readout is automatic. On some DMMs, the range is changed automatically, which further simplifies operation.

1.1.2 Ramp-type digital meters

Figure 1-2 shows the operation of a basic ramp-type digital voltmeter. Such meters measure the time required for a linear voltage ramp to change from a value that is equal to the voltage being measured to zero (or vice versa). The *time period* is measured with an electronic counter (chapter 4) and is displayed on a decade readout. The amp-type meter is essentially a *voltage-to-time converter*, plus a counter/readout.

A ramp voltage is generated at the start of a measurement cycle (there are usually two or three measurement cycles per second). The ramp voltage is compared continuously with the voltage being measured. At the instant the two voltages become equal, a coincidence circuit generates a pulse that opens a gate. The ramp continues until a second coincidence or comparator circuit senses that the ramp has reached 0 V. The output pulse of this comparator closes the gate.

The time duration of the gate opening is proportional to the input pulse. The gate allows pulses to pass to the counter circuit, and the number of pulses counted during the gating interval is thus a measure of voltage. The elapsed time, as indicated by the count on the readout, is proportional to the time the ramp takes to

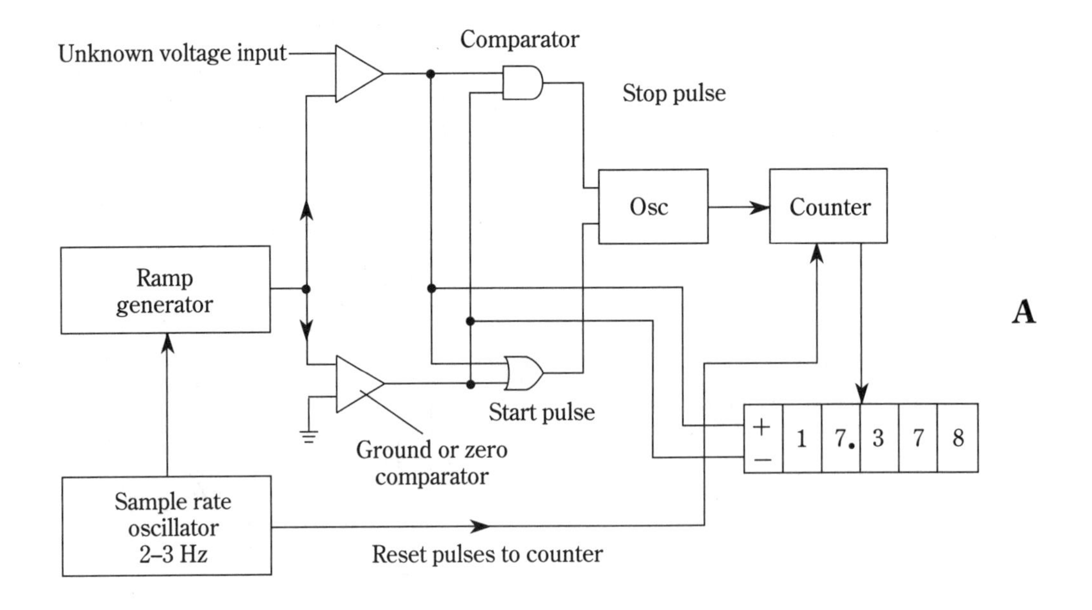

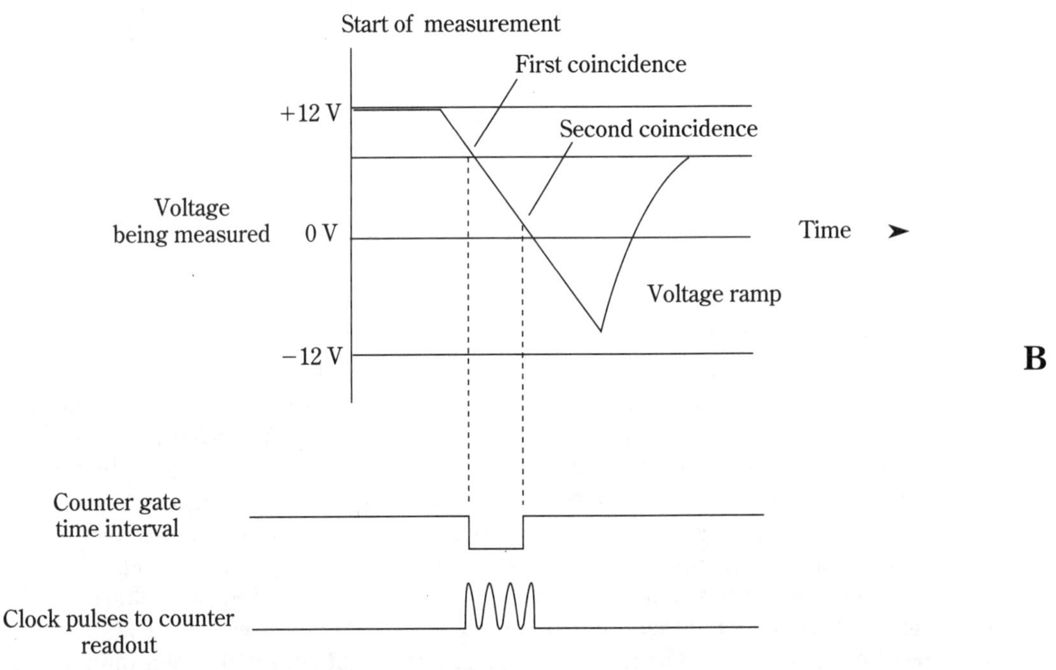

1-2 Basic ramp-type digital voltmeter.

travel between the unknown voltage and 0 V (or vice versa in the case of a negative input voltage to be measured). Therefore, the count is equal to the input voltage.

The order in which the pulses come from the two comparators indicates the polarity of the unknown voltage. This triggers a readout that indicates plus, or minus, as required. Notice that all or most of the circuit components shown in Fig. 1-1 are often contained on a single chip or card in present-day DMMs and DVMs.

A *staircase ramp-type DMM* or *DVM* makes voltage measurements by comparing the input voltage to an internally generated staircase ramp, rather than to a linear ramp. Otherwise, operation of the two ramp-type meters is essentially the same. Staircase ramp-type meters are generally more accurate than the basic linear ramp-type meter.

1.1.2 Integrating-type digital meter

Figure 1-3 shows operation of a basic integrating-type digital voltmeter. One of the problems with a ramp meter is that measurements at the end of the timer interval can occur simultaneously with a noise burst. These noise signals can lengthen (or shorten) the time interval, thus making the count incorrect. This problem can be overcome by an integrating-type meter that makes the measurement on the basis of *voltage-to-frequency conversion*, rather than voltage-to-time, as is used in the ramp meter.

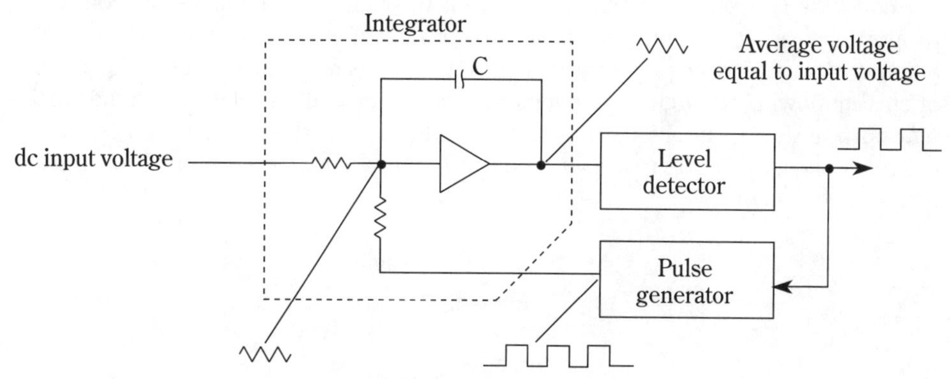

1-3 Basic integrating-type digital voltmeter.

As shown in Fig. 1-3, the circuit functions as a feedback-control system that governs the rate of pulse generation, making the average voltage of the rectangular pulse train equal to the input voltage. The ramp slope is proportional to the input voltage (for example, a higher voltage at the input results in a steeper slope, resulting in a shorter time duration for the ramp). As a result, the pulse repetition rate is higher. Because the repetition rate is proportional to the input voltage, the pulses can be counted during a known time interval to find a digital measure of the input voltage.

The primary advantage of this type of analog-to-digital conversion is that the input is integrated over the sampling period, and the reading represents a true average of the input voltage. The pulse-repetition frequency tracks a slowly varying input voltage so closely that changes in the input voltage are accurately reflected as changes in the repetition rate. The total pulse count during a sampling interval thus represents the average frequency and thus the average voltage. This is important when noisy signals or voltages are measured because the noise can be averaged out during the measurement. Another advantage of the integrating meter is that the measurement circuit can be isolated by shielding and transformer coupling.

An *integrating/potentiometric DMM* or *DVM* combines the continual measurement of true average input with accuracy from precision resistance ratios and stable reference voltages. With this meter, the V/F converter generates a pulse train with a rate that is exactly proportional to the input voltage. The pulses are gated for a precise time interval and fed to the first four places of a counter. The stored (undisplayed) count is transferred to a D/A converter, which produces a highly accurate voltage proportional to the stored count. This voltage is subtracted from the unknown voltage at the input to the V/F converter. After a second gating period, the total count is transferred to the counter readouts, which indicates the integral of the input voltage.

1.2 Nondigital (analog) meters

The simplest and most commonly used movement in a nondigital VOM or multimeter is the *D'Arsonval movement* shown in Fig. 1-4. Current through the armature coil sets up a magnetic field that reacts with the permanent magnet's field to rotate the coil, with respect to the magnet. Current through the coil makes the coil turn a proportional amount. Thus the basic meter movement is an analog device. The amount of travel for the pointer attached to the coil is related directly to the amount of current flowing through the movement. The meter scale is related to some particular current. For example, if 1 mA is required to rotate the coil and pointer across the full scale, a half-scale reading is equal to 0.5 mA, a quarter-scale reading is equal 0.25 mA, and so on.

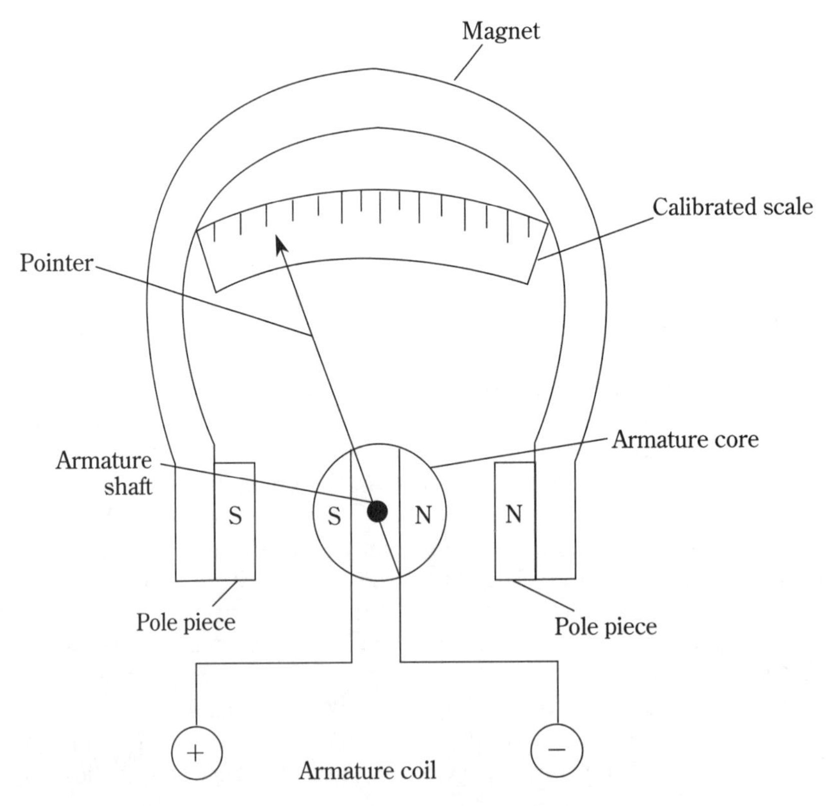

1-4 Basic D'Arsonval meter movement.

1.2.1 Basic ammeter

The basic movement of Fig. 1-4 by itself, forms an *ammeter* (*ampere meter*). A true ammeter measures current in amperes. In electronics, current is more often measured in milliamperes or microamperes. Most movements used in nonelectronic meters produce a full-scale deflection with a few microamperes of current.

A *shunt* must be connected across the movement to measure currents greater than the full-scale range of the basic movement. The shunt can be a precision resistor, a bar of metal, or a simple piece of wire. Meter shunts are usually precision resistors that can be selected by a switch. Shunt resistance is a fraction of the movement resistance. Current divides between the meter and shunt, with most of the current flowing through the shunt. Shunts must be precisely calibrated to match the movement.

Figure 1-5 shows the two typical current range-selection circuits for VOMs. In Fig. 1-5A, individual shunts are selected by the range-scale selector. In Fig. 1-5B, the shunts are cut in or out by the selector. If the selector is in position 1, all three shunts are across the movement, giving the least shunting effect (most current

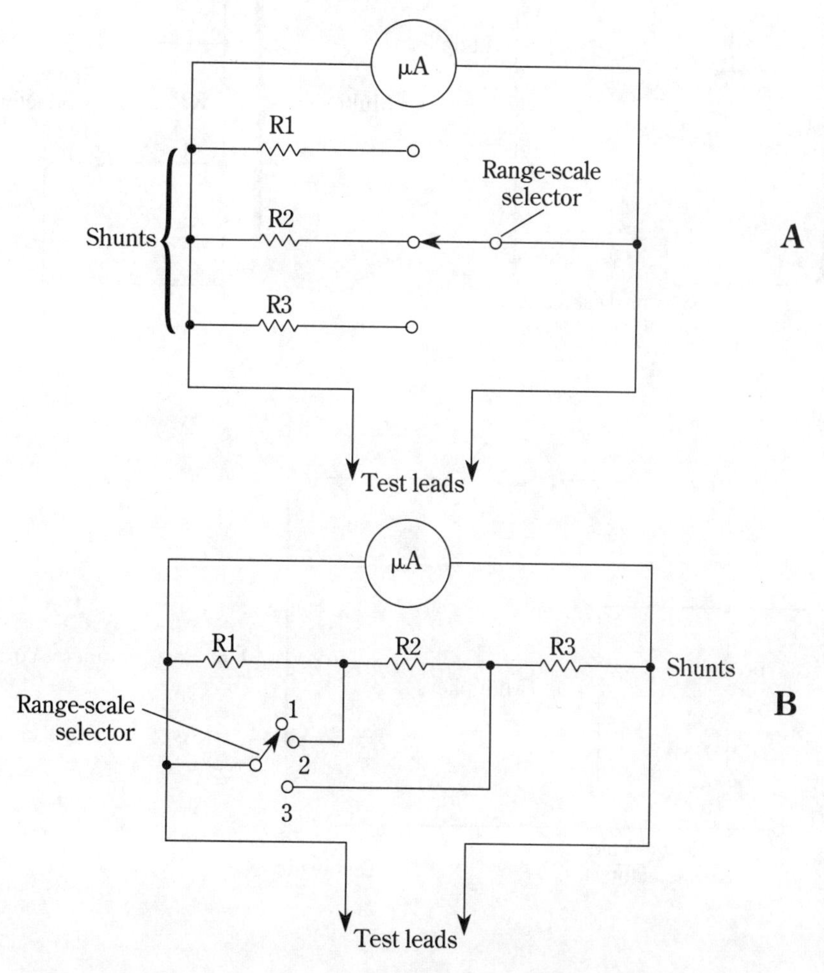

1-5 Two typical current range-selection circuits for VOMs.

through the movement). In position 2, resistor R1 is shorted out, with resistors R2 and R3 shunted across the movement, increasing the meter's current range. In position 3, only R3 is shunted across the movement, and the meter reads maximum current.

1.2.2 Basic voltmeter

Figure 1-6 shows the basic voltmeter circuit, where the movement is connected in *series* with the resistors. The series resistance is known as a *multiplier* because the resistance multiplies the range of the basic movement.

As shown in Fig. 1-6A, the voltage divides across the movement and series resistance. If a 0.5-V full-scale movement is used, and you want to measure a full scale

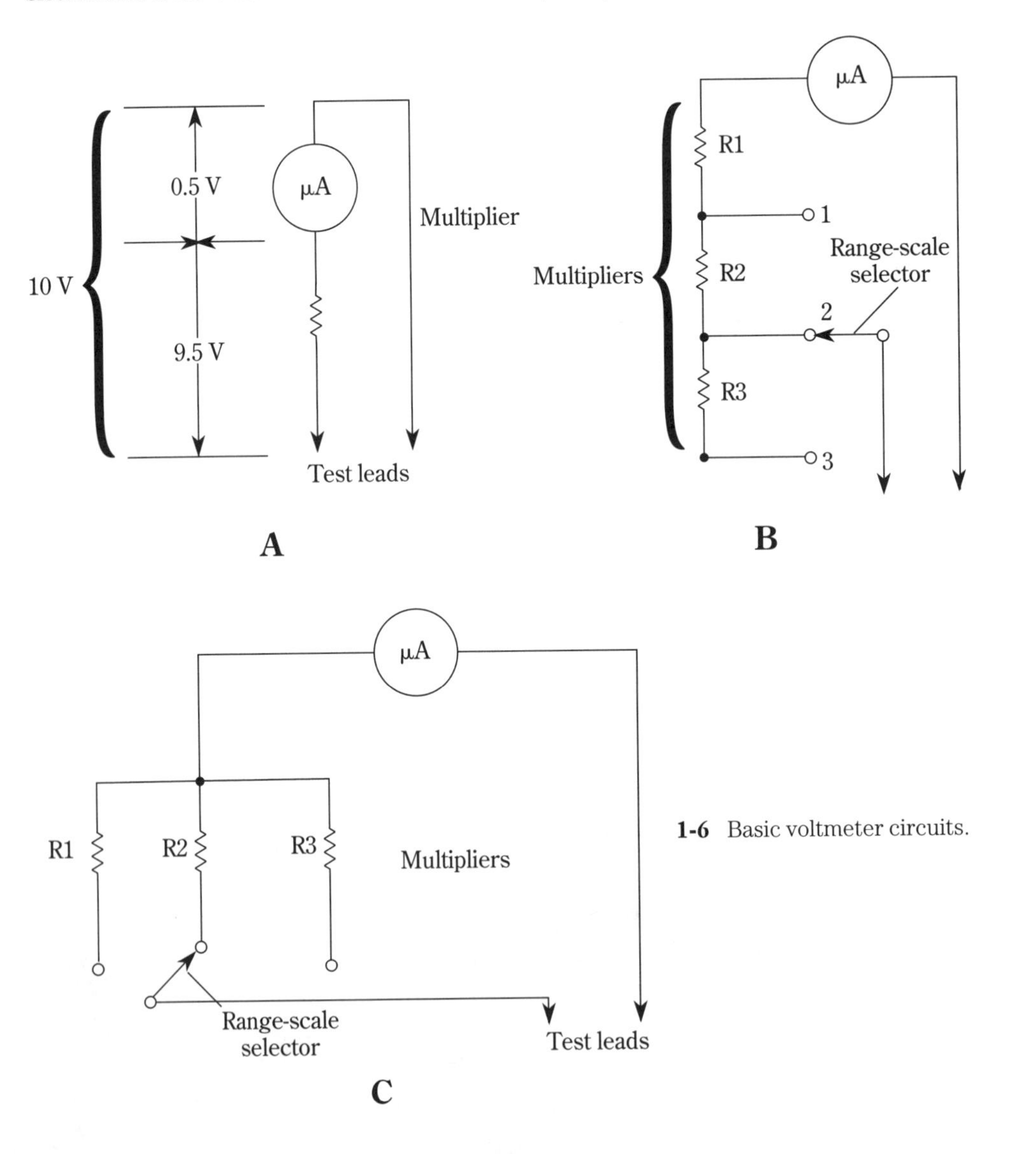

1-6 Basic voltmeter circuits.

of 10 V, the series resistor must drop 0.5 V. If a 100-V full scale is desired, the series resistance must drop 99.5 V, and so on. In Fig. 1-6B, the multipliers are cut in or out by the range-scale selector. In Fig. 1.6C, individual multipliers are selected.

The term *ohms per volt* is a measure of a VOM's sensitivity and represents the number of ohms required to extend the range by 1 V. For example, if the movement requires 50 μA for full-scale deflection, then the VOM has a sensitivity of 20,000 ohms/V (1 V = 20,000 × 50^{-6}). Thus, VOMs with movements requiring the least current have the highest ohms-per-volt rating and are the most sensitive (put less load on the circuit being measured, and have a less disturbing effect on the circuit because less current is drawn from the circuit).

1.2.3 Basic ohmmeter

Figure 1-7 shows the basic ohmmeter circuits, where the movement is connected in series with resistances and a power source (such as a battery in portable meters). This ohmmeter has five range scales that can be selected by a switch. In all but the X1 position, a series multiplier (similar to that of a voltmeter) is connected to the circuit and drops the voltage by a corresponding amount. This reduces current flow through the entire circuit, usually by a ratio of 10, so that the ohmeter scale represents 10, 100, 1000 or 10,000 times the indicated amount.

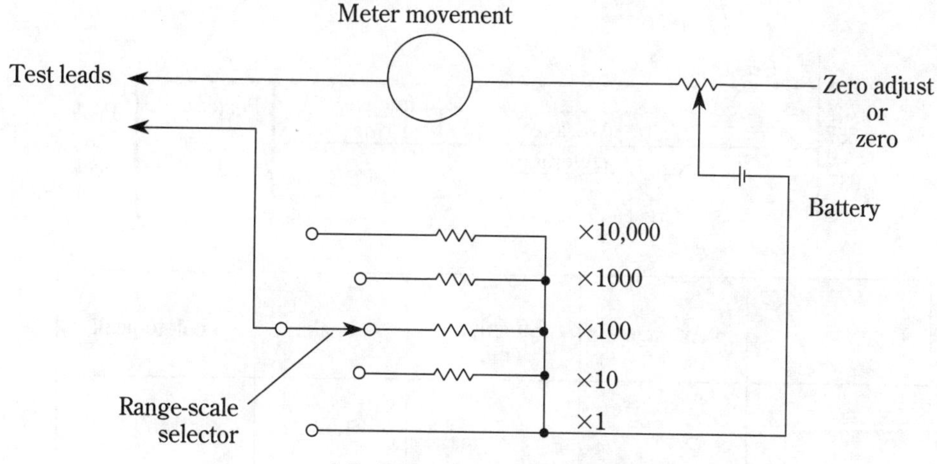

1-7 Basic ohmmeter circuit.

No matter what scale is used, the meter and battery are in series with a variable resistor that allows the circuit to be zeroed. As a battery ages, its output drops. Also, it is possible that with extended age or extreme temperature the resistance values or movement itself could change in value. Any of these conditions would make the ohmmeter scale inaccurate.

In use, the test leads are shorted together, and the variable resistor (usually marked ZERO ADJUST or ZERO) is adjusted until the movement pointer is at zero (at the right-hand side of the ohmmeter scale). When the leads are opened, the pointer then drops back to "infinity" or "open" (left-hand side), and the VOM is ready to read resistance.

1.2.4 Basic alternating-current meter

Because ac reverses direction during each cycle, the basic VOM movement cannot be connected directly to ac. Instead, the movement is connected to the ac voltage through a rectifier, such as shown in Fig. 1-8A. The remainder of the ac meter circuit can be identical to that of a dc meter. The circuit of Fig. 1-8A works well with low-fre-

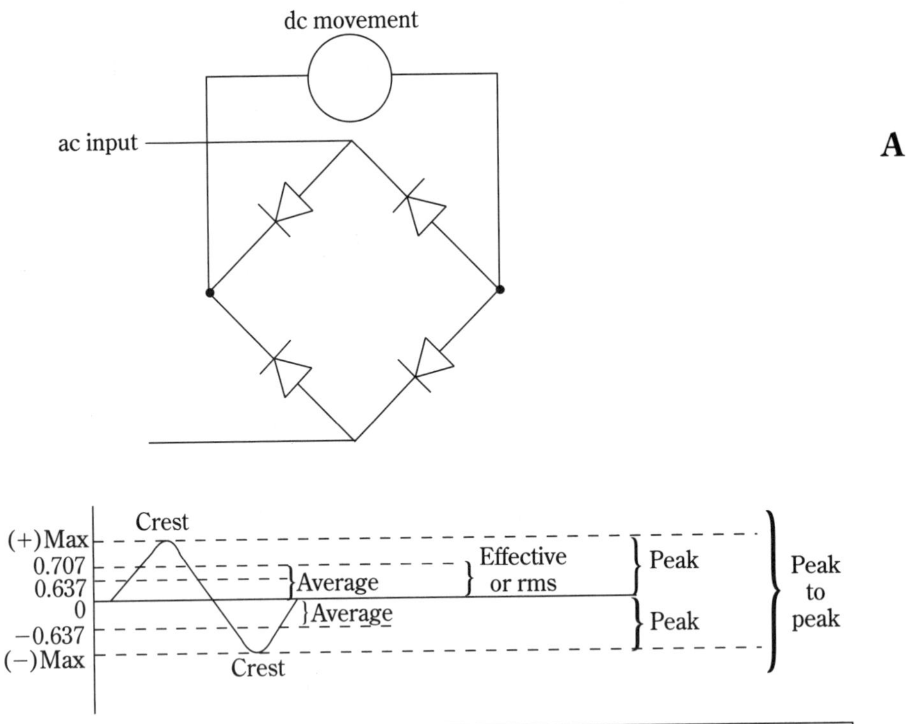

Given	Average	Effective rms	Peak	Peak-to-peak
Average	- - -	1.11	1.57	1.271
Effective (rms)	0.900	- - -	1.411	2.831
Peak	0.637	0.707	- - -	2.00
Peak-to-peak	0.3181	0.3541	0.500	- - -

B

1-8 Basic alternating-current meter relationships.

quency currents, but presents a problem as frequency increases (because the movement and multiplier resistances might load the circuit being tested). An RF probe can be connected ahead of the basic meter circuit to overcome this problem. As covered in chapter 5, the RF probe for most VOMs is a slender metallic prod on the end of an insulated rod or handle connected to the instrument through a flexible insulated lead.

Ac meter scales also present a problem for VOMs. As shown in Fig. 1-8B, there are four ways to measure an ac voltage: *average*, *rms* or *effective*, *peak*, or *peak-to-peak*.

Peak voltage is measured from the crest of one half cycle, and *peak-to-peak* is measured from the crests of both half cycles. However, the direct current to the movement is less than the peak alternating current because the voltage and current drop to zero on each half cycle.

With a full-wave bridge rectifier (Fig. 1-8A), the current or voltage is 0.637 of the peak value (a half-wave rectifier delivers 0.318 of the peak value). This is known as *average value*, and some meters are so calibrated. Most meters have rms (root-mean-square) scales. In an rms meter, the scale indicates 0.707 of the peak value (assuming the usual full-wave rectifier is used). This is the *effective value* of an alternating current.

A direct current flowing through a resistor produces heat. So does an alternating current. The effective value of an alternating current or voltage is that value which produces the *same amount of heat* in a resistor as the direct current or voltage of the same numerical value. The term *rms* is used because it represents the square root of the average of the squares of all instantaneous values in a *perfect sine wave*.

1.2.5 Clip-on meters and current probes

Figure 1-9 shows the basic elements of a clip-on meter or current probe. Alternating currents can be picked up by a coil of wire around the conductor, stepped up through a transformer, and measured by a voltmeter. A *clip-on meter*, complete with built-in coil, transformer, and meter movement, is useful where conductors carry heavy currents, and where it is not convenient to open the circuit to insert an ammeter.

Current probes are similar to clip-on meters, except that probes are generally used with an amplifier to measure small currents. A typical probe clips around the wire carrying the current to be measured and, in effect, makes the wire the one-turn primary of a "transformer" formed by ferrite cores and a many-turn secondary within the probe. The signal induced in the secondary is amplified and can be applied to any suitable voltmeter for measurement. Often, the amplifier constants are chosen so

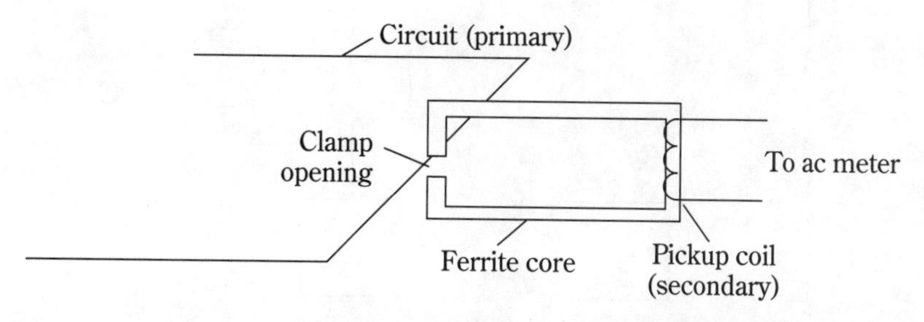

1-9 Basic elements of a clip-on meter or current probe.

that 1 mA in the wire being measured produces 1 mV at the amplifier output. In this way, current can be read directly on the voltmeter.

1.2.6 Basic electronic meters

Figure 1-10 shows basic electronic meter circuits. The amplifier is typically solid state (often with a FET input stage). However, there are still some vacuum-tube voltmeters (VTVM) available for special-purpose applications. A typical VTVM pro-

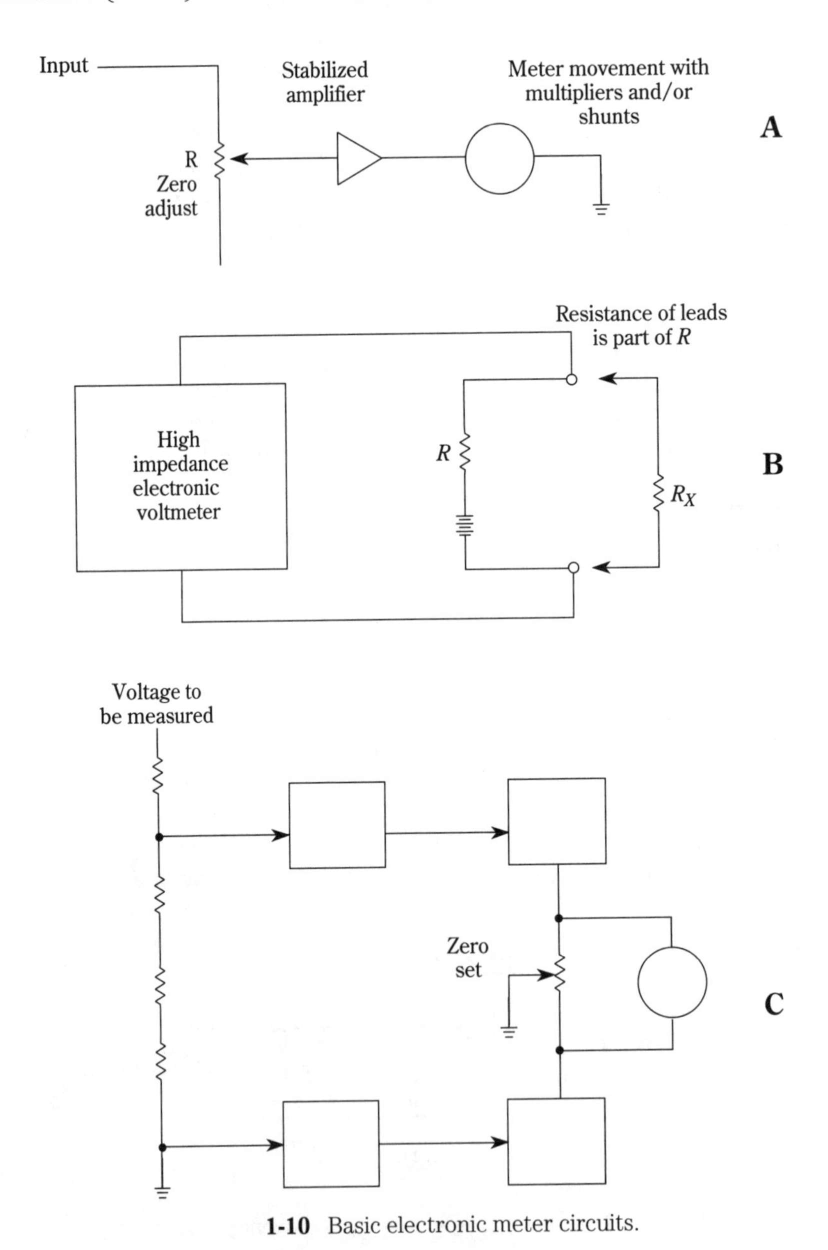

1-10 Basic electronic meter circuits.

vides about 11-MΩ *input impedance*, whereas a FET meter amplifier can provide up to about 100 MΩ (although most shop-type FET meters are in the 15-MΩ range). The amplifier is generally direct coupled. That is, there are no coupling capacitors between the amplifier stages.

When an electronic meter is used as a voltmeter, resistance *R* has a large value (usually several MΩ) and is connected in parallel across the circuit being measured. Because of the high resistance, very little current is drawn from the circuit, and operation of the circuit is not disturbed. The voltage across resistance *R* raises the voltage at the amplifier input from the zero level. This causes the meter at the amplifier output to indicate a corresponding voltage.

When an electronic meter is used as an ammeter, resistance *R* is a low value (a few ohms or less) and is connected in series with the circuit being measured. Because of the low resistance, there is little change in the total circuit current, and operation of the circuit is not disturbed. The current flow through resistance *R* causes a voltage to be developed across R, which raises the amplifier input level from zero and causes the meter to indicate a corresponding (current) reading.

When an electronic meter is used as an ohmmeter, the circuit is similar to that of Fig. 1-10B, where current in the circuit depends on the series combination of the unknown resistor RX and the internal resistor R. Both the voltage and current in the external circuit change according to the value of the unknown. The resistance scales of the meter are calibrated for measurement of this unknown value.

Figure 1-10C shows one of the most common circuits used in electronic meters. This is essentially a *differential amplifier*, in which the voltage to be measured is applied to one input, and the other input is grounded. The zero-set resistance is adjusted so that the meter reads zero when no input voltage is applied. When the voltage to be measured is applied across the input resistance, the circuit is unbalanced, and the meter indicates the proportional unbalance as a corresponding voltage reading.

One of the reasons for using the differential amplifier is to minimize drift because of power-supply changes. A direct-coupled amplifier cannot tell the difference between power-supply change and change in the voltage being measured (especially when measuring small voltages). With a differential amplifier, both halves of the amplifier (grounded input and ungrounded input) change simultaneously with power-supply changes.

Another technique for eliminating drift in electronic meters is to convert the dc voltage being measured to ac, amplifying the corresponding ac value (in an amplifier with input and output coupling capacitors, and not affected by supply changes), and then restoring the signal to a corresponding dc value for readout on the movement.

In high-accuracy laboratory meters, a taut-band suspension is substituted for the pivots and jewels of a shop-type meter. The moving coil in the taut-band meter mechanism is suspended on a platinum-alloy ribbon, eliminating friction and problems concerning repeated measurements. Lab meter faces are custom-calibrated and photographically printed to match exactly the linearity characteristics of each individual movement at all points. This eliminates the tracking error that is found on many mass-produced meters. By combining custom-calibrated scales with taut-band suspension, the possibility of mechanical error is kept to an absolute minimum.

Electronic meters for measuring ac voltages also use an amplifier with the meter movement, but add a rectifier circuit to convert the ac into dc. Notice that most shop-type meters are rms-reading instruments. This is also true of lab meters, although it is possible to use average-, peak-, or peak-to-peak reading meters for special applications. Although a meter might be rms-*reading*, it is usually average- or peak-responding. That is, the scale reads rms values, but the movement operates on an average or peak value. An average-responding (rms-reading) ac electronic meter uses a rectifier between the amplifier and movement. A peak-responding ac electronic meter uses a rectifier at the input. The rectifier charges a small input capacitor to the peak value of the input signal or voltage.

1.3 Differential meters

Figure 1-11 shows basic differential voltage-measurement techniques. The concept of differential voltage measurement is to apply an unknown voltage against one that is accurately known and to measure the difference between the two on an indicating device. If the known voltage is adjusted to the exact potential of the unknown voltage, you can determine the unknown quantity being measured as accurately as the known voltage (or reference standard).

Measurements made by the differential technique (sometimes called a *potentiometric or manual voltmeter measurement*) are recognized as one of the most accurate means of relating an unknown voltage to a known voltage. In practice, these measurements are made by adjusting a *precision resistive divider* to divide down an accurately known reference voltage. The divider is adjusted to the point where the divider output equals the unknown voltage, as indicated by the null voltmeter shown in Fig. 1-11A. The differential measurement techniques are similar to the bridge-circuit measurements of chapter 6.

With the differential technique, the unknown voltage is determined to an accuracy limited only by the accuracies of the reference voltage and the resistive divider. Meter accuracy is of little consequence because the meter serves only to indicate any residual difference between the known and unknown voltages.

A high-voltage standard is required to measure high voltage. This need can be overcome by inserting resistive voltage dividers on both sides of the null meter, as shown in Fig. 1-11B. However, this results in relatively low input resistances for voltages higher than the reference standard. This low input resistance is undesirable because accurate measurements might not be obtained if substantial current is drawn from the source being measured. Many differential meters offer input resistance approaching infinity only at a null condition, and then only if an input voltage divider is not used.

This problem can be overcome with an *input isolation stage*, as shown in Fig. 1-11C. The amplifier (typically an IC op amp) ensures that the high input impedance is maintained, regardless of whether the instrument is adjusted for a null reading or not. A further advantage of the amplifier is that the resistive voltage divider (which permits voltages as high as 1000 V to be compared to a precision 1-V reference) can be placed at the amplifier output, rather than being in series with the measured voltage source. This isolation permits the instrument to have high input impedance on all ranges.

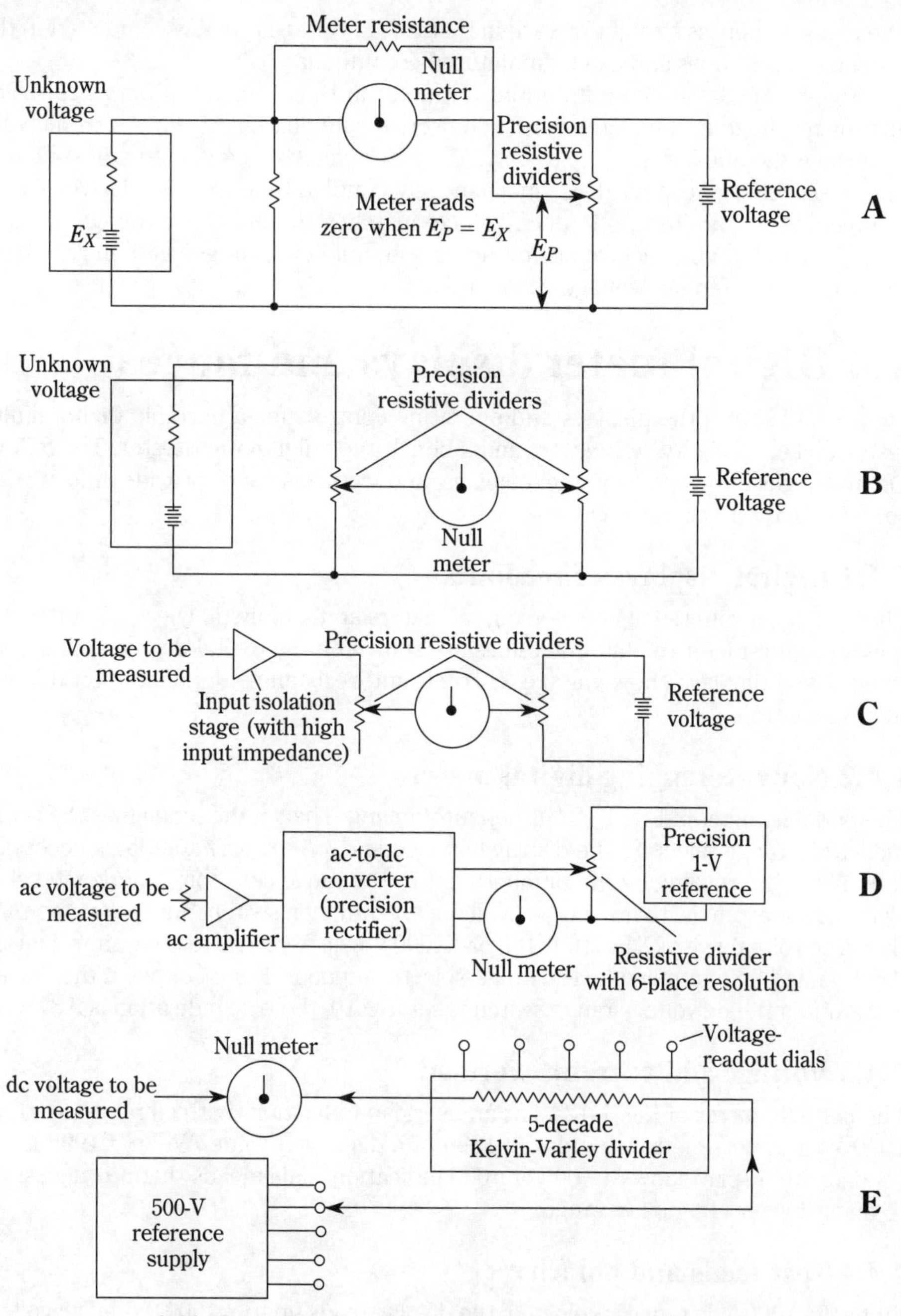

1-11 Basic differential voltage-measurement techniques.

Figure 1-11D shows a differential voltmeter in the *ac mode of operation*. This circuit uses a precision rectifier to convert the unknown ac voltage directly into dc voltage (equivalent to the average value of the ac voltage), and the resulting dc voltage is read to five- or six-place resolution by the potentiometric voltmeter technique.

The measurement is straightforward in that the ac voltage remains connected to the converter at all times and can be monitored continuously.

Figure 1-11E shows a differential voltmeter in the *dc mode of operation*. The unknown voltage is connected across the series combination of the electronic voltmeter and the 500-V reference supply. The reference voltage is then adjusted with five or six voltage-readout dials (mechanically coupled to a precision Kelvin-Varley voltage divider) until the reference voltage matches the unknown voltage, as indicated by a null on the electronic voltmeter. The unknown voltage is then read from the five or six reference voltage-readout dials.

1.4 Digital meter displays and ranges

Figure 1-12 shows the displays and operating controls for a portable digital multitester (called a *digital voltmeter*) and a bench-type digital multitester. The following notes describe each of the displays and controls, and provide information concerning their accuracy and use.

1.4.1 Digital displays or readouts

The first major difference between digital meters and nondigital (Sec. 1.5) meters is that all digital-meter readings or values are shown on the digital displays. Thus, the same digital displays show voltage, current, and resistance, depending on the settings of controls.

1.4.2 Nonautoranging digital meter

The portable meter of Fig. 1-12A is nonautoranging. That is, the range must be set by means of a range selector. The display has a *fixed decimal point*, and it is necessary to multiply the reading by the number next to the range selector (a horizontal slide switch). For example, to measure dc voltage, the function switch must be set to mA/V; then a dc volts range of X1, X10, X100 or X1000 is selected. The reading shown on the display must be multiplied by the range-selector number. For example, if the display is 0.987 and the dc voltage range switch is set to X10, the true indication is 9.87 V.

1.4.3 Voltage and current overload

The portable meter of Fig. 1-12A has an overload indication. With all ranges (except 1000 V) it is possible to use up to ±1999 of a displayed value. When ±1999 is exceeded, the display shows a =000 or ≡000 indication. This means that it is necessary to manually select a higher range.

1.4.4 Test leads and polarity

When the 1000-Vdc range is selected, the display reads up to ±1000. When ac volts is selected, the display reads up to 500. Notice that the same position of the range selector is used for ac volts measurement, and for dc volts X1000. Thus, the choice between ac and dc measurements must be made with the test leads (a common feature on portable digital meters). Normally, the black test lead is connected to the COMMON jack, and the read test lead is connected to dc volts or ac volts, as required.

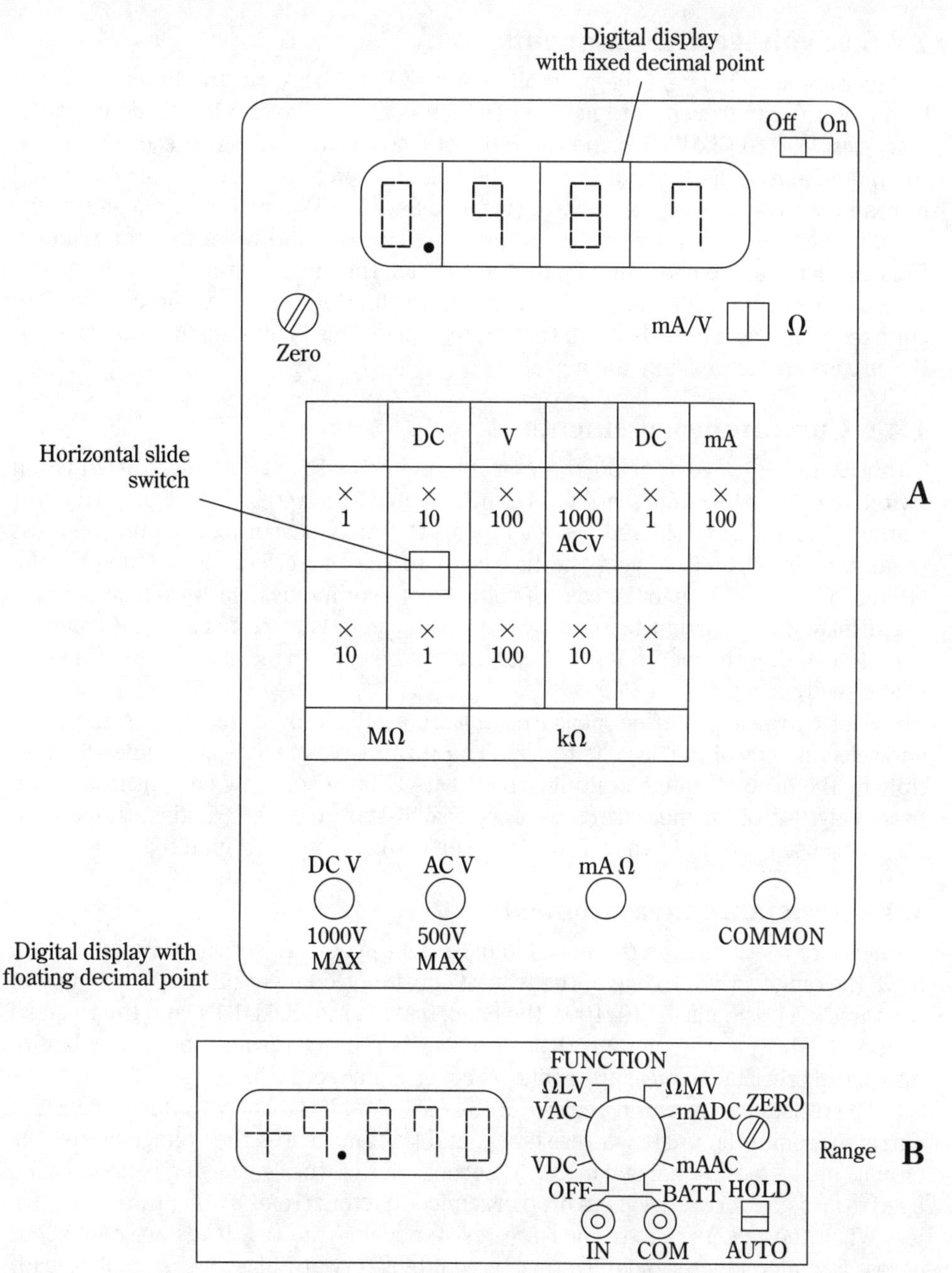

1-12 Typical digital-meter displays and ranges.

When the test leads are connected into a circuit to measure dc voltages, a (–) or minus polarity signal appears in the display if the voltage being measured is negative with respect to the COMMON lead. If no polarity sign appears, the voltage being measured is positive with respect to the COMMON lead.

1.4.5 ac voltage measurements

To measure ac voltage on the portable meter of Fig. 1-12A, set the function switch to mA/V, the range switch to ac volts (which is also dc volts X1000), connect the black test lead to COMMON, and the red test lead to AC V. Under these conditions, the meter senses the average value of the voltage being measured and is calibrated to read the rms value of a sine wave (refer to Sec. 1.2.4). Any direct current in the circuit being tested is added to the alternating current, and will affect the reading. This problem can be overcome by first measuring the circuit for any dc voltage and then subtracting the dc voltage from the ac reading. An alternate procedure is to connect a capacitor in series with the red test lead. This blocks any dc from the meter circuits and leaves only the ac voltage.

1.4.6 Current-measurements

To measure direct current on the portable meter of Fig. 1-12A, set the function switch to mA/V, the range switch to DC mA (either 1 or 100), connect the black test lead to COMMON, and the red test lead to mA Ω. When measuring current, you must break the circuit and connect the test leads to the two circuit-connection points. (Refer to Sec. 1.12.) As in the case of voltage, a (–) or minus polarity sign appears in the display if the current being measured is negative with respect to the common lead. Notice that the meter of Fig. 1-12A cannot be used to measure alternating current directly.

Four extra ranges of dc measurement are available by using the mA Ω socket and selecting one of the dc volts ranges. The current-range value is obtained by multiplying the dc volts range multiplier by 0.1 μA. These very low current ranges are extremely useful for measuring leakage in solid-state devices (diodes, transistors, etc.) because current down to 100 pA can be resolved on the digital readout.

1.4.7 Resistance measurements

To measure resistance on the portable meter of Fig. 1-12A, set the function switch to Ω, the range switch to one of the two MΩ positions connect (1 or 10) or to one of the three kΩ positions (1, 10, 100), the black test lead to COMMON, and the red lead to mA Ω. Always make certain that all power is removed from the circuits before making any resistance measurements, as covered in Sec. 1.10.

The resistance-measuring circuit of the meter applies a known value of constant current through the unknown resistance, and then measures the voltage across the circuit (unlike most nondigital meters). Because of this, the *resistance ranges* can be used to measure forward voltage drop of semiconductors (refer to chapters 7 and 10).

When the meter is set to measure resistance and the test leads are connected across a semiconductor diode so that the diode is forward biased, the display reading is the forward voltage drop, expressed in volts. For example, when the range selector is set to kΩ X1, the constant-current value is 1 mA. If the reading is 0.5 (indicating a diode forward resistance of 500 Ω), the forward voltage drop is 0.5 V.

The resistance ranges of the portable meter are also provided with an *overload indication*. The meter shows flashing bars until a resistance lower than the maximum reading of the range is connected. Both positive overload and negative overload can occur. A positive overload is where the resistance value is greater than the maximum reading of the range. A negative overload is where the resistance value is lower than the resolution provided by the last digit of the scale. For example, on the kΩ X10 resistance range, the extreme right-hand digit equals 10 Ω. If a 5-Ω resistor is connected, there will be a negative-overload indication.

A positive overload always produces flashing bars. A small negative overload also shows flashing bars, but a large negative overload can produce fixed bars. Should these fixed bars appear on the resistance function, indicating a large negative overload, the condition can be corrected by temporarily placing the test leads across a large resistance.

1.4.8 Zero adjustment

Many digital meters are provided with a zero adjustment, such as the front-panel ZERO control in Fig. 1-12A. This control is accessible with a small screwdriver, and does not normally required repeated adjustment. The zero setting of the meter is checked by setting the range switch to X1000, and the function switch to mA/V. With no input and no test leads connected, the display should read 000, or possible -000. If not, adjust the ZERO control until the display is 000 or -000. A slight jitter to a reading of 001 may occur. This is normal and will not affect meter accuracy.

1.4.9 Accuracy and resolution

The accuracy of a digital meter is closely related to the resolution. The accuracy figures or specifications for digital meters (and most other digital readout instruments) is usually given as a percentage, ±1 count (or possibly ±2 counts for ac measurements). Thus, it is necessary to determine what each count represents on each scale to determine accuracy.

As an example, in the meter of Fig. 1-12A, the accuracy of the 1-V range for dc voltage measurements is rated at 1 percent ±1 count (in this case, the 1 percent applies to the reading). The resolution for the 1-Vdc range is 1 mV (the extreme right-hand digit equals 1 mV). As a result, if the display is 0.987, the true voltage is within 1 percent of the reading (within 9.87 mV) ±1 mV. This produces an area of uncertainty of about 22 mV (almost 11 mV above or below the display reading).

1.4.10 Autoranging digital meter

The meter of Fig 1-12B is capable of full autoranging. That is, the range changes automatically to suit the value being measured, once a particular function (voltage, current, or resistance) is selected by a FUNCTION switch. The display has a floating decimal point and is therefore a direct reading. It is not necessary to multiply the reading by a scale factor.

For example, to measure dc voltage, the FUNCTION switch must be set to VDC, the test leads connected to the COM and IN jacks, the RANGE switch is set to AUTO, and the test-lead tips connected to the circuit. The digital display moves automatically to the appropriate one of five dc voltage ranges (±0.2, ±2, ±20, ±200, ±2000). This is, if the display is +9.870, the true indication is +9.87 V (the display is in the ±20 range).

1.4.11 Maximum displays

Each digital range is four-place in the meter of Fig. 1-12B. That is, the maximum display for the 0.2 range is 0.19999, the 2 range is 1.999, the 20 range is 19.99, and the 200 range is 199.9. The maximum display for the 2000-V range is 1000 because this is the maximum voltage capability of the meter.

1.4.12 Range override

It is possible to override the autoranging function in the meter of Fig. 1-12B. When the RANGE switch is set to HOLD, the display remains in the last range selected, or moves to a lower range. For example, if the RANGE switch is moved to HOLD when a reading of +9.870 is obtained in AUTO, the display remains in the ±20 range (or drops to a lower range, if necessary). This feature is very useful when making repetitive voltage measurements in a particular range. When the meter is set to AUTO, the display goes through each range (as necessary) to reach the appropriate range for each range (as necessary) to reach the appropriate range for each measurement. This can be a waste of time if all measurements are in the same range. However, the total response time, including full autoranging, is less than 3 seconds.

1.4.13 Minimum displays

Most digital meters have some minimum display limits. For example, in the ±2 range of the Fig. 1-12B motor, the minimum display limit is 0.180 V. Voltages below this value will not appear on the ±2 range. However, this presents no problem to the operator because, in either HOLD or AUTO mode, the display automatically drops down to the next lower range (±0.2) and produces the appropriate reading.

1.4.14 Overload and overrange indications

Where there is an overload or overrange condition, all display elements of the Fig. 1-12B meter (except the kΩ and MΩ indicators) flash repeatedly. When measuring voltages greater than 1000 (an overload) in the AUTO mode, the display elements continue to flash until the voltage is removed. The display elements also flash to indicate an overrange condition when the RANGE switch is in HOLD and the selected range is exceeded. The flashing numbers displayed during the overload/overrange conditions can be any number from 000 to 1000. These numbers are essentially meaningless and are not to be considered an accurate indication of value.

1.4.15 Test leads and connections

In normal operation of the Fig. 1-12B meter, the red test lead is connected to the IN jack, and the black test lead is connected to the COM jack. Notice the standard international operator warning symbols (covered in the Introduction) located next to the jacks. These symbols indicate that no more than 1000 V should be connected to the jacks and that the manual should be consulted before connecting any leads to the jacks.

Although the meter is supplied with test leads, a number of alternative test leads can be used. The manual recommends "twisted pair", coaxial or shielded-cable test leads when measurements must be made in the presence of strong RF signals. This subject is covered further in Sec. 1.9.

The meter can also be used with an RF probe or high-voltage probe (chapter 5). When a high-voltage probe is used, make certain to multiply the indicated reading by the probe division factor.

1.4.16 Polarity indications

In addition to the four-place readout, the display for the Fig. 1-12B meter is provided with polarity indications or signs. When measuring dc voltage or current (FUNCTION switch set to VAC or mA DC) a (+) or positive-polarity sign appears in the display if the voltage or current being measured is positive with respect to the COM (black) lead. A (–) or negative-polarity sign appears for voltages or currents that are negative with respect to the COM lead. No polarity indications or signs are displayed for ac and resistance measurements.

1.4.17 Zero adjustment

For the most accurate results, the meter of Fig. 1-12B should be zeroed at regular intervals, particularly when the function is changed. For dc voltage and current functions, the meter is zeroed by connecting the test leads, and (if necessary) adjusting the front-panel ZERO control to display +0.0000. An occasional display of +.0001 or –.0001 is normal. The resistance function is zeroed in essentially the same way. However, it is possible for the test leads to show some small resistance, typically 0.1 or 0.2 Ω (which appears as .0001 or .0002 on the kΩ range). The ZERO control is inoperative on the ac voltage and current functions.

1.4.18 Digital-meter dc voltage measurement considerations

The following considerations apply to the autoranging meter of Fig. 1-12B. However, similar considerations apply to a variety of digital meters when measuring dc voltages.

Accuracy specifications The accuracy specifications for dc voltage is ± (0.05 percent of reading + 0.10 percent of the range + 1 count) for low-impedance voltage sources. Measurements of relatively high resistance can cause a significant reading error. This condition applies not only to digital meters, but to any electronic meter with high input resistance or impedance.

The input resistance of the meter is 10 MΩ on all dc voltage ranges. The amount of error because of meter loading can be determined by the equation:

$$\text{Percent error} = -\frac{R_S}{R_S + 10\ \text{M}\Omega} \times 100$$

where R_S is the resistance of the circuit where dc voltage is being measured (called the *source resistance*).

For example, a source resistance of 10 kΩ results in a loading error (reading error because of the meter load) of approximately –0.1 percent. The error has a minus sign because the loading reduces the voltage under "load" from the "unloaded" (meter not connected) value. The loading error becomes very significant for source resistances above 100 kΩ for most digital meters.

Variations in zero reading Over an extended period of operating time (usually hours), there might be some variation in the shorted-input zeroing on the .2-VDC range (to which the meter automatically ranges when the test leads are shorted together). The least-significant (right-hand) digit may, because of ambient temperature changes, vary positive or negative from +.0000. The right-hand digit is equivalent to 100 μV in the .2-VDC range. Because voltage measurements on higher ranges are more common, the effect of this variation is reduced, in decade steps, to the extent that even several hundred microvolts are an insignificant part of values usually measured. Thus, it is typically not an important effect on measurement accuracy when this occurs. For critical measurements on the .2-VDC range, a touch-up of the front-panel ZERO control removes this possible error.

Offset adjustments The ZERO control can be used as an offset adjustment to remove small, residual voltages when making differential or null measurements. However, the range of adjustment decreases in decade steps such that only a very limited zeroing capability is provided on higher ranges. Zero offset is recommended only for the lowest (.2 and 2-VDC) ranges. Readjust the ZERO control when zero-offset measurements are complete.

Open-circuit display counts When the meter inputs are open-circuited on the .2-VDC range, there might be several counts displayed. This is normal (for the meter of Fig. 1-12B and many digital meters) and does not produce significant measurement error when the leads are connected to a low resistance (less than 1 MΩ). The problem is caused by bias currents in the meter circuits, and can generally be ignored (except where very low voltages are to be measured in circuits with high resistance).

1.4.19 Digital-meter ac voltage measurement considerations

The following considerations apply to the autoranging meter of Fig. 1-12B. However, similar considerations apply to a variety of digital meters when measuring ac voltages.

Maximum voltages and complex waves The maximum ac voltage allowable between the IN and COM jacks is 700 Vrms. This corresponds to about 1000 V peak. When measuring ac voltage, any input other than a pure sine wave causes an error because the ac converter is average-sensing and RMS-calibrated. Square waves, sawtooth waves, and so on can best be measured with an oscilloscope (chapter 2). The measurement of complex waves is covered further in Sec. 1.26.

Capacitive coupling When the volts ac function is selected, the meter is capacitively coupled to the circuit so that any dc is blocked from entering the meter. Thus, only the ac voltage present in the circuit appears on the meter.

Accuracy specifications The accuracy specification for ac voltage is ± (0.25 percent of reading +0.15 percent of range + 1 count) over the frequency range from 40 Hz to 20 kHz. The input impedance of the meter on all ac ranges is frequency dependent (as is the case with any ac meter). In this particular meter, the input is 10 MΩ, in parallel with about 90 pF. This corresponds to about 9.47 MΩ, at 60 Hz. As in the case of dc, measurements at relatively high source resistances can cause a significant reading error. The meter input impedance at frequencies other than 60 Hz can be determined by:

$$Z_{IN} = \frac{10\ \text{M}\Omega}{\sqrt{1 + (5.655 \times f)^2}}$$

where Z_{IN} is the effective meter input impedance and f is the frequency in kHz.

Loading error The loading error caused by the impedance of the circuit or device being measured (source impedance) can be determined by:

$$\textit{Percent error} = \left(\frac{Z_{\text{source}}}{Z_{\text{source}} + Z_{\text{IN}}}\right) \times 100$$

The loading error at low frequencies (below 100 Hz) can be very significant for source impedances above 100 kΩ, and at higher frequencies (near 10 kHz) for source impedance even as low as 2000 Ω. The error will always have a minus sign because the leading always reduces the voltage under load from the unloaded value.

Open-circuit display counts When the meter inputs are open-circuited on the .2 or 2 Vac ranges, there might be a significant number of counts displayed, because of stray line-voltage radiation. This is normal (and not unusual for digital meters) and does not produce a measurement error when the leads are connected to a low-impedance source.

Variations in zero reading In the HOLD mode (with the meter input shorted), the display might show a 1- or 2-count "residual" on the higher ranges of the Vac function. The accuracy specification for this function includes the 1- or 2-count offset.

1.4.20 Digital-meter direct-current measurements

The following considerations apply to the autoranging meter of Fig. 1-12B. However, similar considerations apply to a variety of digital meters when measuring direct current.

Meter and circuit protection The meter is fuse-protected (Sec. 1.6) for a 3-A maximum direct current on the mA dc ranges. If this is exceeded, the fuse opens, the display reads zero, and the circuit under test is protected (opened).

Accuracy specifications and insertion loss The accuracy specification for mA DC is ± (0.15 percent of reading + 0.10 percent of range + 1 count). When current is measurement, the meter will, to some degree, affect operation of the circuit being tested and thus affect accuracy. This effect, known as *insertion loss*, causes a voltage drop and reduces the actual circuit current to the current displayed on the meter. The error must be considered if the resistance of the circuit under test is not at least 1000 times the shunt resistor for the range being used.

As an example, on the .2 mA DC range, the shunt resistance is 1000 Ω. Therefore, a source resistance of 100 kΩ results in an insertion-loss error of about 1 percent of the reading. Insertion-loss error for other source resistances can be determined by the following relationship:

$$\textit{Percent error} = \left(\frac{R_{\text{shunt}} + 0.15}{R_{\text{source}} + R_{\text{shunt}} + 0.15} \times 100\right)$$

where 0.15 Ω is approximate fuse and wiring resistance of the meter (Sec. 1.20) and R_{shunt} = 1000, 100, 10, 1, and 0.1 Ω for the .2, 2, 20, 200, and 2000 mA DC ranges, respectively.

The insertion-loss error will always have a minus sign because the inserted current is always less than the not-inserted current. To reduce this effect, use the highest range possible consistent with the measurement resolution required (by means of the HOLD mode).

Zero offset The ZERO control may be used as an offset adjustment to remove small, residual currents when making differential or null measurements. However, because the ZERO control has the same (count) effect on all ranges, the offset technique applies only on the range where the initial offset adjustment is made. Readjust the ZERO control as necessary for differential or null measurements.

Variations in zero readings Over an extended period of time (usually hours), there might be some variation in the zeroing of the .2 mA DC range to which the meter automatically ranges when the test leads are open-circuited. The least-significant (right-hand) digit might, because of ambient temperature changes, vary positive or negative from a display of +0.0000. Periodically check the zeroing (on the VDC mode) and touch up, if necessary.

Measuring combined dc and ac When you attempt to measure direct current with a substantial ac or pulse component superimposed on the direct current, a significant error can result if the peak-to-peak variation in current exceeds three times full-scale direct current of the range being used (1.5 times on the 2000 mA DC range). This error can be minimized by using the HOLD mode and by using a higher range.

1.4.21 Digital-meter alternating-current measurements

The following considerations apply to the autoranging meter of Fig. 1-12B. However, similar considerations apply to a variety of similar digital meters when measuring alternating current.

Meter and circuit protection The meter is fuse-protected (Sec. 1.6) for a 3-A maximum alternating current on the mA AC ranges. If this is exceeded, the fuse opens, the display reads zero, and the circuit under test is protected (opened).

Complex waves When measuring alternating current, any input other than a pure sine wave causes an error because the ac converter is average-sensing and RMS (sine wave)-calibrated. Square waves, sawtooth waves, and so on can best be measured with an oscilloscope (chapter 2). The measurement of complex waves is covered further in Sec. 1.26.

Capacitive coupling When the mA AC function is selected, the meter is capacitively coupled to the circuit so that any dc is blocked from entering the meter. Thus, only the ac present in the circuit appears on the meter.

Accuracy specifications and insertion loss The accuracy specification for mA is ±(0.30 percent of reading + 0.10 percent of range + 2 counts) over the frequency range 40 Hz to 10 kHz (5 kHz on 2000 mA AC range). As in the case of direct current (Sec. 1.4.20), the meter causes some insertion loss when measuring alternating current. The effects of ac insertion loss can be calculated in exactly the same way as for dc insertion loss.

The mA AC ranges are particularly affected by high common-mode voltage, which is produced when making current measurements in circuits that are not directly at earth (power line) levels. This problem is covered further next.

1.4.22 Ground loops, floating measurements and common-mode problems

One of the problems in any meter, particularly differential and digital meters that operate from a power line, is an effect known as *common-mode insertion*. This is essentially a form of noise or undesired signals appearing at the input of an instrument, because of circulating *ground currents* between the meter and circuit under test.

One of the major causes of common-mode signals is *induced ground currents*, usually at the power-line frequency (typically 60 Hz). These signals can generate a potential of several volts between the circuit ground and the meter PC board or case ground. Unless bypassed, these currents cause a voltage to appear at the input. This voltage could be larger than the voltage (or current) being measured. In any event, the ground-loop voltage (or current) is added to the measured voltage, resulting in an improper reading.

Figure 1-13 shows a typical common-mode problem in a digital meter, where a bridge-type transducer is connected to the input of the meter. The output voltage across RH and RL of the bridge is applied to the measuring circuit input and results in the desired reading (bridge circuits are covered further in chapter 6). Notice that there is another circuit (or ac path) through RL, C, and RC. Capacitor C is the input capacitance between the COM terminal of the meter and the board or case ground. *RC* is the resistance between the earth grounds of the transducer and the meter.

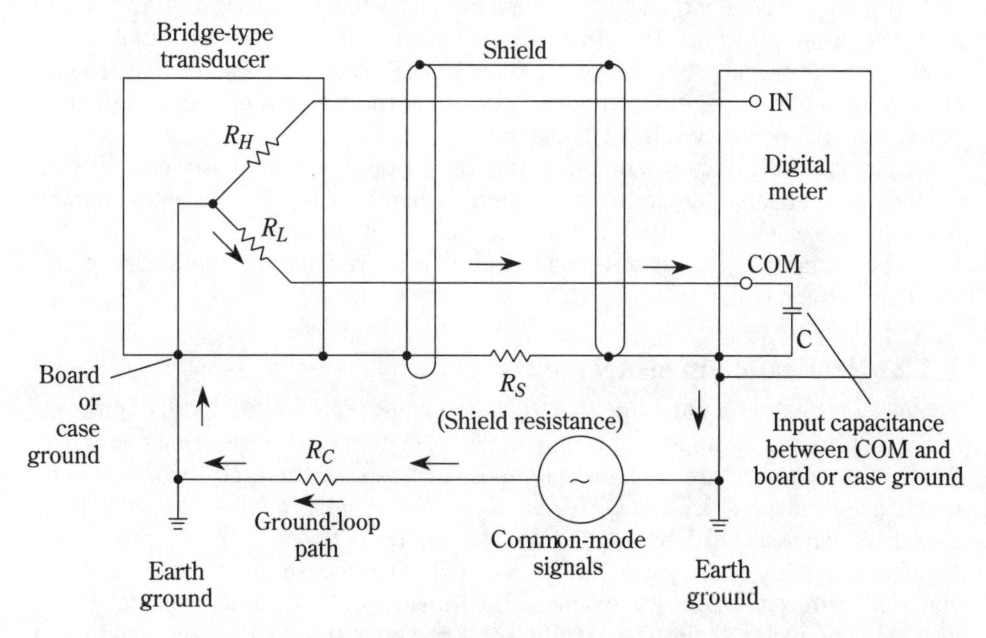

1-13 Typical common-mode problems in a digital meter.

In the circuit of Fig. 1-13, the junction RH and RL (the circuit being measured) is shown connected directly to the board or case ground of the transducer. This connection need not be direct, but it can be reactive (such as the capacitance or inductance between the junction and case or board). Where there is no direct connection to ground, the circuit is said to be *floating*, and the measurement is a *floating measurement*.

Although the values of *C* and *RC* are low, they do exist and provide an ac path for any signals or currents present in the ground line (known as the *ground loop*). These currents or signals cause an undesired voltage to be developed across RL. This voltage is mixed with the desired transducer output and applied to the measuring circuit of the meter. Such voltages or signals are a source of error. The error is most noticed when measuring current (particularly alternating current), but it is also a problem on the voltage ranges.

The ability of a meter to reject a common-mode signal, and thus reduce the undesired signal currents, is called *common-mode rejection* or *common-mode rejection ratio (CMR* or *CMRR)*. The term *CMRR* is usually applied to the rejection capability when both inputs are floating. Neutral-mode rejection ratio (NMRR) is sometimes applied to indicate the rejection capability when one input is grounded. Usually, CMRR and NMRR are specified in decibels (Sec. 1.5.4) at some frequency or range of frequencies. For example, a typical common-mode rejection figure is 130 dB minimum at 60 Hz. This means that the undesired signal effect of a given common-mode signal is reduced by 130 dB or more. Thus, the effect of a 100-V, 60-Hz common-mode signal is reduced to that produced by a 33-μV maximum equivalent signal.

As a point of reference, the meter of Fig. 1-12B has an NMRR rating of greater than 80 dB (over a frequency range of 50 Hz to 20 kHz), a CMRR rating of greater than 100 dB at dc, and greater than 70 dB (over a frequency range of 50 Hz to 1 kHz). These specifications apply to the dc voltmeter function when the meter is operated on line power. The ratings improve to greater than 120 dB at dc and greater than 100 dB (over a frequency range of 50 Hz to 20 kHz) when the meter is operated on battery power (and the power cord is disconnected).

As is typical with many meters, battery operation (with the power cord disconnected) can minimize ground-loop problems when making floating and common-mode measurements. Another way to minimize such problems is to connect the meter power cord to an isolation transformer (as covered under General Safety Precautions in the Introduction).

1.4.23 Resistance measurements

The following considerations apply to the autoranging meter of Fig. 1-12B. However, similar considerations apply to a variety of digital meters when measuring resistance.

Resistance function and test voltages Notice that there are two resistance functions or ranges, Ω LV and Ω HV. The Ω LV function has a full-scale test voltage of 200 mV, whereas Ω HV has a full-scale test voltage of 2 V.

The test voltages are particularly important when measuring resistances in circuits with semiconductors. For example, the threshold of conduction voltage of a silicon semiconductor (transistor, diode, etc.) is about 0.5 V. The threshold for a germanium semiconductor is about 0.2 V. Because the full-scale test voltage on Ω HV exceeds the conduction thresholds on all types of semiconductors, erroneous (usually low) readings can result when resistance measurements are made. The Ω HV function is best suited for measurements that require a higher test voltage, and for very high resistances (up to about 20 MΩ), but should be avoided when semiconductors are included in the circuit. This subject is covered further in Sec. 1.10.

Accuracy specifications and test currents The accuracy specifications for resistance ranges is ± (0.10 percent of reading + 0.10 percent of range + 1 count). When measuring resistance, the voltage developed across the measured resistance is directly proportional to the current applied. For example, a reading of 1.000 kΩ corresponds to a voltage of 1.000 V on Ω HV. The current applied by the meter is determined by the measurement range used. The test current for each range is as follows:

Ω LV	**Ω HV**	**Test current**
0.2 kΩ	2 kΩ	1.0 mA
2 kΩ	20 kΩ	0.1 mA
20 Ω	200 kΩ	10 μA
200 kΩ	2 MΩ	1 μA
2 MΩ	20 MΩ	0.1 μA

Response time considerations The response time for resistance measurements (or time required for an accurate resistance reading) depends on circuit capacitance and test current. As shown by the previous table, the test current on Ω HV is quite low. If there is appreciable capacitance in the circuit under test, the time for the low currents to charge the capacitors can be quite long. For example, on Ω HV, and for measurements in the 20-MΩ range, it can take more than 1 minute for an accurate reading, if the circuit capacitance is 1 μF. Higher capacitances produce even longer response times. When measuring in pure resistive circuits, the response time is less than 3 seconds.

Variations in zero resistance readings Over an extended period of time, there might be some variation in the shorted-input zeroing on Ω LV. The least-significant or right-hand digit might, because of ambient temperature changes, vary from the desired display of .0000 kΩ. On any digital meter, periodically check the resistance-function zeroing. On the meter of Fig. 1-12B, the problem can be minimized using the Ω HV function.

Zero offset in resistance measurements The ZERO control can be used as an offset adjustment to remove small, residual resistance readings (such as the test-lead resistance when using to .2-kΩ range on Ω LV). The ZERO control has an equal effect on all Ω LV ranges and a decade-reduced effect on all Ω HV ranges. Thus, ZERO offset should be used only for the range being used. Readjust the ZERO control when the offset measurements are complete. This is done by setting the RANGE switch to HOLD and connecting the test leads together. Then, adjust the ZERO control until the display indicates .000 MΩ.

High resistance measurements When making measurements in the 20-MΩ range, the input circuit might be susceptible to noise. Voltage-producing noise fields on the test leads can change the display significantly. To keep errors at a minimum under these circumstances, keep the test leads as short as possible and twist the leads to minimize any pickup effects. It is good practice to twist the leads whenever possible to equalize any stray-radiation effects.

1.5 Nondigital meter scales and ranges

Figure 1-14 shows the scales and operating controls for a typical VOM (called a *multitester*). The following notes describe each of the scales and provide information concerning their accuracy and use.

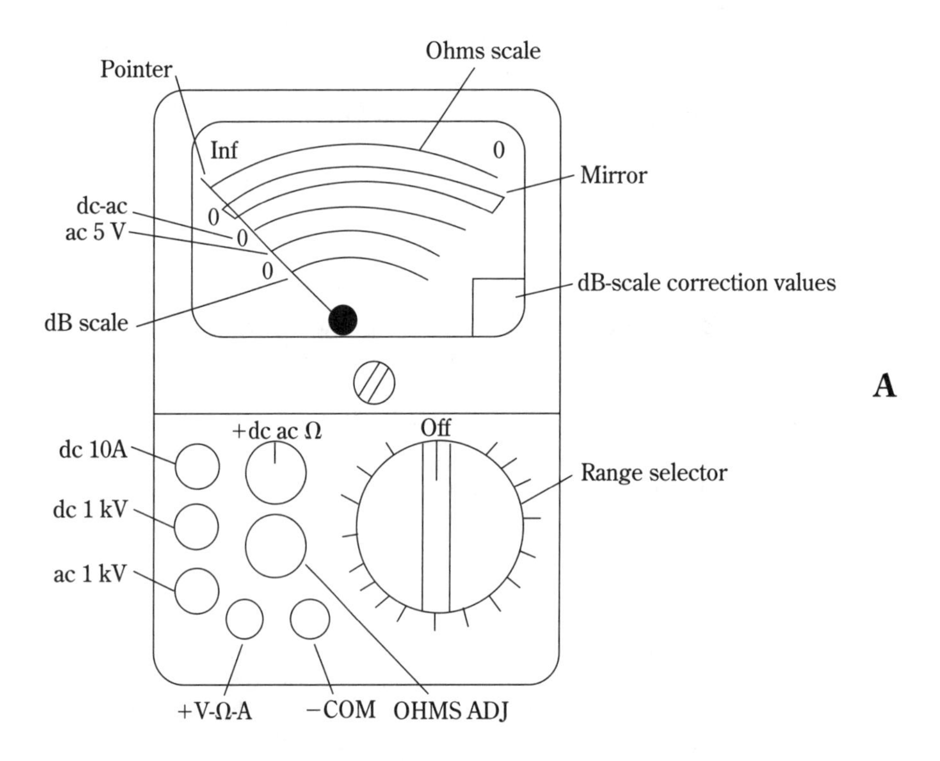

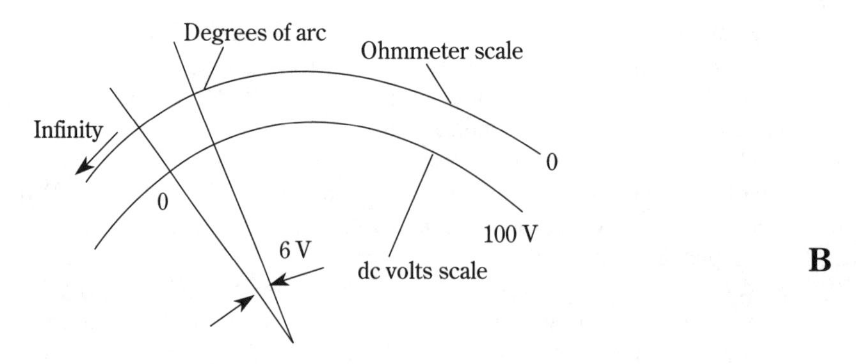

1-14 Typical nondigital (analog) meter scales and range.

1.5.1 Ohmmeter scales and ranges

Notice that the zero indication for the ohmmeter scale is at the right. Although this condition is typical, it is not found on every make and model of nondigital or analog meter. Many times an electronic analog voltohmeter will have the zero indication at the left.

The high-resistance or infinity end of the ohmmeter scale is cramped. Nondigital ohmmeter scales are always nonlinear. As a result, nondigital ohmmeters provide the most accurate indications at midscale or near the low-resistance end.

In general, the ohmmeter scale is considered to be as accurate as the dc voltmeter scale. However, in the case of a battery-operated VOM, the condition of the battery affects accuracy. As a battery ages and the voltage output drops, the resistance indications will be lower than the actual value.

As an example, if a battery voltage drops to 90 percent of the minimum value, a 100-Ω resistor will produce a 90-Ω indication (approximately). This is true even though the ohmmeter is zeroed before making the measurement. Greatest accuracy is obtained if the ohmmeter is zeroed on each range just prior to making the measurement.

In some nondigital meters, the ohmmeter scale is rated in *degrees of arc* rather than percentage of full scale (as are the voltmeter and ammeter scales). For example, the dc voltage accuracy of a meter could be ±3 percent of full-scale. On the 100-V scale, this results in an accuracy of ±3 V. As shown in Fig. 1-14B, an indication of ±3 V (or a total of 6 V) corresponds to a certain number of degrees of arc. In turn, this arc defines the accuracy of the ohmmeter scale. Because the ohmmeter scale is nonlinear, it is possible that the error is not constant over the entire scale. However, the error should not exceed the rated accuracy at any point on the scale.

1.5.2 dc scales and ranges

The zero indication for the dc scales is almost always on the left. Usually, the dc scales are linear, with no cramping or bunching at either end. Notice that there are three basic dc scales (with maximums of 10, 50, and 250) on the meter of Fig. 1-14. These same scales are also used for ac measurements. Each of these scales serves multiple purposes, depending upon the position of the range selector. As a result, you should make note of both the scale reading and the range-selector position when using any nondigital meter.

Also, on basic VOMs such as shown in Fig. 1-14A, it is necessary to connect the test leads to appropriate jacks for different types of measurement (this is usually not necessary when using electronic meters with a probe-type test leads).

In addition to multiple test-lead jacks and a range selector, the meter of Fig. 1-14A has a meter-function switch with two positions (–DC and +DC AC Ω). The –dc position is used only if the voltage to be measured is negative with respect to ground. On simple VOMs, positive and negative voltages are measured by reversing the leads. Usually the black lead is connected to negative and the red lead is connected to positive, when positive dc voltage is to be read. The leads are then reversed for a negative dc voltage.

Always verify the switch positions and lead connections for any meter using the instruction manual (preferably before you ever use the meter). If a manual is not available, check for correct lead connections with a battery or other dc source where polarity is known. While on the subject of instruction manuals, the scales, switches and test leads of nondigital meters are often confusing to the inexperienced operator (or to those who have used only digital meters), so it is a good idea to study whatever instructions are available. The basic procedures for measurements, using both digital and nondigital meters, are described in Sec. 1.10 through 1.13. This will probably convince you to use only digital meters, if you are not already so convinced!

The zero indication for the ac scales is almost always on the left. In some older nondigital models, the ac scales are somewhat nonlinear, with some cramping or bunching on the low end. This is because the rectifier circuits required for ac measurement are nonlinear. Both half-wave and full-wave rectifier circuits are nonlinear. Nonlinearity is more pronounced when the multiplier resistance is small (low-voltage ranges). This condition is known as *swamping effect*.

The ac scales of a nondigital meter are never more accurate and usually less accurate than the dc scales. This is because the inaccuracy of the ac rectifier circuit must be added to the inaccuracy of the dc circuit (multiplier, movement, and scales). Accuracies for a typical nondigital meter are ±3% of full scale for dc and ±5 percent of full scale for ac.

The effects of frequency must be considered in determining the accuracy of ac scales. A typical VOM provides accurate ac voltage indications from 15 Hz to 10 kHz, possibly up to 15 or 20 kHz, but rarely beyond that frequency. This means that VOM readings in the high audio range (chapter 12) might be inaccurate. Notice that an ac meter can provide readings well beyond the maximum rated frequency, but these readings are not necessarily accurate. A typical nondigital electronic meter provides accurate ac voltage indications from 15 Hz up to about 3 MHz. Of course, the frequency range of a meter can be extended with an RF probe. However, the probe and meter must be calibrated together, as described in chapter 5.

Some electronic meters permit ac voltages to be read as an RMS value or a peak-to-peak value, whichever is convenient. Notice that the meter responds to the average value in either case, but that the scales provide for RMS and/or peak-to-peak indications.

The RMS scales are accurate only for a pure sine wave. If there is any distortion or if the voltage contains any component other than a pure sine wave, the readings will be in error. On the other hand, peak-to-peak readings are accurate on any type of waveform, including sine waves.

ac voltage measurements are usually made with a blocking capacitor in series with one of the test leads. On some VOMs, the test lead is connected to a terminal marked OUTPUT or some similar function. This blocks any dc present in the circuit being measured from passing to the meter. Such dc might damage the meter, depending on condition. ac voltage can also be measured on some meters without the blocking capacitor. Always check the instruction manual.

1.5.3 Decibel scales and ranges

Many VOMs are provided with decibel (dB) scales. This is one of the few advantages of a nondigital meter. Actually, the ac voltage circuit is used in the normal manner, except that the readout is made on the dB scales. Inexperienced operators are often confused by the dB scales. The following notes should clarify their use. Decibels are covered further in chapter 12.

The dB scales represent power ratios and not voltage ratios. In most cases, 0 dB is considered as the power of 1 mW (0.001 W) across a 600-Ω pure-resistive load. This also represents 0.775 Vrms across a 600-Ω pure-resistive load. The term *decibel meter (dBM)*, is sometimes used to indicate this system (1 mW across 600 Ω).

The dB scale is related directly to one of the ac scales, usually the lowest scale. The VOM range selector must be set to that ac scale if readings are to be taken directly from the dB scale. If another ac scale is selected, a certain dB value must be added to the indicated dB-scale value. These correction values are printed on the meter face and are applicable to that meter only. Always consult the meter face (or instruction manual) for data regarding scales.

Notice that dB-scale readings are not accurate if (1) the voltages are other than pure sine waves, (2) the load impedances are other than pure resistive, and (3) the load is other than 600 Ω.

If the load is other than 600 Ω, it is possible to apply a correction factor. The decibel is based on the following mathematical function:

$$dB = 10 \log \frac{Power\ output}{Power\ input}$$

Because the power changes by the corresponding ratio when resistance is changed (power increases if resistance decreases and voltage remains the same), it is possible to convert the function to :10 log R_2/R_1, where R_2 is 600 Ω and R_1 is the resistance of the load.

For example, assume that the load resistance is 500 Ω instead of 600 Ω, and a 0-dB indication is obtained (0.775 Vrms).

$$\frac{600}{500} = 1.2, \log 1.2 = 0.0792, 10 \times 0.0792 = 0.792$$

Therefore, 0.792 (or 0.8 for practical purposes) must be added to the 0-dB value to give a true reading of 0.8 dB.

The following table lists correction factors to be applied for some common load-impedance values. This table shows the amount of dB correction to be added to the indicated dB value when the load impedance is other than 600 Ω. For example, if the load impedance is 300 Ω, a +3 dB must be added to the indicated value. This can be verified using the previous equation.

Resistive load at 1 kHz	**dBM**
500	+0.8
300	+3.0
250	+3.8
150	+6.0
50	+10.8
15	+16.0
8	+18.8
3.2	+22.7

Notice that if the load resistance is greater than 600 Ω, the power is reduced, and the correction factor must be subtracted. Notice that dBM values are usually based on a frequency of 1 kHz because this is the frequency where the decibel system most nearly corresponds to the characteristics of the human ear.

dB gain-or-loss There is often confusion in making dB measurements at the input and output of a particular circuit (such as an amplifier) to find gain, particularly with different load impedances. The following rules should clarify this problem.

If the input and output impedances are 600 Ω (or whatever value is used on the meter scale), no problem should be found. Simply make a dB reading at the input and the output (under identical conditions), subtract the smaller dB reading from the larger, and notice the dB gain (or loss). For example, assume that the input shows 3 dB, with 13 dB at the output. This represents a 10-dB gain. If the output is 3 dB, with a 13-dB input, there is a 10-dB power loss.

If the input and output load impedances are not 600 Ω, but are equal to each other, the relative dB gain or loss is correct—even though the absolute dB reading is incorrect. For example, assume that input and output load impedances are both 50 Ω, and that the input shows 3 dB with 13 dB at the output. The previous table shows that 10.8 dB must be added to the input and output readings to get the correct dBM absolute value. However, there is still a 10-dB difference between the two readings. Therefore, the circuit shows a 10-dB gain.

If the input and output load impedances are not equal, the relative dB gain or loss indicated by the meter scales is not correct. For example, assume that the input impedance is 300 Ω, the output impedance is 8 Ω, the input shows +7 dB, and the output shows +3 dB, on the scales of a meter using a 600-Ω reference.

There is an apparent loss of 4 dB (7-dB input, 3-dB output). However, the previous table shows that the 300-Ω input (7 dB) requires a correction of + 3 dB (giving a corrected input of +10 dB), and the 8-Ω output (3 dB) requires a correction of +18.8 dB (giving a corrected output of 21.8 dB). Thus, there is an actual gain of +11.8 dB.

1.6 Meter-protection circuits

Most meters (digital and nondigital) are provided with some form of protection against overloading, accidentally connecting the wrong functions to a particular circuit (such as connecting the ohmmeter to a voltage), and similar occurrences. There are several types of meter-protection circuits. However, they are usually one of the following three types.

1.6.1 Fuse

A fuse can be inserted in the COMMON line of a meter circuit to protect the movement, shunt, and multiplier resistors. A fuse is usually effective only against large surges of current. A delicate meter movement can be damaged—even by small current surges, if they are beyond the maximum capability of the movement. Notice that fuses have resistance values that affect the accuracy of the ohmmeter segment of a VOM. If the fuses are to be replaced, an exact replacement must be used.

1.6.2 Diodes

Diodes are used frequently in meter-protection circuits. The most common are *input clamp diodes* and *meter-movement varistor diodes*.

A clamp-diode network can be placed across the meter input (VOM, electronic, or digital). When the meter is connected to a voltage greater than a bias voltage applied to the network, the diodes conduct and current flows through the network. This drops the input voltage to a safe level within the capability of the meter.

A varistor diode can be placed across the movement of a nondigital meter. A varistor diode (typically silicon) has a high forward resistance until a certain forward voltage is reached. At this voltage (usually a fraction of 1 V), the forward resistance drops to almost zero. Therefore, if the voltage across the meter movement increases to the dangerous level, the forward resistance of the diode drops to near zero, and all the current is passed through the diode.

1.6.3 Relay

Some nondigital meters are provided with a relay that opens the movement circuit should there be an overload. Typically, the relay is actuated when contacts are closed by the pointer being driven off-scale in either direction. When the relay is actuated, the COMMON line (or power line) is opened). Once the relay is actuated, the relay remains tripped until it is reset by a mechanical reset button.

1.7 Parallax problems

In nondigital meters, *parallax* is an error in observation that occurs when the operator's eye is not directly over the pointer. This causes the reading to appear at the right or left of the actual indication. Some manufacturers minimize this problem by placing a mirror behind the pointer on the scale, as shown in Fig. 1-14A. To use a mirrored scale most effectively, close one eye, then move the open eye until the pointer and its reflection appear to coincide (or simply buy a digital meter).

1.8 Movement-accuracy problems

A nondigital meter is no more accurate than the basic movement. As the permanent magnet of a movement ages, the magnetic field weakens, and the indications will be in error (usually the reading is low). This is true even though the precision shunt and multiplier resistors might not have changed. Likewise, as a meter is subjected to shock, vibration, and overloads, the mechanical balance is disturbed, and the pointer moves from the zero position.

Meter movements are provided with maintenance adjustments of various types. Usually both electrical and mechanical adjustments are provided. I recommend that you do not try to calibrate meters (unless you are a qualified instrumentation technician). The one possible exception to this is the *mechanical-zero-adjustment*.

As shown in Fig. 1-14, mechanical-zero is actuated by a screwdriver-type adjustment, which is accessible from the meter front panel (directly below the meter face). This mechanical-zero adjustment is not to be confused with the zero-Ω adjustment or the zero adjustment of an electronic meter. The mechanical-zero adjustment is used to set the pointer at zero when no power is applied to the meter and the test leads are not connected.

1.9 Reading-error problems

There are many causes for inaccurate meter readings, especially with nondigital meters. Some of these are the result of "operator trouble," such as trying to get accuracy greater than the meter's capability. However, it is possible to operate a meter properly and still get inaccurate readings. The following notes summarize the most common causes of reading errors.

Excessive vibration or shock can damage a meter movement, or weaken the permanent magnet.

Overloads can damage meter movements, the multiplier or shunt resistors, and rectifiers, even though the components do not actually burn out. For example, excessive current through a resistor causes heat. In turn, the heat can change the resistance value, making the shunt or multiplier out of tolerance. On those meters with printed-circuit wiring, an overload can cause breakdown between components or burn out part of the wiring board.

When ac is applied to a VOM set to measure dc, the ac passes through the movement. This could cause the pointer to move, depending on meter construction, frequency of the ac voltage, and so on. Even without pointer movement, it is possible for an ac voltage to burn out or damage the meter. The presence of ac on a meter (when the meter is set to measure dc) sometimes shows up as vibration of the pointer.

Static electricity can build up on a meter face. This is often caused by wiping the face with a cloth. Such static electricity can cause the pointer to stick on the inside of the meter face and it could show up as an incorrect reading.

If a nondigital meter is operated in a strong magnetic field, the moving coil might be deflected incorrectly. Also, the permanent magnet might remain damaged even after the field is removed. One particular problem in color-television service is the presence of the degaussing coil. The strong ac field set up by the coil has been known to destroy the permanent magnet of a nondigital meter.

If a meter is operated in the presence of radio-frequency signals, the currents generated by the signals can cause reading errors. Electronic voltmeters and digital meters are most subject to such problems because both contain amplifiers that can amplify even small RF signals (this is one of the few advantages of a basic nondigital VOM). Well-designed electronic and digital meters are shielded against RF signals. The test leads are most vulnerable to RF pickup. A shielded coax or a pair of shielded leads offer the best protection. If neither of these are practical, unshielded test leads can be twisted together to minimize the effects of RF signal pickup.

1.10 Basic ohmmeter (resistance) measurements

The first step in making a resistance measurement with a nondigital meter is to zero the meter on the desired resistance range (refer to Sec. 1.4 for the digital meter procedure). Many nondigital meters can be zeroed on other ranges, and will remain zeroed when the range is changed. However, to be safe, zero the meter on each range, as necessary.

As shown in Fig. 1-15A, a nondigital meter is usually zeroed by touching the two test prods together and adjusting the ZERO Ω or Ω control until the pointer is at "ohmmeter zero." This is usually at the right end of the scale for a VOM and at the left end for an electronic meter.

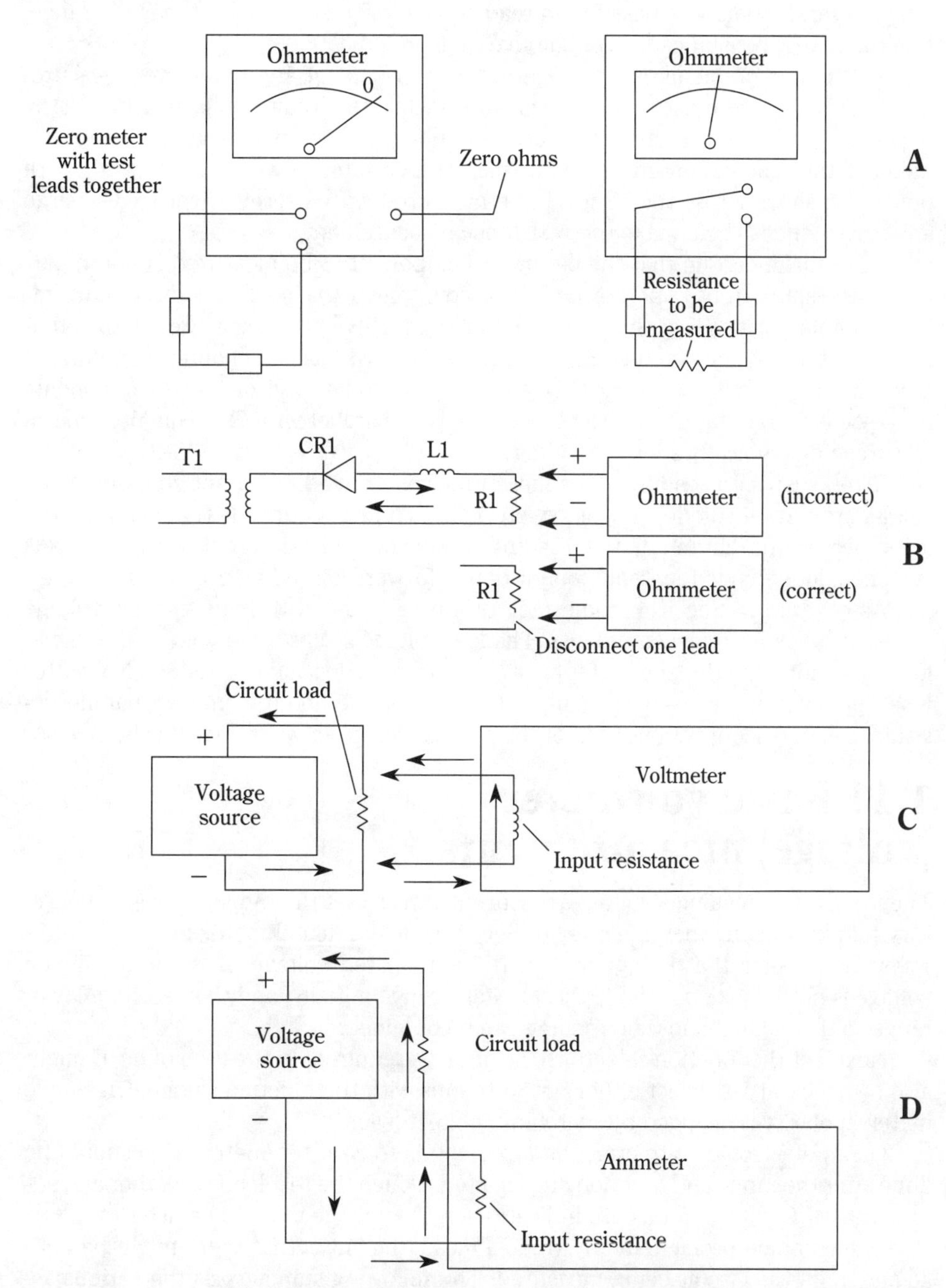

1-15 Basic nondigital-meter measurement techniques.

Once the ohmmeter is zeroed, connect the test prods across the resistance to be measured. Read the resistance from the ohmmeter scale. Make certain to apply any multiplication indicated by the range-selector switch. For example, if an indication of "3" is obtained with the range selector at $R \times 10$, the resistance is 30 Ω. As a practical matter, it should be possible to read such a low resistance on the $R \times 1$ range. Use the lowest resistance range that gives a good indication.

Two major problems must be considered in making any ohmmeter measurements. First, there must be no power applied to the circuit being measured. Any power in the circuit will cause an incorrect reading, and could damage the meter. Remember that capacitors often retain their charge after power is turned off. With power off, short across the circuit to be measured with a screwdriver to discharge any capacitance. Then, make the resistance measurement.

Next, make certain that the circuit or component to be measured is not in parallel with (shunted by) another circuit or component that can pass direct current. For example, assume that the value of R1 in Fig. 1-15B is to be measured. If the ohmmeter battery is connected such that CR1 is forward biased (chapter 10), current flows through the T1 winding, CR1 and the L1 winding. All of these components have some dc resistance, the total of which is in parallel with R1. This produces a false reading (typically a lower reading).

If you suspect a reading when measuring resistance in a circuit containing any solid-state device (diode, transistor, etc.) that can pass current, try switching the leads across the circuit. If the reading is substantially different with the leads switched, look for a circuit component that is forward-biased by the meter voltage.

As covered in Sec. 1.4. some meters are provided with high-and low-voltage scales for resistance measurements. The low-voltage scales use a voltage that is below the conduction threshold of typical solid-state diodes and transistors. No matter what type of meter is used, the simplest and safest method to eliminate parallel resistance is to disconnect one lead of the resistance, as shown in Fig. 1-15B.

1.11 Basic voltmeter (voltage) measurements

The first step in making a voltage measurement is to set the range. This is unnecessary for autoranging meters (refer to Sec. 1.4 for the digital meter procedure). Always use a range that is higher than the anticipated voltage. If the approximate voltage is not known, use the highest voltage range initially, and then select a lower range so that a good midscale reading can be obtained.

Next, set the function selector to ac or dc as required. In the case of dc, it might also be necessary to select either plus or minus with the function switch. On simple meters, polarity is changed by switching the test leads.

On an electronic voltmeter, the next step is to zero the meter. This should be done after the range and function are selected. Touch the test leads together and adjust the ZERO control for a zero indication on the voltage scale to be used.

One common problem in any voltage measurement is that there might be both ac and dc in the circuit being measured. The following summarizes the various aspects of this problem.

To measure ac only but where dc is present, select the OUTPUT or AC ONLY function. This switches a coupling capacitor into the input circuit. On a VOM, this is done by connecting the free test lead to the OUTPUT terminal. In an electronic voltmeter, ac is often selected by a switch on the probe. In some meters, ac is always measured with a coupling capacitor at the input. In any event, the dc is blocked, and the ac is passed.

Use of the OUTPUT function can present problems. The coupling capacitor and the meter resistance form a high-pass filter and might attenuate low-frequency ac voltages. However, most present-day meters provide accurate ac indications above 15 or 20 Hz. It is also possible that the capacitor and meter movement form a resonant circuit that increases the ac signals at some particular frequency (typically at 30 to 60 kHz). Always consider the frequency problem when making any ac voltage measurements.

To measure dc only, but where ac is present, there are several possible solutions. If the ac is of high frequency, it is possible that the meter movement will not respond and there will be no ac indications when the meter is set to measure dc. It is also possible that the meter will not be affected if the ac voltage is low in relation to the dc being measured.

One solution, if the meter is affected by the presence of ac, is to connect a capacitor across the test leads. This provides a bypass for the ac, but it does not affect the dc. However, the capacitor can affect operation of the circuit. Also, remember that the capacitor is charged to the full value of the dc.

In some cases, it is possible to use a high-voltage or attenuator probe (chapter 5) to measure dc in the presence of ac. The series resistance of the probe, combined with the capacitance between the probe's inner and outer conductor or shield, forms a low-pass filter. This filter action has no effect on dc but rejects ac. The fact that electronic voltmeters usually use some form of probe makes these instruments better suited to measure dc (in the presence of ac) than VOMs.

All voltage measurements are made in parallel across the circuit and voltage source, as shown in Fig. 1-15C. This means that some of the current normally passing through the circuit under test is passed through the meter. In a VOM where the total meter resistance (or impedance) is low, considerable current can pass through the meter. This might affect circuit operation. For example, an oscillator that develops a small voltage over a high-impedance circuit can be prevented from oscillating if a VOM is used to measure the voltage (the low-impedance VOM draws excessive current, dropping the voltage to a point where oscillator feedback cannot occur).

This problem of parallel-current drain does not occur in an electronic voltmeter, except when a voltage is measured across a high-impedance circuit. A typical electronic voltmeter has an input impedance of 10 to 15 MΩ. If the circuit impedance is near this value, the current divides between the circuit and the meter, possibly causing an erroneous reading.

1.12 Basic ammeter (current) measurements

The first step in making a current measurement is to set the range. Always use a range that is higher than the anticipated current. If the approximate current is not known, use the highest current range initially, then select a lower range so that a good midscale reading can be obtained.

In many meters, selecting a current range involves more than positioning a switch. A typical VOM requires that the test leads be connected to different terminals, in addition to setting a range switch. Also, it is usually necessary to select ac or dc when measuring current. Many VOMs do not measure ac current.

Notice that when the lowest current scale is selected on a VOM, the meter is actually functioning as a voltmeter. The meter movement (without a shunt) is placed in series with the circuit. Any sudden surges of current can damage the meter movement. There is especially a problem when there is both ac and dc in the circuit being measured. If the ac is of a higher frequency, there is little effect on the meter movement. Lower-frequency ac can combine with the dc and possibly cause reading errors or movement burnout.

All current measurements are made in series with the circuit and power source, as shown in Fig. 1-15D. This means that all of the current normally passing through the circuit under test is passed through the meter. This might or might not affect circuit operation.

1.13 Basic decibel measurements

The procedures for decibel measurements are similar to those for ac voltage measurement, except (1) the OUTPUT function is always used for decibel measurements, and (2) the decibel scales are used instead of the ac scales. When making decibel measurements, use the basic voltage measurement procedures of Sec. 1.11, and observe the precautions concerning decibel scales described in Sec. 1.5.4

1.14 Testing and calibrating meters

All meters should be tested and calibrated periodically against known standards because there can be changes in accuracy because of damage or with normal aging. This applies to both digital and nondigital meters. Meters should also be checked out thoroughly when first placed in use. If the meter is new, check the instrument against the manufacturer's specifications. If the meter is used, and no specifications are available, you can at least establish a reference point for future periodic calibration and test.

Laboratory meters must be checked against precision standards available in the lab, or must be sent out to calibration (metrology) labs. A special problem arises with shop meters. Most shops are not equipped with precision standards, and it is expensive and time-consuming to send meters out for calibration. Therefore, means must be devised to check meters with available standards. The following paragraphs have been included to permit the test and calibration of shop meters against commonly available standards and to make maximum use of these standards.

Notice that the following procedures are for various types of meters. If tests reveal that a meter is not up to standard, the instruction manual for the particular make and model of meter must be checked to find such information as the location of calibration controls or resistors and the actual calibration procedures.

1.15 Ohmmeter test and calibration

The accuracy of an ohmmeter can be checked by measuring the values of precision resistors. If the indicated resistance values are within tolerance, the ohmmeter can be considered as operating properly when making ohmmeter-accuracy checks.

The resistors should have a ±1 percent or better rated accuracy or tolerance. In any event, the resistor accuracy must be greater than the rated ohmmeter accuracy. A typical ohmmeter (VOM or electronic) has a ±2 or ±3 percent accuracy.

Select resistor values that provide midscale indications on each ohmmeter range. Make certain to zero the ohmmeter when changing ranges. Also select precision test resistors that give 25 to 75 percent scale indications. Accuracy is not the same on all parts of the scale because of nonuniform meter movements. This is known as *distribution error*. However, accuracy should be within the rated tolerance on all parts of the scale and on all ranges. Distribution error is generally not a problem in digital meters.

Often, nondigital ohmmeter accuracy is rated in degrees of arc, rather than a percentage of full scale. It is sometimes difficult to relate pointer travel in digress of arc to a percentage. For practical work, remember that the ohmmeter accuracy is approximately equal to the accuracy of the dc scale. For example, assume that the dc scale is rated as accurate to within ±2% accuracy. Then, the ohmmeter scale is also accurate to within the same degree of pointer travel, or arc, as it takes for the pointer to move ±2 small divisions on the dc scale.

1.16 Voltmeter test and calibration

Both the ac and dc scales of a voltmeter must be checked for voltage accuracy. In addition, the ac scales must be checked for accuracy over the entire frequency range.

The obvious method to check the accuracy of a voltmeter is to measure a known voltage or series of voltages, and check that the indicated voltage values are within tolerance. There are two basic methods for making such a test.

1.16.1 Standard voltmeter method

The most convenient method for checking voltmeter accuracy is to compare the voltmeter to be tested against a standard voltmeter of known accuracy. It is common practice in lab work to have one standard voltmeter against which all meters are compared. The standard voltmeter is never used for routine work, but only for test, and is sent out for calibration against a primary standard at regular intervals.

The voltmeter to be tested is connected in parallel with the standard voltmeter, and a variable-voltage source, as shown in Fig. 1-16A. The source is then varied over the entire range of the voltmeter under test, and the voltage indications are compared with those of the standard. In the case of an ac meter, the source is set to a given voltage and the frequency is varied over the range of the meter under test.

With this test method, accuracy depends entirely on the accuracy of the standard meter, and not on the source voltage or frequency. This is very convenient because it is more practical to maintain a meter of known accuracy than a source of known accuracy (especially over a range of voltages).

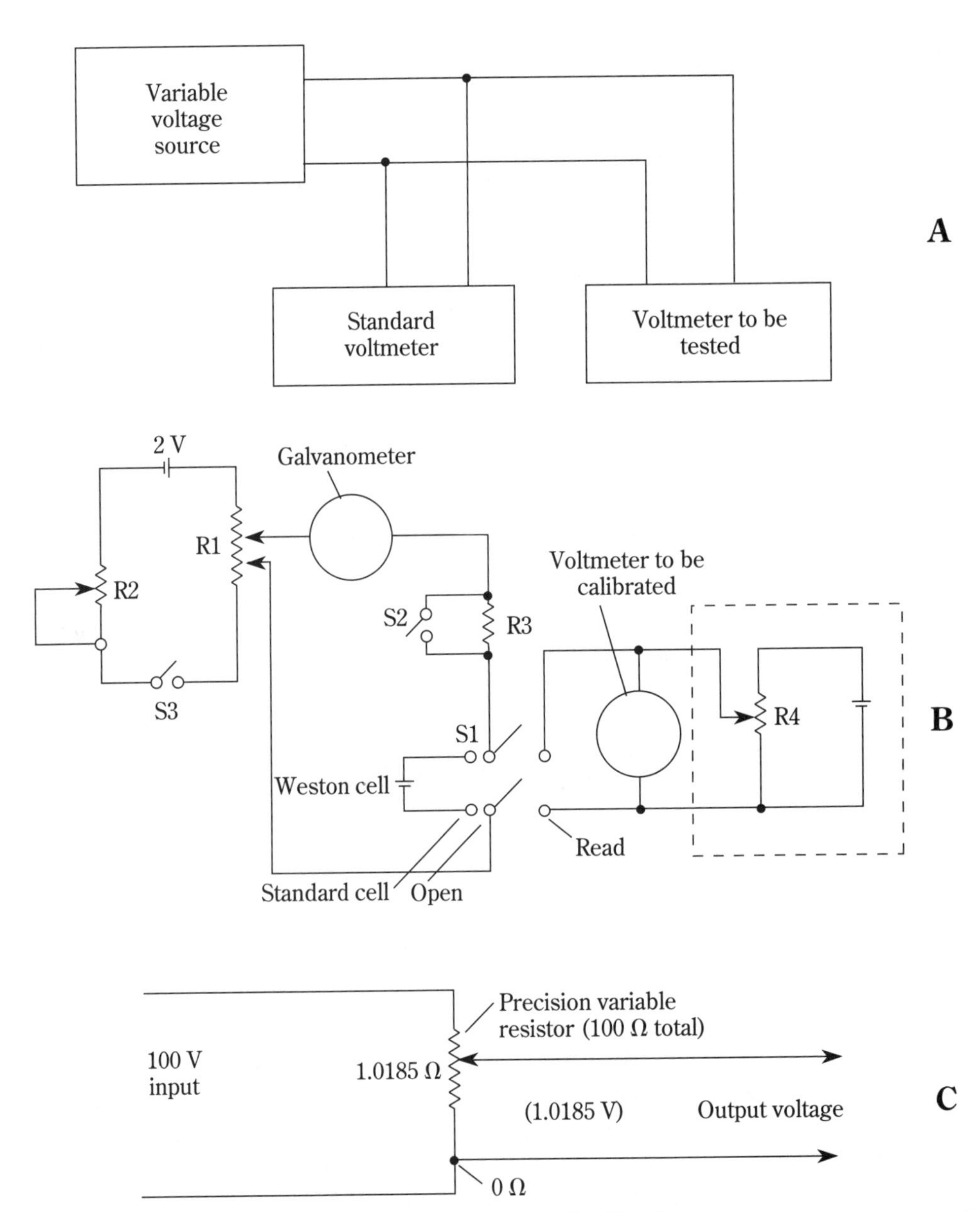

1-16 Voltmeter testing and calibration.

1.16.2 Standard cell method

The next most popular method of checking a meter's voltage accuracy is to measure a voltage source of known accuracy. Usually, this voltage source is a *standard cell*. Inexperienced technicians often check the voltmeter accuracy against a common dry cell or series of dry cells. This is satisfactory for rough shop work, but is not accurate enough for precision work. A single dry cell rarely produces 1.5 V, as is often assumed. The accuracy of a dry cell voltage is usually less than that of a typical VOM or digital meter.

A mercury cell, or series of mercury cells, provides much better accuracy than a common dry cell. The voltage output of a typical mercury cell is 1.35 V. The mercury cell maintains this voltage for a long time, over a wide temperature range, with an accuracy greater than that of a typical shop meter.

The most accurate source is a standard cell, such as the *Weston cell*. There are two types of Weston cells; the *normal cell* and the *student* or *shop cell*. The normal cell is the most accurate, providing of 1.0183 V at 20°C. The student or shop cell provides a voltage from 1.0185 V at 20°C. The student or shop cell provides a voltage from 1.0185 V at 20°C. Thus, the student cell can be off by as much as 0.5 mV. Usually, this is greater accuracy than is required for all but precision lab meters.

Notice that a student cell is less affected by changes in temperature (compared to a normal standard cell). Once the accuracy of a student cell is established, the cell can be considered as remaining at that voltage over the normal range of room-temperature operation.

Inexperienced technicians often connect a meter to be tested directly to a standard cell. Although this provides an accurate indication, it also places a damaging current drain on the standard cell. Laboratories generally use some form of calibration circuit with standard cells. Most of these calibration circuits use the balance method and involve a galvanometer, such as the special bridge method, and involve a galvanometer (such as the special bridge circuits described in chapter 6).

Figure 1-16B shows a calibration circuit using the balance method. Switch S1 is left in the open position. Potentiometers R1 and R2 are adjusted so that the voltage at the tap of R1 is approximately equal to the voltage of the standard cell. S1 is then set to the STANDARD CELL position, and R1 is adjusted until the galvanometer reads zero. At this point, the voltage at S1 is equal to the standard-cell voltage. Initially, S2 is open to place R3 in series with the galvanometer (this protects sensitive galvanometers against large voltage unbalance). Once approximate balance is reached, S2 is closed, and the voltage is adjusted to exact zero. Circuits of this type make it possible to get a balance while drawing less than 0.1 mA from the standard cell.

Once balance is obtained, S1 is set to the READ position. This removes the standard cell from the circuit and places the voltage across the meter under test. If a student or shop Weston cell is used in the circuit, the meter under test should read between 1.0185 and 1.0190 V.

A further variation of the calibration circuit is shown in Fig. 1-16B in dashed lines. This variation is used to protect the sensitive galvanometer. After the galvanometer is balanced against the standard cell and S1 is set to READ, R4 is adjusted until the galvanometer again shows a balance, indicating that the voltage at the tap of R4 is equal to that of R1 (and the standard cell). Then, S1 is set to OPEN, leaving only the voltage from R4 across the meter under test.

1.16.3 Precision voltage-divider method

It is obvious that the circuit of Fig. 1-16B provides a good indication on the 2.5-V scale of a meter, but is of little value on the 250-V range, or any range substantially higher than 2.5 V. Likewise, the circuit can not be used on a lower range such as 250 mV. These problems can be overcome with a *voltage-divider circuit.*

The basic voltage divider can be any form of precision variable resistor from which the exact resistance can be read out, or where the exact ratio of full-resistance to tap-resistance can be shown by an external indicator. Such devices have various names, such as *volt box, decade-ratio potentiometer, ratiometer*, or *decade box*.

Figure 1-16C shows how a voltage divider can be used in meter calibration. In this circuit, a supposed 100-V source is placed across a precision variable-voltage divider of 100 Ω. Each 1-Ω tap or position on the voltage divider should show a corresponding voltage (the 3-Ω position shows 3 V, the 33-Ω position shows 33 V, etc.). The divider can then be set to the 1.0185- or 1.0190-V position, and the resulting voltage balanced against a standard cell. This establishes the accuracy of the voltage source.

For example, assume that the circuit shows a balance against a 1.0190-V standard cell when the resistance shows 1.000 Ω, or about 2% off. This means that the voltage source across the voltage divider is 2% off (2% high), and the voltage readings are 2% off. Notice that the accuracy of the voltage-divider system depends on the accuracy of the voltage divider, not on the accuracy of the voltage source.

Now assume that a 2-V source is placed across a precision variable-voltage divider of 2000 Ω. Each 1-Ω position of the divider is equal to 1 mV with a 2000-mV total. The divider can be set to 1019 mV and balanced against a standard cell to determine accuracy. Once the accuracy is set, the voltage divider can be used to check the entire 250-mV range.

1.17 Ammeter test and calibration

The scales of an ammeter must be checked for accuracy of the current indication. A typical VOM has only dc current ranges, and a typical electronic meter has no current ranges. Most digital meters have both.

1.17.1 Standard ammeter method

The most convenient method of checking current accuracy is to compare the ammeter to be tested against a standard ammeter of known accuracy. As shown in Fig. 1-17A, the ammeter to be tested is connected in series with the standard ammeter and a variable current source. The source is then varied over the entire range of the ammeter, and the current indications are compared. The accuracy of this test depends entirely on the accuracy of the standard ammeter and not on the current source.

1.17.2 Precision voltmeter/resistance method

The next most popular method for checking current accuracy is to use a precision voltmeter and precision resistance. The ammeter to be checked and the precision resistance are connected in series with a variable power source, as shown in Fig. 1-17B. The precision voltmeter is connected across the precision resistance. Current through the circuit is computed using Ohm's law ($I = E/R$).

The value of the precision resistor is chosen so that the voltage indication on the precision voltmeter can be related directly to current. For example, if a 1-Ω resis-

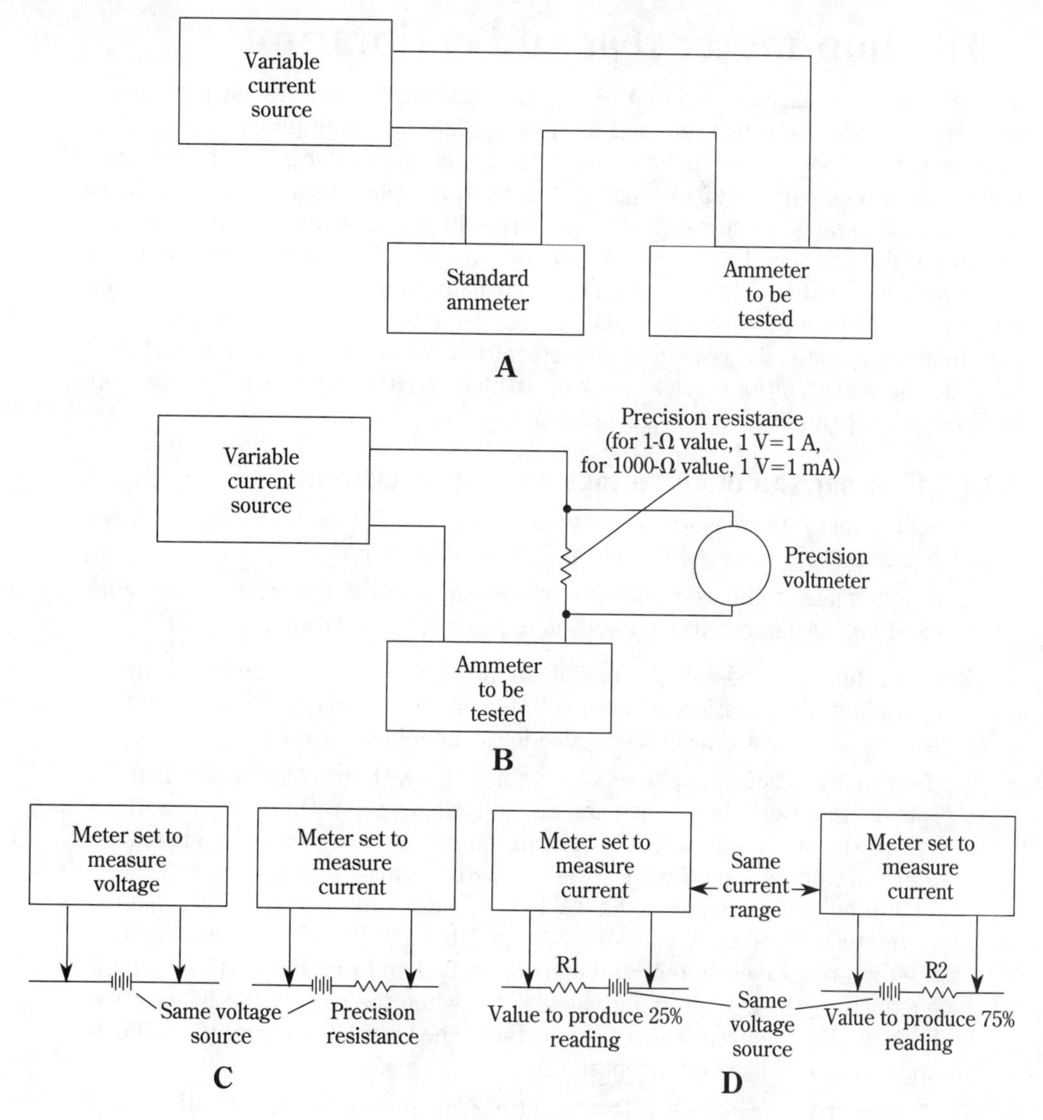

1-17 Meter testing and calibration.

tance is used, the voltage across the resistance can be read directly in amperes (3 V equals 3 A, 7 V equals 7 A, etc.). If a 1000-Ω precision resistor is used, the voltage can be read directly as milliamperes (3 V equals 3 mA, etc.).

The accuracy of this test depends on the accuracy of both the voltmeter and series resistance. The tolerance of both components must be added. For example, assume that both the voltmeter and resistance have a 1 percent tolerance. Then accuracy of the circuit can be no greater than 2 percent. In any event, the combined accuracy must be greater than that of the meter to be tested.

1.18 Shop-meter test and calibration

Most shops do not have a standard cell or even a standard meter of known accuracy against which the shop meters can be checked. The following procedures can be used to test and calibrate shop meters without the use of standards. The procedures are based on comparison of internal scales, such as comparison of the ac scales against the dc scales, or current scales against voltage scales, or one range against another. If the scales and ranges are compared with the same source, and the scales or ranges agree within a given tolerance, it is usually safe to assume that the meter is accurate within a given tolerance. Of course, these procedures do not show such conditions as a gradually weakening movement of a VOM, and should be used only for shop equipment, or for a quick check of lab meters. However, the procedures can be used to locate an inaccurate range or scale.

1.18.1 Comparison of dc voltage against dc current

1. Set the meter to measure dc voltage. Connect the test leads to a fixed-voltage source and measure the voltage as shown in Fig. 1-17C.
2. Set the meter to measure direct current. Connect the test leads to the same fixed-voltage source and a precision resistor, also as shown in Fig. 1-17C.
3. Using the indicated-voltage reading obtained in step 1, calculate the current by adding the precision resistance to the meter internal resistance (on that particular current range). Then, divide by the indicated voltage.
4. If the indicated current agrees within tolerance with the calculated current, it is likely that both the voltage and current ranges are within tolerance. However, if the indicated current is quite different from the calculated current, either the voltmeter multiplier or the ammeter shunt is defective.

   For example, assume that a 3-V battery is measured in step 1 and that the voltmeter indicates 2 V (because of a defective multiplier). Also assume that the total resistance (meter and resistor) is 3 Ω and that the current shunt is good. Using the 2-V figure for calculation (when the battery produces a true 3 V) results in a lower calculated current. The calculated current is 0.666 A, and the true indicated current is 1 A.
5. It is important to make this test with the same power source for both voltage and current readings. Check each voltage range against each current range for a complete test.
6. It is also important to remember that the input resistance of the meter changes for each current range. The input resistances are usually listed in the meter instruction manual. If not, use the following procedure to find the input resistance of a meter on each current range.

1.18.2 Finding input resistance of current ranges

1. Set the meter to measure direct current. Connect the test leads to a fixed-voltage source and precision resistor R1 as shown in Fig. 1-17D. The value of

the resistor should be such that the current reading is about 25 percent of full scale.

2. Replace the precision resistor with another value (R2) that produces a reading 75 percent of full scale, also as shown in Fig. 1-17D. Do not change current ranges.
3. Using the following equation, calculate the input resistance of the current range:

$$\textit{Input resistance} = \frac{I_1R_1 - I_2R_2}{I_2 - I_1}$$

where: R_1 = resistance value that produces a 25 percent reading
R_2 = resistance value that produces a 75 percent reading
I_1 = current indicated when R1 is in the circuit
I_2 = current indicated when R2 is in the circuit

1.19 Testing meter characteristics

In addition to testing a meter for accuracy, it is often helpful to check other characteristics of the meter circuit. The following paragraphs describe the procedures for testing two important meter characteristics.

1.19.1 Checking the ohms-per-volt rating of a VOM

There are two methods for measuring the ohms-per-volt rating of a VOM. One method (with fixed resistor) quickly determines if the ohms-per-volt characteristic is correct (as rated by the manufacturer) but does not show the actual ohms-per-volt rating. The *fixed-resistor method* is as follows:

1. Set the meter to measure dc voltage.
2. Connect the test leads to a fixed-voltage source as shown in Fig. 1-18A that provides an approximate 75 percent full-scale reading.
3. Disconnect the test leads. Insert a precision resistor in the circuit as shown and reconnect the test leads. The value of the fixed resistor that should be equal to the full-scale value is 3 V on a 20,000-ohms-per-volt meter, the precision resistor value should be 60,000 Ω.
4. If the ohms-per-volt characteristics are correct, the voltage indication drops to one-half that obtained in step 2.
5. Repeat the test on each meter scale, using precision resistors of the appropriate value.
6. The test can be made on the ac scale using the same procedure. However, the meter must be set to measure ac and an ac source must be used. Also, it should be noted that the ac Ω-per-volt rating is usually much lower than the dc rating of a VOM.

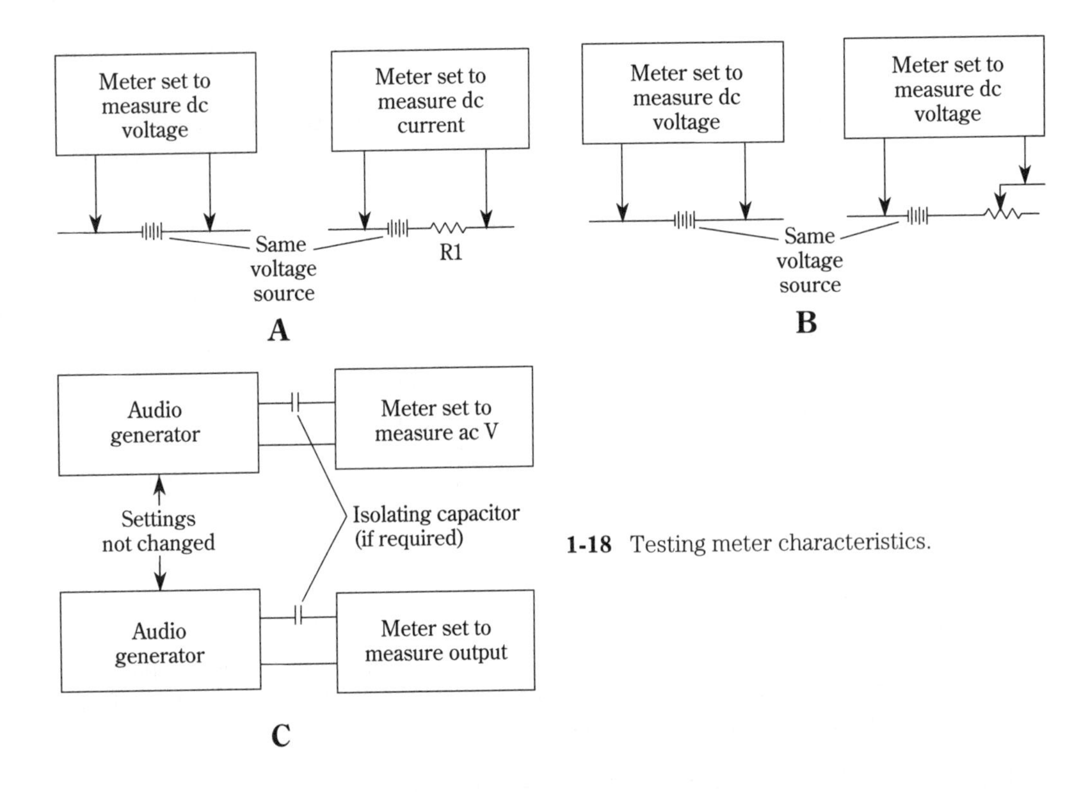

1-18 Testing meter characteristics.

The second method (with variable resistor) shows the actual Ω-per-volt rating. The *variable-resistor method* is as follows:

1. Set the meter to measure dc voltage.
2. Connect the test leads to a fixed-voltage source as shown in Fig. 1-18B that provides an approximate 75 percent full-scale rading. Notice the exact voltage reading.
3. Disconnect the test leads. Insert a variable resistor in the circuit, as shown and reconnect the test leads. Set the variable resistor to the approximate value of the ohms-per-volt rating before connecting the resistor into the circuit.
4. Adjust the variable resistor until the voltage reading is exactly one-half that obtained in step 2.
5. Disconnect the test leads and measure the value of the variable resistor with a precision ohmmeter. The variable-resistor value is equal to the ohms-per-volt rating of the meter.
6. Repeat the test on all scales, including the ac scales if desired.

The ohms-per-volt rating of any meter should be the reciprocal of the current required for a full-scale reading on the basic meter movement. For example, a 100-µA movement produces a 10,000-ohms-per-volt meter (1/0.0001 A = 10,000).

1.19.2 Checking the effect of OUTPUT function on ac voltage readings

As covered in Sec. 1.11, use of the output function might cause an error in readings, particularly at low frequencies. The actual effect can be measured and a scale-factor applied to the readings made in the output mode of operation.

1. Connect the meter to an audio generator as shown in Fig. 1-18C.
2. Set the meter to measure ac voltage directly and not through the OUTPUT terminal.
3. Set the audio generator to some low frequency near the low end of the rated frequency range (20 Hz, 60 Hz, etc.).
4. Adjust the amplitude of the audio-generator signal so that the meter indicates some ac voltage near midscale.
5. Without changing any of the audio-generator or meter controls, apply the generator output through the OUTPUT function.
6. If there is no change in voltage indication, the OUTPUT function has no effect on the circuit. However, if the reading is lower in the OUTPUT function (as is usually the case), a scale factor must be applied. For example, assume that the reading is 7 V on normal ac and 5 V in the OUTPUT mode. Then a 7/5-scale factor must be applied to all readings made in the OUTPUT mode (divide the OUTPUT reading by 5, then multiply by 7).
7. If there is any dc voltage at the audio-generator terminals, use an isolating capacitor, as shown in Fig. 1-18C.

1.20 Calculating multiplier and shunt values

It is possible to convert a basic meter movement into a multirange voltmeter (by adding multiplier) or ammeter (by adding shunts). To calculate the values for multipliers and shunts, it is necessary to know the current required for full-scale deflection and the resistance of the basic meter movement. Usually, the full-scale deflection current is indicated on the scale face. However, the internal resistance is usually not marked.

1.20.1 Finding internal resistance of meter movement and ranges

The internal resistance of a meter movement can be found using the test circuits of Figs. 1-19A or 1-19B. These same test circuits can be used to find the internal resistance of a VOM on its various dc current ranges.

To use the circuit of Fig. 1-19A, disconnect R2 from the circuit by opening switch S1. Adjust R1 until the meter movement is at full-scale deflection. Close switch S1 and adjust R2 until the meter movement is at exactly one-half scale deflection. Disconnect R2 from the circuit and measure the R_2 resistance value with an ohmmeter. The R_2 resistance value is equal to the internal resistance of the meter movement.

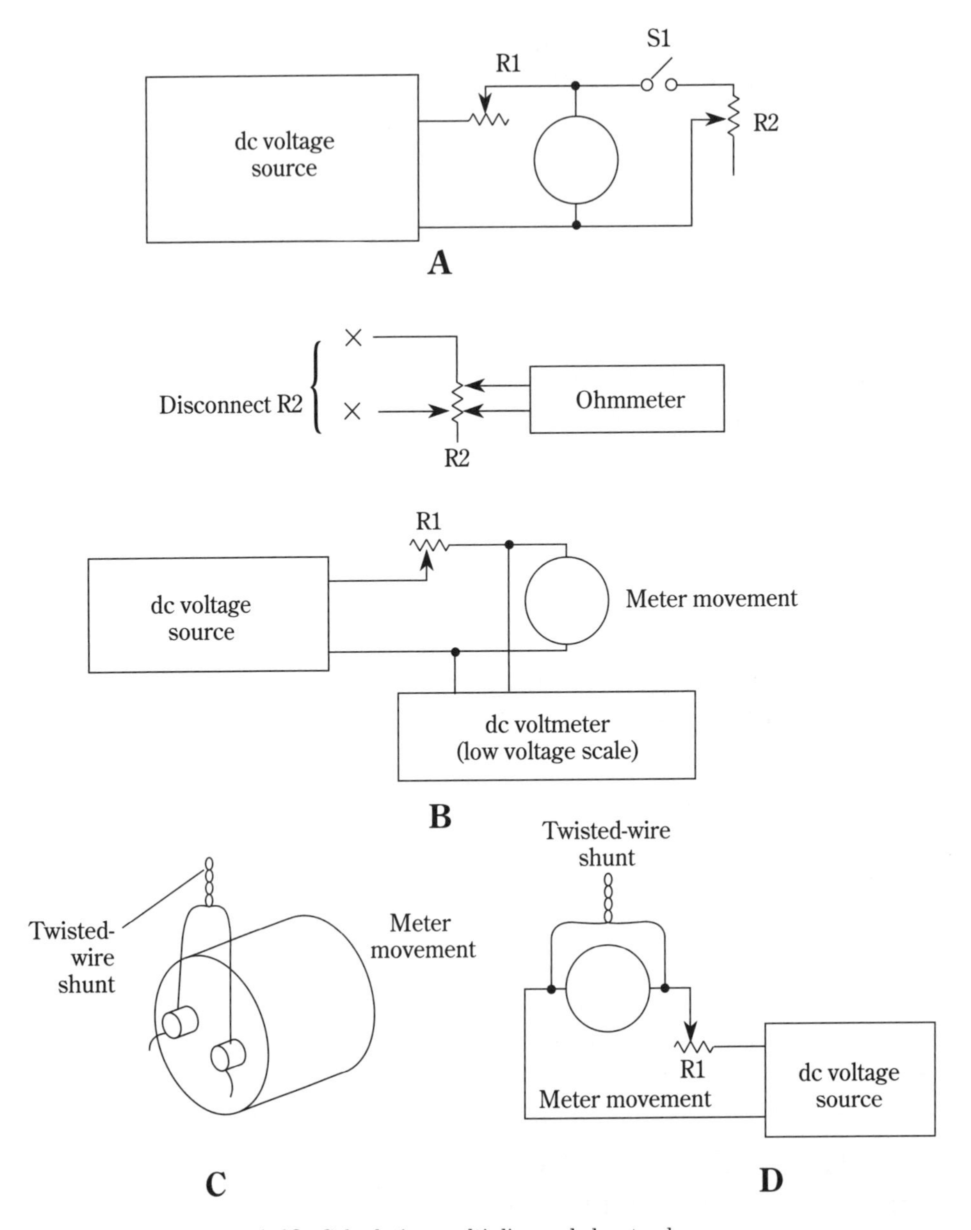

1-19 Calculating multiplier and shunt values.

To use the circuit of Fig. 1-19B, adjust R1 until the meter movement is at approximately 75 percent of full-scale deflection. Adjust R1 for some exact current value. Calculate the internal resistance using Ohm's law ($R = E/I$). Notice that the voltage drop across a typical movement is less than 1 V. Therefore, it is usually necessary to use the lowest scale of the voltmeter.

Never connect an ohmmeter directly across the meter movement. This will damage (and probably burn out) the movement.

1.20.2 Finding multiplier resistance values

The multiplier resistance that is required to convert a basic meter movement into a voltmeter capable of measuring a given voltage range can be calculated using the following equation:

$$R_X = \frac{R_M(V_2 - V_1)}{V_1}$$

where: R_X = multiplier resistance (in series with the meter movement)
R_M = internal resistance of the meter movement
V_1 = voltage required for full-scale deflection of the meter movement
V_2 = voltage desired for full-scale deflection (maximum voltage range desired)

For example, assume that a meter has a 50-µA full-scale movement (with 50 equal divisions on the scale), an internal resistance of 300 Ω, and that it is desired to convert the meter movement to measure 0 to 100 V (each of the 50 scale divisions represents 2 V).

1. Find the voltage required for full-scale deflection of the meter movement: $E = IR$ or $(3 \times 10^2) \times (5 \times 10^{-5}) = 0.015$ V. This is voltage V_1.
2. Use the given equation to find the value of R_X.

$$R_X = \frac{300(100 - 0.015)}{0.015} = 1{,}999{,}700\ \Omega$$

3. To verify this multiplier resistance, add the meter internal resistance (300 Ω) and divide by the full-scale voltage obtained with the multiplier (100 V) to find the ohms-per-volt rating:

$$1{,}999{,}700 + 300 = 2{,}000{,}000; \quad \frac{2{,}000{,}000}{100} = 20{,}000 \text{ ohms/V}$$

4. Divide the full-scale current (50 A) into 1 to find an ohms-per-volt rating of 20,000. Both ohms-per-volt ratings should match.

1.20.3 Finding shunt resistance values

The shunt resistance required to convert a basic meter movement into an ammeter capable of measuring a given current range can be calculated either of two ways:

With method 1, use the following equation:

$$R_X = \frac{R_M}{N - 1}$$

where: R_X = shunt resistance (in parallel with the meter movement)
R_M = internal resistance of the meter movement
N = multiplication factor by which the scale factor is to be increased.

For example, assume that the meter has a 50-μA full-scale movement (with 50 equal divisions on the scale), an internal resistance of 300 Ω, and that it is desired to convert the meter to measure 0 to 100 mA (each of the 50 scale divisions then represents 2 mA).

1. Find the multiplication factor (or N factor) by which the movement is to be increased: N equals the desired scale factor (0.1 A or 100 mA) divided by the movement current (0.00005 A) equals 2000.
2. Use the given equation to find R_X:

$$R_X = \frac{300}{2000 - 1} = 0.150075\ \Omega$$

With method 2, use the following steps.

1. Find the voltage required for full-scale deflection of the meter movement: $E = IR$ or $(3 \times 10^2) \times (5 \times 10^{-5}) = 0.015$.
2. Subtract the meter-movement full-scale current (0.00005 A) from the desired current (0.1 A) to find the current that must flow through the shunt: $0.1 - 0.00005 = 0.09995$ A.
3. Using Ohm's law, find the shunt resistance: $R = E/I$, or $0.015/0.09995 = 0.150075\ \Omega$.

1.21 Fabricating temporary shunts

The internal shunts of commercial meters are precision resistors. External shunts are usually strips or bars of metal connected directly between the meter-movement terminals. Sometimes shunts are strips of metal mounted on insulators. Commercial shunts should be used for permanent operation. However, for temporary use, it is possible to extend the range of any basic meter movement many times using nothing more than a piece of bare wire.

The basic arrangement is shown in Fig. 1-19C, and the calibration schematic is shown in Fig. 1-19D. If a piece of wire is added across the terminals of a basic movement, part of the current passes through the wire. If the wire is twisted as shown, it is possible to adjust the wire's resistance. This controls the amount of current through the wire. The wire is twisted or untwisted as necessary, depending on whether more or less resistance is needed. It is not necessary to know the resistance of the shunt or the internal resistance of the meter movement, but only the full-scale deflection of the movement and the full-scale deflection that is desired.

1. Connect the meter movement to the voltage source, as shown (assume that the meter movement has a 50-μA full-scale rating).
2. Set it to the full value before connecting the meter. Then, gradually reduce the R_1 resistance until the meter reads full scale, 50 μA.
3. Connect the twisted-wire shunt. The meter movement should drop back to zero.
4. Twist or untwist the wire until the movement reads exactly half-scale, 25 μA (the actual current is still 50 μA because the R_1 resistance has not changed).

5. With the shunt wire connected and twisted so that an actual 50-μA flow shows as 25 μA (half scale), full scale is 100 μA. Thus, the 50-μA movement is extended to 100 μA.
6. To extend the range further, adjust R1 until the meter again reads full scale (which is now 100 μA). Then, twist or untwist the shunt wire until the movement again reads half-scale. The actual current flow is still 100 μA, and the movement is extended to a 200-μA full-scale rating.
7. In theory, the process can be repeated any number of times until the movement is extended to any desired range. In practice, a twisted-wire shunt has an obvious drawback. If the wire is exposed to any handling, the shunt resistance changes and throws the calibration off. However, the method is quite accurate for temporary use.

1.22 Extending ohmmeter ranges

It is often convenient to extend the range of an ohmmeter to measure either very high resistances or very low resistances. This is possible using high-ohms and low-ohms adapters. The circuit for a typical high-Ω adapter is shown in Fig. 1-20A. This basic circuit is suitable for use with a VOM or electronic meter. A low-ohms adapter suitable for a VOM is shown in Fig. 1-20B. These adapters can be packaged as a form of probe, if desired. For greatest convenience, the adapter-circuit values should be chosen for a multiplication factor of 10. That is, a high-ohms adapter should increase the ohmmeter reading 10 times, and a low-ohms adapter should decrease the reading 10 times.

Although both circuits use the ohmmeter scales for readout, the low-ohms adapter does not use the VOM internal ohmmeter circuit. Instead, the adapter circuit is connected to the lowest current scale (usually this means a connection directly to the meter movement).

1.22.1 High-ohms adapter

To increase the ohmmeter range by a factor of 10, operate the meter on the highest ohmmeter range (usually $R \times 10{,}000$ or $R \times 100{,}000$) and connect the circuit of Fig. 1-20A.

The external battery should be such that the total voltage (external battery plus internal ohmmeter battery) is 10 times that of the internal ohmmeter battery. For example, if a 4.5-V internal battery is used, the total voltage should be 45 V, and the external battery should be 40.5 V. If it is not practical to get a 40.5-V battery, use a 45-V battery with a 4.5-V external opposing battery in series. This is shown in dashed form in Fig. 1-20A. The total voltage is then the desired 45 V.

The value of the external resistor should be such that the total resistance (external resistor plus input resistance on the highest ohmmeter range) is 10 times that of the input resistance. A good approximation of the input resistance is obtained by noting the center-scale indication on the ohmmeter. 12 and 4.5 are typical center-scale indications on VOMs. These represent 120,000- and 45,000-Ω input resistance, respectively, on the $R \times 10{,}000$ scale.

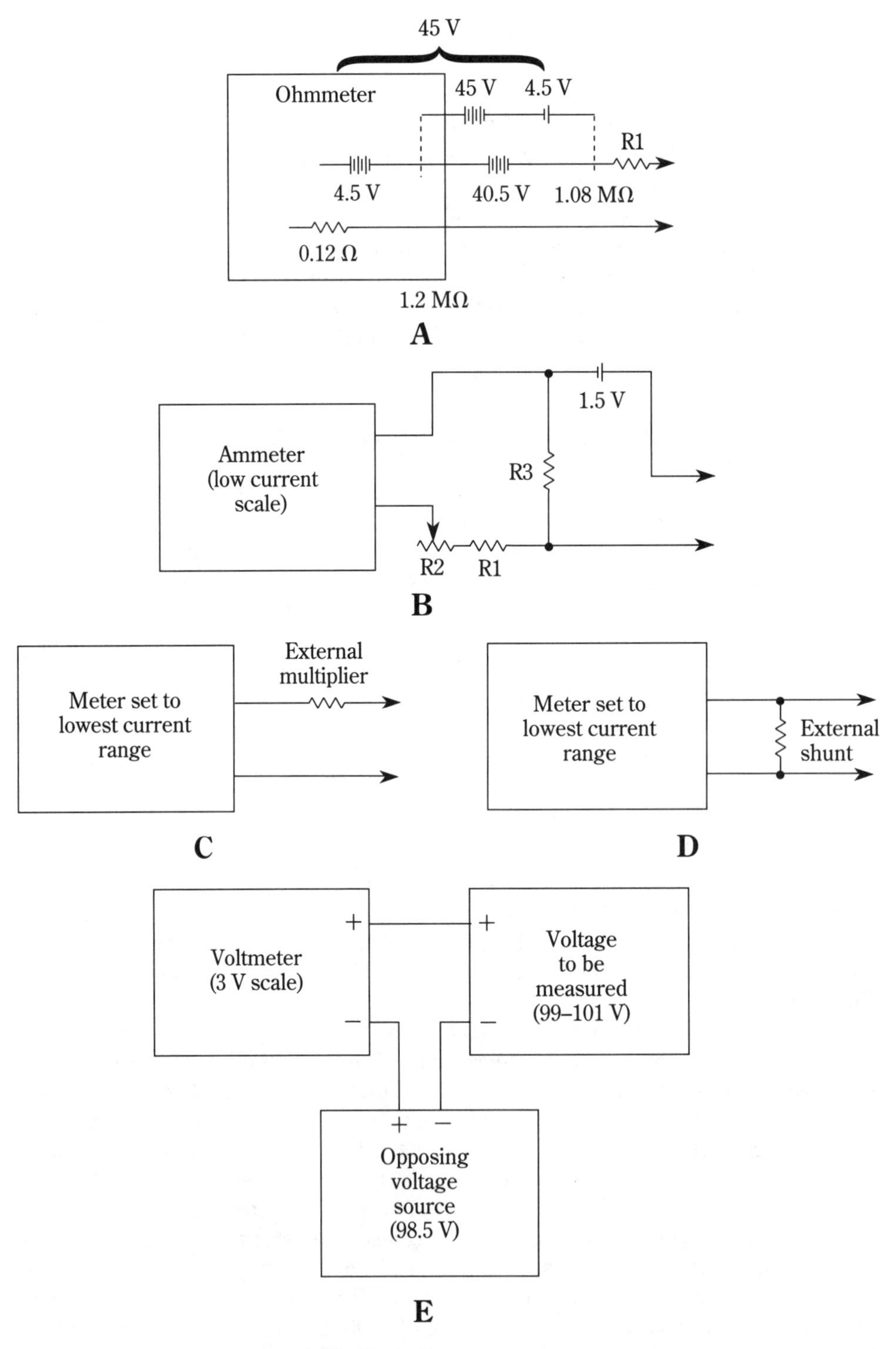

1-20 Extending meter ranges.

As an example, assume that the input resistance is 120,000 Ω. This requires a total resistance of 1.2 MΩ, and an external resistor of 1.08 MΩ (1.1 MΩ for practical purposes). The exact value of the external resistor is not too critical because the entire circuit is adjusted for zero with the ohmmeter ZERO-OHMS control.

The high-ohms adapter is used in the same way as any conventional ohmmeter. The test leads are shorted together, and the meter is zeroed. Then, measurements are made in the normal manner. To check the accuracy of a high-ohms adapter, measure the value of a precision resistor. The accuracy of the high-ohms circuit readings should be within 1 percent of readings made with the ohmmeter. Therefore, if the ohmmeter is rated ±3 percent, then the reading obtained with the high-ohms adapter connected should be ±4 percent.

1.22.2 Low-Ω adapter

To decrease the ohmmeter range by a factor of 10, operate the meter on the lowest current range (usually this is at the basic meter movement, and is 50 μA for a 20,000-Ω/V meter, 20 μA for 50,000 Ω/V, etc.) and connect the circuit of Fig. 1-20B.

The external-battery voltage is not critical. 1.5 V is chosen for convenience. However, the battery should have a high current capacity because there is heavy current drain in the low-ohms circuit.

The value of a fixed resistor R1 and external zero-adjust resistor R2 combined in series should be approximately 10 times the internal resistance of the meter movement. The exact values of R_1 and R_2 are not critical because the entire circuit is adjusted for zero with R2.

The value of shunt resistor R3 should about 0.1 times the center-scale Ω value when the ohmmeter is set to the lowest resistance range (usually $R \times 1$). For example, if the center-scale indication is 12 Ω (on $R \times 1$), the value of R_3 should be 1.2 ohms. A more exact value would be 0.095 times the center-scale reading, or 1.14 Ω for a 12-Ω indication.

R_3 is critical because the shunt resistor determines the 10:1 scale reduction. To check the accuracy of the low-ohms circuit, short the test leads together, zero the circuit with R2, then measure the value of a precision resistor. Use a precision standard resistor in the range 3 to 7 Ω. Remember that a 3-Ω resistor shows a 30-Ω reading on the ohmmeter scale. Then try a 1-Ω (or less) precision resistor. If necessary, select a different value for R3 to obtain a correct low-resistance reading.

Make certain to zero the circuit with R2 before making each resistance measurement. Also, do not keep the low-ohms adapter circuit connected to the circuit under test (or hold the test leads shorted) for a long time. This causes considerable current flow through R3. The heat thus generated may change the resistance value of R3. For best results, use a wire that is the least heat-sensitive (such as Manganin wire).

The low-ohms adapter is used in the same way as any conventional ohmmeter circuit. The test leads are shorted together, and the meter is zeroed with R2. Then, measurements are made in the normal manner. The low-ohms adapter is most effective in measuring values, such as the resistance of a cold-soldered joint, the resistance of a switch contact, and so on.

1.23 Extending voltmeter ranges

The range of a voltmeter can be extended to measure high voltages with a high-voltage probe, as covered in chapter 5, or with external multiplier resistors connected to the basic meter movement, as shown in Fig. 1-20C. In most meters, the basic movement is used on the lowest current range. Values for multipliers can be determined using the procedures of Sec. 1.20.

The basic meter movement can also be used to measure very low voltages. However, great care must be used not to exceed the voltage drop required for full-scale deflection of the basic movement. For example, assume that a 50-μA movement has a 2000-Ω internal resistance. Then, the voltage drop required for full-scale deflection is 0.1 V. Voltages greater than 0.1 V will destroy the meter. To be safe, make a rough check of the voltage to be measured on one of the voltage scales, then work down to the lowest voltage scale. If the indications appear well below the calculated full-scale deflection of the basic movement, it is safe to proceed.

1.24 Extending ammeter ranges

The range of an ammeter can be extended to measure high current values with an external shunt connect to the basic movement (usually the lowest current range), as shown in Fig. 1-20D. Values for shunts can be determined using the procedures of Sec. 1.20.

The range of the basic meter movement cannot be lowered. For example, if a 100-μA movement with 100 scale divisions is used to measure 1 μA, the meter will deflect by only one division. It is not practical to obtain greater deflection.

1.25 Suppressed-zero voltage measurements

Sometimes it is convenient to measure a large voltage source that is subject to small variations on a low-voltage range. This makes it easy to see small voltage differences. For example, assume that a 100-V source is subject to 1-V variations (99 to 101 V). This shows up as one scale division on a 100-V scale. It is possible to use the suppressed-zero technique and measure the difference on a low scale, such as a 3-V scale.

The basic suppressed-zero circuit is shown in Fig. 1-20E. Notice that an opposing voltage is connected in series with the voltage to be measured, and the meter sees only the difference in voltage. For example, assume that a 99- to 101-V source is to be measured on the 3-V scale of a VOM. A 98.5-V opposing voltage is connected as shown (a variable power supply can be used as the opposing voltage). If the source to be measured is at 99 V, the meter reads 0.5 V (99 – 98.5). If the source is at 101 V, the meter reads 2.5 V (101 – 98.5). Both of these indications can be read easily on a the 3-V scale.

Notice that the suppressed-zero technique does not increase accuracy of voltage measurements, but it simply makes small voltage variations easy to read.

1.26 Measuring complex waves

The oscilloscopes described in chapter 2 are the best instruments for measuring complex waves because the waveshape, frequency, and peak-to-peak voltage can be found simultaneously. However, it is possible to measure the voltage of a complex wave with a meter. You get the best results with a peak-to-peak meter because peak-to-peak voltage is independent of actual waveform. Other values of a complex wave (average, RMS, etc.) depend on the waveform (square wave, pulse, sawtooth, half-wave pulsating, full-wave pulsating, etc.). Most service manuals (such as those used in TV) specify the peak-to-peak value of complex waves, so some meters have peak-to-peak scales.

If the waveshape is known, the peak-to-peak value can be converted into the RMS using the data on Fig. 1-21A. For example, assume that a sawtooth wave produces a peak-to-peak reading of 3 V. The RMS value of this voltage is 0.8655 (3×0.2885).

It is also possible to convert an RMS reading to a peak-to-peak value with reasonable accuracy. Remember, when measuring a complex wave with an RMS meter, the indicated value is not the true RMS value because the scales are based on the RMS of sine waves. Again, if the waveshape is known, the RMS value can be converted to a peak-to-peak value using the data on Fig. 1-21B.

For example, assume that you measure a square wave with an RMS meter and get a reading of 3 V. The peak-to-peak value of this is 5.4 V (3×1.8). This value is developed as follows. The average value of a full-wave rectified square wave is equal to the peak value (or one-half of the peak-to-peak value). A conventional RMS meter responds to average, but it indicates RMS, which is 1.11 times average in sine waves (Fig. 1-8B). Therefore, the meter reads average (or peak in the case of a square wave), but indicates this value is 1.11 times the peak. To find the peak, use the reciprocal of 1.11, which is 0.9. To find the peak-to-peak value, multiply the peak value by 2, which is $2 \times 0.9 = 1.8$.

1.27 Matching impedances

One problem often encountered when measuring complex waves is the matching of impedances. To provide a smooth transition between devices of different characteristic impedances, each device must encounter a total impedance equal to its own characteristic impedance. A certain amount of signal occurs in this transition. A simple resistive impedance-matching network that provides minimum attenuation is shown in Fig. 1-21C together with the applicable equations.

For example, to match a 50-Ω system to a 125-Ω system, $Z_1 = 50$ and $Z_2 = 125$, therefore:

$$R_1 = \sqrt{125\,(125 - 50)} = 96.8\ \Omega$$

$$R_2 = 50\sqrt{\frac{125}{125 - 50}} = 64.6\ \Omega$$

Although the network of Fig 1-21C provides minimum attenuation for a purely resistive impedance-matching network, the attenuation, as seen from one end, does

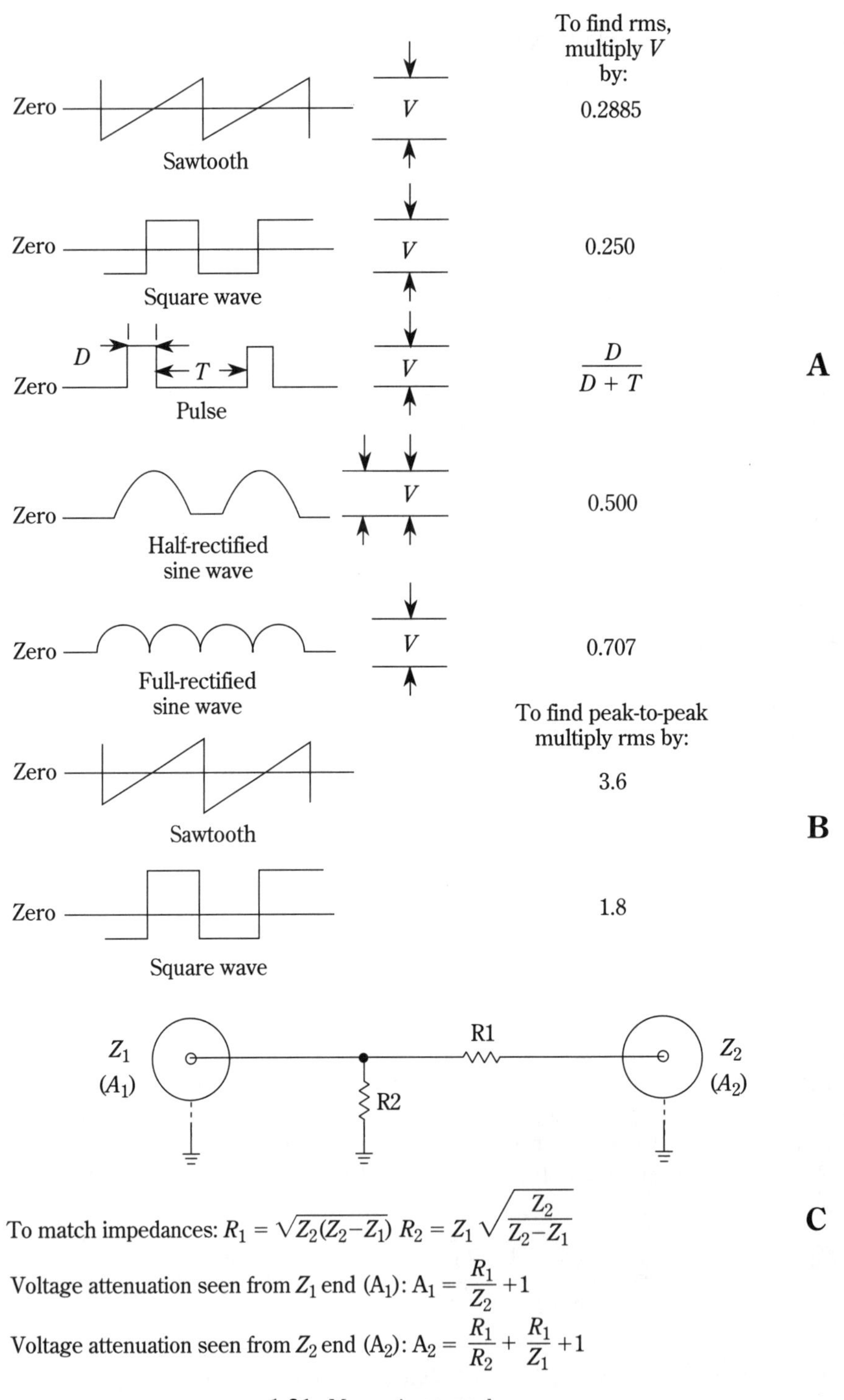

To match impedances: $R_1 = \sqrt{Z_2(Z_2 - Z_1)}$ $R_2 = Z_1\sqrt{\dfrac{Z_2}{Z_2 - Z_1}}$

Voltage attenuation seen from Z_1 end (A_1): $A_1 = \dfrac{R_1}{Z_2} + 1$

Voltage attenuation seen from Z_2 end (A_2): $A_2 = \dfrac{R_1}{R_2} + \dfrac{R_1}{Z_1} + 1$

1-21 Measuring complex waves.

not equal that seen from the other end. A signal applied from Z_1 encounters a voltage attenuation (A_1) that can be determined as follows. Assume that R_1 is 96.8 Ω and Z_2 is 125 Ω.

$$A_1 = \frac{96.8}{125} + 1 = 1.77 \text{ attenuation}$$

A signal applied from Z_2 produces an even greater voltage attenuation (A_2), as follows. Assume that $R_1 = 96.8\ \Omega$, $R_2 = 64.6\ \Omega$, and $Z_1 = 50\ \Omega$:

$$A_2 = \frac{96.8}{64.6} + \frac{96.8}{50} + 1 = 4.43 \text{ attenuation}$$

1.28 Checking individual components

Meters are most often used to check components while the components are connected in the related circuits. However, it is possible to check many individual components with a meter. These tests provide a quick check of component condition, usually on a go/no-go basis. In some cases, it is also possible to find the actual values or characteristics of components. This section describes the procedures for checking individual components out of circuit using a meter. The procedures for checking such components as transistors, FETs, UJTs, PUTs, diodes, thyristors, SCRs, and other control rectifiers are described in chapters 7 through 11.

1.28.1 Photocell tests

Photovoltaic cells produce an output voltage and current in the presence of light and are often called *solar batteries* or *solar cells.* Photoconductive cells are often termed *light-sensitive* or *light-dependent resistors (LDRs)* because they function as resistors and do not generate an output. Instead, photoconductive cells act as a resistance that varies in the presence of light, thus changing the amount of current being conducted through the circuit.

The basic test circuit for a photovoltaic cell is shown in Fig. 1-22A, and the photoconductive test circuit is shown in Fig. 1-22B. In either circuit, the cell is exposed to sunlight or artificial light, and the meter reading is noted. Usually, a single cell produces sufficient output to produce a readable indication on the low-voltage range of most meters. However, some cells might require that the output be read as a current on a low-current range. For example, some cells produce less than 1 mA when exposed to strong sunlight. Always check the manufacturer's data for cell characteristics.

1.28.2 Resistor tests

The obvious test of a resistor is to measure the resistance value with an ohmmeter. The various procedures for resistance measurement are covered in previous sections of this chapter. The following paragraphs describe the procedures for measurement of special resistors and resistance elements.

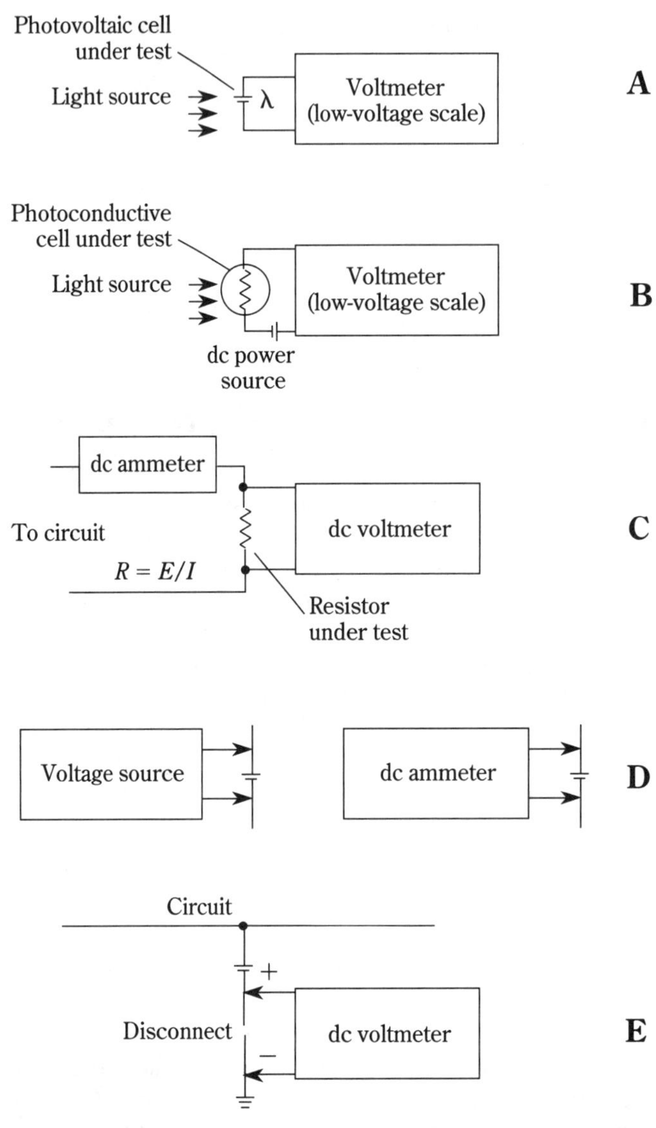

1-22 Photocell, resistor, and capacitor tests.

Thermal or ballast resistors All resistors are subject to variation between hot and cold resistance values, but some resistors are manufactured specifically to produce a large change in resistance for changes in temperature. A simple way to check such components is to heat the resistor (either by operating in the circuit or by placing a voltage across the resistor temporarily) and then measure the hot resistance. The cold resistance can be measured before heating the resistor, and the hot-versus-cold resistance values compared.

Some resistors and resistance elements do not retain their hot resistance and must be measured while operating in the circuit. The filament or heater of a cathode ray (chapter 2) is a good example. The usual tube filament or heater is less than 5 Ω cold and as high as 50 Ω hot.

The in-circuit value of a resistor can be found by measuring the current and voltage and then calculating resistance using Ohm's law ($R = E/I$). The basic test circuit is shown in Fig. 1-22C. Once the hot resistance is found, the resistor can be removed from the circuit, and the cold resistance can be measured with an ohmmeter. It might be more practical to measure the cold resistance first because the resistance element might not cool down immediately.

A ballast resistor can also be checked using the same test circuit of Fig. 1-22C. Most ballast resistors have a positive temperature coefficient (resistance increases with temperature) to maintain constant current flow with changes in voltage. If voltage increases, current flow also increases, and a series-connected ballast resistor becomes hotter. Under these conditions, the ballast resistance increases, increasing the voltage drop across the ballast, and lowering the voltage to the load. This maintains load voltage and current at a constant level.

In practice, there is always some increase in current through a ballast resistor with an increase in voltage. The amount can be checked using the test circuit of Fig. 1-22C and a variable power source. Increase the voltage until the current appears to level off. Then, increase the voltage in small steps and note the change in current for each step. Be careful not to exceed the maximum rated voltage and current for the ballast resistor.

Check potentiometers and variable resistor. In addition to checking the resistance value, it is possible to check the contact of a potentiometer for noise or scratchiness using an ohmmeter. Connect the ohmmeter between the wiper contact and one end of the winding. Then, rotate the contact through the full range of resistance. The ohmmeter resistance reading should vary smoothly through the range. If the meter pointer "jumps" as the resistance is varied (or if a digital readout shows "jitter" at certain points) the potentiometer contact is not riding firmly on the resistance winding or composition element. Usually, this condition is the result of dirt on the contact, and can be cured with contact cleaner.

1.28.3 Capacitor tests

The obvious test of a capacitor is to check for leakage with an ohmmeter. It is also possible to find the approximate value of a capacitor with an ohmmeter. Likewise, it is possible to check operation of a capacitor under various conditions with a voltmeter. The following paragraphs describe these procedures.

Checking capacitor leakage with an ohmmeter The capacitor must be disconnected from the circuit to make an accurate leakage test with an ohmmeter. The basic test is similar to measuring any high-resistance value. The ohmmeter is connected across the capacitor terminals, and the resistance is measured. The following notes should be observed.

1. Use the highest resistance range of the ohmmeter. A typical capacitor can have a resistance in excess of 1000 MΩ.
2. A more accurate test can be made if a high-ohms adapter (Sec. 1.22.1) is used. The higher voltage will show up any tendency of the capacitor to break down.

3. When using higher voltages, make certain not to exceed the voltage rating of the capacitor. This is especially a problem with low-voltage electrolytic capacitors.
4. Another precaution when checking electrolytics is to make certain of the ohmmeter-battery polarity. Usually, the positive terminal is red and the negative is black. However, to make sure, check the polarity against the meter schematic or with an external voltmeter connected to the ohmmeter leads.
5. Usually, capacitors show some measurable resistance when the ohmmeter leads are first connected, but then resistance increases to infinity. This is the result of capacitor charge. If the resistance indication remains below about 1000 MΩ after the capacitor is charged, it is likely that the capacitor is leaking. If the resistance remains at infinity, with no surge or charge indication, it is possible that the capacitor is open. Either of these conditions should be followed up with further tests.
6. Sometimes it is possible to charge a capacitor faster if the ohmmeter leads are first connected with the lowest range ($R \times 1$) in use. Then select each higher ohmmeter range, in turn, until the highest range is in use (the lowest ohmmeter range applies the most voltage).

Checking capacitor leakage with a voltmeter. If a capacitor is suspected of leakage or of being open (as the result of improper circuit operation or by the ohmmeter test just described), the facts can be confirmed using a voltmeter test.

If the capacitor is out of circuit, connect the capacitor leads across a voltage source and hold for approximately 10 seconds, as shown in Fig. 1-22D. For best results, use a voltage near the working voltage of the capacitor. Of course, never exceed the working voltage and always observe polarity for electrolytics.

With the capacitor charged, remove the voltage source and measure the initial capacitor voltage (with a voltmeter, not an ohmmeter). The initial voltage indication should be about the same as the source voltage. If no voltage is indicated, the capacitor is open, if the voltage is very low (in relation to the source) the capacitor is leaking.

This test might not be too effective in testing low-value capacitors—especially with a VOM. The low input resistance of a VOM can discharge a low-value capacitor too quickly to produce a measurable voltage indication. However, most digital and electronic meters will discharge the capacitor slowly because of high input impedance.

If the capacitor is in-circuit, disconnect the capacitor ground lead and then measure dc voltage from the lead to ground, as shown in Fig. 1-22E. Initially, there might be some dc voltage indication because of capacitor charge (the exact amount of initial indication depends on the input resistance of the meter and the capacitor value). However, if the dc voltage indication remains, the capacitor is leaking. It might be necessary to measure the initial charging indication on a high-voltage range of the meter. If the voltage indication drops, move the meter to the lowest voltage range. This makes it possible to measure small voltages caused by capacitor leakage.

If the capacitor shows no voltage indication from the ground lead to ground, the capacitor is definitely not leaking. However, there is still the possibility of an open capacitor. A large-value capacitor will show an indication in most circuits. But, as in the case of the resistance test, a low-value capacitor might charge too quickly to produce a measurable reading.

Checking capacitors by signal tracing. An in-circuit capacitor can be checked quickly and positively for opens using a voltmeter equipped with a signal-tracing probe (chapter 5). Of course, there must be a signal present in the circuit. If necessary, connect a signal generator (chapter 3) to the input of the circuit.

Figure 1-23A shows the basic circuit for checking a coupling or interstage capacitor. Typically, the ac voltage (or signal) should be the same on both the input and output sides of a coupling capacitor. There might be some attenuation of the voltage on the output side (output lower than input). The complete absence of a signal at the output side of the capacitor indicates an open.

Figure 1-23B shows the basic circuit for checking a bypass capacitor. Typically, the ac voltage (or signal) across a bypass capacitor is large if the capacitor is open. The function of a bypass capacitor is to pass ac voltages or signals to ground. Therefore, there should be no ac voltage across the capacitor.

Measuring capacitor values with a voltmeter It is possible to find the approximate value of a capacitor with a meter. The method is based on the time constants of an RC circuit. A capacitor of a given value charges to 63.2 percent of full charge in a given number of seconds through a resistor of given value. Likewise, a capacitor discharges to 36.8 percent of full charge through a given resistance in a given time. If the resistance value is known and the time is measured, the capacitance value can be calculated.

Any type of meter can be used, but the input resistance of the meter must be known. The high input resistance of most digital and electronic meters causes the capacitor to charge and discharge slowly, so such meters should be used with small-value capacitors.

Figure 1-23C shows the discharge method:

1. Move S1 to the charge position and hold for about 1 minute.
2. Move S1 to the discharge (meter) position. Simultaneously start a stopwatch or note the exact time on the second hand of a conventional watch.
3. When the voltage drops to 36.8 percent of the initial value, stop the watch or note the elapsed time interval. For convenience, use an even voltage value, such as 100 V, 10 V, and so on.
4. Divide the elapsed discharge time by the input resistance of the meter to find the capacitance in farads. If the input resistance of the meter is converted to MΩ, the capacitance value will be in microfarads.
5. As an example, assume that the capacitor is charged to an initial value of 10 V, that the capacitor discharges to 3.68 V at the end of 90 seconds, and the input resistance of the meter is 3 MΩ (90/3 = 30 μF).

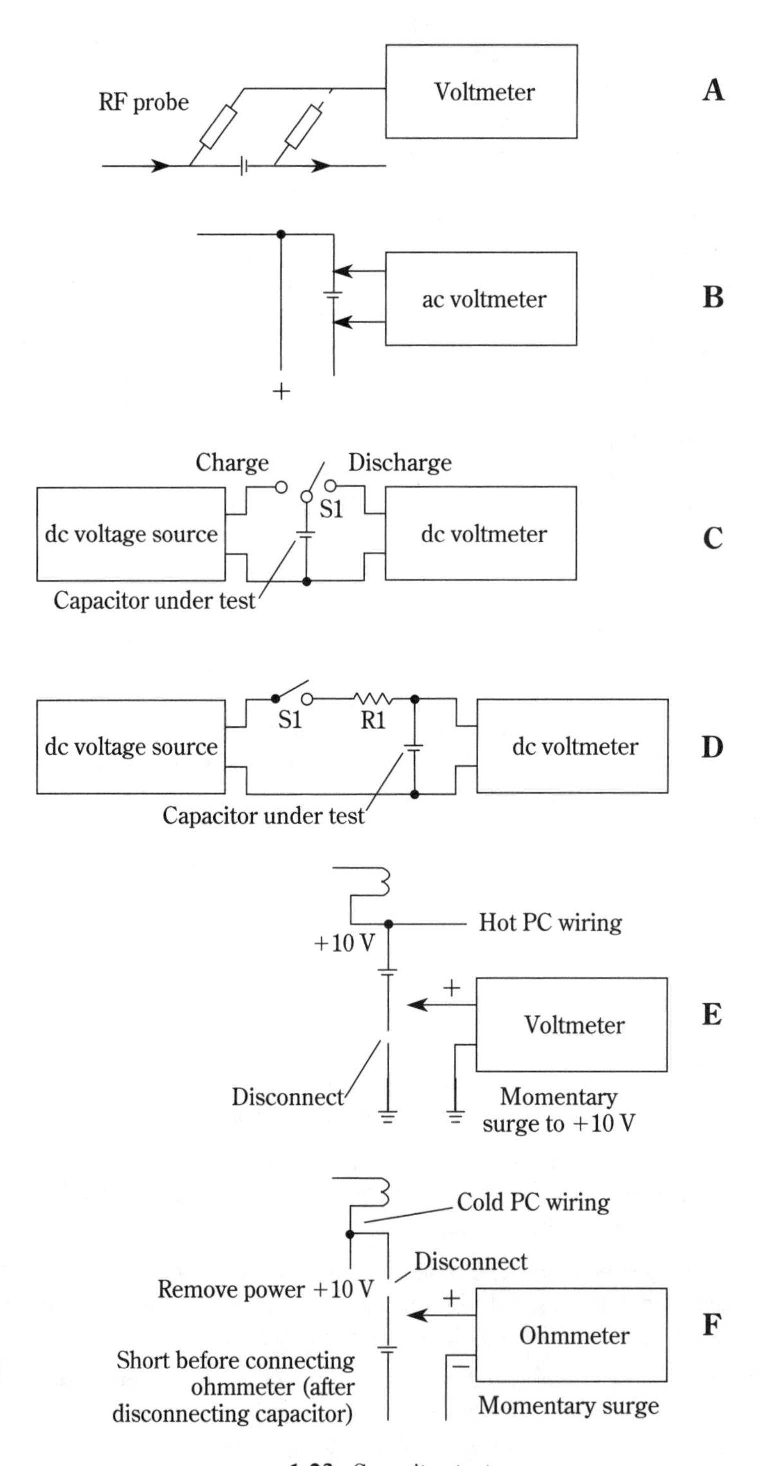

1-23 Capacitor tests.

Figure 1-23D shows the charge method:

1. Make certain that there is no voltage indicated across the capacitor from a previous charge.
2. Close S1. Simultaneously start a stopwatch or note the exact time on the second hand of a conventional watch.
3. When the voltage reaches 63.2 percent of the final value, stop the watch or note the elapsed time interval. For convenience, use an even voltage value such as 100 V, 10 V and so on. Also use an even R_1 resistance value, preferably in MΩ so that the capacitance value can be expressed in microfarads.

   There will be some voltage drop across R1, so the capacitor never fully charges to the source-voltage value. Instead, the capacitor charge depends on the ratio of R_1 to the internal resistance of the meter. For example, if R_1 is 1 MΩ, the meter is 3 MΩ, and the source is 100 V, the capacitor will not charge beyond 75 V (1 + 3 = 4; 100/4 = 25; 100 V – 25 V = 75 V).
4. Divide the elapsed charge time by the value of R_1 to find the capacitance in farads. If R_1 is in MΩ, the capacitance value will be in µF.
5. As an example, assume that the capacitor is charged to an initial value of 100 V (by a source of about 130 V), that the capacitor charges to 63.2 V at the end of 70 seconds, and that the resistance of R1 is 3 MΩ (70/3 = 23.3 µF).

Remember that these tests provide an approximate capacitance value at best, and are not accurate with a capacitor that is leaking. However, the tests show that the capacitor is operating normally and that its approximate value is correct.

Checking capacitors during troubleshooting (circuit voltage method). As shown in Fig. 1-23E, this method involves disconnecting one lead of the capacitor (the lead to the ground of "cold" PC wiring) and connecting a voltmeter between the disconnected lead and ground. In a good capacitor, there should be a momentary voltage indication (or surge) as the capacitor charges up to the voltage at the "hot" end.

If the voltage indication remains high, the capacitor is probably shorted. If the voltage indication remains steady, but not necessarily high, the capacitor is probably leaking. If there is no voltage indication whatsoever, the capacitor is probably open.

Checking capacitors during troubleshooting (ohmmeter method). As shown in Fig. 1-23F, this method involves disconnecting one lead of the capacitor (usually the "hot" end) and connecting an ohmmeter across the capacitor. Make certain all power is removed from the circuit. As a precaution, short across the capacitor to make sure that no charge is being retained after the power is removed. In a good capacitor, there should be a momentary resistance indication (or surge) as the capacitor charges up the voltage of the ohmmeter battery.

If the resistance indication is near zero and remains so, the capacitor is probably shorted. If the resistance indication is steady at some high value, the capacitor is probably leaking. If there is no resistance indication whatsoever, the capacitor is probably open.

1.28.4 Battery tests

The obvious test for a battery is to measure the voltage from all cells together or from each cell on an individual basis. Such a test does not show how a battery maintains the voltage output under load. The following paragraphs describe how to test a battery under dynamic or load conditions.

To measure battery output under load:

1. Connect the battery, load, and meter, as shown in Fig. 1-24A. If the battery is to be tested out-of-circuit, use a load resistance that produces the maximum-rated current flow.

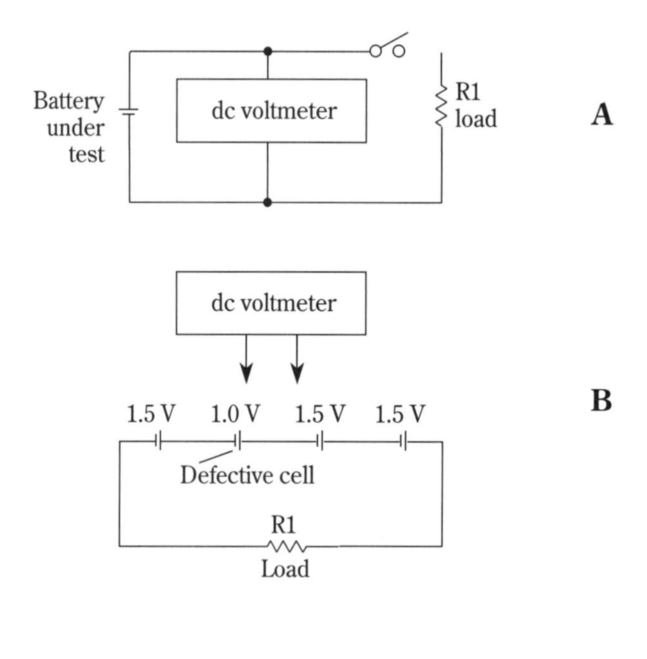

1-24 Battery and quartz-crystal tests.

2. Measure the battery voltage both without a load and with a load. Note any drop in voltage when the load is applied. If the battery is to be tested in-circuit, measure the battery voltage with the equipment turned off. Then, turn the equipment on and measure the full-load voltage.

3. Normally, there is no more than a 10 to 15 percent drop in voltage output when a full load is applied (the exact amount of voltage drop depends on the type of battery). Also, the output voltage of a good battery returns to full value when the load is removed. A defective battery (or cell) drops in output voltage when a load is applied and remains low.
4. A typical lead-acid storage battery produces an output of about 2.1 V without a load, and 1.75 V under full load. The condition of a storage battery can also be checked using a hydrometer (to measure specific gravity).
5. Always use the lowest practical voltage scale to measure battery voltage. This is necessary because a 0.1-V difference can be important in the single cell of a battery.

Locating a defective battery cell To locate a suspected cell in a group of many identical cells, connect all cells in series across a load (Fig. 1-24B). Remove the load. Then, measure the voltage across each cell. The defective cell will show a lower output than the remaining cells, or will possibly show zero output. In some cases, the polarity of the defective cell might reverse.

1.28.5 Quartz crystal tests

The approximate resonant frequency of a quartz crystal can be found using the test circuit of Fig. 1-24C.

1. Set the meter to measure ac voltage.
2. Adjust the generator output to the supposed frequency of the crystal. Then, adjust the generator frequency for maximum indication on the meter. Read the crystal frequency from the counter. Be careful not to increase the generator output voltage beyond the maximum rated limits of the crystal. Excess voltage can crack or otherwise damage the crystal.

1.29 Checking circuit functions

Meters are most often used to check circuit functions by measuring voltage, current, and resistance at various points in the circuit. This is the basis for all troubleshooting. If an expected reading is absent or abnormal, the related circuit components can then be checked on an individual basis. This section describes the procedures for checking general circuit conditions and specific problems that might arise. Chapters 12 through 15 describe the use of meters in checking specific circuits, such as audio, power-supply, RF and communications equipment.

1.29.1 Measuring alternating current in circuits

In-circuit current measurements are always a problem because the circuit must be interrupted (unless a clamp-on adapter is used). ac measurements with nondigital meters are a particular problem because most VOMs do not measure alternating current beyond 60 Hz, and electronic meters often do not measure any form of current.

There are two basic solutions to the problem. Both ac and dc can be measured by inserting a low-value noninductive resistance in series with the circuit, measuring the voltage across the resistance, and then calculating the current using Ohm's law ($I = E/R$), as shown in Fig. 1-25A. If a 1-Ω resistor is used, the voltage indication can be converted directly into current (3 V = 3 A, 7 mV = 7 mA, etc.).

Alternating current can be measured with a transformer, as shown in Fig. 1-25B. The current to be measured is passed through the transformer primary. The voltage developed across the secondary and load is measured by the meter. Commercial versions of this arrangement are used in heavy power equipment. These are known as *current transformers* or *current adapters*. In the commercial equipment, there are several taps on the primary winding. Each tap is designed to carry a particular current and produces a given voltage across the load.

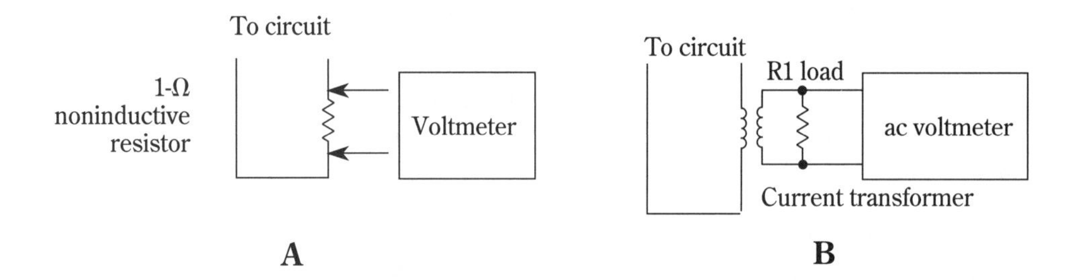

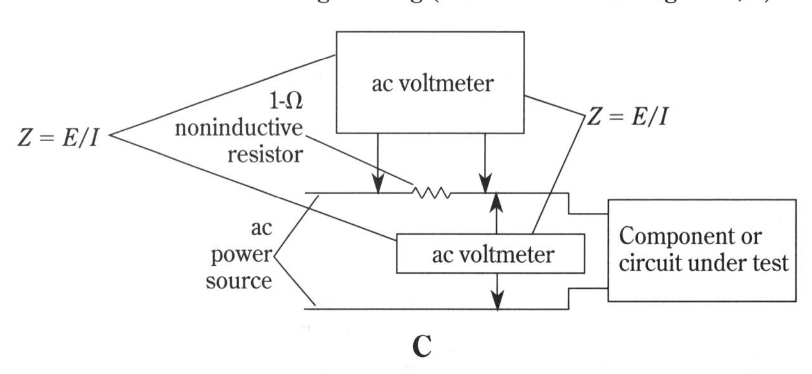

1-25 Measuring current, impedance, and power consumption in circuits.

1.29.2 Measuring impedance and power consumption in ac circuits

The current-measuring circuit of Fig. 1-25A can also be used to measure circuit impedance and power consumption. If the current and voltage are known, the impedance can be calculated using the equation: $Z = E/I$.

Power consumption (in volt-amperes) can be calculated with the equation: $V_A = E \times I$. Of course, true ac power (expressed in watts) is found by multiplying the volt-amperes by the cosine of the phase angle (the volt-ampere figure can be used for most practical purposes).

Figure 1-25C shows the basic test circuit for measuring impedance. The impedance of individual components (headphones, coils, transformers, etc.) or the impedance presented by a complete circuit or equipment (radio receiver, test instrument, etc.) can be measured using the same basic circuit. The following precautions must be observed when using the voltage-current method to find impedance:

1. Accuracy of the measurements depends on accuracy of the resistance value in series with the circuit.
2. A 1-Ω resistor should be used wherever possible. This simplifies the calculations and presents the least disturbance to the circuit. If a 1-Ω resistor is not practical (because the voltage drop is too small to read), use a 1000-Ω or 10,000-Ω resistor. Remember that a 1000-Ω series resistor converts voltage indications into milliamperes (7 V = 7 mA, etc.).
3. A noninductive series resistance must be used. If the resistor has any inductance (as do most wire-wound resistors), the inductive reactance is added to the circuit and produces an error in measurement.
4. The impedance found by the voltage-current test method applies only to the frequency used during test. This is no problem for equipment operated from line power. However, there is a problem for headphones, coils, transformers, and so on, that are operated over a wide range of frequencies. Therefore, these components should be tested over the anticipated frequency range. The component or circuit can be checked at various points throughout the frequency range, and any differences in impedance can be noted.
5. The composition of the waveform also affects the impedance found using the voltage-current test method. For example, a sine wave that contains many harmonics produces a different impedance reading than a pure sine wave. From a practical standpoint, it is usually not essential that circuits or components be tested for impedance with a pure sine wave. However, it is essential that the test be made with waveforms equivalent to those used in actual operation.
6. If the impedance and dc resistance of a component or circuit are known, it is possible to find the reactance, inductance, or capacitance, and power factor. These require considerable calculation or vector analysis and are usually of little practical value. However, the relationship between impedance and dc resistance can be used to determine the presence of reactance in a supposedly resistive load.

For example, a T or L pad used in audio work is considered as a pure resistive load (at audio frequencies). However, it is possible that such a pad could present some reactive load—especially at the high end of the audio range (15 kHz and above). This condition can be determined by measuring the pad impedance at audio frequencies, as described next in Sec. 1.29.3. Then, compare the impedance against the dc resistance. Both should be substantially the same. A large difference indicates some reactance present (probably inductive reactance in the case of L or T pads).

1.29.3 Measuring pad circuits

L and T pads, whether considered as circuits or components, require special measurement procedures. Generally, pads can be checked using an ohmmeter. However, if you suspect that a pad is producing some reactance (inductive or capacitive) the pad must be checked with an audio-signal source.

Measuring L pads for impedance. Usually, the input impedance of an L pad remains constant, but the output impedance varies as the pad setting is changed (or the pad can be reversed so that the output impedance is constant with the input impedance varying).

Figure 1-26A shows the input-impedance test circuit. The load resistance R_1 should be equal to the normal load, as seen from the output side of the pad. Vary the pad throughout the entire range and note the resistance at the input circuit. Assuming no reactance, the dc input-resistance values equal the input-impedance values.

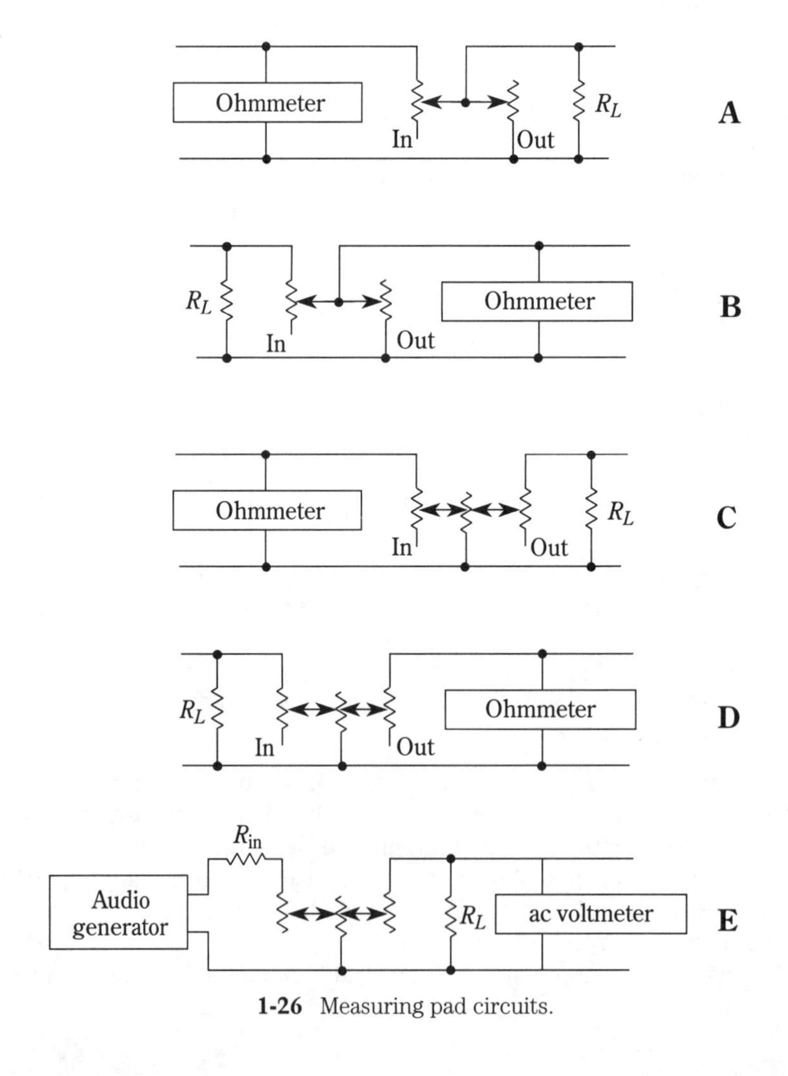

1-26 Measuring pad circuits.

Figure 1-26B shows the output-impedance test circuit. The load resistance R_1 should be equal to the normal load, as seen from the input side of the pad. Vary the pad throughout the entire range and note the resistance at the output circuit. Assuming no reactance, the dc output-resistance values equal the output-impedance values.

Measuring T pads for impedance The input and output impedance of a T pad should remain constant as the pad setting is changed (within a specified tolerance). This is one of the advantages of a T pad over an L pad. However, the input and output impedances of a T pad are not necessarily the same. Therefore, a T pad can also be used to match impedances, as well as to vary signal strength.

The impedance test circuits are shown in Figs. 1-26C and 1-26D. To measure input impedance, Fig. 1-26C, R_L (equal to the normal load, as seen from the output side of the pad) is connected at the output with the ohmmeter at the input. The circuit is reversed in Fig. 1-26D to measure output impedance. In either circuit, vary the pad throughout the entire range and note that the resistance remains constant (within tolerance). Assuming no reactance, the resistance values equal the impedance values.

Measuring pads for reactive effect Figure 1-26E shows the circuit for measuring reactive effect in both L and T pads. The load resistance should be equal to the rated output impedance of the pad (or the impedance that is seen from the output side). The input resistance, R_{IN} should be equal to the rated input impedance of the pad. However, if the pad impedance is close to that of the audio-generator output, then R_{IN} should be added to the generator impedance to equal the pad input impedance.

For example, if the pad input impedance is 10 kΩ, and the audio-generator output impedance is 50 Ω, R_{IN} should be about 10 kΩ. If the pad input is 100 Ω with a 50-Ω generator output, R_{IN} should be 50 Ω.

1. Adjust the audio-generator output voltage for a good reading on the meter.
2. Vary the audio-generator output over the entire frequency range with which the pad is used. Do not change the generator output voltage.
3. If the pad has any reactive effect, the meter reading will not be constant. The meter reading will rise (or fall) at some particular frequency or over some particular frequency range.
4. Repeat the test at various settings of the pad. One point to remember when making this test or any test that involves varying the output of a generator. As covered in chapter 3, all generators do not have a flat output (constant-voltage output over the entire frequency range). The same is true of meters. This can lead to an error in judgment. The ideal remedy is to calibrate the generator over the entire frequency range and record any variations.

A quick check of the generator can be made by connecting the meter directly to the generator output, varying the generator over the entire frequency range, and then noting any variation in output voltage. These variations can then be compared with any variations found with the meter connected at the pad output. If the variations are the same, it is likely that the generator or meter is at fault, not the pad.

1.29.4 Measuring internal resistance of circuits

It is sometimes convenient to measure the internal resistance of a circuit while the circuit is operating. For example, it might be desired to measure the collector resistance of a transistor. Obviously, the circuit resistance cannot be measured with an ohmmeter while the circuit is energized. However, it is possible to measure circuit resistance with a voltmeter and a potentiometer.

1. Connect the equipment, as shown in Fig. 1-27A.
2. Set the potentiometer to zero and measure the circuit voltage. This is the full circuit voltage.

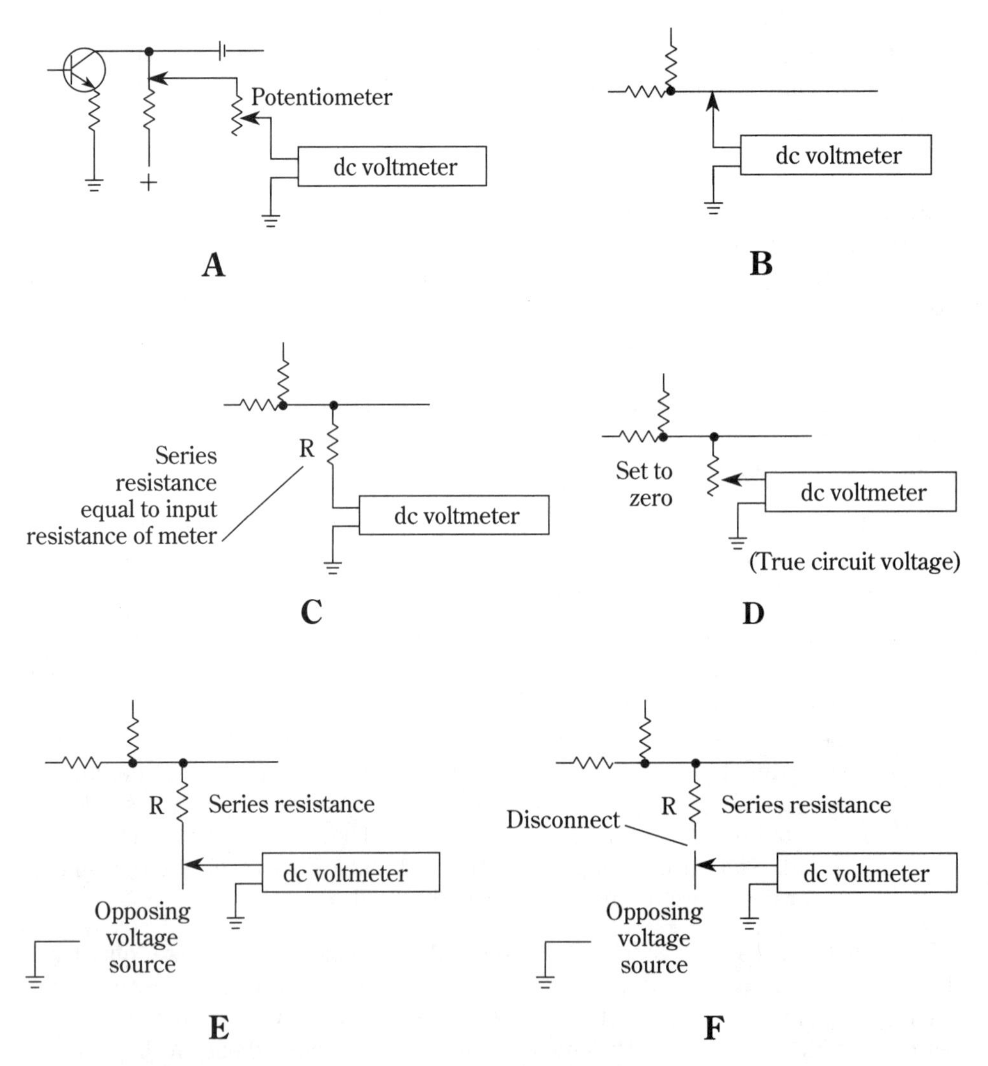

1-27 Measuring internal-resistance and voltage-sensitivity of circuits.

3. Increase the potentiometer resistance until the voltage is one-half of the full-circuit voltage (one-half that obtained in step 2).
4. Disconnect the potentiometer from the circuit. Measure the dc resistance of the potentiometer.
5. Subtract the input resistance of the meter from the dc resistance of the potentiometer. The remainder is equal to the internal resistance of the circuit.

For example, assume that the meter reads 100 V when the potentiometer is set to 0 and 50 V with the potentiometer set to 100,000 Ω. Also assume that the input resistance of the meter is 30,000 Ω. The circuit internal resistance is 70,000 Ω.

1.29.5 Measuring voltage-sensitive circuits

Many circuits are *voltage sensitive*. That is, the voltage varies with changes in load. These circuits are difficult to measure because the load presented by the meter changes the voltage from the normal operating value. Because a VOM has a lower input resistance than an electronic voltmeter, the VOM presents a greater load on the circuit and thus changes the voltage by a greater amount. This is one of the advantages of an electronic meter (and digital meters). However, even electronic or digital meters can cause circuit loading in certain cases.
Determining voltage sensitivity Figure 1-27B and 1-27C show how the voltage sensitivity of a circuit can be determined. The test can be applied to any circuit (the AGC line of a receiver is a typical voltage-sensitive circuit).

1. Measure the voltage directly, as shown in Fig. 1-27B.
2. Insert the series resistor, R, and measure the voltage at the same point in the circuit as shown in Fig. 1-27C. The value of resistor R should be equal to the input resistance of the meter.
3. The second voltage indication (with resistor R in series) should be approximately one-half that obtained by direct measurement. If the second voltage indication is much greater than one-half, the circuit is voltage-sensitive.

Either of the following two methods (potentiometer and opposing-voltage methods) can be used to measure voltage-sensitive circuits:

Potentiometer method:

1. Connect the equipment, as shown in Fig. 1-27D.
2. Set the potentiometer to zero and measure the circuit voltage. This is the initial voltage.
3. Increase the potentiometer resistance until the voltage is one-half the initial voltage (one-half of that obtained in step 2).
4. Disconnect the potentiometer from the circuit. Measure the dc resistance of the potentiometer
5. Find the true circuit voltage with the following equation:

$$\text{True voltage} = \frac{\textit{Initial voltage} \times \textit{Potentiometer resistance}}{\textit{Meter input resistance}}$$

For example, assume that the initial circuit reading is 3 V, the potentiometer resistance needed to reduce the reading to 1.5 V is 140,000 Ω, and the meter input resistance is 60,000 Ω.

$$3 \times 140{,}000 = 420{,}000 \qquad \frac{420{,}000}{60{,}000} = 7 \text{ V}$$

Therefore, the true circuit voltage is 7 V, and the meter caused a 4-V drop.

Opposing-voltage method:

1. Connect the equipment, as shown in Fig. 1-27E. The value of series resistance R is not critical, but it should be near that of the meter input resistance.
2. Adjust the opposing-voltage source until the meter reads zero.
3. Disconnect the meter and opposing-voltage source.
4. Reconnect the meter to measure the opposing-voltage source, as shown in Fig. 1-27F.
5. The true circuit voltage is equal to the opposing-voltage source.

1.29.6 Measuring static voltages of integrated circuits

The static (direct-current) voltages of an IC are measured in essentially the same way as those of transistor circuits, with one major exception. Many ICs require connection to both a positive and negative power source, typically with equal supply voltage (such as +9 V and –9 V). However, this is not the case with the circuit of Fig. 1-28, which requires +9 V at pin 8 and –4.2 V at pin 4.

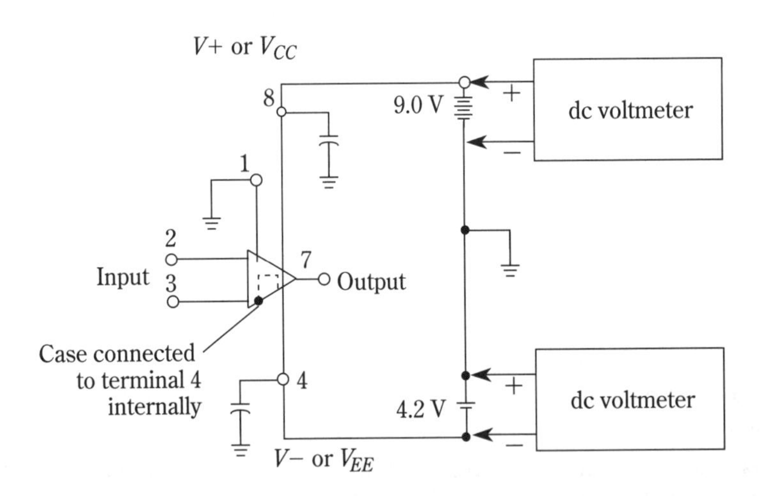

1-28 Measuring static voltages of ICs.

Unlike most transistor circuits, where it is common to label one power-supply lead positive and the other negative, without specifying which (if either) is common or ground, it is usually necessary that all IC power-supply voltages be referenced to a common or ground.

Manufacturers do not agree on power-supply labeling for ICs. For example, the IC of Fig. 1-28 uses V+ to indicate the positive voltage and V– to indicate the negative voltage. Another manufacturer might use the symbols V_{EE} and V_{CC} to represent negative and positive, respectively. As a result, the IC datasheet and/or equipment schematic should always be studied carefully before measuring the power-source voltage.

No matter what labeling is used, the IC typically requires two power sources, with the positive lead of one and the negative lead of the other, tied to ground. Each voltage must be measured separately, as shown in Fig. 1-28.

The IC case (such as a TO-5) of the Fig. 1-28 circuit is connected to pin 4. This is typical for many ICs. Therefore, the case is below ground (or "hot") by 4.2 V.

2
Oscilloscopes

The formal name *cathode-ray oscilloscope* (*CRO*) is usually abbreviated to *oscilloscope*, or, simply, *scope*. No matter what the name, a scope is an extremely fast X-Y plotter capable of plotting an input signal versus another signal, or versus time, whichever is required. A luminous spot acts as a "stylus" and moves over the display area in response to input voltages. In most applications, the Y-axis (vertical) input receives the signal from the voltage being examined, moving the spot up or down in accordance with the instantaneous value of the voltage. The X-axis (horizontal) input is usually an internally generated linear ramp voltage, which moves the spot uniformly from left to right across the display screen. The spot then traces a curve, which shows how the input voltage varies as a function of time.

If the signal under examination is repetitive, at a fast enough rate, the display appears to stand still. The scope is thus a means of visualizing time-varying voltages. As such, the scope has become a universal tool in all kinds of electronic investigations. Oscilloscopes operate on voltages. It is possible, however, to convert current, strain, acceleration, pressure, and other physical quantities into voltages by means of transducers, and thus present visual representations of a wide variety of dynamic phenomena on oscilloscopes.

2.1 The cathode-ray tube

All circuits of a scope are built around a *cathode-ray tube* (*CRT*), such as shown in Fig. 2-1A. As with other vacuum tubes, the filament heats the cathode to the degree where electrons are emitted. The control grid sets the amount of current flow, as in conventional vacuum tubes. Two anodes receive positive dc potentials to accelerate the electrons and form an electron beam. The intensity of the electron beam is regulated by the potential (intensity voltage) applied to the grid. The beam is focused into a sharp pinpoint by controlling the voltage (focus voltage)

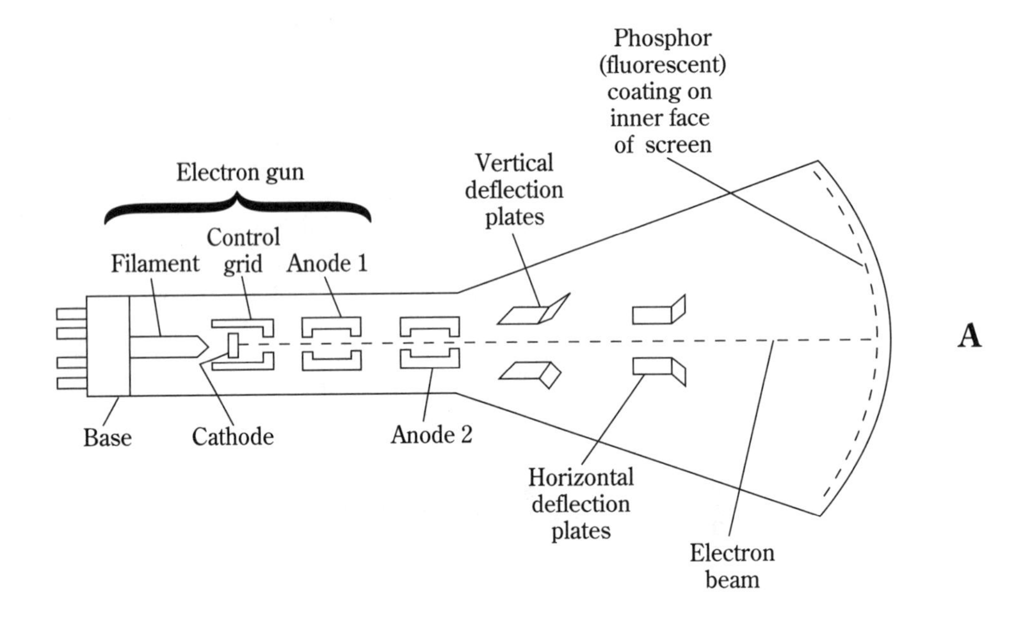

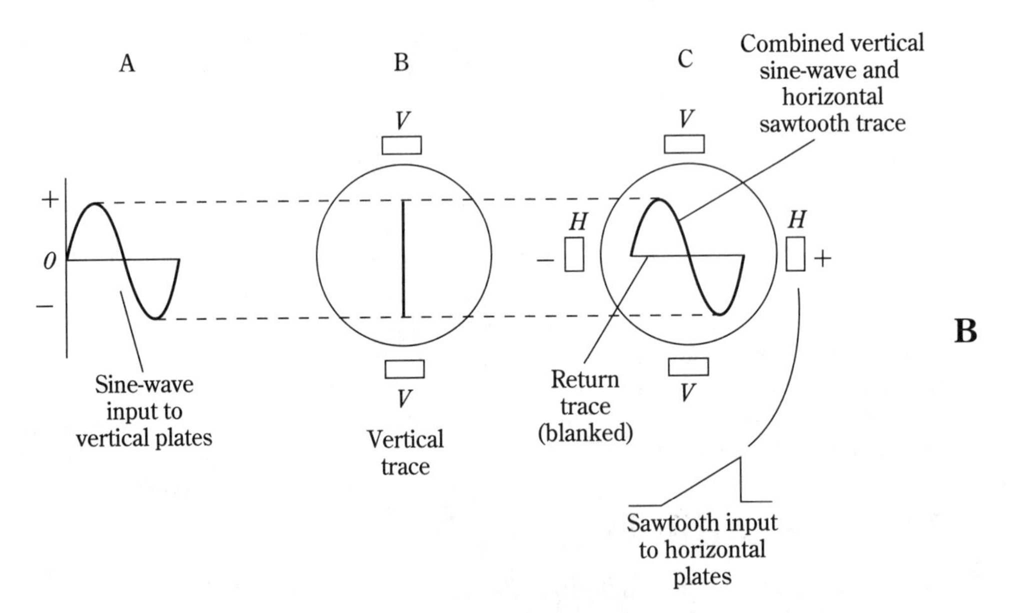

2-1 Cathode-ray tube basics.

on the first anode. A high voltage is applied to the second anode so that the electrons attain high velocity for increased intensity and visibility when the beam strikes the tube face. The beam-forming section of the tube is known as an *electron gun*.

The inside of the tube face is coated with phosphor so that when electrons strike the coating the face will fluoresce and emit light. After the exciting beam leaves the area, the light decays rapidly. However, the composition of the coating can be such that the emitted light persists for an appreciable interval. Because the beam is swept across the tube at a rapid rate, the light must persist for a time sufficient to leave a trace of the waveform drawn on the tube face. At the same time, the persistence must be sufficiently short so that the pattern will disappear if the beam stops.

A standard numbering system is used to identify CRT characteristics. For example, the designation 3AP1 shows that the tube has a 3-in. face diameter. The P1 indicates a medium-persistence phosphor, which has a green glow. The designation A refers to the internal construction and shows that the tube has some structural changes with respect to a 3P1 CRT.

2.2 Beam-deflection system

As shown in Fig. 2-1A, there are two sets of electrostatic deflection plates between the second anode and inner face of the tube. These plates deflect the electron beam both horizontally and vertically so that the beam "writes out" information applied to the deflecting system. For example, a negative potential on one horizontal plate repels the electron beam, whereas a positive potential on the other horizontal plate attracts the beam. If the voltage is a sawtooth, the gradually rising potential gradually pulls the beam toward the positive plate, causing the beam to scan across the tube face. Any potential applied to the vertical deflection plates causes the beam to move vertically.

Figure 2-1B shows how a sine wave is traced out by the deflection system. If the horizontal sawtooth voltage is off, the rapid rise and fall of the sine-wave input causes the electron beam to move up and down the tube face, leaving a vertical-line trace. If the horizontal voltage is on, the sawtooth also pulls the beam from left to right, thus tracing out the input waveshape in visual form. If one sawtooth occurs for each cycle of the sine-wave signal, the beam will trace out a sine wave, as shown.

2.3 Basic frequency measurement

Waveshapes of square waves, pulses, or any other type of signals can be observed. If the input signal has a frequency twice that of the sawtooth, two cycles appear on the screens because the beam is pulled across the screen only once for each two cycles of the input signal. By regulating the ratio of the input signal frequency to the sawtooth sweep frequency, portions of the input signal, or a number of cycles of the input signal, can be made visible on the screen. By calibrating the frequency of the horizontal sweep so that the exact frequency is known, the frequency of the input signals can be calculated. For example, if four cycles of a sine wave appear on the screen and the sawtooth is set for 100 Hz, the frequency of the signal applied at the input is 400 Hz.

2.4 Basic voltage measurement

Oscilloscopes can be used to measure the peak-to-peak voltages of ac signals, pulses, square waves, and so on, as well as dc voltages. Once the peak-to-peak voltage is found, the value can be converted to peak, rms, average, etc. using the information of Fig. 1-8B. A transparent plastic screen is attached to the oscilloscope face to aid in reading voltages (as well as frequency). Such a *screen* (also known as a *grid*, *grid mask*, *grating*, and *graticule*) is found in many forms. Usually, a screen is a transparent scale with vertical and horizontal lines spaced one division apart. The screen is fitted against the CRT face. This allows time (horizontal) and amplitude (vertical) to be read directly. The graduated scales often have small markings that subdivide the major divisions to assist in making accurate measurements. Most laboratory scopes are calibrated in centimeters. A few shop scopes (particularly older models) are calibrated in inches. Still other scopes use no particular standard of measurement, but are simply equal-spaced "divisions."

A low-voltage ac signal and a voltmeter of known accuracy are used to calibrate the scope screen. Some scopes have a front-panel terminal that supplies the calibrating voltage. For example, if the reference voltage is 5 V, this voltage is applied to the vertical input of the scope. The internal horizontal sweep is turned off so that a vertical trace, as shown in Fig. 2-1B, is visible on the screen. This vertical line represents the peak-to-peak voltage of the input signal. The vertical-height control is then adjusted so that the line is five divisions high, and the control is left in this position after calibration. Knowing the rms value of the applied calibrating voltage, the peak-to-peak voltage can be calculated using Fig. 1-8B.

2.5 Basic oscilloscope circuits

Figure 2-2 is a block diagram of a typical oscilloscope. It is beyond the scope of this book to describe all circuits in modern oscilloscopes. Many of these circuits are special purpose. The basic circuits are used in various combinations. Instead of attempting a description of every circuit and combination, the following covers a composite oscilloscope.

2.5.1 Vertical, (Y-axis) channel

Signals to be examined are usually applied to the vertical, or Y, deflection plates through the vertical amplifier. An amplifier is required because the signals are usually not strong enough to produce measurable vertical deflection on the CRT. A typical CRT requires 50 V to produce a deflection of 1 in. The high-gain amplifier in a lab scope permits a 1-in. deflection with about 0.5 m/mV. The average shop scope requires at least 10 to 20 mV for a 1-in deflection.

Another difference between lab and shop amplifiers is the frequency response. As in the case of any amplifier, the frequency response must be wide enough to pass faithfully the entire band of frequencies to be measured. For simple audio work, an upper limit of 200 kHz is usually sufficient. Television service requires a passband of at least 5 MHz, but preferably 10 MHz, if you service S-VHS, HDTV, and so on. Some lab scopes provide up to 1000 MHz (1 GHz), while 100 MHz is common for lab scopes.

Notice that some scopes provide for direct connection to the vertical plates. This is particularly useful when checking modulation patterns of communications equipment (chapter 15).

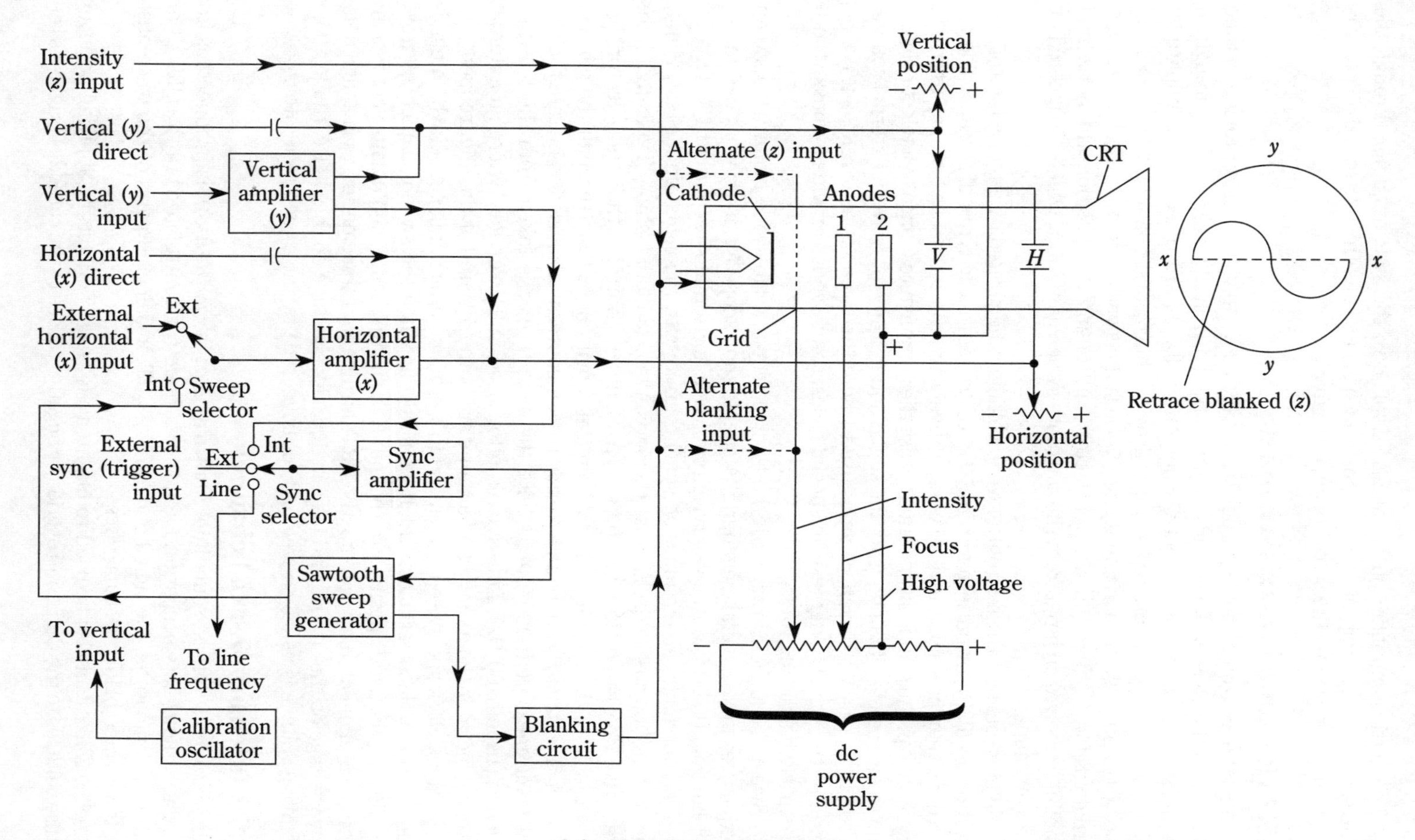

2-2 Basic oscilloscope circuits.

The vertical-amplifier output is also applied to the sync amplifier through the sync selector switching the internal position. This permits the horizontal-sweep circuit to be triggered by the signal being examined.

2.5.2 Horizontal (X-axis) channel

Usually, the horizontal-deflection plates are fed a sweep voltage that provides a time base. As in the case of the vertical channel, the horizontal plates are fed through an amplifier, but can be fed directly when the voltages are of sufficient amplitude. When external signals are to be applied to the horizontal channel, the signals can also be fed through the horizontal amplifier, via the sweep-selector switch (in the external position). When the sweep selector is in the internal position, the horizontal amplifier receives an input from the sawtooth sweep generator, which is triggered by the sync amplifier.

In most scopes, the sawtooth-sweep voltage is generated by a multivibrator, relaxation oscillator or pulse generator. There are four basic types of *sweeps*. The recurrent sweep presents the display repetitively and the eye sees a lasting pattern. In single sweep, the spot is swept once across the screen in response to a trigger signal. The trigger can be obtained from the signal under study, or from an external source.

In most cases, a driven sweep is used where the sweep is recurrent, but is triggered by the signal under test. In special cases, some scopes provide a nonsawtooth sweep, such as a sine wave.

Sweep frequencies vary with the type of scope. A lab scope might have sweep frequencies up to several megahertz. A simple shop scope for audio work has an upper limit of 100 kHz. Most TV service requires a horizontal sweep up to 1 MHz.

The horizontal sweep must be synchronized with the signal under test. If not, the pattern appears to drift across the screen in a random fashion. Typically, there are three ways to synchronize the sweep, all under control of the sync selector; *internal*, where the trigger is taken from the signal under test (through the vertical amplifier); *external*, where an external trigger source is also used to initiate the signal being measured; and *line*, where the sync trigger is taken from the line frequency (usually 60 Hz). Line sync is often used in TV service, where an external sweep generator (chapter 3) and a scope are both triggered at the line frequency.

The oscilloscope sweep is also used to produce a blanking signal, which eliminates retrace. Such retraces, which would occur when the sweep snaps back from the final (right-hand) position to the initial or starting point, could cause confusion. Retrace is eliminated by blanking the CRT during the retrace period with a high negative voltage on the control grid (or a high positive voltage on the CRT cathode). The blanking voltage is usually developed (or triggered) by the sweep generator.

2.5.3 Intensity (Z-axis) channel

Intensity modulation, sometimes known as *Z-axis modulation*, is produced by inserting a signal between ground and the cathode (or control grid) of the CRT. If of sufficient amplitude, the signal cuts off the CRT on selected parts of the trace, just as the retrace-blanking signal does.

The Z-axis can also be used to brighten the trace. Periodically applying positive-pulse voltages to the CRT control grid (or negative pulses to the cathode) brightens the electron beam throughout the trace to give a third, or Z, dimension. These periodically brightened spots can be used as markers for time calibration of the main waveform.

2.5.4 Positioning controls

For many measurements, it is necessary to provide some means of positioning the trace on the CRT face. Such centering provisions are accomplished by applying small, independent, internal dc potentials to the deflection plates. The centering or position controls are particularly useful during voltage calibration, and during enlargement of a waveform for examination of small characteristics (the portion of interest might move off the CRT screen so that positioning controls are necessary to bring it back on again).

2.5.5 Focus and intensity controls

The CRT electron beam is focused in a manner similar to that of an optical lens, but the focal length is altered by changing the ratio of potentials between the first and second anode. This ratio is changed by varying the potential on the first anode with the focus control (a potentiometer on the front panel). The potential on the second anode remains constant.

The intensity of the beam is varied by the intensity control, which changes the grid potential with respect to the cathode. Because the potentials applied to the control grid and anodes are taken from a common voltage-divider network, any change in the intensity control usually requires a compensating change in the setting of the focus control, and vice versa.

2.5.6 Calibration circuit

Most lab scopes have an internally generated and stabilized waveform of known amplitude, which serves as a calibrating reference. Usually, a square wave is used, with the calibrating signal accessible on a front-panel connector. On some scopes, the calibrating voltage is applied to the vertical input through a front-panel control switch. Older shop scopes often have a 60-Hz line terminal at the front panel. The calibrating voltage is applied to the vertical channel by running a lead between the line terminal and the vertical input.

2.5.7 Other oscilloscope circuits

Those familiar with oscilloscope circuits will notice that this section has not covered many basic controls, such as amplifier gain or attenuation, astigmatism, sweep frequency, time, sync polarity, trigger level, and slope, to name a few. These controls, as well as all other controls of typical oscilloscopes, are covered from the operator's viewpoint in Sec. 2.8.

2.6 Storage oscilloscope

The *storage oscilloscope* is especially useful for one-shot displays, where it is not practical to photograph the display. The storage-oscilloscope principle is also used in computer graphics. A storage scope holds a display on the screen for an indefinite time. Several waveforms can be superimposed for comparison.

Several types of storage CRTs are available, and can be generally classified as either *bistable* or *halftone tubes*. The stored display on a bistable tube has one level of brightness. A halftone tube has the capacity of displaying a stored signal at different levels of brightness. The brightness of a halftone tube depends on beam current and the time the beam remains on a particular storage element. A bistable tube either stores or does not store an event. All stored events have the same brightness. With either tube, the operator can remove the display with an erase button. Most storage scopes can be used as a conventional scope when the storage function is disabled.

Storage CRTs can also be classified as either *direct-viewing* or *electrical-readout tubes*. An electrical readout tube has an electrical input and output. A direct-viewing tube has an electrical input but a visual output.

Figure 2-3 is a simplified diagram of a typical storage CRT. The focus, intensity, and accelerator electrodes are omitted for clarity. In addition to the *electron gun* (known as the *writing gun*), the storage CRT also has a *flood gun* (or guns). Unlike the conventional CRT, a storage CRT has four screens behind the display area.

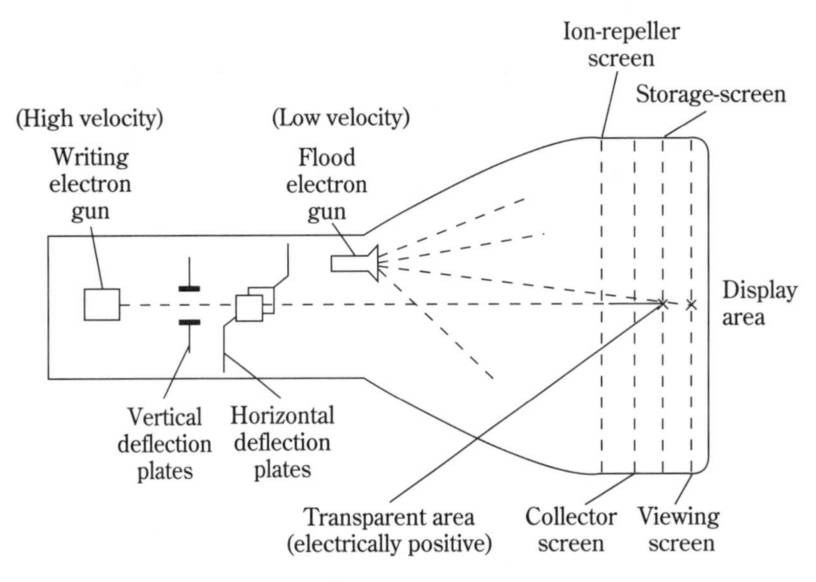

2-3 Storage cathode-ray tube basics.

The viewing screen is a phosphor screen that is similar to that of a conventional CRT. Directly behind the viewing screen is a *storage screen*, usually in the form of a fine metal mesh coated with dielectric. A *collector screen* and an *ion-repeller screen* are located behind the storage screen. Not all storage CRTs have the ion-repeller screen.

When signals are applied to the horizontal and vertical deflection plates, the electron beam from the writing gun is moved to trace out the corresponding waveform in the normal manner. When this beam strikes the storage screen, an electrically positive trace is written out by removing electrons from the storage screen. This leaves a positive trace stored on the storage screen after the signal is removed. The excess electrons removed from the storage screen are collected by the collector screen.

The removal of electrons from the storage screen makes the affected areas (of the dielectric) transparent to low-velocity electronics (that is, low-velocity electrons pass through the areas written out by the high-velocity electron beam). Any time after the trace is written on the storage screen, the flood gun can be turned on to spray the entire storage screen with low-velocity electrons. These electrons pass through the transparent areas, but do not penetrate other areas of the storage screen.

Electrons passing through the storage screen reproduce the trace on the phosphor viewing screen. In effect, the storage screen acts as a "stencil" to the low-velocity electrons from the flood gun. When the display is to be erased, the voltage at the collector is changed, causing the storage screen to discharge (all of the electrons are restored to the dielectric atoms in the storage screen).

2.7 Sampling oscilloscope

Conventional oscilloscopes are limited in bandwidth to frequencies in the megahertz region. *Sampling oscilloscopes* have bandwidths up to about 12 GHz (and above). This permits the sampling oscilloscope to measure signals of extremely high frequency or to monitor waveforms of very fast rise times (0.5 ns or faster is typical).

The sampling oscilloscope uses a stroboscopic approach to reconstruct the input waveform from samples taken during many recurrences of the waveform. This technique is shown in Fig. 2-4. The sampling pulse "turns on" the sampling circuit for an extremely short interval, and the waveform voltage at that instant is measured. The CRT spot is positioned vertically to correspond to this voltage amplitude.

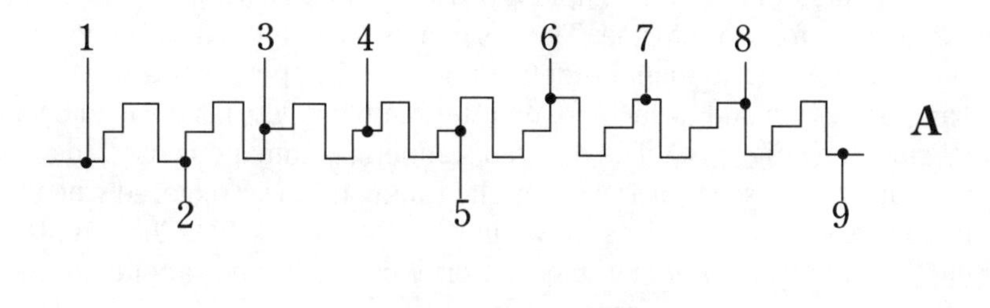

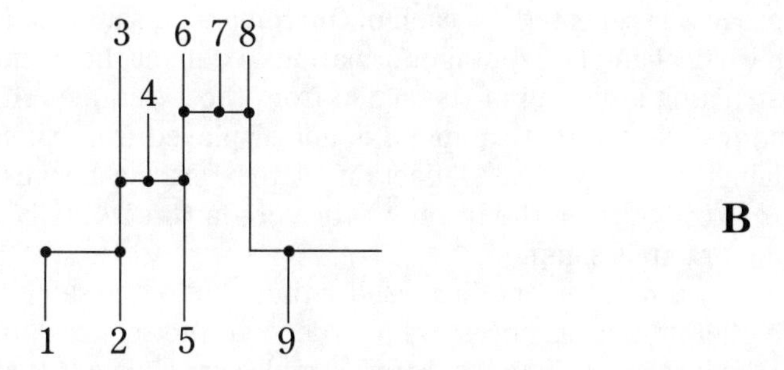

2-4 Sampling-oscilloscope technique to reconstruct a waveform.

The next sample is taken during a subsequent cycle at a slightly later point on the input waveform. The CRT spot moves horizontally and is repositioned vertically to the new voltage. In this way, the scope plots the waveform point by point (or dot by dot). As many as 1000 samples are used to reconstruct a single, recurrent waveform.

A bright trace is obtained (regardless of sampling rate, sweep speed, or waveform duty cycles) because each CRT spot remains "on" during the full interval between samples. For example, assume that a pulse with an extremely fast rise time is monitored. If the rise time is at the maximum speed of the electron beam in a conventional scope, the beam moves so quickly that a weak trace is produced on the leading and trailing edges of the waveform. In a sampling scope, the leading and trailing edges are sampled many times at different points from top to bottom, producing a bright, steady trace.

The *random-sampling oscilloscope* is an improved version of the sampling scope. Operation of a random-sampling scope can be divided into two parts: (1) timing the sample to fall within a time window (displayed portion of the waveform or signal period), and (2) constructing the display from a series of such samples. In a nonrandom-sampling scope, the samples are taken at a specific time interval. In random sampling, there is considerable time uncertainty between the signal being sampled and the sample-taking process.

Figure 2-5 shows how the X and Y deflection signals are generated in a random-sampling scope. In this example, five randomly placed samples are taken from a signal. These samples are taken of successive repetitions of the signal. This is known as *random equivalent-time sampling*. There is also a method of random real-time sampling, where many randomly placed samples are taken during a single, relatively slow, signal occurrence.

The Y component of the first sample is passed through a sampling gate and held in a memory circuit for a brief time. The Y signal is then applied to the vertical deflection plates in the normal manner, deflecting the CRT spot vertically.

The sampling command (which opened the sampling gate) is delayed by a fixed interval, as shown by Fig. 2-5C. The delayed sampling command is used to sample a horizontal timing ramp (sawtooth sweep). The timing ramp is triggered when the input signal reaches a certain level, as shown in Figs. 2-5A and 2-5D. The sample of the timing ramp is held in a memory circuit for a brief period, then is applied to the horizontal deflection plates in the normal manner.

The process is repeated for each of the remaining samples. Each sample supplies both vertical and horizontal information to deflect the beam from dot to dot, thus constructing a display of the signal from those samples that fall within the "time window." Notice that sample 3 is not displayed even though it is sampled and available at the vertical deflection plates. This is because the horizontal sweep has already driven the beam off the screen (horizontally) before the horizontal sweep ramp is sampled.

If the delay-time interval is increased, a time shift of the sampling distribution to the left (earlier in time) is necessary in order that the required information be collected for the display. Although only five samples are shown, a typical random-sampling scope might use 1000 samples to reconstruct such a waveform.

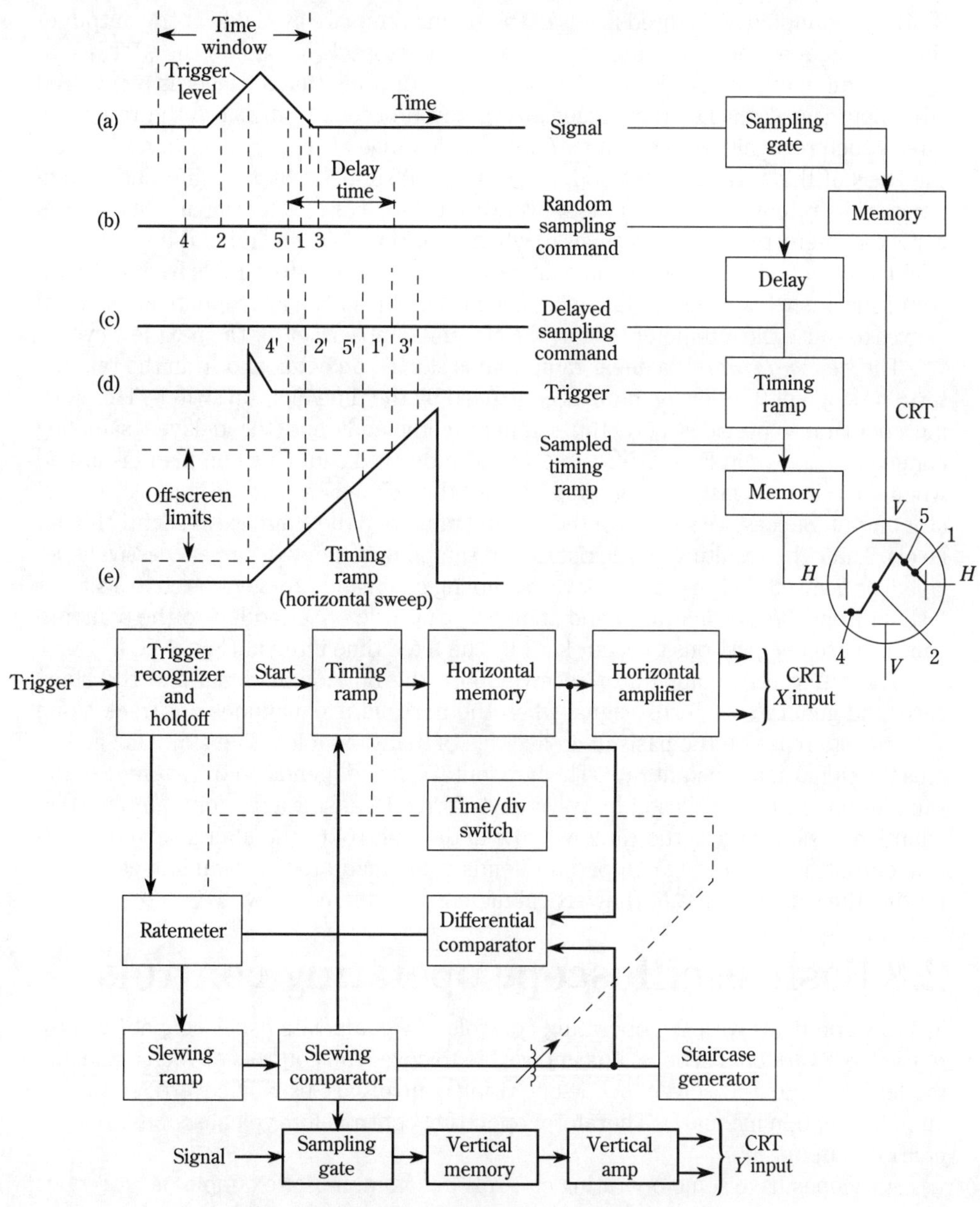

2-5 Random sampling-oscilloscope techniques.

As shown in Fig. 2-5F, the trigger recognizer and hold-off block responds to the presence of a suitable trigger (when the input signal reaches a certain level, Fig. 2-5A) and immediately starts the timing ramp. The hold-off function prevents restarting the timing ramp until the ramp has sufficient time to complete the previous cycle.

The *timing ramp* is a linear ramp with a slope that is controlled by a time/division switch. The starting and subsequent sampling of this ramp is shown in Fig.

2-5E. The sampled and stored levels of the timing ramp are available at the output of the horizontal memory, and are applied to the horizontal (X-axis) of the CRT.

The ratemeter block also receives an input indicating that a trigger is recognized (the input signal reaches the trigger level), and proceeds to measure the repetition rate of such recognitions over an extremely wide range of trigger repetition rates. On the basis of this measurement and an error signal supplied by the differential comparator, the ratemeter supplies the slewing ramp with a pretrigger signal. This pretrigger is the ratemeter's "best guess" as to when to start the slewing ramp. The pretrigger might also contain considerable time uncertainty on a sample-to-sample basis if a number of successive samples are taken later than desired, and an appropriate error signal arrives to cause the ratemeter to start the slewing ramp earlier for the next few cycles.

The *slewing ramp* is a linear ramp that is started on command from the ratemeter. The slope of the slewing ramp is controlled by the time/division switch. The slewing comparator provides both the sampling commands and the delayed sampling commands, shown in Figs. 2-5B and 2-5C. The delayed sampling command is issued when the relatively fast slewing ramp reaches the voltage level of the staircase generator output. Successive sweeps of the slewing ramp find the staircase at slightly higher levels. Thus, the resulting comparisons and sampling commands are successively delayed, or "slewed" in time. The delayed sampling commands are generated in a similar fashion from the slewing ramp and staircase, but a dc offset added to the staircase causes these comparisons to occur later by the fixed time interval (Fig. 2-5C).

The differential comparator receives both the horizontal signal and the staircase, and generates an error signal when the horizontal signal does not track along with the staircase on the basis of an average of many samples. The staircase generates the signal at a constant rate. The horizontal signal depends on the average of the samples taken. Thus, a closed-loop system is formed, causing a random sampling distribution to slew across the time window under control of the staircase generator. The output of the slewing comparator sends a command to the sampling gate, permitting the signal to pass to the vertical memory circuit, as shown.

2.8 Basic oscilloscope operating controls

This section deals with the operating controls of a composite oscilloscope. Because of the large variety of scopes, it is impossible to cover the controls of each make and model (also, the functions of scope operating controls are usually covered in the related instruction manuals). Therefore, operating controls for typical scopes are concentrated upon.

All scopes have some operating controls in common. For example, most scopes have controls that position the trace vertically and horizontally. Not all controls, however, are found in any one scope. Most simple, inexpensive scopes do not have a single-sweep control or an operating mode where a single sweep can be displayed. Controls and connectors are not always found in the same location and in the same form on all scopes. On one instrument, the astigmatism control is a front-panel operating knob. On another scope, astigmatism control is a side-panel adjustment.

Figure 2-6 shows the front-panel operating controls of a composite scope. This instrument is a dual-trace scope with both vertical channels capable of displaying

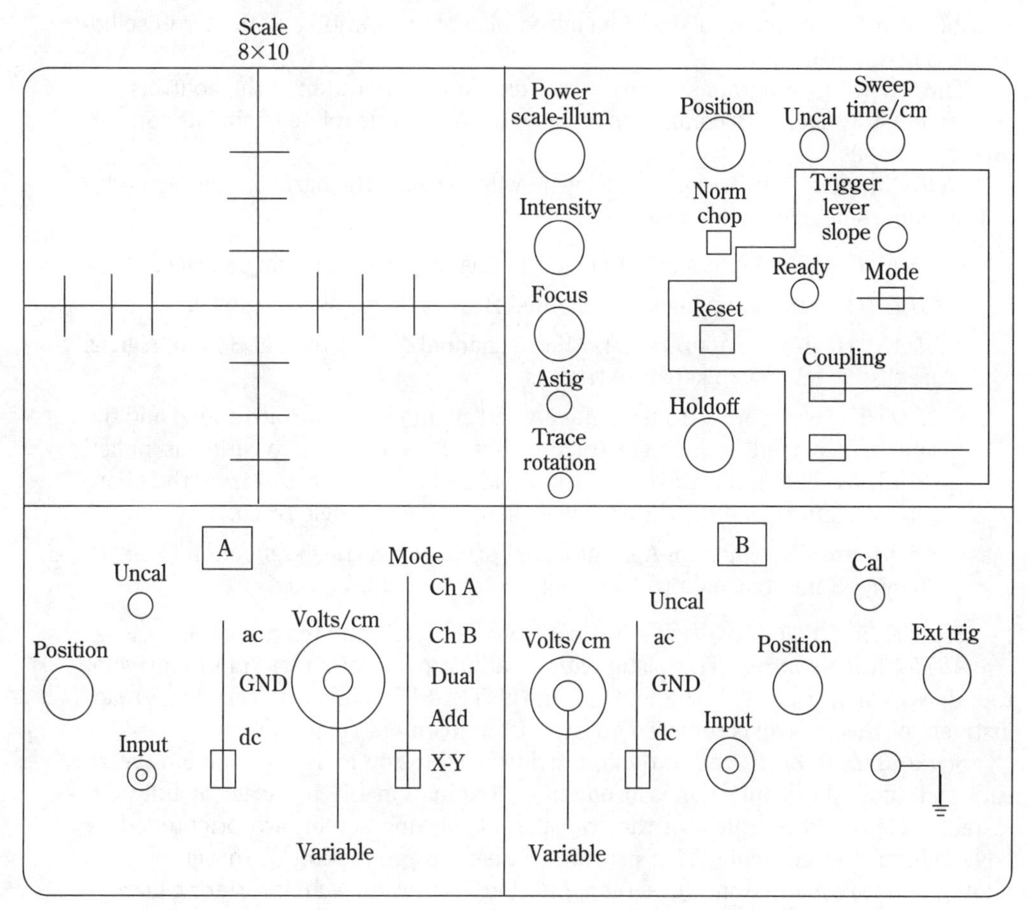

2-6 Front-panel operating controls of a composite scope.

signals from dc up to 35 MHz. The vertical deflection factor is 1 mV/division for signals from dc to 10 MHz, and 3 mV/division for signals from 10 to 35 MHz.

There are four modes of sync, including single sweep and fixed. The single-sweep mode allows a push button to enable a single sweep, which begins at the next sync trigger. The fixed mode automatically fixes the sync-level threshold at the center of the sync waveform. Five choices of sync coupling are available including a sync separator for video signals.

The sync signal is selectable from five sources, including an alternate for dual-trace operation (channel A trace is triggered by the channel A input signal, and the channel 3 trace is triggered by the channel 3 input signal). Variable hold-off allows triggering to be inhibited beyond the sweep duration. A built-in signal delay line allows the leading edge of the high-speed signals to be observed.

UNCAL lights remind the operator that controls are not properly set for calibrated voltage or time measurements. For dual-trace operation, alternate or chop sweep modes (Sec. 2.10.2) can be selected automatically with sweep time or can be manually overridden. The instrument also provides for electrical trace rotation (adjustable from the front panel), and a slotted bezel for mounting a standard oscilloscope camera (Sec. 2.11).

The following paragraphs describe the function or operation of the controls and connectors, together with comments on how these controls relate to similar controls on other scopes.

MODE switch This five-position lever switch selects the basic operating modes of the scope. The five modes are:

- *CH A* Only the input signal to channel A is displayed as a single trace.
- *CH B* Only the input signal to channel B is displayed as a single trace.
- *DUAL* Dual-trace operation. Both the channel A and channel B input signals are displayed on two separate traces.
- *ADD* The waveforms from channel A and channel B inputs are added and the sum is displayed as a single trace. When the channel B position is pulled (PULL INVERT), the waveform from channel 3 is subtracted from the channel A waveform and the difference is displayed as a single trace.
- *X-Y operation* Channel A input signal produces vertical deflection (Y-axis). Channel B input signal produces horizontal deflection (X-axis).

TRACE ROTATION control This control electrically rotates the trace.

ASTIG adjustment The astigmatism adjustment provides optimum spot roundness when used with the FOCUS and INTENSITY controls. Very little readjustment of this control is required after initial adjustment (usually).

Scale This 8- by 10-cm graticule provides calibration marks for voltage (vertical) and time (horizontal) measurements. Illumination of the scale is fully adjustable. The engraved lines of the transparent viewing screen are brightened by edge-lighting the graticule. This provides a sharp reproduction of the lines when photographs are made from the screen, but does not produce an interfering glare.

FOCUS control This control provides adjustment for a well-defined display. If a well-defined trace cannot be obtained with the FOCUS control, it might be necessary to adjust the ASTIG control. To check for proper setting of the astigmatism adjustment, slowly turn the FOCUS control through the optimum setting. If the astigmatism adjustment is correctly set, the vertical and horizontal positions of the trace will come into sharpest focus at the same position of the FOCUS control. This relationship between focus and astigmatism should be the same for all displays. However, it may be necessary to reset the FOCUS control slightly when the INTENSITY control is changed.

INTENSITY control This control regulates the brightness of the display (from very bright to total darkness). The setting of the INTENSITY control can also affect the correct focus of the display. Slight readjustment of the FOCUS control might be necessary when the intensity level is changed.

To protect the CRT phosphor of any scope, do not turn the INTENSITY control higher than necessary to provide a satisfactory display. Also, be careful that the INTENSITY control is not set too high when changing from a fast to a slow sweep rate.

POWER/SCALE-ILLUM control Full counterclockwise rotation of this control (OFF position) turns off the scope. Clockwise rotation turns on the scope. Further clockwise rotation increases the illumination level of the scale.

NORM-CHOP switch This push-button switch operates with the triggering SOURCE switch to provide automatic or manual selection of alternate or chop method of dual-trace sweep generation.

With the SOURCE switch in the ALT position, the alternate sweep is selected regardless of sweep time and the NORM-CHOP switch has no effect.

POSITION control Rotation of this control adjusts the horizontal position of traces (both traces when operated in the dual-trace mode). Also called *horizontal-centering control*, *horizontal-position control*, or *X-position control* on some scopes, this function moves the trace left and right to any desired horizontal position on the screen. This particular POSITION control also has a push-pull function. When this POSITION control is pulled out, the display is magnified five times (5×). The display is normal when the POSITION control is pushed in.

The 5× magnification permits closer observation of a part of the signal. As shown in Fig. 2-7A, the center division of the unmagnified display is the portion visible on the screen in magnified form. Different portions of the display can be viewed by rotating the POSITION control to bring the desired portion into the viewing area (*sweep-magnifier functions* are usually found only on lab scopes).

Sweep-time UNCAL indicator This LED glows when the sweep time VARIABLE control is not set to CAL position, and reminds the user that time measurements are not calibrated.

SWEEP TIME/CM switch This switch is the horizontal coarse sweep-time selector. The switch selects calibrated sweep times of 0.1 μs/cm to 0.5 s/cm in 21 steps when the VARIABLE control is set to CAL (fully clockwise). On some shop-type scopes, this switch is called the *sweep-range selector* or the *coarse frequency control.* In such instruments, the switch positions are calibrated in terms of frequency, rather than time (typically 10 to 100 Hz, 100 to 1000 Hz, 1 to 10 kHz, 10 to 100 kHz, and so on). In lab scopes, the sweep rates are expressed in time for convenience because the time interval of the of the display is of greater importance than the sweep frequency for most scientific measurements. If time is not given directly, it must be calculated from frequency, as is true for most older shop scopes.

Sweep-time VARIABLE control This control is the fine sweep-time adjustment. In the extreme clockwise (CAL) position, the sweep time is calibrated as set by the SWEEP TIME/CM switch. In other positions, the variable control provides a continuously variable sweep rate. On shop scopes, the variable control is usually called the *sweep frequency control*, *fine frequency control*, or *frequency vernier*, and permits continuous variation of sweep frequency within any range provided by the sweep range selector.

HOLDOFF control This control adjusts holdoff time (trigger inhibit period beyond sweep duration), and is usually set to the NORM position because no holdoff period is necessary. The HOLDOFF control is useful when a complex series of pulses appears periodically, such as in Fig. 2-7B. Improper sync might produce a double image, as shown. Such displays can be synchronized VARIABLE with the sweep-time VARIABLE control, but this is impractical because time measurements are then uncalibrated. An alternative method of synchronizing the display is with the HOLDOFF control, which adjusts the duration of a period (after the sweep) in which triggering is inhibited. The sweep speed remains the same,

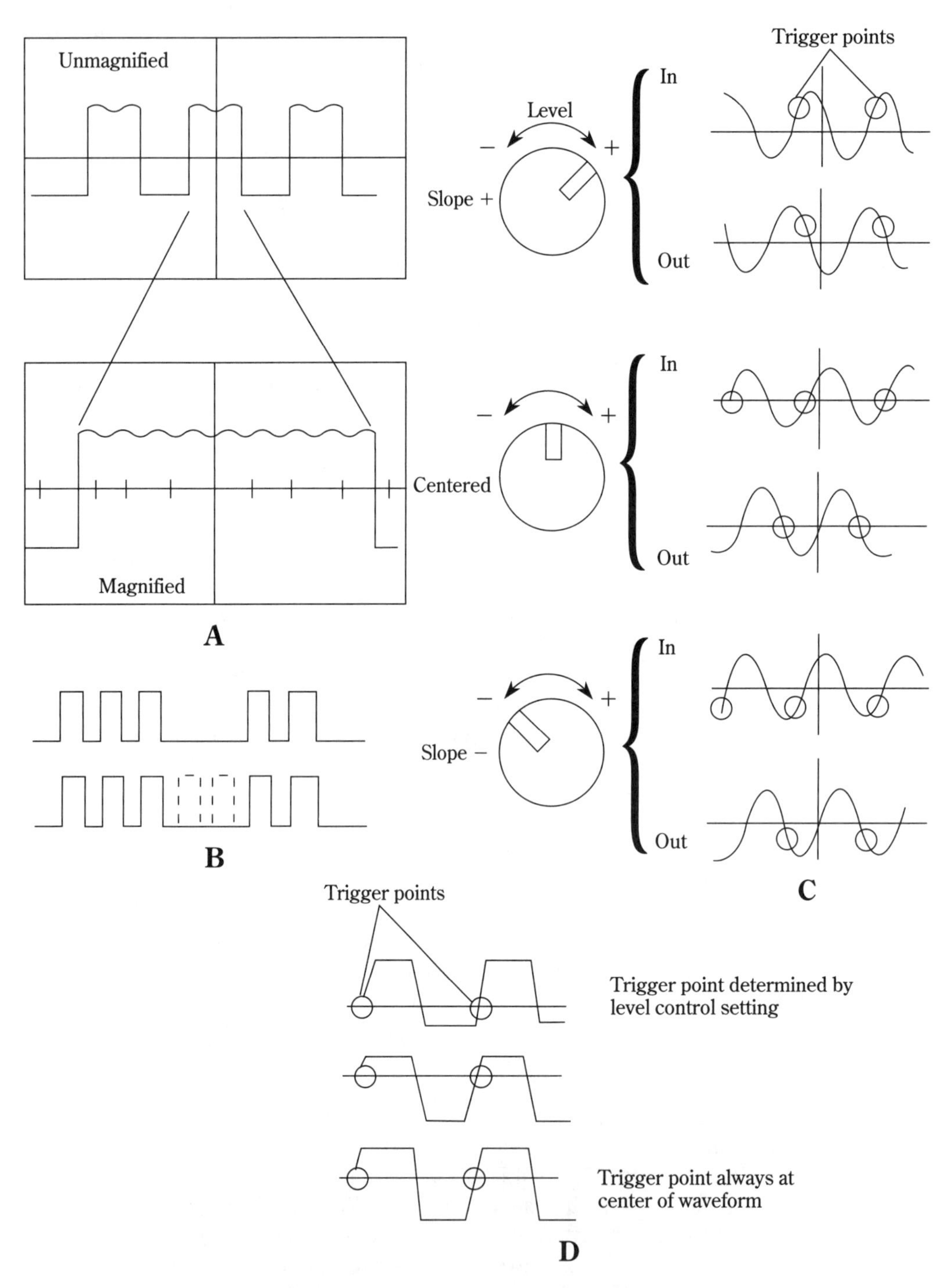

2-7 Sweep-magnification, level-control, and trigger-point displays.

but the triggering of the next sweep is "held off" for the duration selected by the HOLDOFF control. At the MAX setting, the holdoff period is about 10 times greater than at the lowest setting.

EXT TRIG jack This jack is the input terminal for external trigger signals.

CAL terminal This terminal provides a calibrated 1-kHz, 0.1-V peak-to-peak square-wave signal. The signal can be used for calibration of the vertical-amplifier attenuators, and to check frequency compensation of probes used with the scope (chapter 5).

RESET button When the triggering MODE switch is in SINGLE, pushing the RESET button initiates a single sweep that begins when the next sync trigger occurs.

READY/TRIGGER indicator In the single triggering mode, this indicator lights when the RESET button is pressed, and goes off when sweep is completed. In the NORM, AUTO, and FIX triggering modes, the indicator turns on for the duration of the triggered sweep. The indicator also shows when the LEVEL control is properly set to obtain triggering.

LEVEL control Rotation of the triggering LEVEL control varies the triggering threshold level as shown in Fig. 2-7C. In the slope + position (clockwise), the triggering threshold shifts to a more positive value, and moves to a more negative value in the slope position. When the LEVEL control is centered, the threshold is set at the approximate average of the signal used as the triggering source. The READY/TRIGGER indicator turns on when the sweep is triggered, indicating that the triggering LEVEL control is within the proper range. If the READY/TRIGGER indicator does not turn on at the center setting of the LEVEL control, there is insufficient sync signal for triggered operation, or one of the sync selection switches might be improperly set.

When the triggering LEVEL control is pushed in, the sync trigger is developed from the positive-going slope of the sync signal, and when the control is pulled out (PULL SLOPE NEG), the sync trigger is developed from the negative-going slope, as shown in Fig. 2-7C. The push-pull action of the LEVEL control also selects positive (in) or negative (out) polarity of composite video pulses when the COUPLING switch is in the video position.

Figure 2-7C shows slope + and slope-triggering from a sine wave. However, any waveshape produces a trigger when the signal is above the trigger threshold setting. Rotation of the LEVEL control adjusts the phase for which sweep triggering begins on sine-wave signals. For any type of triggering signal, the LEVEL control can be used to set an amplitude threshold to prevent triggering by noise or undesired portions of the waveform.

Triggering MODE switch This four-position switch selects the triggering mode as follows:

- *SINGLE* This position enables the RESET switch for triggered single-sweep operation. When the signal to be displayed is not repetitive or varies in amplitude, shape, or time, a conventional repetitive display can produce an unstable presentation. The single-sweep mode is used to avoid this condition. The single sweep can also be used to photograph a nonrepetitive signal.

- *NORM* This is the most commonly used setting for triggered sweep operation. The triggering threshold is adjustable with the triggering level control. No sweep is generated in the absence of triggering signal or when the LEVEL control is set so that the threshold exceeds the amplitude of the triggering signal.
- *AUTO* This position selects automatic sweep operation where the sweep generator free-runs to generate a sweep without a trigger signal (this is called *recurrent sweep operation* on many scopes). In AUTO, the sweep generator automatically switches to triggered sweep operation if an acceptable trigger signal is present. The AUTO position is handy when first setting up the scope to observe a waveform because AUTO provides sweep for waveform operation until other controls can be properly set. AUTO sweep dc for operation must be used for dc measurements, and for signals of such low amplitude that the sweep is not triggered. With this scope, signals that produce even 0.5 cm of vertical deflection are adequate for normal triggered-sweep operation.
- *FIX* This mode is the same as the AUTO mode, except that triggering always occurs at the center of the sync trigger waveform, regardless of the LEVEL control setting, as shown in Fig. 2-7D. This makes it unnecessary to constantly readjust the LEVEL control when making waveform measurements at several points that might be of different amplitude and waveshape. The NORM or AUTO modes must be used to trigger at any point other than the waveform center.

COUPLING switch This five-position switch selects input coupling for the sync trigger signal as follows:

- *AC position* This is the most commonly used position. AC position permits triggering from 10 Hz to over 35 MHz, and blocks any dc component of the sync trigger signal.
- *LF REJ position* In the LF REJ position, dc signals are rejected, and signals below about 10 kHz are attenuated. Therefore, the sweep is triggered only by the higher-frequency components of the signal. This position is particularly useful for providing stable triggering if the trigger signal contains line-frequency components, such as 60-Hz hum, as shown in Fig. 2-8A.
- *HF REJ position* The HF REJ position attenuates signals above 100 kHz. Dc signals are rejected. This position is used to reduce high-frequency noise and is especially useful to trigger on modulation, rather than RF signals, when observing amplitude-modulated waveforms.
- *VIDEO position* This position is used primarily to view composite video signals in television and VCR service. In VIDEO, the sync trigger signal is routed through a sync-separator circuit. The VIDEO position operates in conjunction with the SWEEP TIME/CM switch to automatically select horizontal or vertical sync pulses at appropriate sweep speeds for viewing each type of waveform.

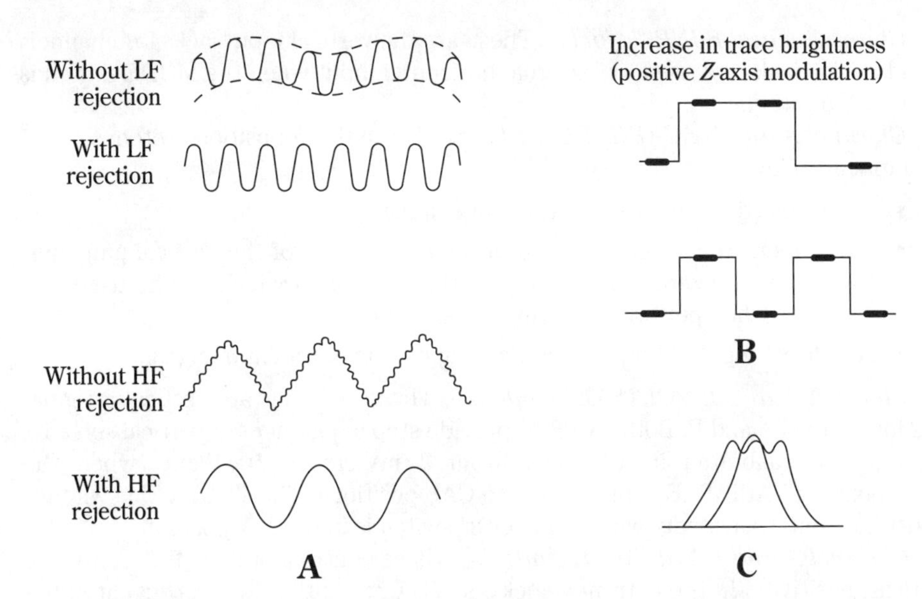

2-8 Noise-rejection, Z-axis modulations, and double-image displays.

Vertical-sync pulses (FRAME) are automatically selected at sweep speeds of 0.1 ms/cm and slower because these sweep speeds are appropriate for viewing vertical frames of video. Horizontal-sync pulses (LINE) are automatically selected at sweep speeds of 50 μs/cm and faster because these sweep speeds are appropriate for viewing horizontal lines of video.

- *DC position* This position permits triggering from dc to over 35 MHz. DC can be used to provide stable triggering with low-frequency signals which would be attenuated in the AC position, or with low-repetition rate signals. The LEVEL control can be adjusted to provide triggering at the desired dc level on the waveform. When using internal triggering, the setting of the vertical position controls can affect the dc trigger level.

Z-AXIS INPUT jack (rear panel) This connector is for intensity modulation of the CRT display (Sec. 2.5.3). The trace displayed can be intensity modulated (Z-axis input), where frequency or time-scale marks are required. A 5-V peak-to-peak or greater signal applied to the Z-axis input jack provides alternate brightness and blanking of the trace. A positive voltage input increases brightness. Figure 2-8B shows typical displays with Z-axis modulation.

Channel A and B POSITION controls These controls provide vertical-position adjustment for the channel A and B traces. The controls become horizontal controls when the MODE switch is in the X-Y position. The controls are also push-pull, and select normal or inverted polarity of the channel A and B displays (display is inverted when the control is pulled).

Channel A and B UNCAL indicators These indicators glow when the A VARIABLE and B VARIABLE controls are not set to CAL. This reminds the user that the channel A or B measurements are not calibrated.

Channel A and B INPUT jacks These are the vertical input jacks for channels A and B. The jacks are used for external horizontal input when the MODE switch is in the X-Y position.

Channel A and B AC-GND-DC switches These three-position switches select the input as follows:

- *AC* blocks dc component of the input signal.
- *GND* opens the signal path and grounds the input of the vertical amplifier. This provides a zero-signal base line, the position of which can be used as a reference when performing dc measurements.
- *DC* direct input of both ac and dc components of the input signal.

Channel A and B VOLTS/CM switches These switches are vertical attenuators for channel A and B. Both switches provide step adjustment of vertical sensitivity, and are calibrated in 12 steps, from 2 mV/cm to 10 V/cm (when the corresponding VARIABLE control is set to CAL position). The VOLTS/CM switches adjust horizontal sensitivity when the MODE switch is in the X-Y position.

Channel A and B VARIABLE controls These controls provide fine control of vertical sensitivity. In the extreme clockwise (CAL) position, the vertical attenuator is calibrated (as determined by the corresponding VOLTS/CM switch). The variable controls become the fine horizontal-gain control when the MODE switch is in the X-Y position.

2.8.1 Miscellaneous operating controls

In addition to the controls covered thus far, other controls must be adjusted for proper scope operation. The following additional controls or adjustments are found on many scopes. On some instruments, the additional functions are adjusted by front-panel controls. On other instruments, the function is set by a screwdriver adjustment.

VOLTAGE REGULATION sets the output voltage of regulated power supplies for the scope circuits.

CALIBRATION VOLTAGE sets the calibration voltage output.

SWEEP FREQUENCY sets the horizontal-sweep oscillator frequencies to match the sweep-selector calibrations.

LINEARITY sets linear horizontal and vertical deflection on each side of the screen center (Sec. 2.9.3).

FREQUENCY COMPENSATION sets amplifier and attenuator components (vertical and horizontal) for wideband response.

HUM BALANCE cancels power-supply hum.

One front-panel control, found on older scopes used for television and VCR service, is a *PHASING control*. Adjustment of this control varies the horizontal sweep when being driven by the line voltage at the line frequency. The PHASING control is especially important when the scope is used with a sweep generator (driven at line frequency) to observe a response pattern (chapter 3). If there is a phase shift between the sweep generator and horizontal sweep, even though both are at the same frequency, a double pattern appears (Fig. 2-8C). This condition can be corrected by shifting the scope sweep-drive signal phase.

2.9 Oscilloscope specifications and performance

The following is a summary of the most important scope specifications.

2.9.1 Sweeps and scales

In most cases, scopes have built-in sawtooth sweep generators that produce constant-speed horizontal beam deflection. In early scopes, the generators ran continuously, and horizontal calibration was based on repetition frequency. In present-day scopes, sweeps are usually calibrated in terms of a direct unit of time for a given distance of spot travel across the screen. This accounts for the term *time base*.

A scope with the widest range of sweeps is usually the most versatile. However, primary usefulness of a scope is as a high-speed device. Fastest sweeps are usually considered adequate if the sweep can display one cycle of the upper passband frequency across the full horizontal scale.

When accurate rise-time measurements must be made (with the fastest sweep) a useful figure of merit for sweep adequacy is: $M = T_R/T_D$, where M is the figure of merit, T_R is the vertical-system rise time, and T_D is the time per division of the fastest sweep.

Using this system, figures of merit greater than 1 are seldom found in scopes having rise times less than about 30 ns. Figures of merit greater than about 6 exceed the ideal and offer no further advantage. In any event, do not attempt rise-time measurements where the rise time of the signal to be examined exceeds the scope vertical-deflection system rise time.

Time-base accuracy is usually specified in terms of the permissible full-scale sweep-timing error for any calibrated sweep. That is, an accuracy of 3% means that the actual full-scale period of any sweep should not be more than 3% greater or lesser than indicated.

Magnified sweeps might have poorer accuracy ratings than unmagnified sweeps because magnification is usually achieved by reducing amplifier feedback. In some cases, portions of sweeps can be magnified by increasing the gain of the horizontal amplifier, allowing either or both ends of the sweep to go off-screen, and then positioning the display so that the desired portion is on-screen. This method delays the presentation of a sweep portion.

In another method of sweep magnification, fast sweeps are triggered just before the time when the signal to be examined occurs. This method is generally preferred (greater time-base accuracy, less jitter, and so on).

2.9.2 Rise time and high-frequency response

For fast scopes, rise time is the more important specification; with slower scopes, passband (or bandwidth) is the more frequently used specification. Generally, transient response is optimum when the product of rise time and frequency response produces a value between about 0.33 and 0.35.

Transient response is the faithfulness with which a deflection system displays fast-rising step signals. The most common transient-response distortions are overshoot, ringing, and reflections from impedance discontinuities in the vertical-signal delay line. These forms of distortion make clean step signals appear to have spikes, squiggles, or bumps, and make unclean signals appear worse.

In an example of combining rise time and frequency response to determine optimum transient response, the product of 0.023 μs rise time (0.023×10^{-6} s) and 15 MHz (15×10^{6} Hz) equals 0.345. Factors less than 0.35 probably indicate overshoot in excess of 2%. Factors greater than 0.4 probably indicate overshoot in excess of 5%.

Ideally, oscilloscopes should have a vertical system that is capable of rising in about one-fifth the time that it takes the fastest step signal to rise. To find signal rise time, use the equation $T_S = T_I/T_A$, where T_S is signal rise time, T_I is the indicated rise time on the display, and T_A is the vertical-system (usually the vertical amplifier) rise time. Of course, the accuracy of this calculation falls off sharply for signals that rise faster than the scope vertical system.

2.9.3 Display geometry

Here are some common limitations and application pitfalls that apply to almost all scopes. Some easily made performance checks are also included.

Trace and scale alignment A horizontal trace should coincide with the horizontal scale markings on the graticule. Misalignment, as shown in Fig. 2-9A, usually indicates a need to reorient the CRT (or the scale), but might be caused by inadequate CRT shielding or the presence of a strong magnetic field. CRTs operated with low accelerating potentials are most susceptible. Even the earth's magnetic field can alter trace alignment.

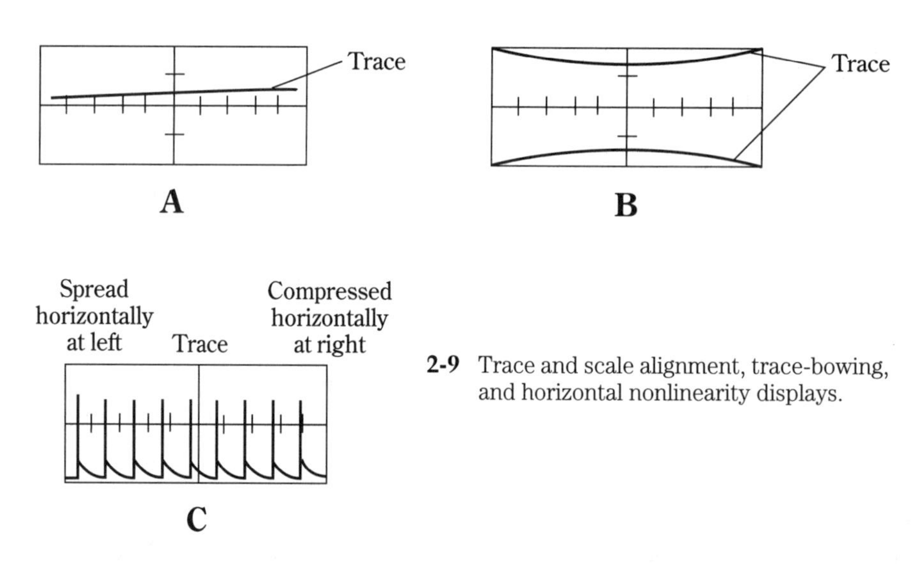

2-9 Trace and scale alignment, trace-bowing, and horizontal nonlinearity displays.

CRT deflection perpendicularly The horizontal plates should deflect the beam in a direction perpendicular to that of the vertical deflection plates. A small deviation (and generally acceptable) is less than 1° near the center of the screen (a 1° error is a displacement from perpendicular of 1 mm in 5.7 cm). Note that axis inaccuracy affects the whole display geometry (and is more noticeable at the screen edges).

Trace-bowing Beam deflection can deviate from a straight line when a trace appears near the outer limits of the useful screen area, as shown in Fig. 2-9B (exaggerated trace bowing). A CRT can be tested for trace bowing with pairs of horizontal or vertical lines. Another method is to use a single beam trace, and manually position the beam rapidly back and forth (horizontally for a vertical trace, and vertically for a horizontal trace) near all four sides of the useful screen area. Bowing tolerance depends on the CRT type and the operating voltages (first and second anodes, etc.). A typical high-quality CRT will not deviate from the parallel lines by more than 1 mm on the edges, or by more than 0.5 mm at the top and bottom.

To minimize bowing, some systems have adjustable voltages on special electrodes in the CRT. Excessive bowing can be caused by improper accelerating voltage, a poor CRT, or an appreciable difference between the dc levels of the vertical and horizontal deflection plates (with spot centered).

Horizontal nonlinearity Equal changes in voltage on the horizontal plates should produce equal changes in spot position. Any CRT nonlinearity, such as shown in Fig. 2-9C, cannot usually be corrected. Other nonlinearities depend on the quality (and adjustment) of the sweep generator and/or horizontal amplifier.

Amplitude compression Amplitude linearity of a horizontal or vertical amplifier with dc coupling can be checked by observing any change in amplitude of a small signal while positioning the display through the useful screen area. In a perfect system, there should not be compression (or expansion) when the signal is positioned at all points throughout the useful area. For amplification with ac coupling, change the input signal in measured steps, and check if the displayed signal amplitude changes by exactly corresponding amounts.

2.9.4 Position drift

High-gain amplifiers with dc coupling tend to drift. In some cases, as much as an hour or more is required for the rate of drift to reduce to a minimum. After warm-up, the maximum amount of drift to be expected is often specified in terms of millivolts (or microvolts) per hour. The amount of position change that such drift represents depends on the deflection factor selected. For example, if the deflection factor is 1 mV/cm, and the drift specification is 1 mV/hr, the drift in any 1-hour period should not exceed 1 cm. In most cases, the drift per hour is of little significance because measurements generally take no more than 1 minute or so. Of course, this is not the case where some precision instruments (precision power supplies, amplifiers, etc.) are to be monitored with a scope over long periods.

2.10 Oscilloscope accessories

As in the case of meters (chapter 1), an oscilloscope can perform most functions without accessories. One major exception that applies to both instruments is a probe. In fact, many of the probe types used with rectors can also be used with scopes. For that reason, all information concerning probes is covered in chapter 5. The following paragraphs describe a few other accessories that can be used effectively with both shop-type and lab scopes.

2.10.1 Electronic switch

In many applications, it is convenient to display two signals simultaneously (phase measurement by the dual-trace method described in Sec. 2.13 is such an application). Many lab scopes have a dual-trace provision (or multiple-trace capability), where two (or more) signals can be applied directly to separate scope input and will be displayed simultaneously, as shown in Fig. 2-10. The same type of operation can be accomplished with a shop scope when an electronic switch or "chopper" is used at the input.

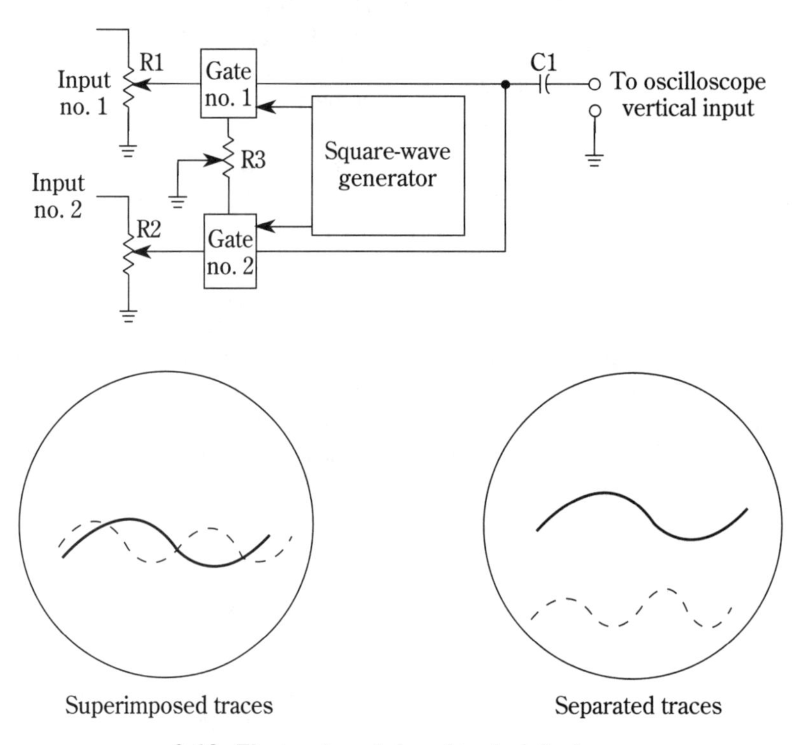

Superimposed traces   Separated traces

2-10 Electronic switch and typical displays.

Whether built-in, or part of an accessory, the electronic switch acts as two gates (one for each input signal), which open and close on alternate half-cycles at a predetermined frequency. One gate is open while the other gate is closed. The output of both gates is applied to the scope vertical input so that both signals are displayed (note that the gate-switching frequency must be much greater than either signal frequency).

In the circuit of Fig. 2-10, each signal is applied to a separate gain control and gate stage. The gate stages are alternately biased to cutoff by square-wave signals from the square-wave generator. Only one gate stage is a condition to pass the signal at any given time. The output of both gates is applied directly to the scope vertical input.

Figure 2-10 also shows both superimposed and separated traces (each trace is composed of tiny bits of the corresponding signal. Because the switching rate is fast, the traces appear as solid lines). Most electronic switches have some form of positioning control so that the traces can be superimposed or separated, whichever is more convenient (the positioning control is shown as R3). Also, so that the amplitude or height of both signals can be made to appear approximately the same on the display, despite an actual difference in signal strength, the electronic switch has separate gain controls (R1 and R2).

2.10.2 Other accessories

Many accessories besides probes and an electronic switch are available for use with lab scopes. Some of these are highly specialized units, such as diode or transistor switching-time testers. Other accessories are essentially extensions of the scopes, such as wideband amplifiers and time-base units with extended ranges. Still other instruments that are sometimes classified as accessories are actually test devices that can be used on their own (such as square-wave generators, pulse generators, power supplies, time-mark generators, calibrators, and spectrum analyzers).

The basic operating principles, as well as detailed operating procedures, for these specialized accessories are described in the instruction manuals for the accessories, or in the manuals for the scopes with which the accessories are to be used. Therefore, scope accessories are not covered in further detail here. When accessories are required to perform the procedures discussed elsewhere in this book, the operating procedures and test connections are described in the relevant chapters.

2.10.3 Plug-in accessories

Plug-in devices are often used to extend the usefulness of lab scopes. In this case, the basic scope consists of a CRT and power circuits. The remaining circuits (such as amplifiers, sweep generators, time-base generators, trigger circuits, etc.) are contained in plug-in units. These plug-ins can be changed to perform specific tests. In laboratory work, it is usually easier to use the basic scope and those accessory plug-ins required for routine or essential operation (vertical and horizontal amplifiers, and horizontal-sweep generator). Then, as new work areas or requirements are developed, additional plug-ins are used to meet the needs.

2.11 Basic operating procedures and recording methods

This section describes the basic operating procedures for various scopes, and covers the basic methods for recording scope displays. Typical scopes are described. As covered in the Introduction, it is assumed that you will become familiar with your particular scope. This is absolutely unessential because the procedures given in this section can be used only as a general guide to operating scope controls. You must understand each and every control on your particular equipment to follow the instructions of later sections and chapters.

2.11.1 Placing an oscilloscope in operation

The first step in placing a scope in operation is to read the instruction manual. Although most instruction manuals are weak in applications data, they do describe turn-on, turn-off, and the logical sequence for operating the controls. After the manual setup instructions are digested, compare the procedures with those described here.

Throughout the following sections and chapters, the direction "place oscilloscope in operation" is used frequently. This direction refers to the following procedure, which has been put in this chapter to avoid repetition.

1. Set the power switch to off.
2. Set the internal recurrent sweep to off.
3. Set the focus, gain, intensity, and sync controls to their lowest position (usually full counterclockwise).
4. Set the sweep selector to external.
5. Set the vertical-and horizontal-position controls to their approximate midpoint.
6. Set the power switch to on. It is assumed that the power cord is plugged in!
7. After a suitable warm-up period (as recommended by the instruction manual), adjust the intensity control until the trace spot appears on the screen. If a spot is not visible at any setting of the intensity control, the spot is probably off-screen (unless the scope is defective). If necessary, use the vertical- and horizontal-position controls to bring the spot into view. Always use the lowest setting of the intensity control needed to see the spot. This prevents burning of the scope screen.
8. Set the focus control for a sharp, fine dot.
9. Set the vertical- and horizontal-position controls to center the spot on the screen.
10. Set the sweep selector to internal (use the linear internal sweep if more than one sweep is available).
11. Set the internal recurrent-sweep to on. Set the sweep frequency to any frequency, or recurrent rate, higher than 100 Hz.
12. Adjust the horizontal-gain control (if any) and check that the spot is expanded into a horizontal trace or line. The line length should be controllable by adjusting the horizontal-gain control.
13. Return the horizontal-gain control to zero (or to the lowest setting). Set the internal recurrent-sweep to off.
14. Set the vertical-gain control (if any) to the approximate midpoint. Touch the vertical input with your finger. The stray signal pickup should cause the spot to be deflected vertically into a trace or line. Check that the line length is controllable by adjustment of the vertical-gain control.
15. Return the vertical-gain control to zero (or to the lowest setting).
16. Set the internal recurrent sweep to on. Advance the horizontal-gain control to expand the spot into a horizontal line.
17. If required, connect a probe to the vertical input.
18. The scope should now be ready for immediate use. Depending on the test to be performed, and on the type of scope, it might be necessary to calibrate the scope. Voltage and current calibration procedures are described in Sec. 2.12.

2.11.2 Basic operating precautions

The following precautions apply to all scopes, and should be observed each time the scope is operated (in addition to the precautions listed in the Introduction, and in the scope instruction manual).

1. Even if you have had considerable experience with scopes, always study the instruction manual of any scope with which you are not familiar.
2. Use the procedures of Sec. 2.11.1 to place the scope in operation. It is good practice to go through the procedures each time that the scope is used. This is especially true when the scope is used by other persons. You cannot be certain that position, focus, and (especially) intensity controls are at safe positions, and the scope could be damaged by switching the power in immediately.
3. As in the case of any CRT device (such as a TV), the CRT spot should be kept moving on the screen. If the spot must remain in one position, keep the intensity control as low as possible.
4. Always use the minimum intensity necessary for good viewing.
5. If possible, avoid using a scope in direct sunlight, or in a brightly lighted room. This permits a low-intensity setting. When the scope must be used in a bright light, use the viewing hood.
6. Make all measurements in the center area of the screen. Even if the CRT is flat, there is a chance of reading errors caused by distortion at the edges.
7. Avoid operating a scope in strong magnetic fields. Such fields can cause distortion of the display. Most good-quality scopes are well shielded against magnetic interference. However, the face of the CRT is still exposed, and is subject to magnetic interference.
8. Never operate the scope with the shield or case removed. Besides the danger of exposing high-voltage circuits (several thousand voltages are used on the CRT, even though the remaining circuits are low-voltage solid state), there is the hazard of the CRT imploding and scattering glass at high velocity.
9. Do not attempt to repair a scope unless you are a qualified instrument technician. If you must adjust any internal controls, follow the instruction manual.
10. Study the circuit under test before making any test connections. Try to match the capabilities of the scope to the circuit. For example, if the circuit has a range of measurements to be made (ac, dc, RF, pulse), you must use a wideband, dc scope with a low-capacitance probe and possibly a demodulator probe. Do not try to measure 6-MHz TV signals with a 100-kHz bandwidth scope. On the other hand, it is wasteful to use a dual-trace 100-MHz lab scope to check out the audio sections of transistor radios.

2.11.3 Recording an oscilloscope trace

In many applications, a scope trace must be recorded. Recording is often done with a Polaroid Land camera equipped with a special lens and a mounting frame (bezel) attached to the scope. It is also possible to record with a conventional camera, with a moving-film camera, or even by hand in some situations. The following sections briefly describe the basic recording methods (instruction manuals for scope cameras and accessories are quite detailed, and are not duplicated here).

2.11.4 Oscilloscope camera systems

Figure 2-11 shows the basic operating principles of a typical oscilloscope camera system. The viewing system consists of a viewing hood and two mirrors. Light from the scope screen strikes the beam-splitting mirror, where a portion of the light is transmitted to the camera lens and another portion is reflected back to the second mirror.

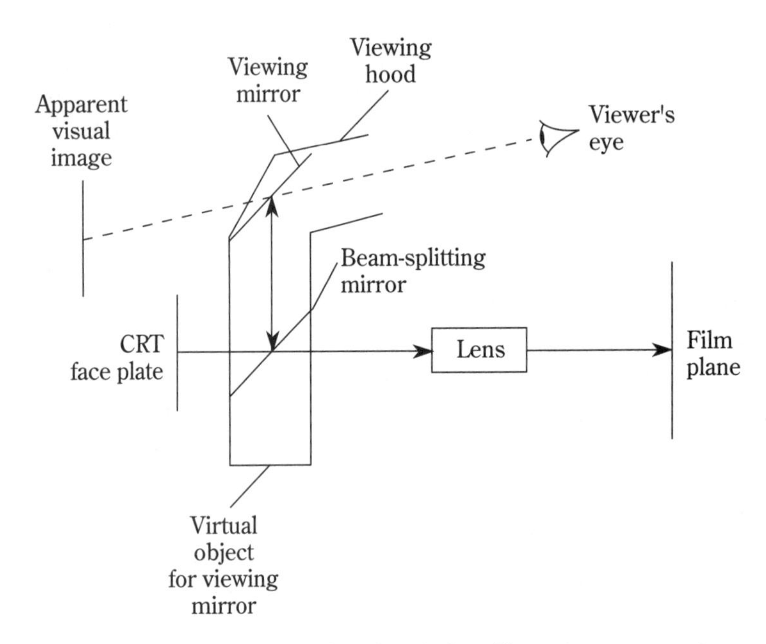

2-11 Basic operating principles of typical oscilloscope camera system.

Because of the 45° arrangement of the beam-splitting mirror, the observer sees the scope display as if looking directly toward the scope screen. The view is full-size, but the image appears about 20 in. away even though the lens is considerably closer to the screen. The difference in the two distances produces a small amount of parallax between the viewed and photographed images. The small parallax can generally be ignored.

Writing rate The term *writing rate* is often used in scope photography. Writing rate is a figure of merit that roughly describes the ability of a particular camera system mounted on a particular scope to photograph fast-moving traces. The writing-rate figure expresses the maximum spot rate (usually in cm per μs) that can be photographed satisfactorily. The faster the scope spot moves, the dimmer the trace becomes because the electron beam strikes each point on the phosphor coating for a shorter period of time.

Because of many variables (type of film, CRT, camera optics) it is not practical to assign an absolute value of writing rate to any scope or camera. However, it is possible to compare the effectiveness of two films by measuring the writing rate under the same conditions. Although ASA rating of film is not the final determining factor, it is practical to assume that a film with very high ASA ratings has a higher maximum writing rate than films with lower ASA speed ratings, all other factors being equal.

CRT phosphors If a scope is to be used for photograph work, selection of the CRT phosphor can be important. For low sweep rate or repetitive-sweep applications, where a high writing rate is not required, almost any type of phosphor is satisfactory. However, selection of the CRT phosphor becomes important for single-sweep or low-repetition rate applications using the fast sweep rates. In such applications, use of the proper phosphor can mean the difference between getting a good photograph and getting none at all.

When a scope trace is to be photographed, the most important single characteristic of a phosphor is the color of the emitted light. A blue or violet fluorescence is most suitable. In general (all other factors being equal), the shorter the wavelength of the emitted light, the better the phosphor for photographic applications.

Most scope users are concerned with both viewing and photographing the display. In this case, the P2 or P31 phosphors provide a good writing rate, and have sufficient persistence to permit easy viewing. The P11 phosphor is generally considered to be the best when the only concern is photographing the trace. However, the medium-short persistence of P11 is somewhat undesirable for general-purpose work (direct viewing of the scope screen). The following is a summary of CRT phosphor characteristics, as applied to writing rate and scope brightness.

2.11.5 Photographing repetitive oscilloscope traces

The following procedures can be used to obtain an exposure for both Polaroid and conventional film when photographing repetitive scope traces.

1. Adjust the graticule lighting for clear lines. Graticule lighting should not be so high as to produce glare, but high enough to make the lines shine. This is true for almost all photographic recording, except where the exposure is very short. Medium graticule lighting for short exposures might not produce a sharp reproduction. Sometimes, this condition can be corrected by double exposure. The preferred technique is to expose the film first with the graticule only, then with the graticule and trace combined. As with most photography, the exposure times must be found by experiment (much experiment). Notice that a dim, thin trace requires longer exposure, but gives better reproduction. A bright trace can produce a halo or afterglow.
2. Mount the camera bezel on the scope, following the instruction manual procedures.
3. Apply a signal to the scope input and adjust the controls for the desired display.
4. Attach the camera to the bezel and secure the camera against the scope.
5. Adjust the scope focus, astigmatism, and intensity controls for a sharp trace.

6. Set the camera aperture selector for the largest lens opening (smallest f-stop number) and carefully focus the camera on the trace. If you are using an external graticule, and want both the trace and graticule lines to be clear, focus the camera halfway between the trace and graticule.
7. Set the scope intensity to midrange, graticule lighting about 75% of full range, shutter speed to ⅕ second, and aperture selector to f/5.6. These control settings should be reasonably close for film speeds of 400 ASA, and a waveform near 1 kHz. For film with a 3000 ASA rating and a waveform frequency of 1 kHz use a shutter speed of ⅕ second and an aperture setting of f/4.5.

2.11.6 Photographing single-sweep displays

The following procedures can be used to obtain an exposure for both Polaroid and conventional film when photographing single-sweep displays. Notice that single-sweep displays are formed when the scope spot sweeps across the screen only once. Actual exposure time is thus determined by the duration of the sweep plus the phosphor persistence, and not by the shutter setting (provided that the shutter is open sufficiently long). In one type of single-sweep photography, the graticule exposes the film for the time set by the shutter, whereas the spot on the screen exposes the film for only the duration of the sweep. Therefore, it is not usually possible to adjust the trace and graticule for the same intensity (and good pictures) because the effective exposure times for the two are different. Success in getting good photographs of single-sweep displays comes only with experience (an unbelievable amount of wasted film). However, the following tips can reduce the amount of experimenting and film required.

1. Select a shutter speed longer than the event to be photographed. If necessary, use the "time" or "bulb" camera function and hold the shutter open while manually triggering the sweep.
2. Use the highest possible intensity without causing defocus of the trace (do not use this intensity setting for normal operation, but only when photographing single-sweep traces).
3. Where practical, use f-stops higher than f/4, if an external graticule is used. This allows both the trace and graticule to be in focus (hopefully).
4. Remember that because the shutter speed has already been determined, the selection of lens openings determines how well the trace photographs. In single-speed applications, the camera setting must be made for trace intensity and duration. Graticule intensity cannot be used as a reference.

2.11.7 Conventional cameras for oscilloscope recording

Conventional cameras can be used, with certain limitations, in place of Polaroid cameras for oscilloscope recording. The adapters used with Polaroid cameras can also be used with some conventional cameras, including 35 mm. The film must be processed in the normal manner.

It is also possible to use a conventional camera without an adapter, if several precautions are taken and if the inconvenience can be tolerated. Two basic problems are to be considered: focal length of the camera and ambient light.

Most cameras have a minimum focal length of 2 or 3 ft. This produces a very small image on the negative, requiring blow-up of the print (with the usual distortion problems). This problem can be minimized using a camera fitted with a close-up lens.

The ambient light problem can be overcome with a cardboard tube between the scope screen and camera to exclude the light. The author has successfully used an ancient twin-lens reflex, with 400 ASA film, focusing from a distance of about 12 in., through a hood or light shield. Of course, any such arrangement is makeshift, and should be used as a temporary or emergency method (try it if you have nothing else to do).

2.11.8 Moving-film cameras for oscilloscope recording

Moving-film cameras are sometimes used in highly specialized oscilloscope work, such as studying lightning, noise, electrical breakdown, cosmic rays, metal fatigue, or any random occurrence. Moving-film cameras are similar to motion-picture cameras in that both have film which is drawn across the lens. In a moving-film camera, the film is drawn continuously by an adjustable-speed motor, and there is no opening and closing of a shutter for each frame, as in a motion-picture camera.

The moving film itself provides the horizontal sweep and the time base. Usually, the signal to be measured is applied to the scope vertical input in the normal manner, but the horizontal sweep is switched off. Because operating a moving-film camera for oscilloscope recording is so specialized, the instruction manuals provide a great amount of detail, which is not repeated here.

2.11.9 Hand-recording oscilloscope traces

When a camera is not available, it is possible to "hand-record" oscilloscope traces. Of course, such recording is limited to traces of long duration, which remain stationary. Also, hand-recording requires considerable skill and should be used only as a temporary or emergency technique.

The oscilloscope trace can be recorded on a transparent plastic overlay cut to the same size as the oscilloscope screen. Thin paper can also be used, but this requires that the intensity be advanced.

If the plastic overlay is used, the actual trace is made with a well-sharpened grease pencil, permitting the trace to be rubbed off and the overlay reused when the particular trace is no longer needed. Attach the overlay to the screen with two-sided adhesive tape, or hold firmly in place. It might be convenient to describe a graticule on the overlay, and to align this graticule with that of the scope screen. The plastic overlay should be thick enough so that it does not wrinkle, but not so thick that it produces parallax (0.01 in. is usually satisfactory).

If paper is used, the actual trace is made with a medium pencil, being careful not to scratch the screen. Also be careful not to advance the intensity to a point where the screen could be burned. It might be convenient to use graph paper, with a graticule ruled on one side. If possible, use a graph paper on which the divisions correspond to the scope screen divisions.

The static electricity present on plastic or paper sheets might be a problem when placed against the scope screen. The static electricity charge can distort the display. Discharge the paper or plastic before attaching it to the scope. This can be accomplished by touching the sheet to a good ground.

2.12 Measuring voltage and current

The oscilloscope has both advantages and disadvantages when used to measure voltage and current. The most obvious advantage is that the scope shows waveform, frequency, and phase simultaneously with the amplitude of the voltage or current being measured. The meter shows only amplitude. Similarly, most meters are calibrated in relation to sine waves. When the signals being measured contain significant harmonics, the calibrations are inaccurate. With the scope, the voltage is measured from the displayed wave, which includes any harmonic content. In certain applications, the high-speed response of a scope makes it the only instrument capable of transient voltage measurement.

Another problem, although not a disadvantage, is that voltages measured with a scope are peak-to-peak, whereas most voltages specified in troubleshooting manuals are rms (except where scope waveforms are given in the manual).

To sum up, if the only value of interest is voltage or current amplitude, use the meter because of simplicity. Use the scope where waveshape characteristics are of equal importance.

This chapter describes the procedures for measuring both ac and dc voltages and currents with a scope, including calibration, if necessary. The vertical amplifier of a lab scope usually has a step attenuator where each step is related to a specific deflection factor (such as a mV/cm). These scopes need not be calibrated for voltage or current measurements because calibration is an internal adjustment performed as part of routine maintenance.

The vertical amplifiers of some scopes have a step attenuator and a variable attenuator or gain control. The steps might or might not have a specific volts/cm factor. Such scopes must be calibrated before being used to measure voltage and current. The following paragraphs describe such calibration, where necessary.

2.12.1 Peak-to-peak-measurements: ac laboratory oscilloscope

1. Connect the equipment, as shown in Fig. 2-12.
2. Place the scope in operation (Sec. 2.11).
3. Set the vertical step attenuator to a deflection factor that allows the expected signal to be displayed without overdriving the vertical amplifier.
4. Set the input selector to measure ac. Connect the probe to the signal being measured.
5. Switch on the scope internal recurrent sweep.
6. Adjust the sweep frequency for several cycles on the screen.
7. Adjust the horizontal gain control to spread the pattern over as much of the screen as desired.

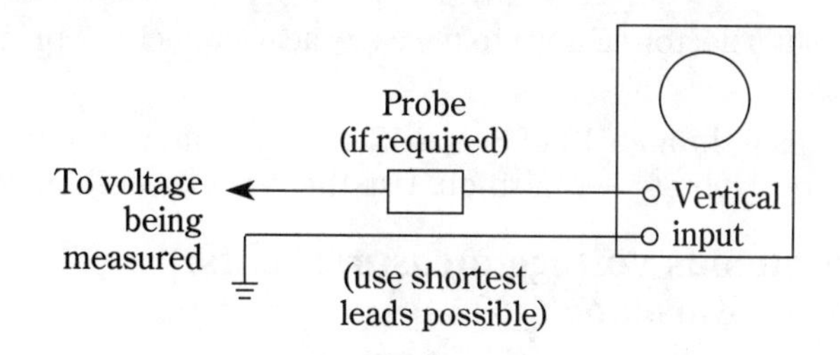

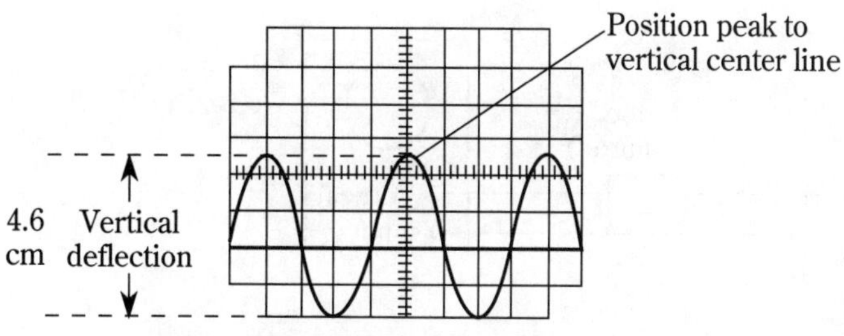

2-12 Measuring peak-to-peak voltages.

8. Adjust the vertical position control so that the downward excursion of the waveform coincides with one of the graticule lines below the graticule centerline, as shown in Fig. 2-12.
9. Adjust the horizontal position control so that one of the upper peaks of the signal lies near the vertical centerline (Fig. 2-12).
10. Measure the peak-to-peak vertical deflection in cm. Notice that this technique can also be used to make vertical measurements between two corresponding points on the waveform, other than peak-to-peak. However, the peak-to-peak points are usually easier to measure.
11. Multiply the distance measured in step 10 by the vertical attenuator setting. Also include the attenuation factor of the probe, if any. For example, assume a peak-to-peak vertical deflection of 4.6 cm (Fig. 2-12) using a 10× attenuator probe and a vertical-deflection factor of 0.5 V/cm. The peak-to-peak value is: $4.6 \times 0.5 \times 10 = 23$ V.
12. Notice that if the voltage being measured is a sine wave, the peak-to-peak value can be converted to peak, rms, or average, as shown in Fig. 1-8B. Similarly, if a peak, rms, or average value is given and must be measured on a scope, Fig. 1-8B can be used to find the corresponding peak-to-peak value.

2.12.2 Peak-to-peak measurements: ac shop oscilloscope

1. Connect the equipment, as shown in Fig. 2-12.
2. Place the scope in operation (Sec. 2.11).
3. Set the vertical gain control to the calibrate-set positions, as determined during the calibration procedure (Sec. 2.12.5).

4. Set the input selector (if any) to measure ac. Connect the probe to the signal being measured.
5. Repeat steps 5 through 12 of Sec. 2.12.1. Do not move the vertical-gain control from the calibrate-set position. Use the vertical-position control only.

2.12.3 Instantaneous voltage measurements: dc laboratory oscilloscope

1. Connect the equipment, as shown in Fig. 2-13.

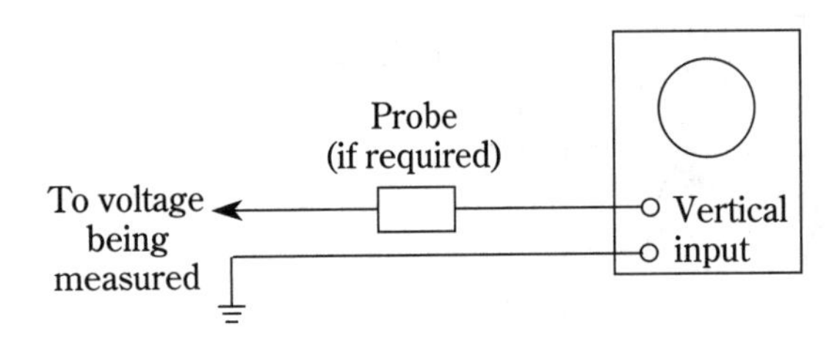

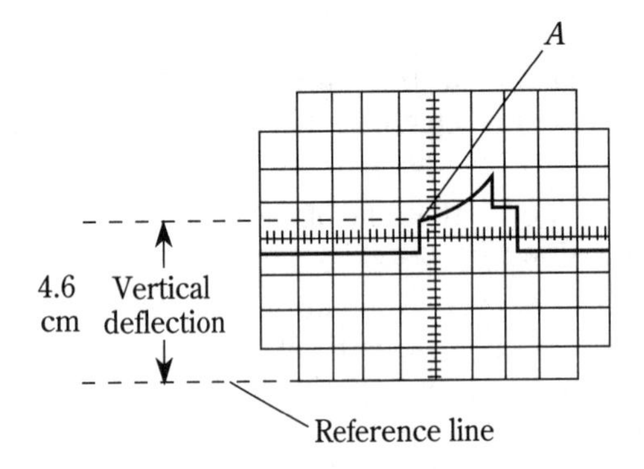

2-13 Measuring instantaneous voltages.

2. Place the oscilloscope in operation (Sec. 2.11).
3. Set the vertical step attenuator to a deflection factor that allows the expected signal, plus any dc, to be displayed without overdriving the vertical amplifier.
4. Set the input selector to ground. On most lab scopes, the input switch that selects either ac or dc measurement also has a position that connects both vertical-input terminals to ground (or shorts them together). If no such switch position is provided, short the vertical input terminals by connecting the probe (or other lead) to ground.
5. Switch on the oscilloscope internal recurrent sweep. Adjust the horizontal gain control to spread the trace over as much of the screen as desired.
6. Using the vertical-position control, position the trace to a line of the graticule below the centerline, as shown in Fig. 2-13. This establishes the reference line. If the average signal (ac plus dc) is negative with respect to ground, position the trace to a reference line above the graticule centerline. Do not move the vertical-position control after this reference line is established.

To measure a voltage level with respect to a voltage other than ground, make the following changes in steps 4 and 6. Instead of shorting the vertical-input terminals, apply the reference voltage to the vertical input, then position the trace to the reference line.

7. Set the input selector switch to dc. The ground reference line, if used, can be checked at any time by switching the input selector to the ground position. Connect the probe to the signal being measured.
8. If the waveform is outside the viewing area, set the vertical step attenuator so that the waveform is visible.
9. Adjust the sweep frequency and horizontal gain controls to display the desired waveform.
10. Measure the distance in centimeters between the reference line and the point on the waveform at which the dc level is to be measured. For example, in Fig. 2-13, the measurement is made between the reference line and point A.
11. Establish polarity of the signal. Any signal-inverting switches on the scope must be in the normal position. If the waveform is above the reference line, the voltage is positive (and vice versa).
12. Multiply the distance measured in step 10 by the vertical attenuator setting. Also include the attenuation factor of the probe, if any. For example, assume the vertical distance measured is 4.6 cm (Fig. 2-13). The waveform is above the reference line, using a 10× attenuator probe and a vertical-deflection factor of 2 V/cm. The instantaneous voltage is: $4.6 \times +1 \times 2 \times 10 = +92$ V.

2.12.4 Instantaneous voltage measurements: dc shop oscilloscope

1. Connect the equipment, as shown in Fig. 2-13.
2. Place the oscilloscope in operation (Sec. 2.11).
3. Set the vertical-gain control to the calibrate-set position, as determined during the calibration procedure (Sec. 2.12.5).
4. Set the input selector (if any) to measure dc (notice that all shop scopes do not have a dc input, and are capable of measuring only ac voltages or signals). Connect the probe to the signal being measured.
5. Repeat steps 5 through 12 of Sec. 2.12.3. Do not move the vertical-gain or vertical-position controls to bring the waveform into view. Use the vertical step attenuator only.

2.12.5 Calibrating the vertical amplifier for voltage measurements

The following procedure applies only to those scopes that have a variable gain for the vertical amplifiers. The basic calibration consists of applying a reference voltage of known amplitude to the vertical input and adjusting the vertical-gain control for a specific deflection. Then, the reference voltage is removed and the test voltages are measured, without changing the vertical-gain setting. The calibration remains accurate as long as the vertical-gain control is at this calibrate-set position.

The vertical amplifier can be voltage-calibrated by several methods. For example, the calibrating voltage can be ac or dc (on scopes capable of dc operation), variable or fixed, internal or external. The method depends on the type of scope and the available calibrating voltage. The following procedure assumes that there is a source of square waves, with known accuracy as to amplitude and frequency. Notice that some scopes have such square waves available for calibration of probes (chapter 5).

1. Place the oscilloscope in operation (Sec. 2.11).
2. With the calibrate square waves removed, use the vertical-position control to set the trace at the graticule horizontal centerline.
3. Apply the calibrating square waves to the vertical input.
4. Set the horizontal-sweep frequency to some frequency lower than that of the calibrating square waves. Several cycles of square waves should appear when the sweep frequency is lower than the calibrating square-wave frequency. If the sweep frequency is considerably lower than the calibrating frequency, the flat peaks of the square waves will blend and appear as two horizontal lines. If the horizontal sweep is not on, the trace appears as a vertical line when the square waves are applied. Two square waves should appear if the sweep frequency is one-half that of the square-wave frequency. All three patterns are shown Fig. 2-14.

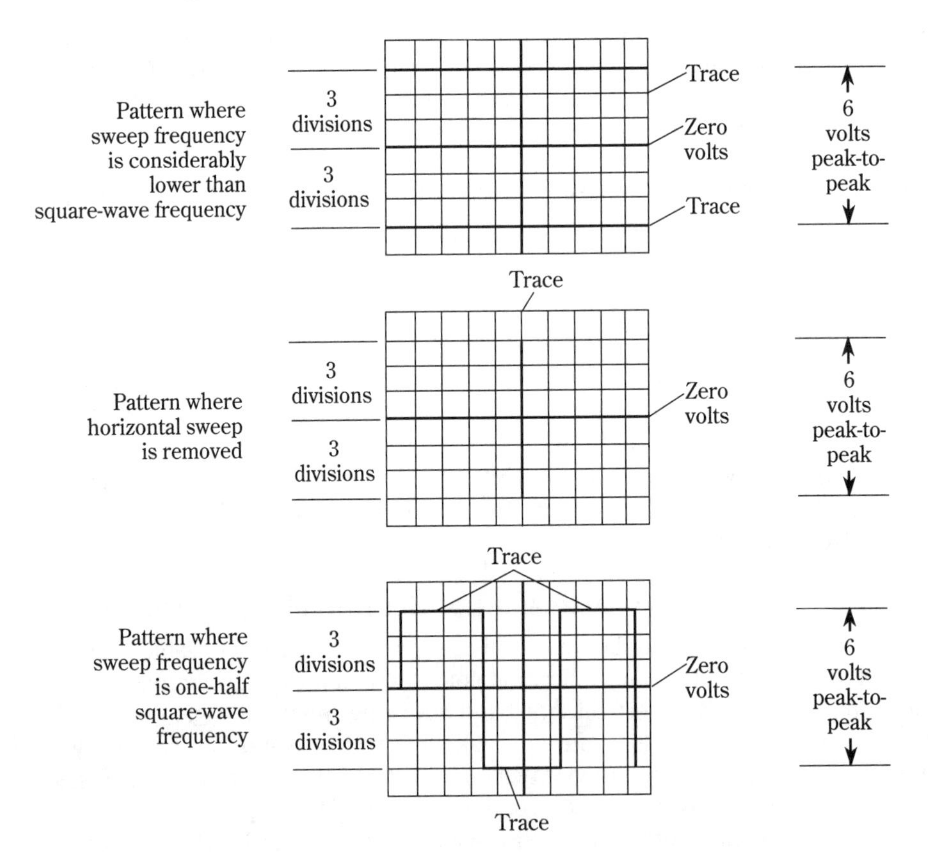

2-14 Voltage calibration with internal square waves.

5. Without touching the vertical-position control, adjust the vertical-gain and align the flat peaks of the square waves with the desired scale divisions. For example, assume square waves with a peak-to-peak amplitude of 6 V and a scale such as shown in Fig. 2-14. Set the vertical gain so that the square waves are spread a total of six divisions (three divisions up from the centerline and three divisions below). Each division then equals 1 V peak-to-peak, giving the scope a vertical-deflection factor of 1 V/cm, with the step attenuator set to 1×. If the step attenuator is then moved to 10×, the factor is increased to 10 V/cm.

   When the graticule horizontal centerline is used as a reference, as described in step 2, the square-wave peaks above the centerline indicate positive voltage, and vice versa. The amplitude of the square waves above (and below) the centerline is the peak voltage. This is equivalent to one-half the peak-to-peak voltage (Fig. 1-8B).
6. If desired, the position of the vertical-gain control should be noted and recorded as the calibrate-set position. Use this same position for all future voltage measurements. It is recommended that the calibration be checked at frequent intervals. Once the calibrate-set position is established, the trace can be moved up or down as required by the vertical-position control without affecting the volts/cm factor.

2.12.6 Composite and pulsating voltage measurements

In practice, most voltages measured are composites of ac and dc, or are pulsating dc. For example, a transistor amplifier that is used to amplify an ac signal has both ac and dc on the collector. This can be measured on a dc scope. The procedures are essentially a combination of peak-to-peak measurements and instantaneous dc measurements.

Figure 2-15 shows the measurement displays for some typical composite and pulsating voltages. The procedures are essentially the same as described in Sec. 2.12.3.

1. Connect the equipment as shown in Fig. 2-15.
2. Repeat the procedures of Sec. 2.12.3, except as follows. Using the vertical-position control, position the trace to a convenient location on the graticule. If the voltage to be measured is pulsating dc, the horizontal centerline should be convenient. If the voltage is a composite, and the average signal (ac plus dc) is positive, position the trace below the centerline. If the average is negative, position the trace above the centerline. Do not move the vertical-position control after this reference line is established.

   Notice that if the voltage to be measured is pulsating dc, the trace remains on one side of the reference lines, but starts and stops at the reference line, as shown in Fig. 2-15A. If the voltage is a composite of ac and dc, the trace can be on either side of the reference line (or possibly cross over the reference line), but it usually remains on one side, as shown in Fig. 2-15B. If the voltage is a non-sine wave (such as a sawtooth, pulse spike, etc.), the trace might appear on both sides of the reference line, or it might be displaced above or below the line, as shown in Fig. 2-13C.

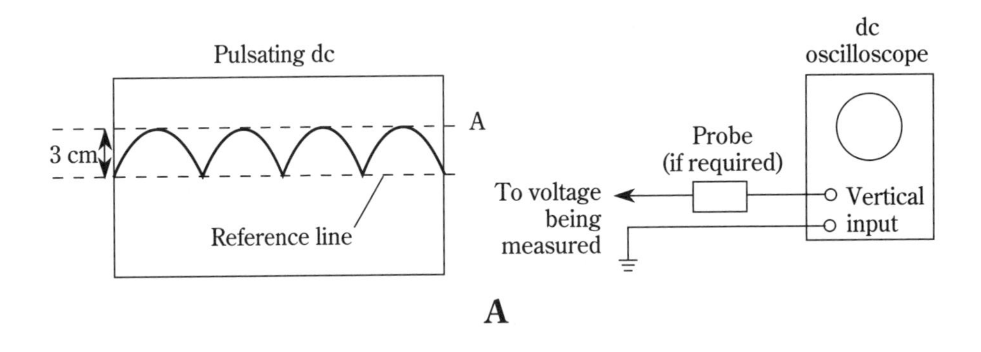

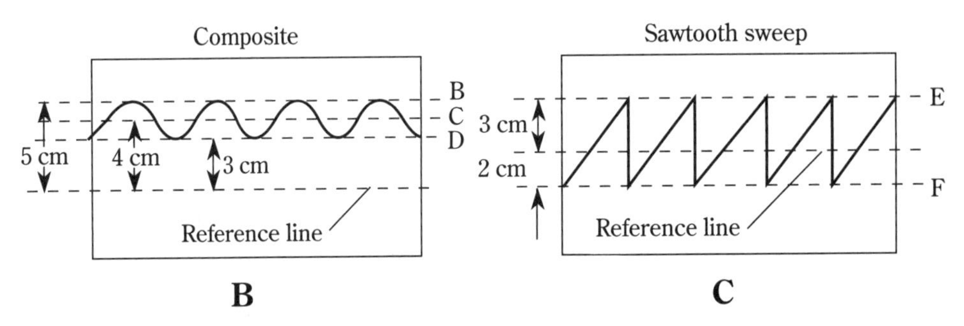

2-15 Measuring composite and pulsating voltages.

3. Calculate the voltages shown on Fig. 2-15, as follows:

   Assume that the vertical distance measured is 3 cm from the reference line to point A of Fig. 2-15A. The waveform is above the reference line (pulsating dc), using a 10× attenuator probe and a vertical-deflection factor of 2 V/cm. The voltage is 3× + 1 × 2 × 10 = 60 V (peak).

   Assume that the vertical distance measured is 3 cm from the reference line to point D (Fig. 2-15B), 4 cm to point C (Fig. 2-15B), and 5 cm to point B (Fig. 2-15B). The waveform is above the reference line (ac combined with positive dc), using a 10× attenuator probe and a vertical-deflection factor of 2 V/cm. The dc component (reference line to point C) is 4*N* + 1 × 2 × 10 = +80 V, and the peak-to-peak of the ac component (point B to D) is 2 × 2 × 10 = 40 V (peak to peak). The 2-cm value is obtained by subtracting the point-D value (3 cm) from the point-B value (5 cm).

   Assume that the vertical distance measured is 3 cm from the reference line to point E (Fig. 2-15C), and 2 cm from the reference line to point F. The waveform is above and below the reference line (sawtooth sweep), using a 10× attenuator probe and a vertical-deflection factor of 2 V/cm. The positive peak of sweep (point E1 is 3× + 1 × 2 × 10 = +60 V (peak), the negative peak of sweep (point F) is 2× –1 × 2 × 10 = –40 V (peak), and the peak to peak of sweep (point E to F) is 60 + 40 = 100 V (peak to peak).

2.12.7 Current measurements with a test resistor

The most common method of measuring an unknown current is passing the current through a resistance of known value, and then measuring the resultant voltage. This is the basic principle of most voltmeters. Because the scope can be used as a voltmeter, scopes can be adapted to measure current. A resistor of known value and accuracy is the only other component required for the procedure. Once the voltage is measured, the current can be calculated using basic Ohm's law: $I = E/R$.

1. Connect the equipment, as shown in Fig. 2-16.

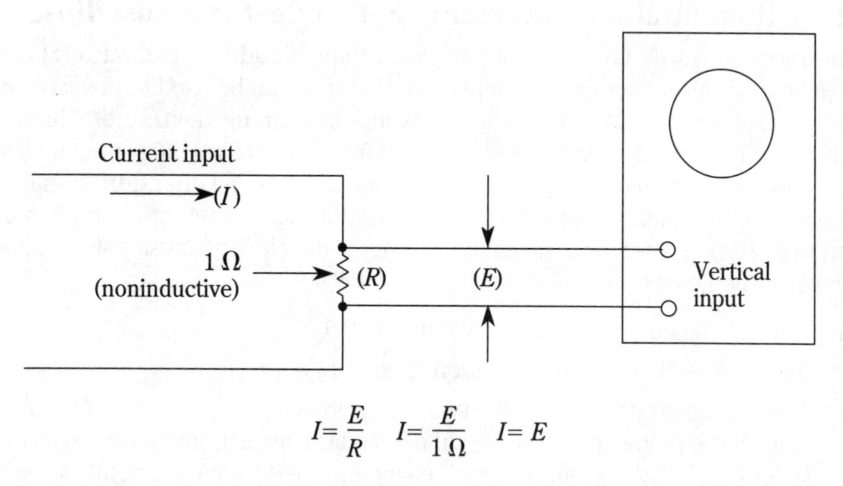

$$I = \frac{E}{R} \quad I = \frac{E}{1\,\Omega} \quad I = E$$

2-16 Measuring current with a test resistor.

2. Place the oscilloscope in operation (Sec. 2.11).
3. Apply the current to be measured through the resistor. Note that the calculations are simplified if a 1-Ω resistor is used. The wattage of the resistor must be at least double the square of the maximum current (in amperes). For example, if the maximum anticipated current is 10 A, the minimum wattage of the resistor should be $10^2 \times 2 = 200$ W.
4. Measure the voltage drop across the resistor using the procedures of this section, whichever is applicable to the type of voltage being measured. The unknown current equals the voltage measured (provided that a 1-Ω resistance is used). If peak-to-peak voltage is measured, the resistant current value is also peak-to-peak. To determine the rms or average value, convert the measured voltage to rms or average using Fig. 1-8.
5. If the current being measured is the result of a composite voltage (ac plus dc), both the ac and dc voltages should be measured separately, as described in Sec. 2.12.6. The ac voltage (peak to peak) should then be converted to rms (using Fig. 1-8). The dc and ac currents (which are equivalents to the corresponding voltages) should be combined to find the composite or total current as follows:

$$\textit{Total current} = \sqrt{\textit{ac current (rms)}^2 + \textit{dc current}^2}$$

2.12.8 Current measurements with a current probe

As covered in chapter 1, measuring current with a test resistor has two disadvantages: (1) the current must be opened so that the resistor can be inserted during testing, and (2) operation of the circuit can be affected by the additional resistance. Both of these problems can be eliminated with a current probe. Such probes are similar (if not identical) to the current probes described for meters. Because current probes are used with lab scopes, and are provided with detailed instructions, no operating procedures are given here.

2.12.9 Differential measurements with a dual-trace oscilloscope

A dual-trace scope with an ADD function (both channels added algebraically) can be used to make differential measurements. Such scopes can be used to observe waveforms and measure voltages between two points in a circuit, neither of which is at circuit ground. Figure 2-17 shows the connections for a typical differential measurement (to measure the output signal of a push-pull amplifier). Other differential measurements include monitoring the inputs or outputs of a differential amplifier, the output of a phase splitter, and the amount of signal developed across a single section of a voltage divider or attenuator.

1. Connect the equipment, as shown in Fig. 2-17.
2. Place the oscilloscope in operation (Sec. 2.11).
3. Connect the ground clips of the two scope probes to the PC board ground, and connect the probe tips to the circuit point where measurements are to be made. Never connect the ground clip of a scope probe to a circuit point other than ground. The ground clip of each probe is an earth ground, and will short any circuit point to ground. Unless that circuit is already at ground, the PC board under test (and possibly other boards in the equipment) can be damaged.
4. Set the vertical step attenuators of both channels as necessary to a deflection factor that allows the expected signal to be displayed without overdriving the vertical amplifier. The sensitivity/attenuation controls of both channels must be identical.
5. Set the input selector to measure ac. Operate the controls as necessary to select the ADD function.
6. Operate the scope controls as necessary to display a single waveform, as shown in Fig. 2-17.
7. If the channel A and channel B signals are in phase, the displayed waveform is the sum of the amplitudes of the signals. Any imbalance or difference between the signals can be checked by setting the scope controls to the invert mode. Typically, this inverts channel B. The displayed waveform then becomes the difference between the two signals.
8. If the channel A and channel B signals are 180° out of phase (as in the case of a push-pull amplifier, shown in Fig. 2-17), it is necessary to invert one channel (channel B) by operating the scope invert control. With one channel inverted, the waveform then becomes the sum of the two 180° out-of-phase

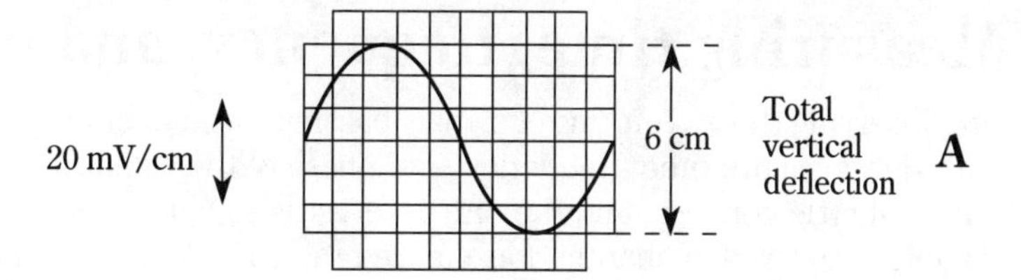

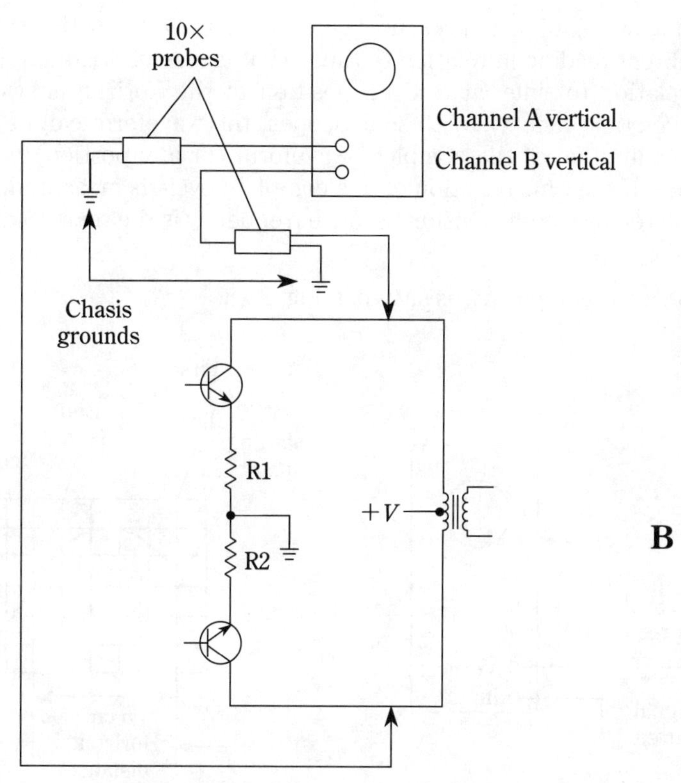

2-17 Measuring differential voltages with a dual-trace oscilloscope.

signals. Any imbalance or difference between the two signals can be checked by setting the scope invert control to normal mode. The waveform then becomes the difference between the two signals.

9. With the controls set to get a sum of the two signals, position the waveform (as necessary) to measure peak-to-peak voltage as covered in this section.
10. Measure the peak-to-peak vertical deflection in centimeters. Multiply the measured distance by the vertical attenuator settings. Include the attenuation factors of both probes. It is assumed that both probes have an identical attenuation factor. For example, assume a peak-to-peak vertical deflection of 6 cm (Fig. 2-17) using 10 × probes and a vertical deflection factor of 20 mV/cm. The voltage is 6 × 29 mV × 20 = 2.4 V (peak to peak).

2.13 Measuring time, frequency, and phase

In addition to observing the waveforms of signals passing through circuits, the scope can also be used to measure time, frequency, and phase of the signals or voltages. If the waveform is of little concern, and the only interest is signal frequency, then the frequency counters covered in chapter 4 are more practical.

2.13.1 Time-duration measurements

The horizontal sweep of most present-day scopes is provided with a selector control that is direct reading in relation to time. That is, each horizontal division has a definite relation to time, at a given position of the horizontal-sweep switch (such as ms/cm and μs/cm). With such scopes, the waveform can be displayed, and the time duration of the complete waveform (or any portion) can be measured directly. If the time duration of one complete cycle is measured, frequency can be calculated by simple division because frequency is the reciprocal of the duration of one cycle.

1. Connect the equipment, as shown in Fig. 2-18.

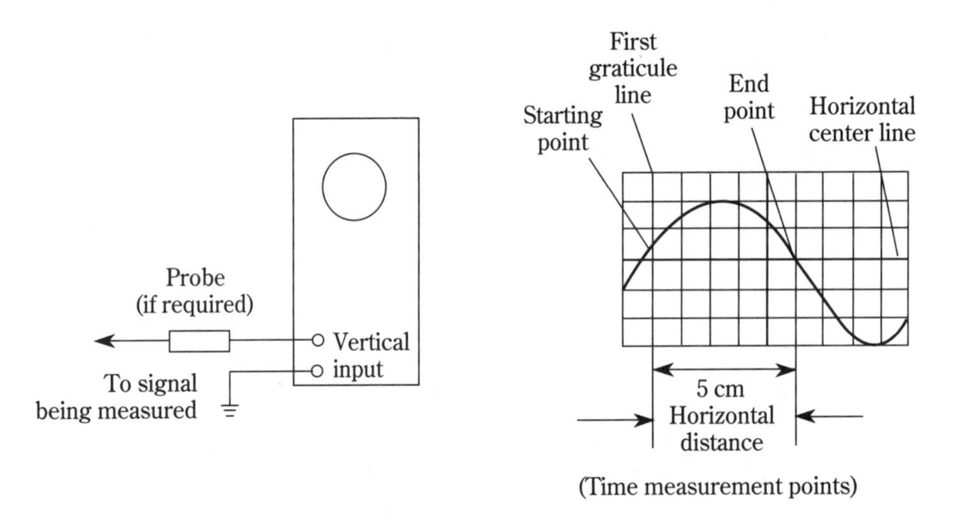

2-18 Measuring time duration.

2. Place the oscilloscope in operation (Sec. 2.11).
3. Set the vertical step attenuator to a deflection factor that allows the expected signal to be displayed without overdriving the vertical amplifier.
4. Connect the probe (if any) to the signal being measured.
5. Switch on the scope internal recurrent sweep. Set the horizontal sweep control to the fastest sweep that displays a convenient number of divisions between the time-measurement points, as shown in Fig. 2-18. On most scopes, it is recommended that the extreme sides of the screen not be used for time-duration measurements. There might be some nonlinearity at the beginning and end of the sweep.

6. Adjust the vertical-position control to move the points between which the time measurement is made to the horizontal centerline.
7. Adjust the horizontal-position control to move the starting point of the time measurement area to the first graticule line.
8. Measure the horizontal distance between the time-measurement points (Fig. 2-18). If the horizontal sweep is provided with a variable or vernier control, make certain that the variable function is off or at calibrate.
9. Multiply the distance measured in step 8 by the setting of the horizontal-sweep control. If sweep magnification is used, divide the answer by the multiplication factor. For example, assume that the distance between time-measurement points is 5 cm, as shown in Fig. 2-18, the horizontal-sweep control is set to 0.1 ms/cm, and there is no sweep magnification. The time duration is $5 \times 0.1 = 0.5$ ms. Now, assume that there is sweep magnification of 10×. This results in a time duration of $(5 \times 0.1)/10 = 0.05$ ms.

2.13.2 Frequency measurements

The frequency measurement of a periodically recurrent waveform is essentially the same as time-duration measurement, except that an additional calculation must be performed. In effect, a time-duration measurement is made, then the time duration is divided into 1. Because frequency of a signal is the reciprocal of one cycle:

1. Connect the equipment, as shown in Fig. 2-19.
2. Place the oscilloscope in operation (Sec. 2.11).
3. Set the vertical step attenuator to a deflection factor that allows the expected signal to be displayed without overdriving the vertical amplifier.
4. Connect the probe (if any) to the signal being measured.
5. Switch on the scope internal recurrent sweep. Set the horizontal sweep to a rate that displays one complete cycle (Fig. 2-19). Avoid using the extreme sides of the screen, if practical. There might be some nonlinearity at the beginning and end of the sweep.
6. Adjust the vertical-position control so that the beginning and end points of one complete cycle are located on the horizontal centerline. Any two points representing one complete cycle can be used. It is usually more convenient to measure at points where the waveform swings from negative to positive (or vice versa), or where the waveform starts the positive (or negative) rise.
7. Adjust the horizontal-position control to move the selected starting point of the complete cycle to the first graticule line.
8. Measure the horizontal distance between the beginning and end of a complete cycle. If the horizontal sweep has a variable control, make certain that it is off or in the calibrate position.
9. Multiply the distance measured in step 8 by the setting of the horizontal-sweep control. If sweep magnification is used, divide the answer by the multiplication factor. Then, divide the measured time into 1 to find frequency. For example, assume that the distance between the beginning and end of a complete cycle is 8 cm (Fig. 2-13A), the horizontal-sweep control is set to 0.1 ms/cm, and there is no sweep magnification. The time duration is $8 \times 0.1 = 0.8$ ms (0.0008 s), and the frequency is $1/0.0008 = 1250$ Hz.

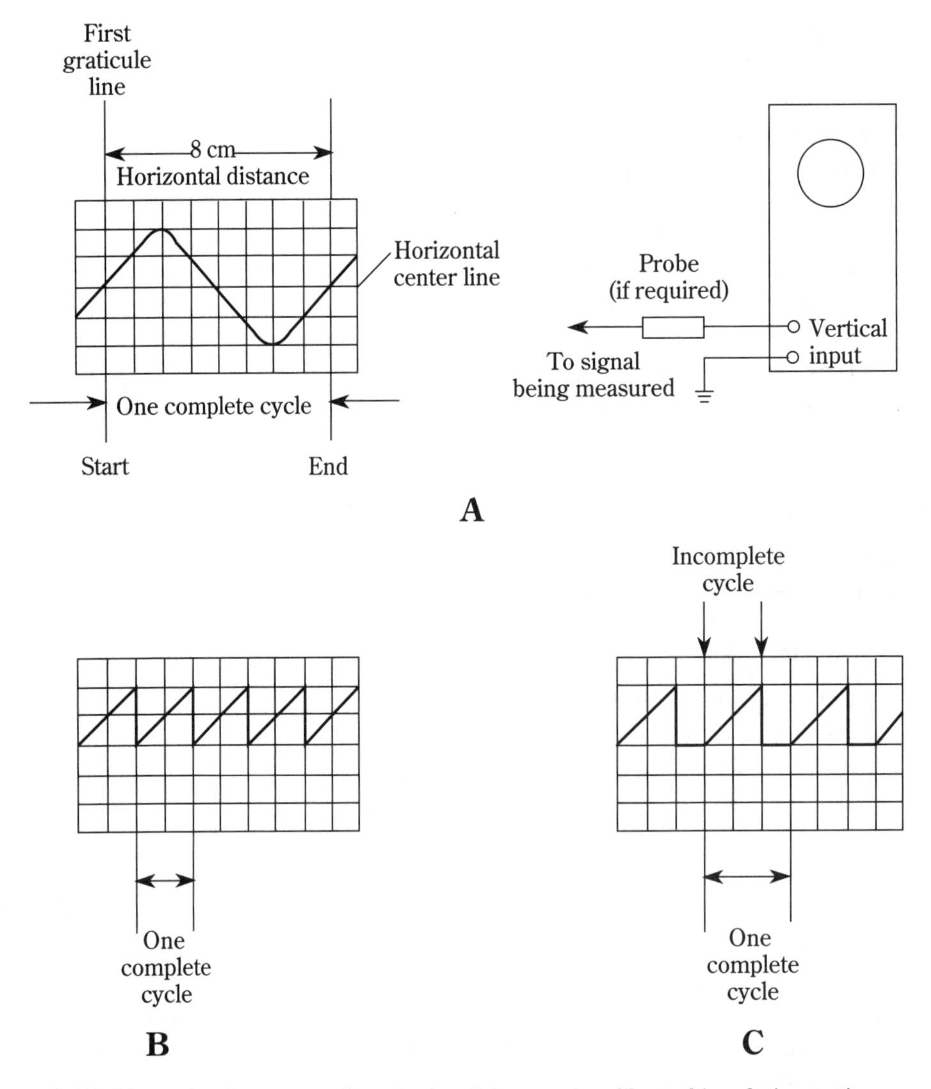

2-19 Measuring frequency where horizontal sweep is calibrated in relation to time.

2.13.3 Frequency and time measurements with shop oscilloscopes

The horizontal sweep of some inexpensive scopes (such as used in schools, production lines, field service, hobby, etc.) is provided with controls that are direct reading in relation to frequency. Usually, there are two controls: a step selector or range control and a vernier. The frequency of the horizontal trace is equal to the scale settings of both controls.

When a signal is applied to the vertical input, and the horizontal controls are adjusted until one complete cycle occupies the entire length of the trace, the vertical signal is equal in frequency to the horizontal-sweep settings. If desired, the frequency can then be converted to time using: *Time* = 1/*Frequency*.

1. Connect the equipment, as shown in Fig. 2-22.
2. Place the oscilloscope in operation (Sec. 2.11).
3. Set the vertical step attenuator to a deflection factor that allows the expected signal to be displayed without overdriving the vertical amplifier.
4. Connect the probe (if any) to the signal being measured.
5. Switch on the internal recurrent sweep. Set the horizontal-sweep controls (step/range and vernier) so that one complete cycle occupies the entire length of the trace (Fig. 2-20).

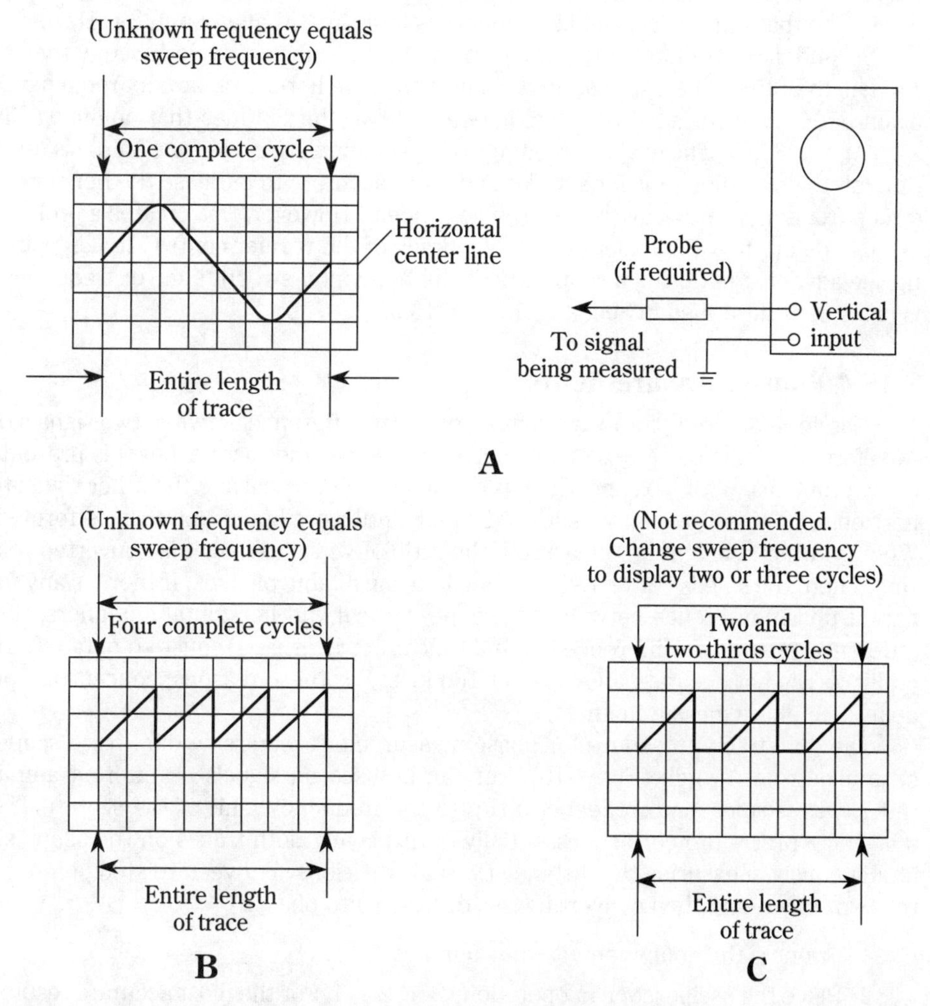

2-20 Measuring frequency where horizontal sweep is calibrated directly in units of frequency.

6. Read the unknown signal frequency directly from the horizontal-sweep controls. Assume that the step/range control is set to the 10-kHz position, and vernier indicates 5 (on a total scale of 10). This indicates that the horizontal-sweep frequency is 5 kHz. If one complete cycle of vertical signal occupies the entire length of the trace, the vertical signal is also at a frequency of 5 kHz.
7. If it is not practical to display only one cycle on the trace, more than one cycle can be displayed, and the resultant horizontal-sweep frequency indication multiplied by the number of cycles. Two important points must be remembered when using more than one cycle.

First, multiply the indicated sweep frequency by the number of cycles. For example, assume that the step/range control is set to 10 kHz, and the vernier indicates 5 (on a total scale of 10) when four complete cycles of vertical signal occupy the entire length of the trace (Fig. 2-20B). This indicates that the horizontal-sweep frequency is 5 kHz, and that the vertical-signal frequency is four times that amount, 20 kHz.

Second, it is absolutely essential that an exact number of cycles occupy the entire length of the trace. For example, assume that the horizontal-sweep frequency is again 5 kHz, and the vertical-signal is two and two-thirds times that amount (Fig. 2-20C), 13.33 kHz. The exact percentage of the incomplete cycle (one-third) is quite difficult to determine. It is far simpler and more accurate to increase the horizontal-sweep frequency until exactly three cycles appear. However, this creates a problem in that you are now dependent on the accuracy of the vernier control. It also points up the advantage of using a scope where the horizontal-sweep time (or frequency) is precise, as described in Sec. 2.13.1 and 2.13.2.

2.13.4 Phase measurements

The classic X-Y method for measurement of phase difference between two signals or waveforms on a scope screen can be used on a single-trace scope, but this presents many problems. With X-Y, one signal is applied to the vertical and the other signal to horizontal. The scope controls are adjusted until an elliptical pattern is formed. When width and height are measured, the ratio of width to height is converted to a sine. Then, the sine is converted to an angle using a table of sines. If there is any inherent phase difference between the scope vertical and horizontal amplifiers, this difference (as well as difference in amplitude and waveshape) must be noted. Generally, X-Y is not accurate above about 100 kHz. For these reasons, concentrate on dual-trace phase measurement.

The dual-trace procedure for phase measurement requires a dual-trace scope, or an electronic switch (Fig. 2-10), but can be used on signals of different amplitudes, waveshapes, and frequencies (up to the frequency limit of the scope). The dual-trace procedure consists essentially of displaying both traces on the scope simultaneously, measuring the distance (in scale divisions) between related points on the two traces, and then converting this distance into phase.

1. Connect the equipment, as shown in Fig. 2-21.
2. Place the oscilloscope in operation (Sec. 2.11). For the most accurate results, the cables connecting the two signals to the scope input should be of the

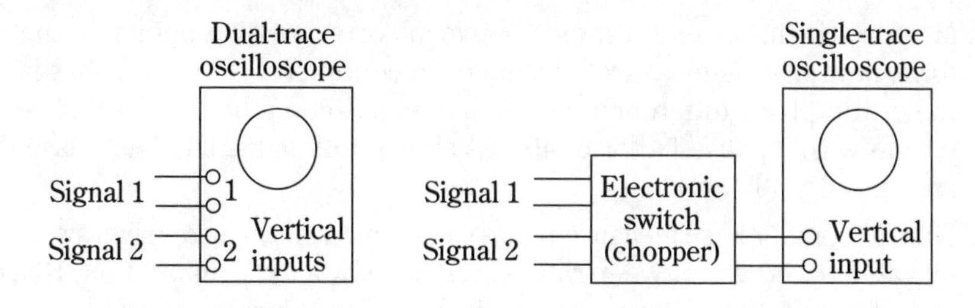

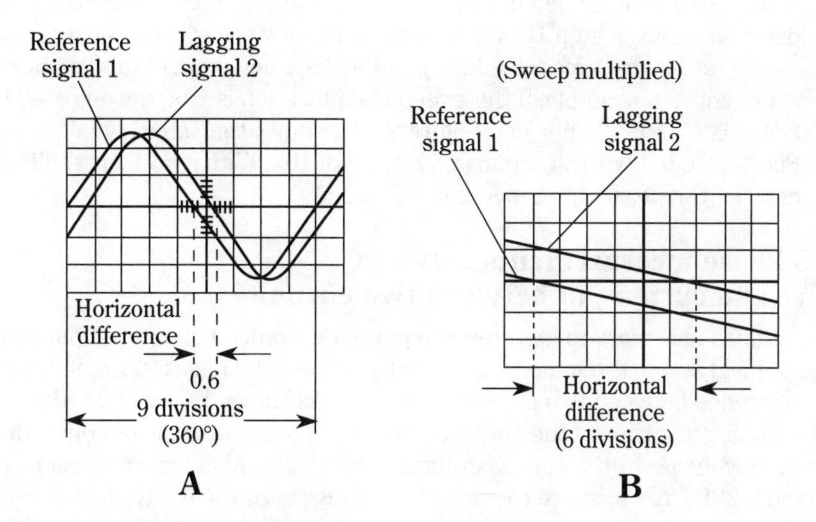

2-21 Measuring phase difference with dual traces.

same length and characteristics. At higher frequencies, a difference in cable length or characteristics could introduce a phase shift.

3. Set the step attenuators to deflection factors that allow the expected signals (both traces) to be displayed without overdriving the amplifiers.
4. Switch on the scope internal recurrent sweeps.
5. Set the position controls until the pattern is centered on the screen.
6. Set the gain controls to spread the pattern over as much of the screen as desired.
7. Switch on the dual-trace function or electronic switch.
8. Adjust the sweep control until one cycle of the reference signal occupies exactly nine divisions (9 cm) horizontally, as shown in Fig. 2-21A. Either of the two signals can be used as the reference signal, unless otherwise specified by requirements of the particular test. It is usually simpler if the signal of the lowest frequency can be used as the reference signal.
9. Determine the phase factor of the reference signal. For example, if 9 cm represents one complete cycle, or 360°, then 1 cm represents 40°.

10. Measure the horizontal distance between corresponding points on the waveform. Multiply the measured distance (in centimeters) by 40° (phase factor) to get the phase difference. For example, assume a horizontal difference of 0.6 cm with a phase factor of 40°, as shown in Fig. 2-21A. The phase difference is $0.6 \times 40° = 24°$.
11. More accurate phase measurements can be made if the scope has sweep magnification (where the sweep rate is increased by a fixed amount 5×, 10×, etc.), and only a portion of one cycle is displayed. In this case, the phase factor is determined as described in step 9. Then, the approximate phase difference is determined, as in step 10. Without changing any other controls, increase the sweep rate (sweep magnification) and make a new horizontal-distance measurement. For example, if the sweep magnification is 10×, the adjusted phase factor is 4°(40°/10) per cm. Figure 2-21B shows the same signal as used in Fig. 2-21A, but with sweep magnification of 10×. With a horizontal difference of 6 cm, the phase difference is $6 \times 4° = 24°$.

2.13.5 Phase measurements between voltage and current, or between two currents

It is sometimes necessary to measure the phase difference between a voltage and a current applied across the same load. Likewise, it might be necessary to measure the phase difference between two separate currents. As shown in Fig. 2-22, this can be done by passing a portion of the current through a fixed resistor, thus converting the current to a voltage. In the case of voltage/current, the phase measurement (Figs. 2-22A and 2-22B), the phase of the resultant voltage is compared to the load-voltage phase. In the current/current phase measurement (Figs. 2-22B and 2-22D), the phase of the resultant voltages are compared.

Once the test connections are made, the phase differences between voltage and current (or current and current) are determined using the procedures of Sec. 2.13.4. The actual resistance values of R_1 and R_2 are not critical, but the wattage must be at least double the square of the maximum current. Also, where current through a load is measured, the value of R_1 should be low in relation to the load resistance.

2.14 Checking components and circuits

Oscilloscopes are most often used to check components while the components are connected in the related circuits. However, it is possible to check many individual components with a scope, both in and out of circuit. The scope is particularly useful in checking those components (and circuits) where response curves, transient characteristics, phase relationships, and time are of special importance. Of course, any component (or circuit) that can be checked with a meter can also be checked with a scope because the scope can function as an ac or dc voltmeter (Sec. 2.12). If a component or circuit requires only a voltage check, it is more practical to use a simple meter or multitester.

This section describes the procedures for checking those components or circuits where a scope can provide an alternate or faster method. The procedures for checking such components as transistors, FETs, UJTs, PUTs, diodes, thyristors, SCRs, and

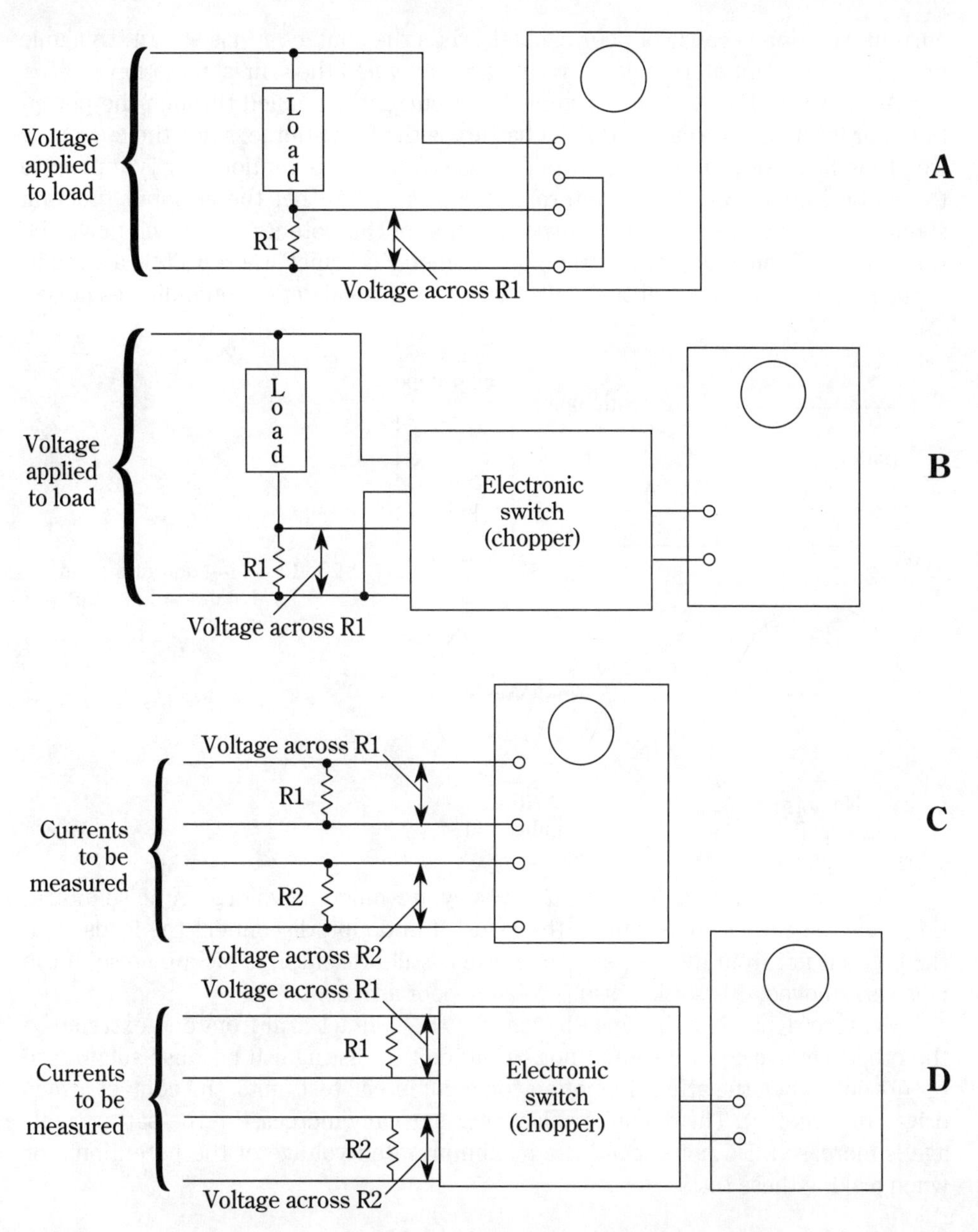

2-22 Phase measurement between voltage and current.

other control rectifiers are described in chapters 7 through 11. Chapters 12 through 15 describe the use of scopes in checking specific circuits such as audio, power-supply, RF, and communications equipment.

2.14.1 Checking potentiometers

A scope can be used to check the noise (both static and dynamic) of a potentiometer, rheostat, variable resistor, or slider resistance. Static noise is the result of any current variation because of poor contact, when the contact arm is at rest. Dynamic noise is the amount of irregular current variation when the arm is in motion.

As shown in Fig. 2-23, a constant direct current is applied through the potentiometer from an external source. (A battery is the best source, since the current is free from noise or ripple.) The output voltage from the potentiometer is applied to the vertical channel, with the internal sweep on. Note that the ac input function should be used since the voltage divider action of the potentiometer will move the trace vertically on a dc scope. If the potentiometer is "quiet" there will be a straight horizontal trace, with no vertical deflection. Any vertical deflection indicates noise.

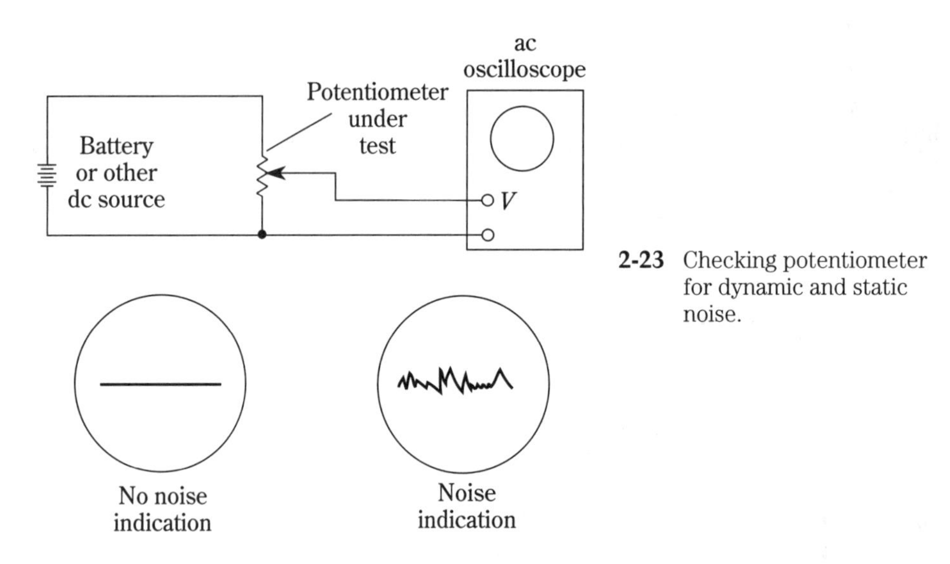

2-23 Checking potentiometer for dynamic and static noise.

Measure the static noise level, if any, as you would any voltage. A noise indication could be caused by pickup in the leads. If in doubt, disconnect the leads from the pot, but not from the scope. If the noise is still present, it is pickup noise, If the noise is removed, it is static noise (probably poor arm contact).

Measure dynamic noise level by varying the contact arm from one extreme to the other. Dynamic noise should not be difficult to distinguish because such noise occurs only when the arm is in motion (on commercial test units, the contact arm is driven by a motor). The dynamic noise level is usually increased if the battery voltage is increased. Do not exceed the maximum rated voltage of the potentiometer when making these tests.

2.14.2 Checking relays

A scope can be used to check the make and break of relay contacts. The presence of contact "bounce," as well as the actual make time and break time of the contacts can be displayed and measured. To be effective, the scope should be capable of single-sweep operation. Also, because of the instantaneous nature of the trace, the display should be photographed (unless a storage-type scope is used).

Figure 2-24A is the connection diagram for testing dc relays. When S1 is closed, a dc voltage from the external battery is applied to the sweep-trigger input to initi-

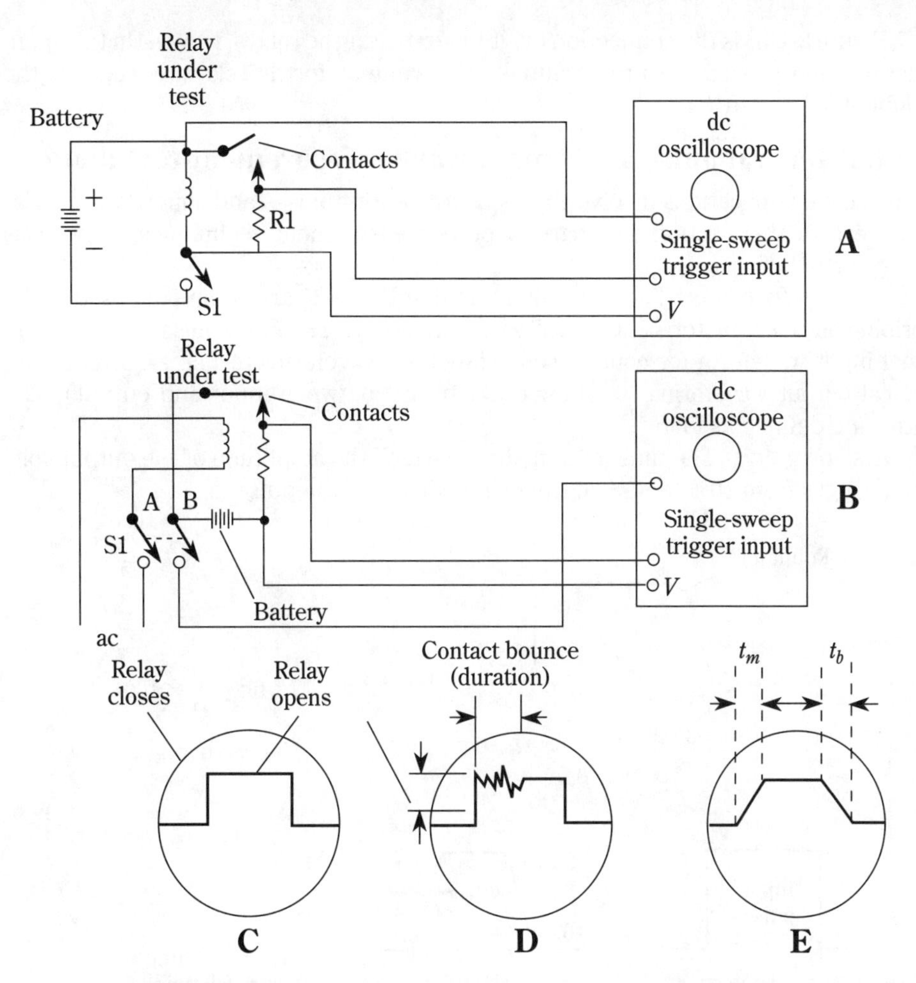

2-24 Checking relays for make time, break time, and contact bounce.

ate a single sweep. Current is also applied to the relay, causing the relay contacts to close. When the contact closed, the positive voltage across R1 is applied to the vertical channel. When S1 is opened, the dc voltage is removed and the relay contacts open, removing the positive voltage from the vertical input.

Ideally, the opening and closing of the relay contacts produces a rectangular trace similar to that of Fig. 2-24C. If the contacts are bouncing, the display is similar to that of Fig. 2-24D. The make time and break time can be measured on the time-calibrated horizontal axis (Fig. 2-24E). The value of R1 is chosen to limit the relay contact current to the value specified in the relay manufacturers data.

If the trace is to be photographed, hold the camera shutter open, close and open S1, then close the camera shutter and develop the picture. Using the developed photo, measure the bounce (if any) amplitude along the vertical axis of the scope screen. Measure bounce duration, as well as make time (t_m) and break time (t_b), along the horizontal axis.

Figure 2-24B is the connection diagram for testing ac relays. Notice that the connections and procedures are essentially the same as for dc relays, except for the double-pole switch, S1.

2.14.3 General pulse and square-wave measurement techniques

Many scope applications involve the measurement of pulses and square waves. For that reason, this section summarizes pulse-measurement techniques, as well as pulse characteristics.

Pulse definitions The terms illustrated in Fig. 2-25 are commonly used in describing pulse characteristics, as well as square waves. The input pulse represents an ideal input waveform for comparison. The other waveforms in Fig. 2-25 represent typical output waveforms (to show relationships between input and output). The terms are defined as follows:

Rise time (t_R) The time interval during which the amplitude of the output voltage changes from 10% to 90% of the rising portion of the pulse.

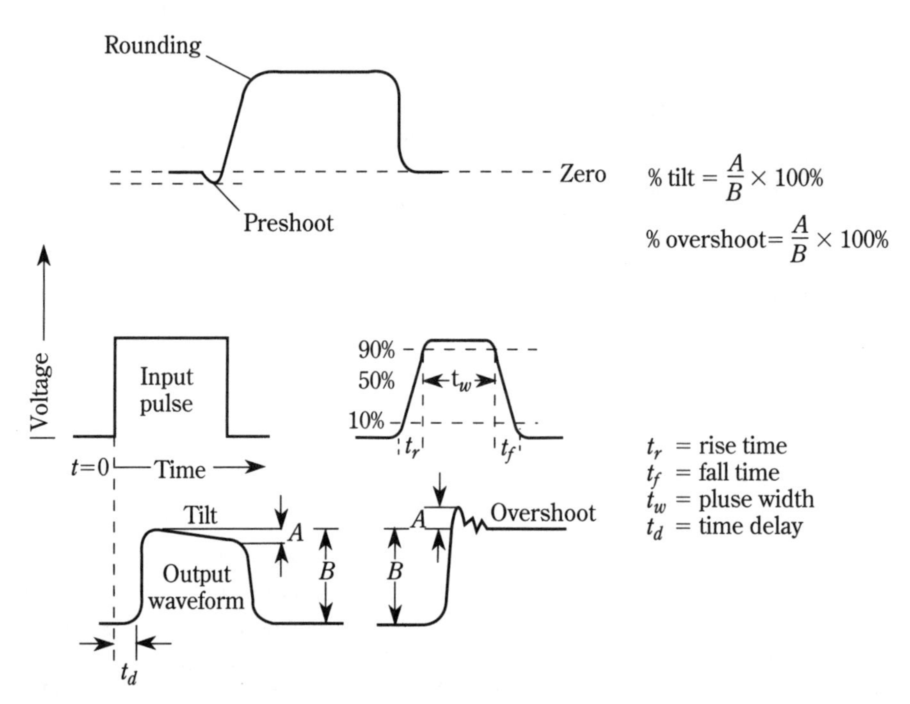

2-25 Basic pulse and square-wave definitions.

Fall time (t_f) The time interval during which the amplitude of the output voltage changes from 90% to 10% of the falling portion of the waveform.

Pulse width or duration (t_w) The time duration of the pulse measured between the 50% amplitude levels of the rising and falling portions of the waveform.

Time delay (t_d) The time interval between the beginning of the input pulse (t = 0), and the time when the rising portion of the output pulse attains an arbitrary amplitude of 10% above the baseline.

Tilt A measure of the tilt of the full amplitude (flat top) portion a pulse. The tilt measurement is usually expressed as a percentage of the amplitude of the rising portion of the pulse.

Overshoot A measure of the overshoot occurring generally above the 100% amplitude level. This measurement is also expressed as a percentage of the pulse rise.

These definitions are guides. When the actual pulses are very irregular (such as excessive tilt, overshoot, etc.), the definitions might become ambiguous.

Rule of thumb for rise-time measurements: because rise-time measurements are of special importance in pulse testing, the relationship between the scope rise time and the rise time of the circuit under test must be taken into account. Obviously the accuracy of rise-time measurements can be no greater than the scope rise time. Also, if the circuit is tested with an external pulse, the rise time of the pulse source must be taken into account.

For example, if a scope with a 20-ns rise time is used to measure the rise time of a 15-ns device, the measurements will he hopelessly inaccurate. If a 20-ns pulse generator and a 15-ns scope are used to measure the rise time, the fastest times for accurate measurement are something greater than 20 ns. Two basic guidelines can be applied to rise-time measurements.

The first method is known as the *root of the sum of the squares*. It involves finding the squares of all rise times associated with the test, adding these squares together, and then finding the square root of this sum. For example, using the 20-ns generator and the 15-ns scope, the calculation is $20 \times 20 = 400$; $15 \times 15 = 225$; $400 + 225 = 625$; $\sqrt{625} = 25$ ns. This means that the fastest possible rise time capable of measurement is 25 ns.

One major drawback to this rule is that the coax cables required to interconnect the test equipment are subject to "skin effect." As frequency increases, the signals tend to travel on the outside or skin of the conductor. This decreases conductor area, and increases resistance. In turn, this increases cable loss. The losses of cables do not add properly to apply the root-sum-squares method, except as an approximation.

The second rule or method states that if the signal being measured has a rise time 10 times slower than the test equipment, the error is 1%. This amount is small and can generally be ignored. If the signal has a rise time 3 times slower than the test equipment, the error is slightly less than 6%. By keeping these relationships in mind, the results can be interpreted intelligently.

Matching pulse-generator impedance to device under test One problem often encountered when testing pulsed circuits is the matching of impedances (as covered in Sec. 1.27). To provide a smooth transition between circuits of different characteristic impedance, each circuit must encounter a total impedance equal to its own characteristic impedance. A certain amount of signal attenuation usually occurs during this transition. A simple resistive impedance-matching network that provides minimum attenuation is shown in Fig. 1-21C, together with the applicable equations.

2.14.5 Measuring pulse time with external timing pulses

On those scopes where the horizontal-sweep circuits are calibrated in units of time, pulse width (duration), or spacing between pulses can be measured directly by counting the number of screen divisions along the horizontal axis (Sec. 2.13). It is also possible to use an external *time-mark generator* to calibrate the horizontal scale of a scope for time measurement. A time-mark generator (sometimes called a *marker generator* or a *time-base generator*, chapter 3) produces a series of sharp pulse spikes, spaced at precise time intervals. These pulses are applied to the scope vertical input and appear as a wavetrain (Fig. 2-26).

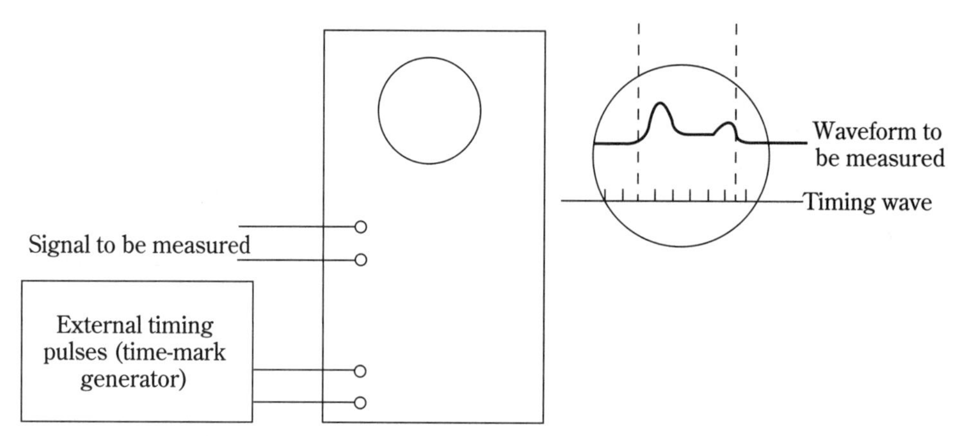

2-26 Measuring pulse time with external timing pulses.

If the scope is single trace, the horizontal gain and positioning controls are adjusted to align timing spikes with screen lines, until the screen divisions equal the timing pulses. The accuracy of the scope timing circuits is then of no concern because the horizontal channel is calibrated against the external generator. The timing pulses are removed, and the signal to be measured is applied to the vertical input (provided that the horizontal controls are not touched). Duration or time is read from the calibrated screen divisions in the normal manner.

If the scope is dual trace, the generator can be connected to one vertical input while the other vertical input receives the signal to be measured. The two traces can be superimposed or aligned, whichever is convenient. No matter what type of scope timing is used, the main advantage of an external timing generator is greater accuracy and resolution.

2.14.6 Measuring pulse delay

The time interval (or delay) between an input pulse and output pulse introduced by a delay line, digital circuit, etc., can be measured with a dual-trace scope. If the delay is exceptionally short, the screen divisions can be calibrated with an external time-mark generator.

1. Connect the equipment, as shown in Fig. 2-27.
2. Place the oscilloscope in operation (Sec. 2.11). Switch on all remaining equipment.
3. Operate the controls to produce stationary input and output pulses, as shown.
4. Count the timing spikes (or screen divisions) between the input and output pulses to determine the delay interval. Count the timing spikes or divisions between the beginning and end of each pulse to determine pulse width or duration.

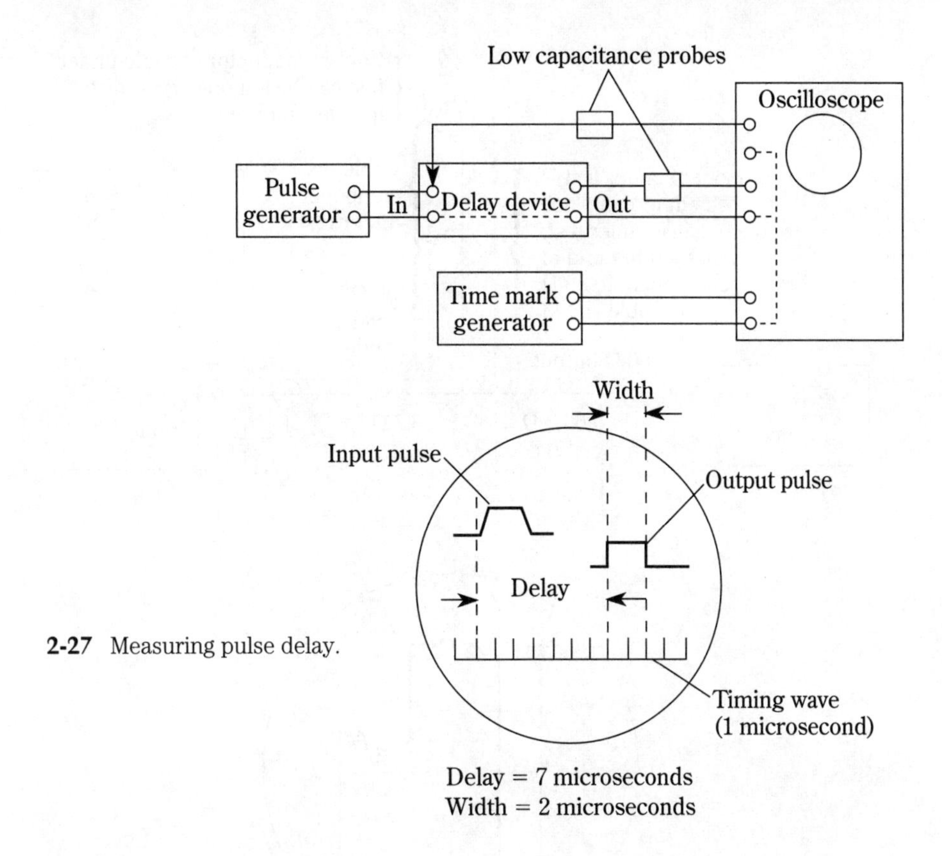

2-27 Measuring pulse delay.

2.14.7 Measuring impedance with a scope and pulse generator

A scope/pulse-generator combination can be used to measure impedance of an unknown device by comparison of the reflected pulse with the output pulse from the generator.

1. Connect the equipment, as shown in Fig. 2-28.
2. Place the scope in operation (Sec. 2.11). Switch on the internal recurrent sweep. Set the sweep selector and sync selector to internal.
3. Switch on the pulse generator, as described in the instruction.
4. Set the scope controls to display the output pulse, and the first reflected pulse.

   This measurement technique is based on comparison of reflected pulses with output pulses. As a signal travels down a transmission line, a reflection is generated and sent back along the line to the source each time the signal encounters a mismatch or difference impedance. The amplitude and polarity of the reflection are determined by the value of the impedance encountered in relation to the line impedance. If the mismatch impedance is higher than that of the cable, the reflection is of the same polarity as the applied signal. If the mismatch is lower, the reflection is of opposite polarity.

   The reflected signal is added to or subtracted from the amplitude of the pulse if the signal returns to the source before the pulse has ended. For a cable with an open end (no termination), the impedance is infinite and the

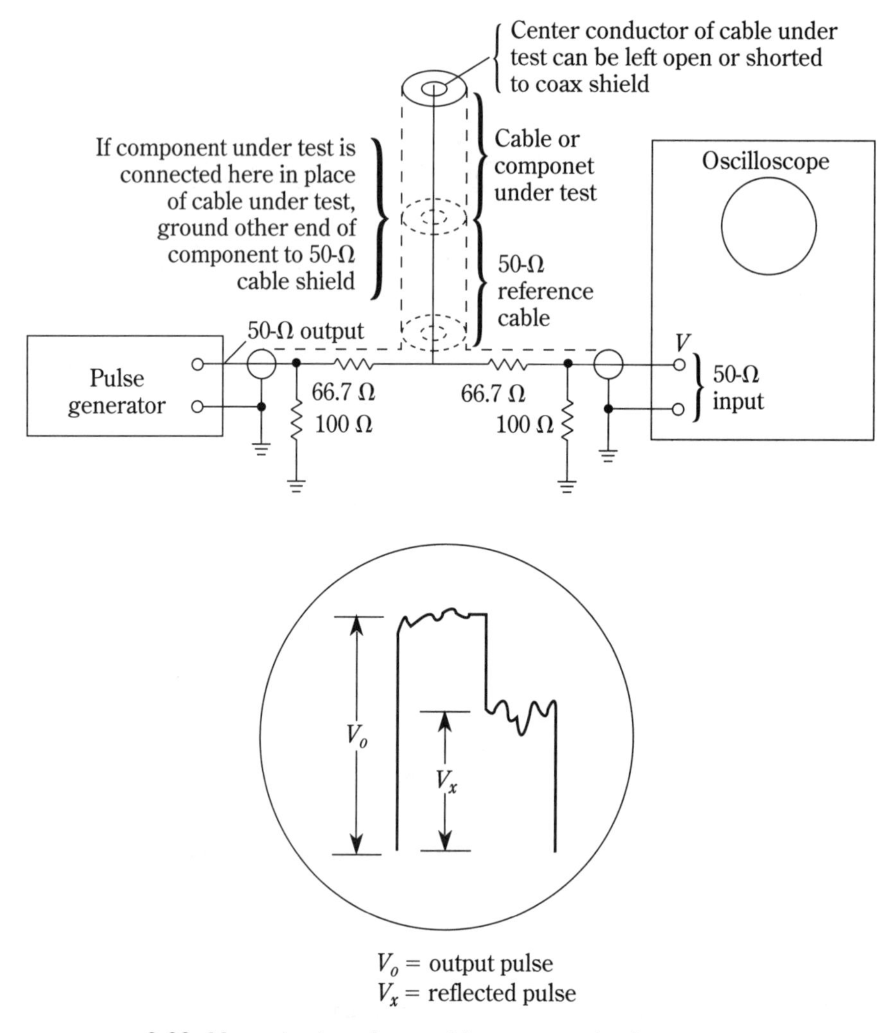

2-28 Measuring impedance with a scope and pulse generator.

pulse amplitude is doubled. For a cable with a shorted end, the impedance is zero and the pulse is canceled.

5. Observe the output and reflected pulses on the scope. Using Fig. 2-28 as a guide, determine the values of V_O (output voltage amplitude) and V_X (reflected voltage amplitude).
6. Calculate the unknown impedance using the equation $Z = 50/(2V_O/V_X) - 1$, where Z = unknown impedance, 50 = the reference impedance (50-Ω coax), V_O = the peak amplitude produced by the 50-Ω reference impedance, and V_X = peak amplitude at the time of the reflection.

2.15 Using an oscilloscope in digital testing

Because of the high speeds involved, and the multiple-signal nature of digital circuits, the logic analyzer is the most effective tool for test and troubleshooting of digital circuits (computers, microprocessor-based systems, etc.). The logic analyzer can test many digital circuits, and can display each step of a digital program. Of course, a scope is an excellent companion for the logic analyzer, and the following paragraphs describe some classic uses for the scope in digital testing. For coverage of digital-equipment circuits, as well as the related test and troubleshooting techniques, read the author's best-selling *Lenk's Digital Handbook* (McGraw-Hill, 1993).

2.15.1 Monitoring digital pulses with an oscilloscope

Figures 2-29 through 2-38 summarize the most common applications of the scope in digital testing. In all cases, the internal recurrent sweep is used, and the sweep/sync controls are set to display a suitable number of pulses. Although the amplitude and duration (or frequency) of pulses can be measured with the connections shown, the relationship of pulses (input to output, memory to clock, etc.) is the factor of major interest.

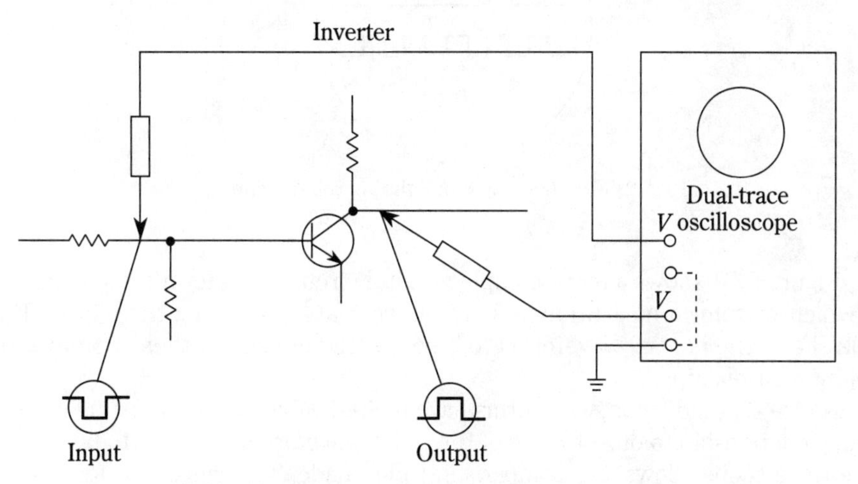

2-29 Checking pulse input/output relationships.

Checking pulse input/output relationships Figure 2-29 shows a scope used to check the input and output of an inverter. If the scope has an invert function, the output pulse can be inverted (so that both pulses appear positive or negative), and the traces can be superimposed to show a comparison of input to output (in relation to amplitude, duration, shape, and timing). Similarly, if the scope has an ADD function (both channels added algebraically) the pulses can be superimposed and the ADD function selected. Under these conditions, any pulse trace that remains is the result of a difference between input and output.

For example, if the pulses are as shown in Fig. 2-29, and are added algebraically with the ADD function, the input should cancel the output, resulting in a straight horizontal line. If the output pulse is not identical to the input pulse (for example, the output is higher), there is partial cancellation, and the trace is a positive pulse of limited amplitude.

Checking clock timing relationships Figure 2-30 shows a scope used to check the output of a memory circuit in relation to the clock pulse. In digital equipment, it is common for a large number of circuits to be synchronized, or to have specific time relationships to each other.

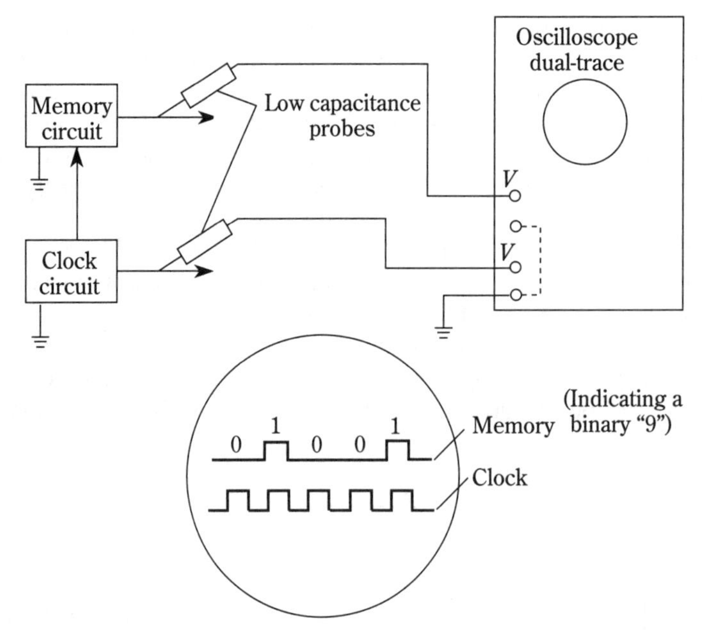

2-30 Checking clock timing relationships.

Figure 2-31 shows a more complex digital circuit and identifies several points at which waveform measurements are appropriate (including the clock). Figure 2-32 shows the normal waveforms to be expected at each of these points and the timing relationships.

Notice that individual waveforms have limited value in digital testing unless the timing relationship to one or more of the other waveforms is known to be correct. A dual-trace scope allows this comparison to be made. Typically, waveform 3 is displayed on one channel, and waveforms 1, 2, and 4 through 10 are successively displayed on the other channel.

In the time-related waveforms shown in Fig. 2-32, waveform 8 or 10 is an excellent sync source for viewing all waveforms because there is but one trigger pulse per frame. For convenience, external sync using waveform 8 or 10 as the sync source may be desirable. With external sync any of the waveforms can be displayed without readjustment of the sync controls. Waveforms 4 through 7 should not be used as the sync source because the signals do not contain a trigger pulse at the start of the frame.

It is not necessary to view the entire waveform (Fig. 2-32) in all cases. There are many times when a closer examination of a portion of the waveforms is more practical. In such cases, it is recommended that the sync remain unchanged, while remagnification (a change in sweep speed) is used to expand the waveform displays.

Checking frequency-division relationships Figure 2-33 shows the waveforms involved in a basic divide-by-two circuit. Notice that there is an output pulse for each two input pulses. Also, the output can be synchronized with the leading edge (positive-going) or trailing edge (negative-going) of the input pulses.

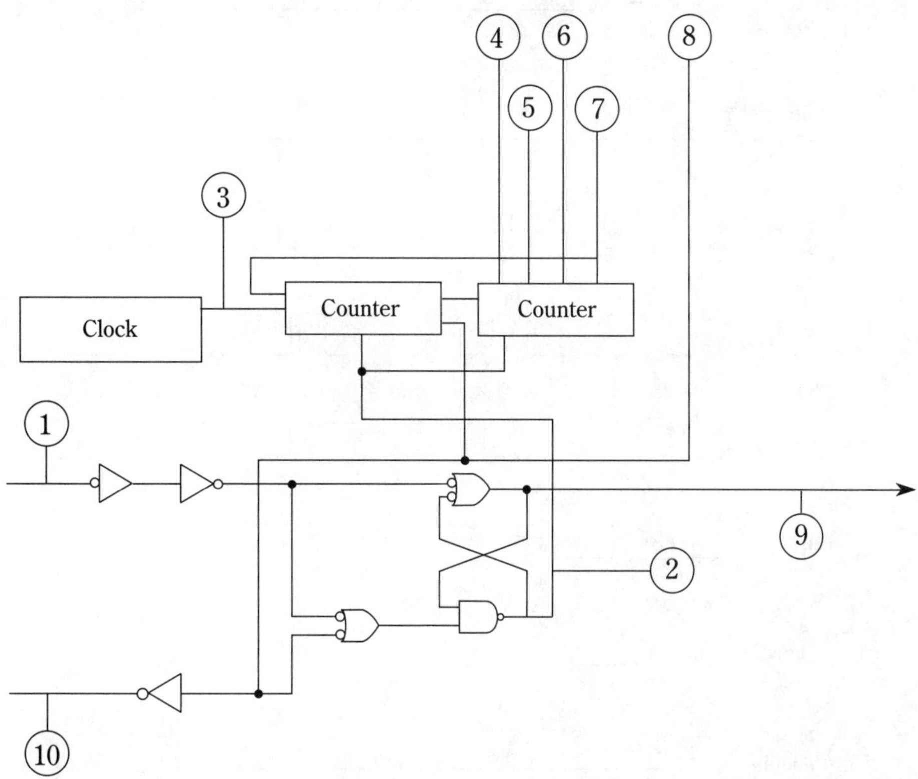

2-31 Typical digital circuit with identification of waveform test points.

Figure 2-34 shows the waveforms involved in a basic divide-by-eight circuit. There is one output pulse for eight input pulses; both ideal and practical time relationships are shown. In most digital circuits, the accumulated effects of components in consecutive states produce a propagation time delay that can be significant in a critical circuit.

Checking propagation time delay Figure 2-35 shows how to measure propagation time delay when the scope has both the invert and ADD functions. In Fig. 2-35A, the input to a divide-by-eight circuit is applied to channel A, and the output of the divide-by-eight circuit is applied to channel B in the normal manner. The ADD function is not used, and neither channel is inverted, In Fig. 2-35B, the ADD mode is selected and neither channel is inverted. In Fig. 2-35C, the ADD mode is selected and channel B is inverted. In all three cases, propagation time T_P is measured from the leading edge of the ninth reference pulse (channel A). As shown, a more precise measurement can be obtained if the T_P portion of the waveform is expanded or magnified.

Checking digital gates Figures 2-36 and 2-37 show scopes used to check an OR gate and an AND gate, respectively. Notice that with an OR gate, the circuit produces an output pulse when either or both input pulses are present. An AND gate can produce an output pulse only when there is a coincidence of two input pulses. The procedures shown in Figs. 2-36 and 2-37 can be applied to other digital gates, such as NAND, NOR, EXCLUSIVE-OR, and EXCLUSIVE-NOR.

2-32 Typical digital-circuit waveforms.

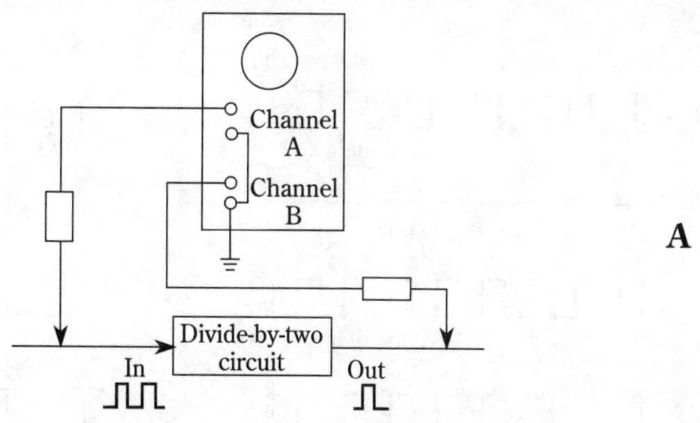

A

2-33 Checking frequency-division relationships (divide by two).

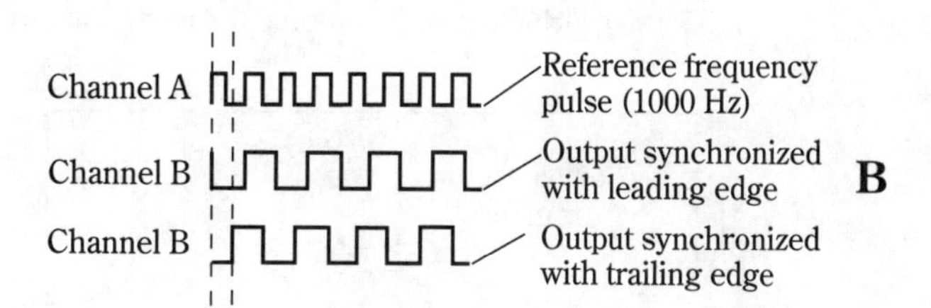

B

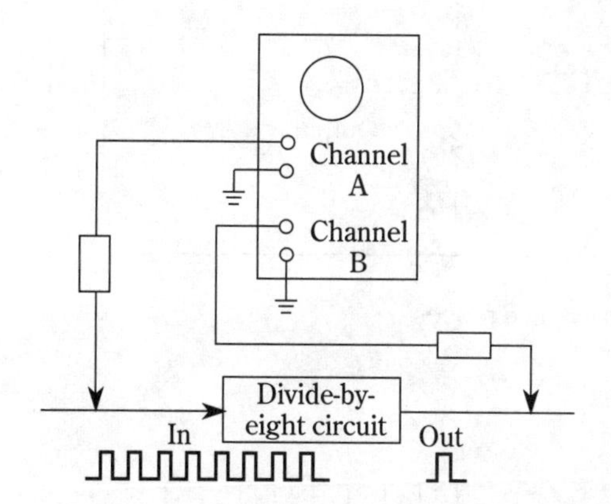

2-34 Checking frequency-division relationships (divide by eight).

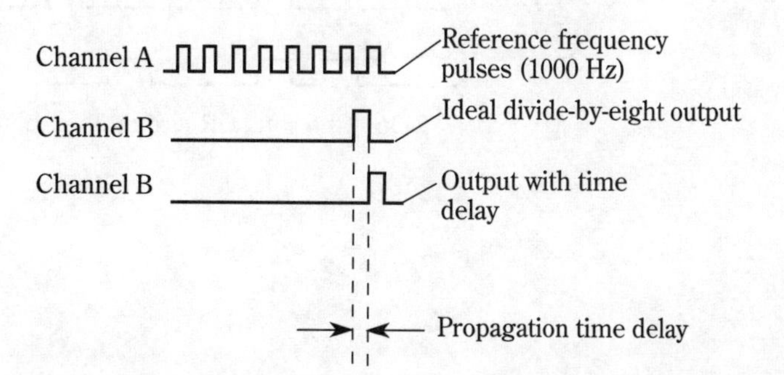

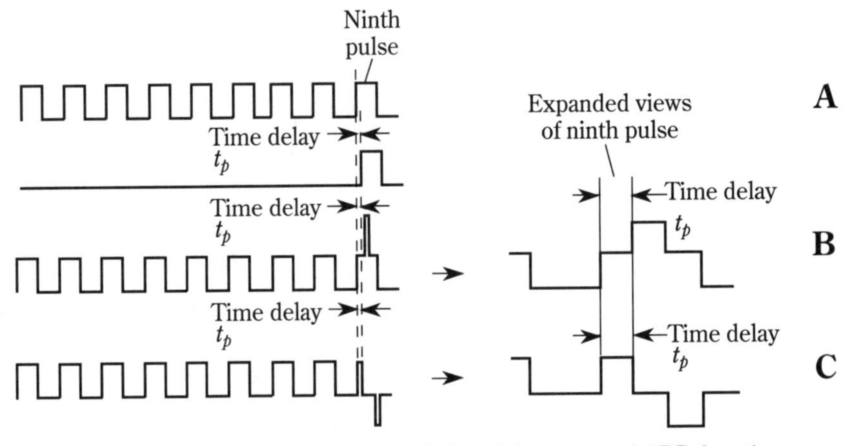

2-35 Measuring propagation time delay with invert and ADD functions.

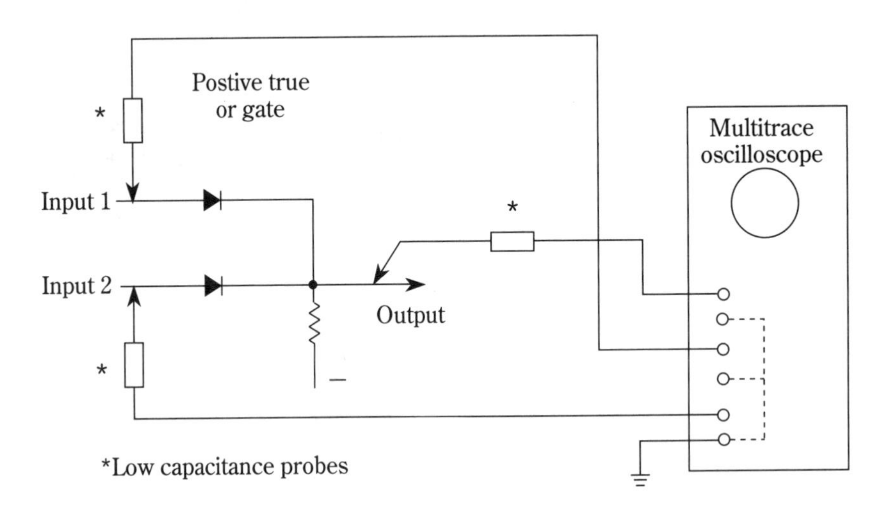

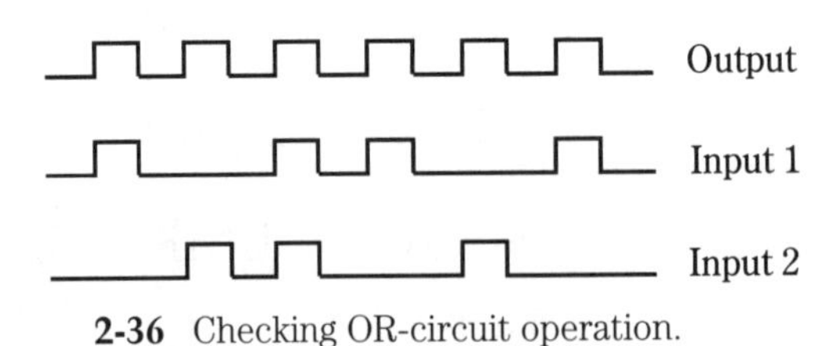

2-36 Checking OR-circuit operation.

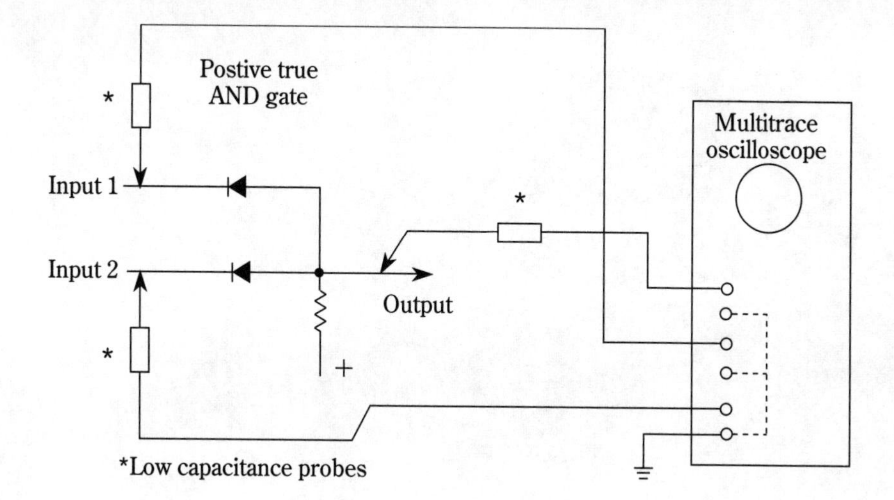

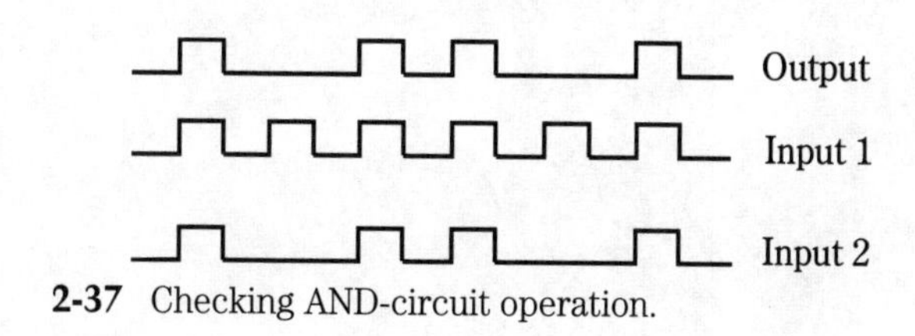

2-37 Checking AND-circuit operation.

3
Generators

A generator is an indispensable tool for circuit test and troubleshooting. Without a generator, you depend entirely on signals produced by the equipment being tested, and you are limited to signal tracing only. This means that you have no control over frequency, amplitude, shape, or modulation of the signals, and have no means of signal injection.

With a generator of the appropriate type, you can duplicate signals or produce special signals required to test, adjust, and troubleshoot all circuits. Also, the frequency, amplitude, shape, and modulation characteristics of the signals can be controlled so that you can check operation of the circuits being tested under various signal conditions (weak, strong, normal, abnormal, etc.).

3.1 Generator basics

At one time, most generators were single purpose. That is, a separate generator was required for RF, audio, sweep, sine, triangle, etc. Today, all or most of these functions are combined in one instrument called a *function generator*. However, when a generator is used exclusively to produce the pulses required in digital testing, the instrument is called a *pulse generator*. Likewise, when a generator produces only RF (but over a much wider frequency range), the instrument is called an *RF generator*.

In addition to these general-purpose generators, there are special-purpose generators used to test and adjust such specific circuits as FM stereo, TV stereo, and color TV. The features of such generators are described in this chapter, and some simple techniques to check the accuracy of the features are provided.

3.1.1 Oscillators

An *oscillator* (audio, RF, pulse, ramp, etc.) is the simplest form of signal generator. At the most elementary level of testing, an oscillator can provide a signal source. Except for the simplest test/troubleshooting, most electronic service requires multiple signals beyond the capability of a basic oscillator.

3.1.2 Probe, pencil generators

Probe generators (also known as *pencil-type noise generators*, *signal injectors*, and various other names) are essentially pulse generators or oscillators with a fast-rise waveform output and no adjustments. The fast-rise output produces simultaneous signals over a wide frequency range. The output signals can be used to used to test or troubleshoot the receiver, audio, and modulation sections of communications sets (chapter 15), as well as the RF sections of AM/FM tuners. However, except in basic test situations, such an instrument has many obvious drawbacks.

For example, to check the selectivity of receiver RF circuits, the signal source must be variable in amplitude (as described in chapter 15). To check the detector or audio portion of receiver circuits, the signal source must be capable of internal and/or external modulation. These characteristics are not available in the pencil-type instrument. As a result, even the least expensive shop-type (or even kit-type) generators have many advantages over the pencil generators.

3.1.3 Function generators

There is no standardization for the features found on present-day function generators. For example, the top-of-the-line function generator for a certain manufacturer generates sine, square, triangle, positive pulse, negative pulse, positive ramp, and negative ramp waveforms over frequencies from 0.001 Hz to 5 MHz in 8 ranges with a 1000:1 frequency control. The outputs can be produced in five modes: continuous, linear sweep, logarithmic sweep, tone burst, and single cycle. For test applications that require an unusual waveform, a variable-symmetry control can be used to alter the shape of any output selected.

At the other end of the line (for the same manufacturer) is a function generator that produces sine, triangle, and square waveforms over frequencies from 2 Hz to 200 kHz (the range most suitable for audio test and service). The outputs can be produced in three modes: linear sweep, log sweep, and continuous. The internal linear/log sweep rate is selectable in three ranges: slow (25 s), medium (250 ms), and fast (2.5 ms).

From these brief descriptions, you can see that all function generators have certain characteristics in common. That is, the instruments produce signals capable of being varied in frequency and amplitude, and capable of internal and external modulation or control. However, the top generators have several outputs (and refinements) not found in the lesser equipment, as well as a number of quality features (this accounts for the wide difference in price).

Function-generator operating procedures are not dwelled upon here. The instruction manuals for function generators are generally well written. That is, the manuals describe how to operate the controls to produce a specific output signal. For the tests described in this book, it is assumed that you have read the manual for your function generator, and can push button A to produce a sine output, and then can adjust dial B to set the output level.

Most of the tests in this book can be performed with a function generator. The tests described in chapters 14 and 15 are exceptions (these tests require an RF generator, and possibly a sweep/marker generator). Also, for digital tests, a pulse generator can be of considerable benefit. The characteristics of these generators are described in the following paragraphs.

3.1.4 RF generators

The following is a summary of the differences found in shop-type and laboratory RF generators, as well as some simple techniques to ensure accuracy of the output signals.

Output meter In most shop generators (as well as most function generators), the output amplitude is either unknown or approximated by means of dial markings. The lab RF generator incorporates an *output meter*. This meter is usually calibrated in microvolts so that the actual output can be read directly. If you must have maximum accuracy in output amplitude, monitor the output with a digital meter. With the possible exception of a precision lab-type RF generator, the output-amplitude shown by an external digital meter is more accurate than any dial marking or output meter.

Percentage of modulation meter Most shop RF generators have a fixed percentage of modulation (usually about 30%). Lab RF generators provide for a variable percentage of modulation, and a meter to indicate this percentage. Some RF generators have two meters (one for output amplitude and one for modulation percentage).

Output uniformity Shop generators vary in output amplitude from band to band. Also, shop generators cover part of the frequency range with harmonics. Lab generators have a more uniform output over the entire operating range, and cover the range with pure fundamental signals.

Wideband modulation Often, the oscillator of a shop generator is modulated directly. This can result in undesired frequency modulation. The oscillator of a lab generator is never modulated directly (unless it is designed to produce an FM output). Instead, the oscillator is fed to a wideband amplifier where the modulation is introduced. Thus, the oscillator is isolated from the modulating signal.

Frequency, or tuning, accuracy The accuracy of the tuning dials for a typical shop generator is about 2 or 3 percent, whereas a lab generator has from about 0.5 to 1 percent accuracy. However, neither instrument can be used as a frequency standard for test or adjustment of RF equipment because the required accuracy is usually much greater. For example, the FCC requires an accuracy of 0.005 percent or better (preferably 0.0025 percent) for CB equipment.

There are laboratory instruments generally described as *communications monitors*, or *frequency meters*, and *signal generators*, that provide signals with accuracies up to 0.00005 percent. These instruments are designed for commercial communications work (radio and TV broadcast, etc.), are quite expensive, and are thus not usually found in a typical service shop.

Combined frequency counter and generator To overcome the accuracy problem in practical work, the simplest approach is to monitor the generator (RF, function, audio, etc.) with a frequency counter (chapter 4). Using this technique, the frequency of the signal is determined by the accuracy of the counter, not the generator.

Frequency range Obviously, any generator used must be capable of producing signals at all frequencies used in the equipment under test. It is also convenient if the generator can produce signals at both harmonic and subharmonic frequencies (this is particularly true in RF work).

Frequency drift Because a generator must provide continuous tuning across a given range, some type of VFO must be used. As a result, the output is subject to drift, instability, modulation (by noise, mechanical shock, or power-supply ripple), and other problems that are associated with any type of variable oscillator. Frequency instability does not present too great a problem in most RF tests, provided that you monitor the generator output with a frequency counter. Of course, continuous drift can be annoying. For this reason, lab generators have temperature-compensated capacitors, frequency synthesizers, and PLL circuits to minimize drift. Similarly, the effects of line-voltage variations are offset by regulated power supplies.

Shielding The better generators have more elaborate shielding—especially for the output-attenuator circuits, where RF signals are most likely to leak. The leakage of RF from generators is something of a problem during receiver-circuit sensitivity tests (chapters 14 and 15) or any tests involving low-amplitude (microvolt range) RF signals from the generator.

Bandspread Less-expensive generators usually have a minimum number of bands for a given frequency range. This makes the tuning dial or frequency-control adjustments more critical and also more difficult to see. Lab generators usually have a much greater bandspread. That is, precision generators cover a smaller part of the frequency range in each band.

3.1.5 Pulse generators

Some manufacturers recommend a pulse generator for test and troubleshooting digital circuits. Ideally, the pulse generator should be capable of duplicating any pulse that is present in the circuits being tested. So, the generator output should be continuously variable (or at least adjustable in steps) in amplitude, pulse duration (or width), and frequency (or repetition rate) over the same range as the circuit pulses.

The simplest method for checking pulse-generator output is to monitor the output with a scope. Pulse amplitude, shape, width, repetition rate, rise time, fall time, and so on can be measured directly on the scope screen, as described in chapter 2. This is most practical because most tests described in this book (that involve pulses) also require a scope.

The following is a summary of the features found on a typical general purpose pulse generator (some of these features may also be available on high-end function generators).

Pulses are produced over a frequency range from 1 Hz to 5 MHz. The three output levels are: TTL level (5 V), 50 Ω, or 600 Ω. A separate control is provided to adjust pulse width, pulse period, and pulse delay. The generator also provides a trigger output (TTL), and can be triggered from an external source (–18 V to +18 V). Pulse-width range is from 100 ns to 1 s in 7 decade steps. The trigger delay (or pulse) delay is the same as for pulse width. The repetition-rate range is from 200 ns to 1 s in 7 decade steps. The pulse duty-cycle can be adjusted up to 100% (or dc). The rise and fall times for the three output levels are: less than 25 ns for TTL, less than 15 ns for 50 Ω, and less than 75 ns for 600 Ω. The 50-Ω output is 0.25 to 10 V open circuit, and 0.15 to 5 V with a 50-Ω load. The 600-Ω output is –15 V to +15 V adjustable with dc offset.

There are 10 operating modes: *free running*, where pulses are produced on a continuous-operation basis; *triggered*, where one pulse is produced for each trigger input; *pulse amplifier*, where one pulse is produced when an input crosses a selected threshold; *external gate*, where pulses are produced at the selected frequency when a TTL input is present; *trigger delay (post delay)*, where trigger output is a pulse, variable in width independently of pulse output; *pulse delay*, where the output pulse is delayed by an amount selected; *manual*, where one output pulse is produced when a manual button is pushed (one-shot mode); *fixed delay*, where the trigger pulse is set at a fixed width of about 1 μs; *invert*, where all pulse outputs are inverted; and *square wave*, where all pulse outputs have a 50% duty cycle.

3.1.6 Sweep/marker generator

Most present-day RF tuned circuits (such as TV/VCR front ends) have *frequency-synthesis (FS) tuning*, and do not require a sweep/marker generator for test or alignment. However, sweep/marker alignment is an excellent method for test/alignment of older non-FS receiver circuits (including RF front end, IF and sound traps). The sweep/marker technique can also be applied to other circuits (such as checking the frequency response of filters). The following paragraphs summarize the sweep/marker-generator features and operating procedures.

Sweep/marker generator basics A sweep/marker generator is used with a scope to display the bandpass characteristics of the circuit under test. The sweep portion of the generator is essentially an FM generator. When the generator is set to a given frequency, this is the center frequency. The sweep width, or the amount of variation from the center frequency, is determined by a control, as is the center frequency. The marker portion is essentially an RF generator with highly accurate dial markings. The marker signals necessary to pinpoint frequencies when making sweep-frequency alignments are usually produced by a built-in marker adder. Sweep/marker generators designed specifically for TV service often have additional features, such as a variable bias to disable AGC circuits.

Basic sweep-frequency measurement procedure Figure 3-1 shows the relationship between the sweep/marker generator and the scope during sweep-frequency test. If the equipment is connected as shown in Fig. 3-1A, the scope sweep is triggered by a sawtooth output from the generator. The scope internal sweep is switched off, and the scope sweep-selector and sync-selector controls are set to external. Under these conditions, the scope sweep represents the total sweep spectrum, as shown in Fig. 3-1C, with any point along the horizontal trace representing a corresponding frequency. For example, the midpoint on the trace represents 15 kHz. If you want a rough approximation of frequency, adjust the horizontal gain control until the trace occupies an exact number of scale divisions on the scope screen (such as 10 cm for the 10- to 20-kHz sweep signal). Each centimeter division then represents 1 kHz.

If the equipment is connected as shown in Fig. 3-1B, the scope sweep is triggered by the scope internal circuits. Certain conditions must be met to use the test connections of Fig. 3-1B. If the scope has a triggered sweep, there must be sufficient delay in the vertical input, or part of the response curve can be lost. If the scope sweep is not triggered, the generator must be swept at the same frequency as the scope, and there must be some means of adjusting the phase between the sweeps. If

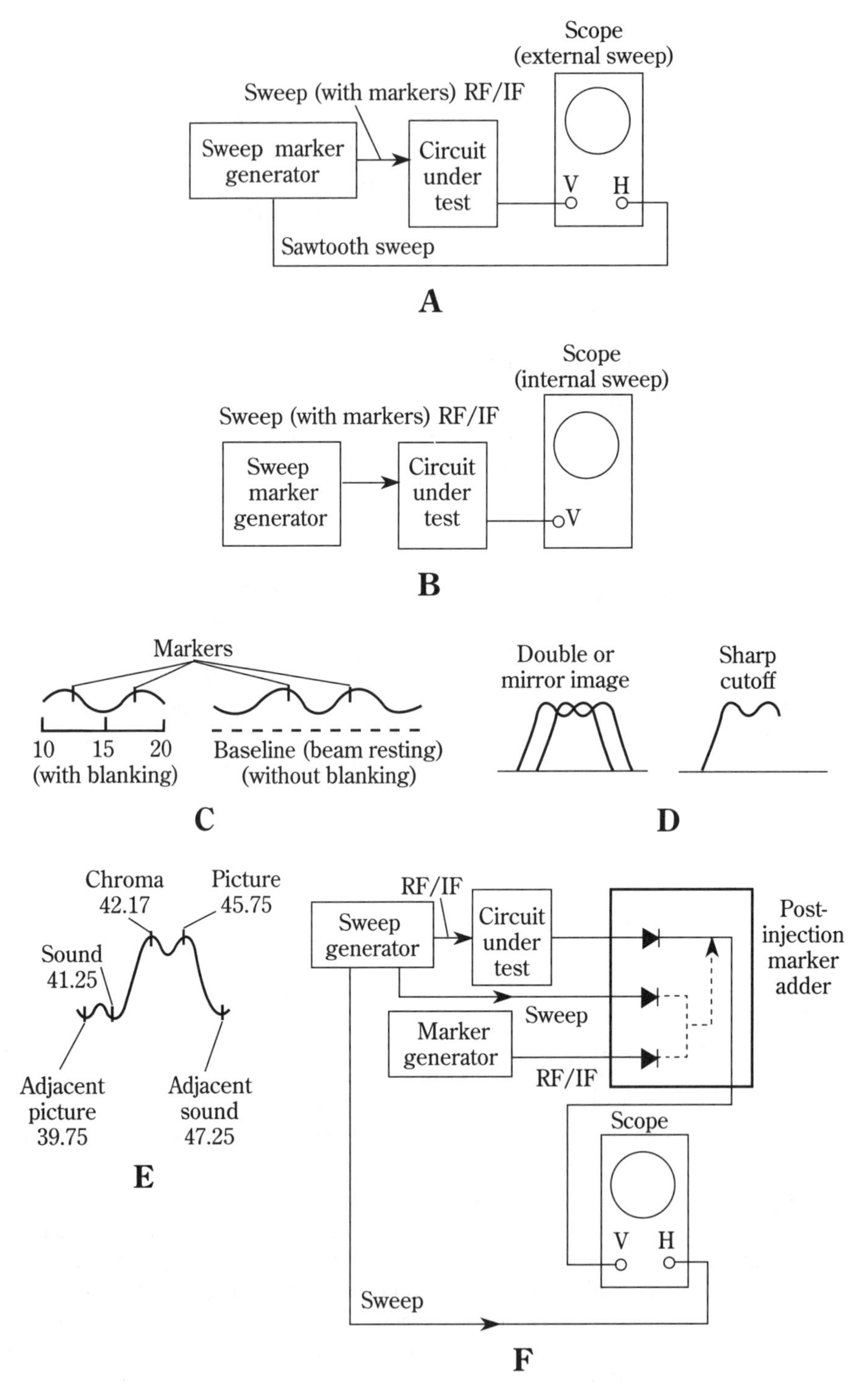

3-1 Basic sweep-frequency measurement procedure.

not, the curve can be prematurely cut off, or the curve might appear as a double image (or mirror image), as shown in Fig. 3-1D.

Sweep/marker generators for TV service have a number of markers at precise, crystal-controlled frequencies. Such markers are illustrated in Fig. 3-1E, which shows the bandpass response curve of a typical RF circuit (a VCR tuner and VIF package). The markers can be selected (one or several at a time), as needed. The response curve depends on the circuit under test. If the circuit has a wide bandpass characteristic, the generator is set so that the sweep is wider than that of the circuit. Under these conditions, the trace starts low at the left, rises toward the middle, and then drops off at the right. This method of test tells at a glance the overall bandpass characteristics (sharp response, irregular response at certain frequencies, and so on). The exact frequency limits of the bandpass can be measured with the markers.

Most sweep/marker generators have some form of postinjection marker-adder circuit, where a portion of the sweep output is also mixed with the marker output, as shown in Fig. 3-1F. The mixed and detected signal from both generators is then mixed with the detected output from the circuit. The scope vertical input represents the detected values of all three signals (sweep, marker, and circuit output).

Basic sweep/marker and scope test procedure The following steps describe the basic procedures for using a sweep generator with a scope. For a thorough discussion of sweep-frequency test procedures, read the author's best-selling *Lenk's RF Handbook*, McGraw-Hill, 1992.

1. Connect the equipment, as shown in Fig. 3-1F.
2. Place the scope and generator in operation.
3. Switch off the scope internal sweep, and set the scope sweep-selector and sync-selector controls to external so that the horizontal sweep is taken from the generator sweep, and the length of the sweep represents the total sweep-frequency spectrum.
4. Use markers (built-in or from an adder) for the most accurate frequency measurement.
5. Switch the generator blanking control (if any) on or off as desired. With the blanking function in effect, there is a zero reference line on the trace (Fig. 3-1C). With blanking off, the horizontal baseline does not appear. The generator blanking function is not to be confused with the scope blanking (which is bypassed when the sweep signal is applied directly to the scope horizontal oscillator).

3.2 FM-stereo generator

An FM-stereo generator is essential for troubleshooting the FM portion of an AM/FM tuner. This is because an FM generator simulates the very complex modulation system used by FM-stereo broadcast stations. Without an FM stereo modulation system, you must depend on constantly changing signals from the stations, making it impossible to adjust the FM portion of the tuner or to measure frequency response after adjustment.

3.2.1 FM-stereo modulation system

Before the characteristics of an FM-stereo generator are described, review the basic-FM stereo modulation system (which is quite complex when compared to that of the AM broadcast system). It is essential that you understand the FM system to test or adjust the FM portion of any AM/FM tuner.

Figure 3-2A shows the composite audio-modulating signal used for FM stereo. This FM system permits stereo tuners and receivers to separate audio into left and right channels and permits mono FM tuners to combine left- and right-channel audio into a single output.

Figure 3-2B shows the block diagram of an FM-stereo modulator and transmitter. Left- and right-channel audio signals are applied through preemphasis networks to a summing network that adds the two signals. A low-pass filter limits this signal to the 0- to 15-kHz audio band, which is the maximum authorized for FM broadcast service. The (L + R) signal contains both left- and right-channel audio in a 0- to 15-kHz baseband.

Summing and suppression The left-channel audio is applied to another summing network, along with the right-channel audio, which is inverted. The summing network effectively subtracts the two signals. The 0- to 15-kHz (L – R) signal is fed to a balanced modulator along with a 3.8-kHz sine wave. The balanced modulator produces a *double-sideband suppressed-carrier (DSBSC) subband*, centered around 38 kHz. A bandpass filter limits the signal to the 23- to 53-kHz range (±15 kHz of the 38-kHz carrier). The resulting L – R signal is a 23- to 53- subband, with the 38-kHz subcarrier fully suppressed.

Pilot signal When FM stereo is broadcast, a low-level 19-kHz pilot signal is transmitted simultaneously. The pilot signal is generated by a stable, crystal-controlled oscillator operating at 19 kHz. The 19-kHz pilot oscillator output is also applied to a frequency doubler, providing the 38-kHz carrier for the balanced modulator.

Composite modulating signal The 0- to 15-kHz baseband (L + R), 23 to 53-kHz subband (L – R), and 19-kHz pilot signal are applied to a summing network, resulting in a composite audio signal consisting of three components shown in Fig. 3-2A. The composite audio signal is applied to the modulator, which FM-modulates the RF carrier of the transmitter. For a fully modulated RF carrier, the (L + R) signal accounts for 45 percent, the (L – R) signal for 43 percent, and the pilot signal for 10 percent.

Decoding Figure 3-2C shows the FM section of a typical AM/FM tuner, from the discriminator to the audio-amplifier inputs. The discriminator output is a composite audio signal similar to that shown in Fig. 3-2A. Stereo tuners decode the signal shown in Fig. 3-2A and then separate the audio into original left and right channels. Because mono FM tuners have only a 0- to 15-kHz audio response, the 19-kHz pilot signal and the 23- to 53-kHz (L – R) subband are rejected. However, the (L + R) baseband signal is accepted and combines left- and right-channel audio to produce mono audio.

A PLL locks the sine-wave output of the 38-kHz VCO in phase with the received 19-kHz pilot signal as follows. A divide-by-2 circuit converts the 38-kHz VCO output to a 19-kHz sine wave, which is one of the inputs to a phase comparator. The other input is the received 19-kHz pilot signal. The phase comparator produces an error voltage to lock the VCO in-phase with the 19-kHz pilot signal.

The composite audio from the discriminator is applied through a 23- to 53-kHz bandpass filter, which blocks the 0- to 15-kHz (L + R) bandpass and 19-kHz pilot sig-

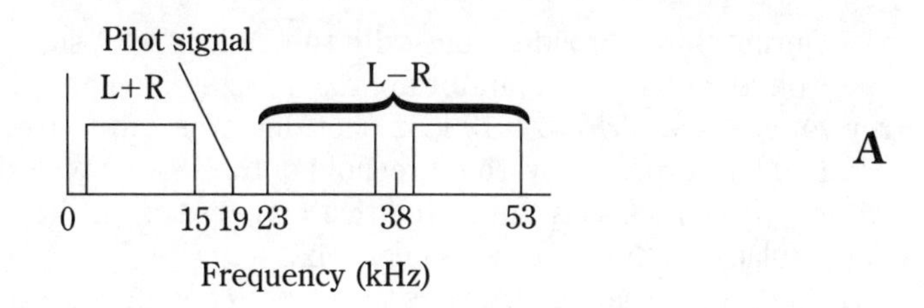

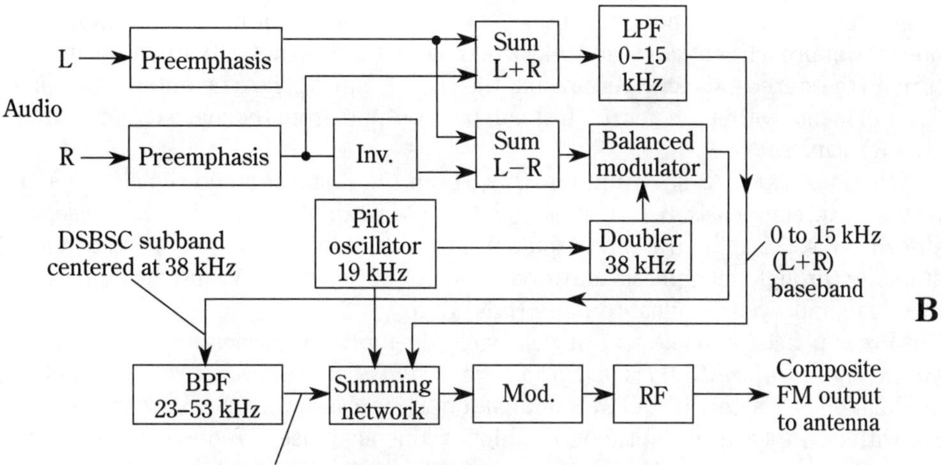

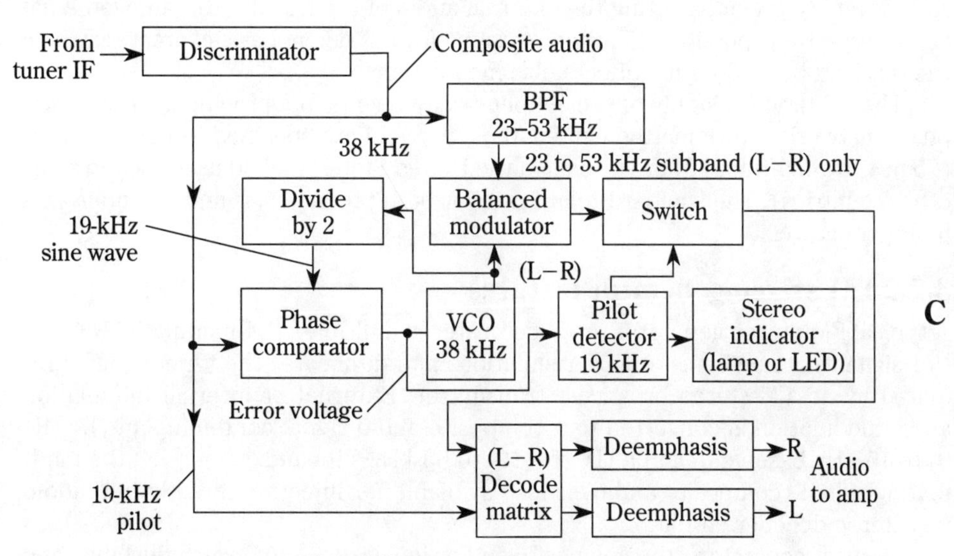

3-2 FM-stereo modulating system.

nals, while passing only the 23- to 53-kHz subband (L – R) signal. The 23- to 53-kHz subband signal is applied to a balanced demodulator. The other input to the demodulator is a 38-kHz carrier from the VCO. The VCO carrier is locked in phase to the 19-kHz pilot signal by the PLL.

During stereo broadcasting, with the 19-kHz pilot signal present, the error voltage from the phase comparator approaches zero as phase-lock is achieved. This error voltage is applied to a 19-kHz pilot detector and to the 38-kHz VCO. When the error voltage drops below the threshold of the 19-kHz pilot detector, the stereo indicator is turned on, as is the switch that couples the (L – R) signal from the balanced demodulator to the L – R decode matrix.

The decode matrix separates the L and R components of the (L + R) and (L – R) signals into independent left- and right-channel outputs. When the (L – R) signal is absent, such as during mono reception, the error signal from the phase comparator does not approach zero, and the 19-kHz pilot detector does not actuate the switch or turn on the stereo indicator. As a result, the (L + R) input is not separated into L and R components within the matrix. Instead, both outputs from the matrix are identical (L + R) signals.

Preemphasis As shown in Fig. 3-2C, both the L and R audio signals are subjected to preemphasis before reaching the FM-modulator circuits. Likewise, as shown in Fig. 3-2C, the L and R outputs from the decode matrix are applied to audio amplifier through deemphasis networks. The use of preemphasis and deemphasis is mostly to improve the signal-to-noise (S/N) ratio.

Preemphasis increases the highs, while deemphasis increases the lows. *Frequency modulation (FM)* is usually a form of *phase modulation (PM)*. The noise-modulation characteristic of PM is not flat, but it increases with the noise frequency. So, with deemphasis in the tuner, the high-frequency noise is reduced to the same level as the low-frequency noise, thus improving the S/N ratio.

Remember that noise modulation is essentially an internal modulation of constant level. By preemphasizing the external audio of the transmitter, an overall flat audio response is possible. This compensates for the deemphasis characteristics of the receiver, but it does not affect noise modulation.

The RC time constants of the coupling components in a preemphasis or deemphasis network determine the center frequency and are specified in microseconds (75 μs is standard for FM broadcasts in the United States, and 50 μs is used in some other countries). The standard rolloff rate for both preemphasis and deemphasis is 6 dB per octave.

3.2.2 FM stereo generator features

A typical FM-stereo generator produces a stereo multiplex FM-modulated RF-carrier signal that conforms to FCC regulations and duplicates the type of signal radiated by an FM-stereo broadcast transmitter. External or internal modulation audio modulation is converted to a composite audio signal containing an (L + R) 0- to 15-kHz baseband, and a (L – R) 23- to 53-kHz subband. Generally, the modulating signal (composite audio) is also available for injection directly into audio and stereo decoder circuits.

On most generators, the composite audio signal is continuously adjustable, and a modulation meter is calibrated to measure the rms value of composite audio. The RF output might be internally or externally modulated. In either case, modulation is continuously adjustable up to 75 kHz. A calibrated modulation meter reads FM deviation in kilohertz.

For external modulation, independent left and right input jacks permit stereo modulation through 75- or 50-μs preemphasis networks, or with no preemphasis (on most generators). The 50-Hz to 15-kHz audio input bandwidth equals that of an FM broadcast transmitter, thus permitting full audio-range frequency-response tests of tuners.

The typical internal-modulation frequency is 1 kHz, which can be selected in one of five combinations: left-channel only, right-channel only, (L + R) baseband, (L – R) subband, and left and right with 1-kHz signal applied to the left channel, and a 50- to 60-Hz line-voltage signal applied to the right channel. These internal-modulation combinations permit complete testing of stereo decoder circuits, including channel-balance and channel-separation characteristics (if required).

During either internal or external modulation, a highly stable 19-kHz pilot signal is generated and combined with the composite audio. The pilot signal can be switched off when desired, such as for the test of the pilot-detector circuit. Some FM-stereo generators have a detachable telescoping antenna to simulate an FM-stereo broadcast transmitter. On most generators, the RF output can be turned off whenever desired, thus generating the composite audio only.

3.2.3 FM IF adjustments with sweep/marker generator

Figure 3-3A shows the connections when the sweep/marker technique (described in Sec. 3.1.6) is used to test and adjust the IF circuits of the FM section of an AM/FM tuner. The procedure can be used at any time, but it is of most value when performed before any extensive troubleshooting (the procedure just might cure a number of problems). A preferred procedure is described in Sec. 3.2.4.

With the sweep generator connected to TP2 (signal), adjust the sweep generator for 10.7 MHz with a 200-kHz sweep width. Adjust the generator output amplitude for a weak signal that produces noise patterns, as shown on the waveforms in Fig. 3-3A. Using a nonmetallic alignment tool, adjust the IF tuning (IFT) of the MD101 FM tuner so that the IF waveform (S curve) in Fig. 3-3A is maximum. Move the sweep generator to TP3 (audio), but leave the frequency and output as previously set. If necessary, reduce the scope gain. Alternately, adjust T201 for a symmetrical S curve, and T202 for linearity between the positive and negative peaks of the IF waveform.

3.2.4 FM IF adjustments with FM generator

Figure 3-3B shows the connections when an FM-stereo generator is used to test and adjust the IF circuits of the FM section of an AM/FM tuner. Some instruction manuals recommend that the sweep/marker technique (Sec. 3.2.3) be performed as a preliminary to this IF alignment.

Adjust the FM generator for 97.9 MHz, modulated by 1-kHz with 75-kHz deviation (100 percent modulation, mono). Adjust the generator output amplitude for 65.2 dBf. The term *dBf* (found in some FM tuner specifications) is the power level measured in dB, referenced to 1 femtowatt (10^{-15}). If you are fortunate, the service literature will spell out a voltage reading (usually in the microvolt range).

Adjust the tuner to 97.9 MHz (as indicated on the tuner digital display), and adjust T201 for 0 V ±50 mV on the null meter (connected between TP4 and TP5). Null

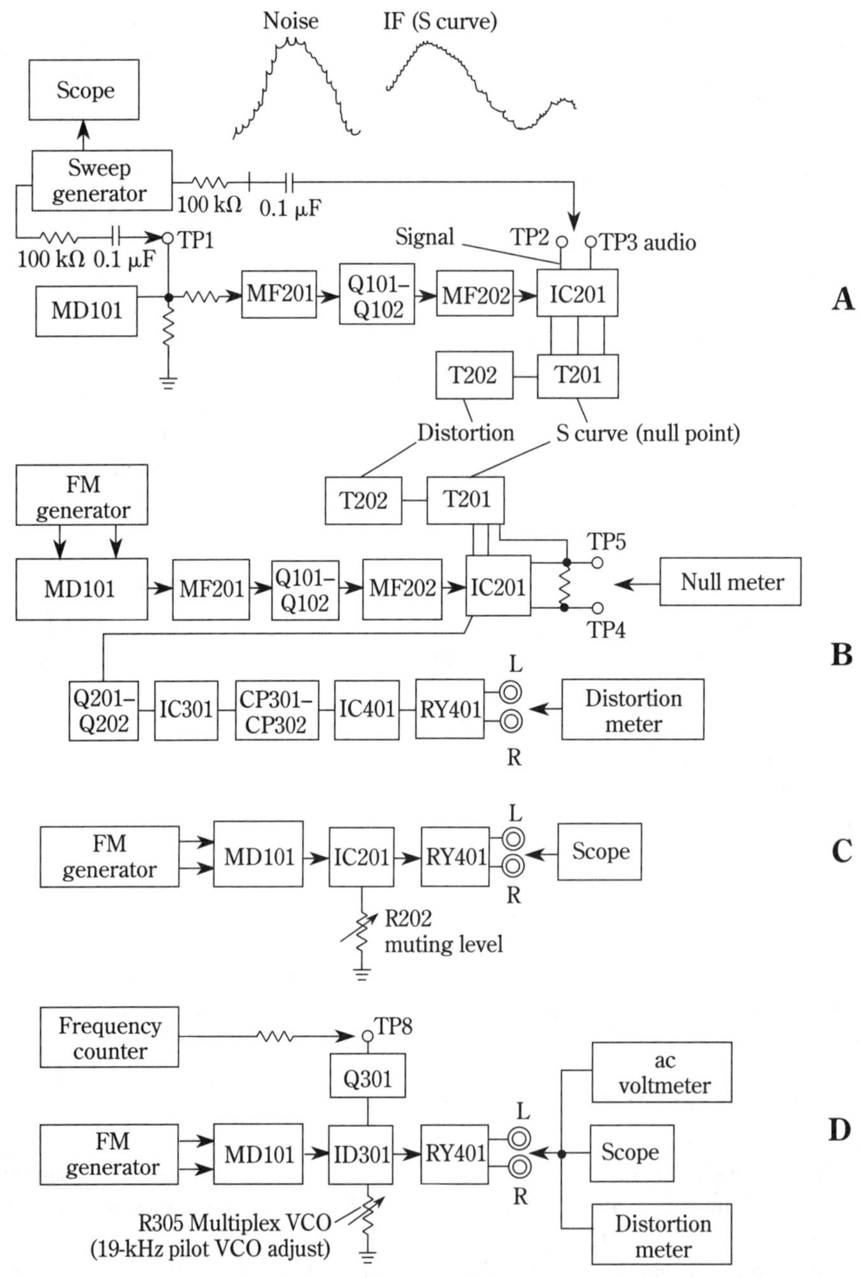

3-3 Tuner FM test and adjustment points.

meters are covered in chapter 6. Adjust T202 for minimum distortion on the distortion meter (connected to the left-channel audio-output jack). Distortion meters are covered in chapter 12. Work between T201 and T202 until you get minimum distortion and (it is hoped) 0 V ±50 mV on the null meter.

3.2.5 FM muting adjustment

Figure 3-3C shows the test and adjustment points. This procedure sets the signal threshold for audio muting in the FM auto mode (where the audio should be muted in between stations and on weak stations).

Adjust the FM generator for 97.9 MHz, modulated by 1 kHz with 75-kHz deviation. Adjust the generator output amplitude for 33 dBf (or the equivalent output voltage). Adjust the tuner to 97.9 MHz (as indicated on the tuner digital display). Place the tuner in FM auto mode (press auto mode and check that the auto-mode indicator, if any, turns on). Adjust R202 until the audio is muted. Then slowly readjust R202 until the audio is just unmuted.

3.2.6 FM multiplex VCO adjustment and distortion test

Figure 3-3D shows the test and adjustment points. This procedure sets the 19-kHz pilot VCO in IC301, and checks the resultant distortion.

Adjust the FM generator for 97.9 MHz with no modulation. Adjust the generator output amplitude for 65.2 dBf (or the equivalent output voltage). Adjust the tuner to 97.9 MHz (as indicated on the tuner digital display). Place the tuner in FM audio mode (press auto mode and check that the auto-mode indicator, if any, turns on). Adjust R305 for 19-kHz ±50 Hz, as indicated on the frequency counter (chapter 4) connected to TP8. Leave the tuner and FM generator set at 97.9 MHz. Apply 1-kHz modulation to the FM generator left channel (with a pilot carrier deviation of 6-kHz and a total deviation of 75 kHz).

Adjust the IFT of MD101 (using a nonmetallic alignment tool) for minimum distortion (as indicated by the distortion meter connected to the left-channel audio output jack). Do not adjust the IFT of MD101 more than one-quarter turn from the setting established in Sec. 3.2.3.

Remove the left-channel modulation from the FM generator, and apply the same modulation to the right channel. Monitor the right-channel audio output with the distortion meter, and check that the distortion is about the same for both channels.

For a thorough discussion of FM receiver test and adjustment, read the author's best-selling *Lenk's Audio Handbook* (McGraw-Hill, 1991).

3.3 TV/VCR stereo generator

A special-purpose generator is required to test and adjust the stereo circuits of a TV set or VCR. There are inexpensive generators that use spot modulation to stimulate dbx encoding at specific frequencies (rather than providing encoding across the entire TV frequency spectrum, as do expensive broadcast-studio instruments). Before the generator characteristics are described, the following is a review of the basic stereo-TV system (which is also very complex and quite different from that for FM-stereos found in AM/FM tuners, Sec. 3.2).

3.3.1 Basic signal flow in stereo TV

Figure 3-4 shows the TV transmitter and receiver for the MTS/MCS (multichannel television sound) system, which delivers audio for a stereo TV program and separate

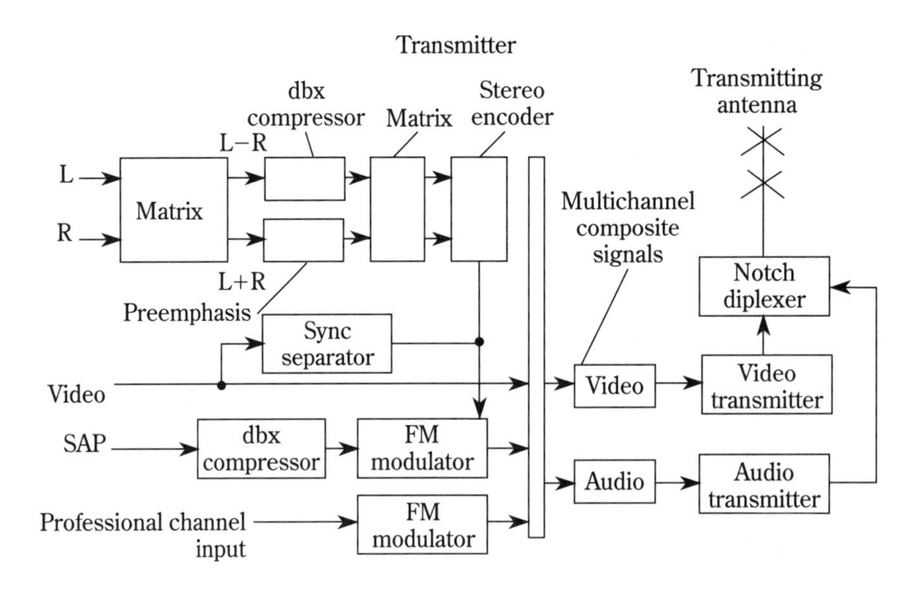

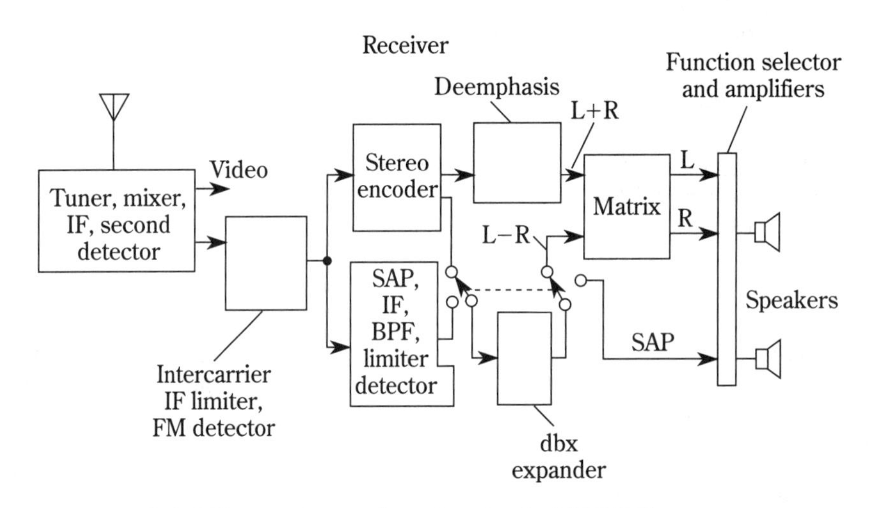

3-4 TV transmitter and receiver for the MTS/MCS system.

audio program (SAP). Stereo-TV broadcasts are made in accordance with the dbx Noise Reduction (NR) System and the Zenith Transmission System. Figure 3-5 shows the signal spectrum used in MTS/MCS.

The main-channel signal is composed of the sum (L + R) signals and is the same as in the conventional TV sound specification. As a result, the same TV sound can be received by a conventional TV set (or VCR) when stereo TV broadcasting is in effect. When the difference between the L and R signals (L – R) is transmitted, the L and R signals are reproduced from the main-channel signal (L + R) and the stereo signal (L – R) by the TV set. The same signals can be recorded and played back by a VCR equipped for stereo TV.

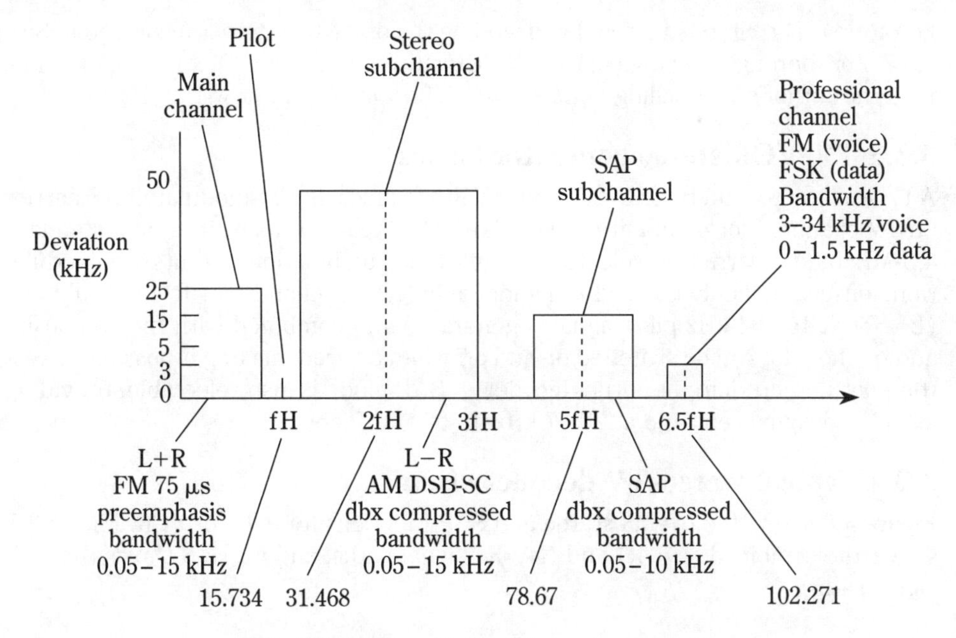

3-5 MTS/MCS multichannel signal spectrum.

The L – R signal is produced by amplitude modulating a subcarrier with a frequency double of the horizontal-scanning frequency (f_H) using the double-sideband suppressed-carrier (DSBSC) system. As a result, the frequency division of the main-carrier signal is double (50 kHz) that of the main channel signal (L + R).

The carrier of the L – R signal is suppressed, so it is necessary to transmit a reference signal to demodulate the L – R signal correctly in the TV or VCR. To do this, a signal with a frequency equal to the horizontal-scanning frequency (f_H), called the *pilot signal*, is inserted between the main-channel signal (L + R) and the stereo signal (L – R). The pilot signal is also used to indicate the presence or absence of the stereo signal (no pilot, no stereo).

The SAP-channel signal frequency-modulates a subcarrier with a frequency of $5f_H$, and is locked to $5f_H$ when there is no SAP. The nonpublic channel (also called the *professional channel signal* frequency-modulates the subcarrier with a frequency of $6.5f_H$. This channel is scheduled to be used for business, not for TV broadcasting.

The sum (L + R), difference (L – R), pilot, SAP channel, and the nonpublic-channel signals are added to produce the multichannel composite signal, which, in turn, frequency modulates the transmitted TV broadcast signal. A noise-reduction system (NR system), which specifically encodes and transmits the L – R and SAP signals, is used to reduce noise in the TV set and/or VCR.

3.3.2 dbx NR system

The dbx noise reduction system is an essential part of the stereo TV system. In the simplest of terms, portions of the transmitted signal are compressed (or encoded) at the TV transmitting station. These same signals are expanded (or decoded) at the TV set or VCR. The process is known as *companding* (or as *compandoring* in some literature). The circuits involved in decoding the dbx NR system are covered in Sec. 3.3.7. For thorough coverage of the TV/VCR stereo system and dbx noise reduction, read the author's best-selling *Lenk's Audio Handbook* (McGraw-Hill, 1991).

3.3.3 TV/VCR stereo generator basics

A typical stereo generator for TV and VCR produces an FM-modulated RF carrier on TV channel 3 or 4, and at the standard TV IF carrier, as well as on a standard 4.5-MHz audio carrier. A selector permits four combinations of internal modulation: left-channel only (L), right-channel only (R), baseband (L + R), and subband (L – R). A 15.734-kHz pilot signal is generated and combined with the composite audio. The pilot can be switched on and off when desired, making it possible to test the pilot-detector circuit in the decoder. A SAP mode is also selectable, providing a subband signal centered at 78.67 kHz, to test SAP operation.

3.3.4 Typical stereo-TV decoder circuits

Figure 3-6 shows the overall stereo and SAP functions for a TV or VCR. Figure 3-7 shows the combined (L + R) and (L – R) signal paths, and Fig. 3-8 shows the SAP signal paths.

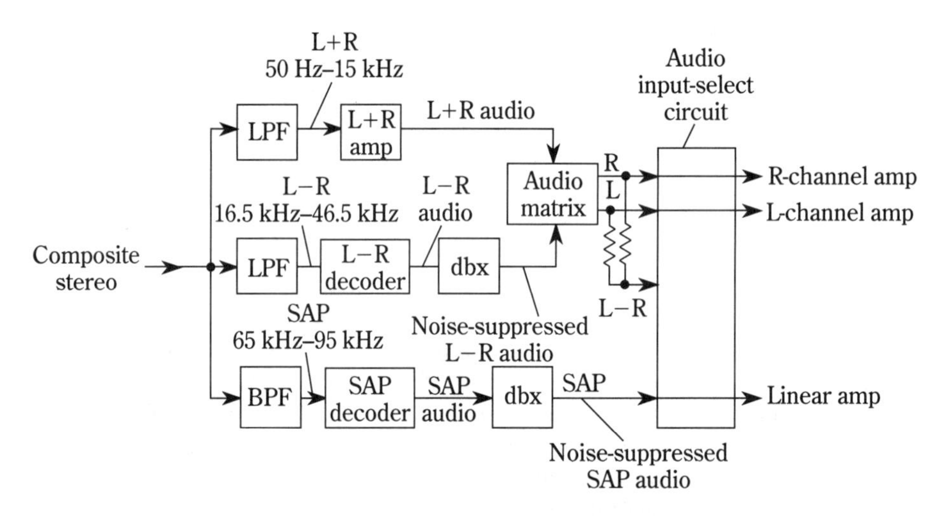

3-6 Overall stereo and SAP functions.

Individual components of the composite stereo signal can take any of three different paths. Which path is determined by low-pass and band-pass filters (LPFs and BPFs) in the input circuit. The L + R audio is selected by an LPF designed to pass only those L + R signals below 15 kHz. The L + R (mono) signal is amplified and applied to the audio matrix. The second LPF passes all frequencies below 46 kHz, which includes the L + R, L – R sidebands, and the pilot signal. These signals are all applied to the L – R decoder.

The L + R signal (because of the low frequency) is rejected by the L – R decoder (only the L – R sidebands and pilot signal are used in the L – R decoder). The pilot

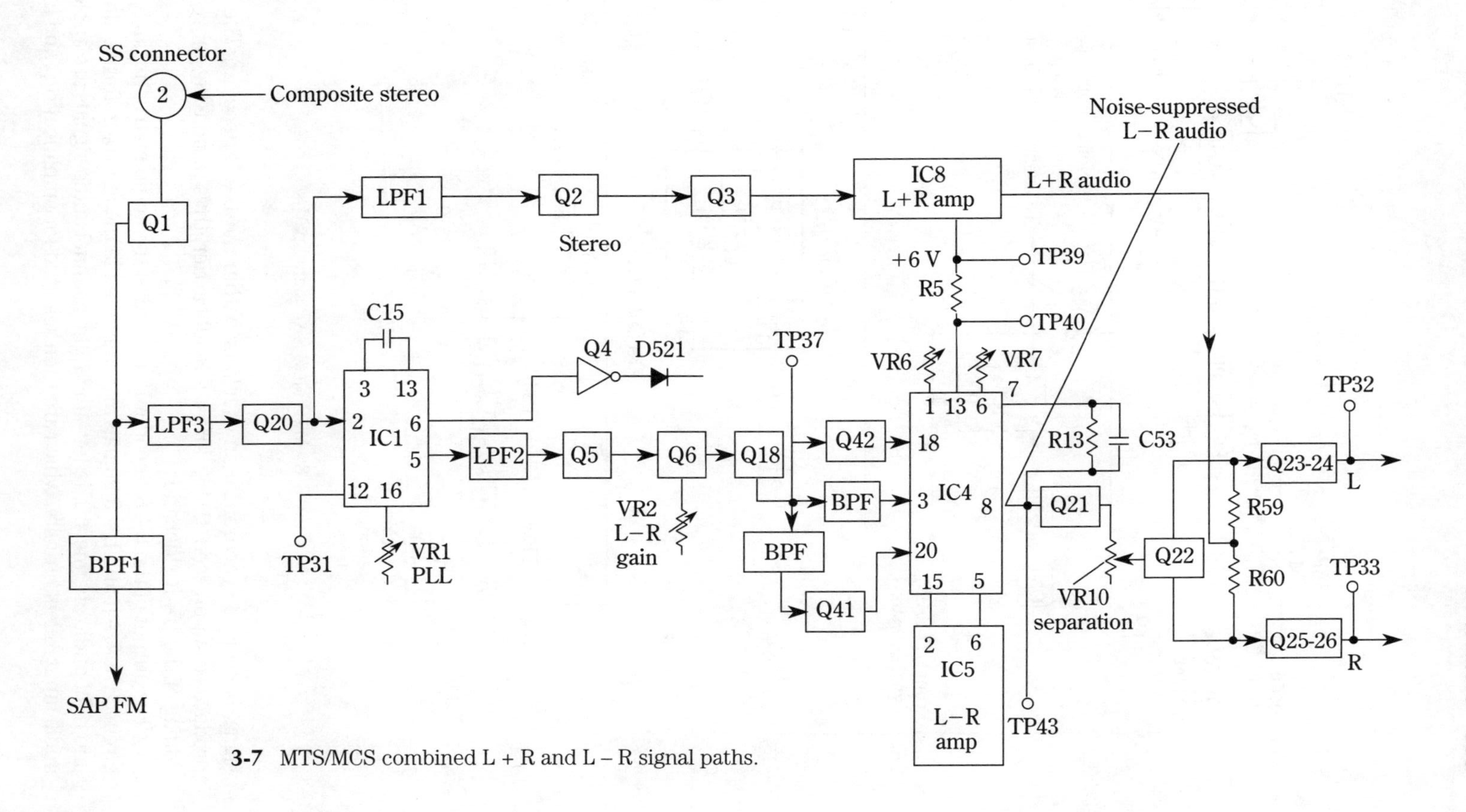

3-7 MTS/MCS combined L + R and L − R signal paths.

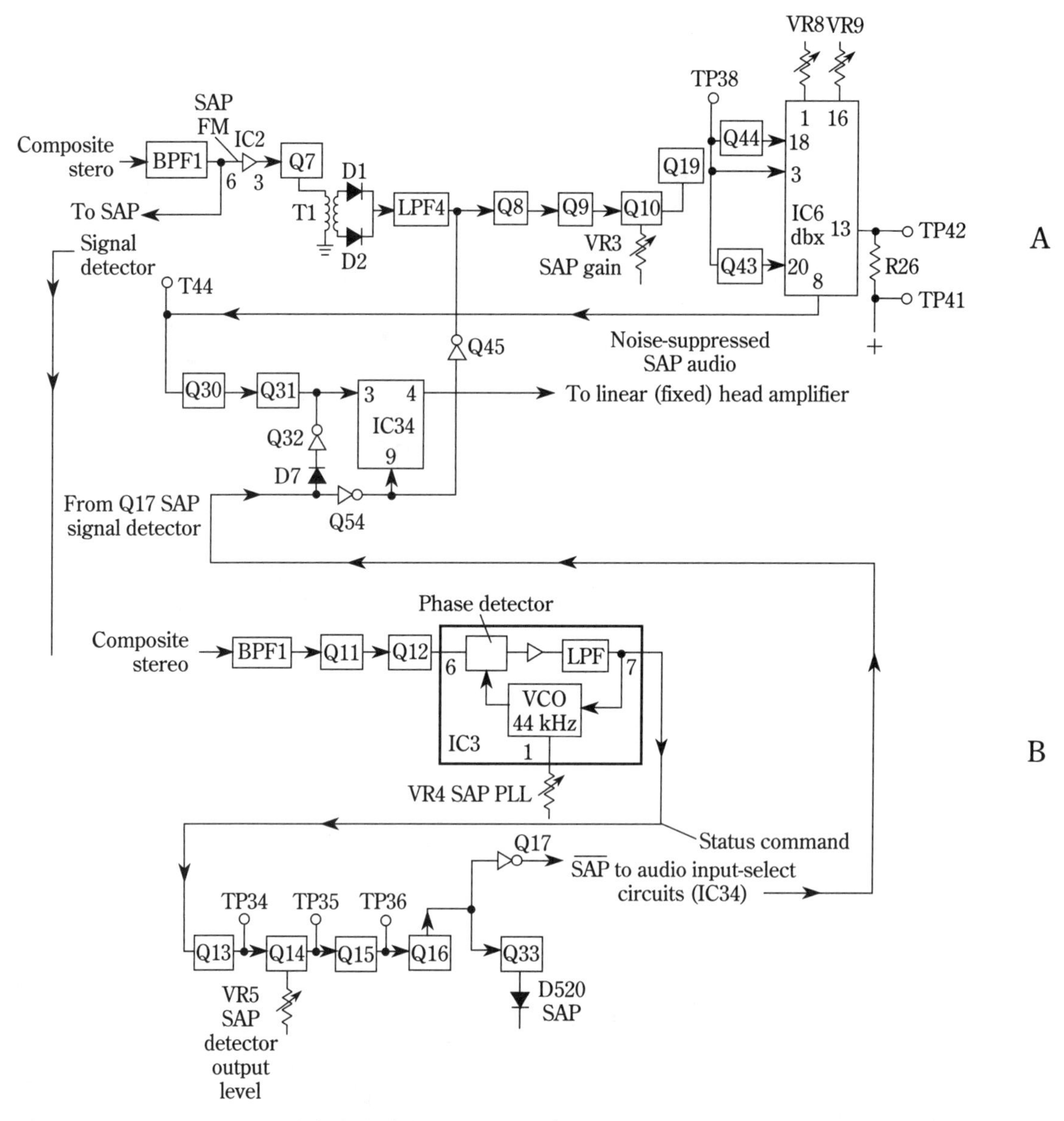

3-8 MTS/MCS SAP signal paths.

signal is used to synchronize an oscillator within the L – R decoder. The oscillator produces a suppressed 31.463-kHz L – R subcarrier, thus permitting subsequent decoding of the L – R sidebands.

The output of the L – R decoder is applied to the audio matrix through a dbx NR circuit. The L – R and L + R signals are combined in the matrix to form individual left and right audio channels. The left- and right-channel audio signals are applied to the audio input-select circuits, which are then selected for amplification (and in the case of a VCR, for recording on tape by the rotating hi-fi audio heads). Samples of each audio channel are added to form the L + R mono audio signal. In a VCR, this signal is recorded by the rotating hi-fi heads if no stereo is broadcast. The L + R signal is also applied to the stationary or linear head when SAP is not broadcast (or selected).

If SAP is present, the SAP audio is passed through the BPF. This BPF passes signals from 65 to 95 kHz, and thus rejects the L + R and L – R signals. The SAP signal is decoded by the SAP decoder and applied to the audio input-select circuit through a separate dbx NR circuit. In many VCRs, the SAP signal can be recorded on the stationary linear head, in place of the L + R, when so selected.

3.3.5 L + R circuits

As shown in Fig. 3-7, the composite stereo signal is amplified by Q1 and applied through filters to the stereo (L – R) and and SAP signal paths. The L + R and L – R signals are rejected by BPF1, but are passed to Q20, where both signals are amplified. The L + R signal is applied to amplifier IC8 through LPF1, Q2, and Q3. The L + R audio from IC8 is applied to the junction of R59 and R60. Both resistors are part of a matrix circuit used to combine the L + R and L – R signals.

3.3.6 L – R decoder

Most of the L – R functions are performed within IC1. These functions include detecting the presence of a stereo signal (a signal with pilot) and decoding the L – R sidebands to produce L – R audio. If the broadcast is in stereo (pilot present), IC1 produces a command at pin 6 to turn on the stereo indicator D521. The L – R sidebands and the pilot are applied to pin 2 of IC1. The L – R sidebands pass to an L – R decoder within IC1, while the pilot is amplified by a preamp. The amplified pilot signal is applied to two comparators through C15.

A VCO in IC1 operates at 4 times the pilot frequency (62.936 kHz). The VCO output is divided twice by 2, resulting in a 31.468-kHz output (applied to the L – R coder as a substitute carrier) and a 15.734-kHz output (applied to both comparators as a second input). When a stereo signal is transmitted, the pilot and the divided-comparison signals from the VCO are compared in both comparators. The PCL comparator (phase comparator lamp) generates a signal that is filtered by an LPF and amplified by a lamp driver to produce a low at pin 6 of IC1. If there is no pilot, pin 6 of IC1 remains high.

A low at pin 6 of IC1 turns on Q4. In turn, Q4 applies 12 V to stereo indicator D521. As a result, D521 turns on when a pilot signal is present (indicating a stereo broadcast to the listener or viewer). If pin 6 is high (no pilot, no stereo), Q4 and D521 remain off.

The *PCV comparator (phase comparator VCO)* develops a correction voltage if a frequency or phase error exists between the pilot signal and the VCO comparison signal. The correction voltage is filtered, amplified, and applied to the VCO to correct any frequency or phase error. VR1, at pin 16 of IC1, is used to set the free-running frequency of the VCR to 4 times that of the pilot frequency.

The VCO output is applied to a divide-by-2 circuit, reproducing a phase- and frequency-corrected duplicate of the original L – R subcarrier. The resulting 31.468-kHz signal is applied to the L – R decoder, along with the L – R sidebands from pin 2 of IC1. The L – R sidebands are decoded, and the resulting L – R audio is available at pin 5 of IC1. The L – R audio is passed through LPF2, where all unwanted signals above

15 kHz are removed. The L – R audio is then amplified by Q5, Q6, and Q18, and is applied to the dbx circuit.

3.3.7 dbx circuits

Although the dbx format used in stereo TV is the most complex part of the system, virtually all of the dbx circuits are contained within IC4. Figure 3-9 shows these circuits in block form.

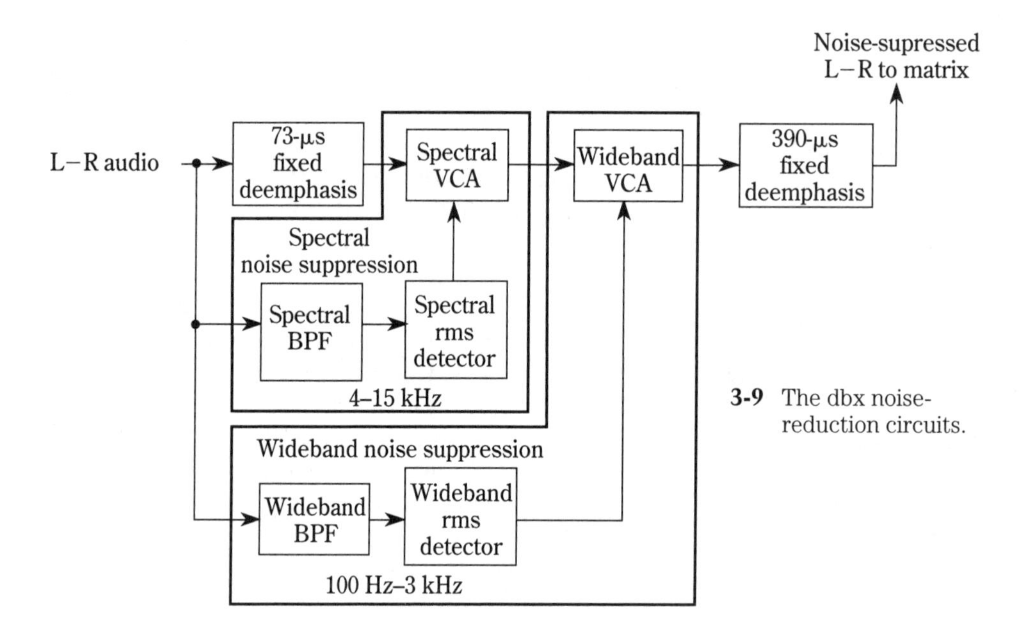

3-9 The dbx noise-reduction circuits.

A BPF passes only those L – R signals within the spectral band (4 to 15 kHz) and applied signals to the spectral rms detector (which detects the signal amplitude). Simultaneously, a second BPF passes those L – R signals within the wideband (100 Hz to 4 kHz) and applies the signals within this band to the wideband rms detector (which also detects the signal amplitude). The complete L – R audio is deemphasized by a fixed 73-μs deemphasis network. The deemphasized L – R signal is then applied to the spectral *voltage-controlled amplifier (VCA)*. Notice that in some dbx ICs, the VCAs are called *current-controlled amplifiers (CCAs)* just to confuse you.

The spectral-VCA gain is controlled by the detected level of the spectral rms detector. The signals from 4 to 15 kHz are either compressed or expanded (depending on the respective amplitude) by varying the spectral-VCA gain. The L – R signal with spectral compression is then directed to the wideband VCA.

The wideband-VCA gain is controlled by the wideband rms detector. The signals from 100 Hz to 3 kHz are compressed by varying the gain to the wideband VCA. The L – R signal from the wideband VCA is deemphasized by a fixed 300-μs deemphasis network, and is applied to the matrix circuit for mixing with the L + R audio signal.

dbx deemphasis and BPF As shown in Fig. 3-7, the L – R audio from Q18 is applied to the spectral VCA at pin 18 of IC4. Q42 and the associated parts form the fixed 73-μs deemphasis network. Q41 and the associated parts form the spectral BPF. L – R signals in the range from 4 to 15 kHz are applied to the spectral rms detector at pin 20 of IC4. L – R signals in the range from 100 Hz to 3 kHz are applied to the wideband rms detector at pin 3 of IC4.

dbx processing Two rms detectors within IC4 receive power from a constant-current generator (also within IC4). The generator output is adjusted by VR6 (the so-called *L – R timing control*), connected at pin 1 of IC4. The setting of VR6 determines the amount of control output from the rms detectors for a given signal amplitude.

Spectral processing is performed by the spectral VCA at pin 18 of IC4, controlled by the spectral rms detector output. The L – R signal is then applied to an L – R amplifier IC5 and an op amp within IC4. VR7 sets the gain of the op amp and is sometimes called the *variable deemphasis (VD) control*. The amplified L – R signal is then applied to another op amp within IC4. This op-amp output at pin 8 of IC4 is applied to Q21 in the matrix circuit. C53 and R13, between pins 8 and 7 of IC4, produce the required 390-μs fixed deemphasis.

3.3.8 Typical stereo TV tests and adjustments

Compare the following adjustment procedures with those shown in the service literature for the stereo TV or VCR being serviced. The test and adjustment points are shown in Figs. 3-7 and 3-8. The following notes supplement the specific test/adjustment procedures in this section.

The L, R, L + R and L – R modes should produce equal audio outputs in the TV or VCR. That is, the audio levels in the left and right channels should be equal for all modulation modes. The L + R or L – R mode should produce equal audio level in each channel. Similarly, the L or R mode should produce the same level in the active channel as in the L + R or L – R modes (this can be used in troubleshooting).

As an example, under a given set of conditions (same measurement points, same audio frequencies) the L and R channels should produce the same audio output voltage. If they do not, an unbalance condition is indicated. This unbalance condition is not necessarily limited to the stereo-decoder circuits, but could as likely be in the audio-amplifier circuits.

The SAP function should produce equal SAP audio-output levels at all frequencies. The SAP function is monaural in our generator, but has dbx encoding. This is not necessarily the case with a TV-station signal, where both stereo and SAP can be broadcast simultaneously (even though you can listen to only one at a time).

L – R adjustment The purpose of the L – R IC1 is to reinsert a carrier into the AM stereo L – R sidebands at pin 2, and produce corresponding audio output at pin 5. The missing L – R carrier is at 31.468 kHz and is produced by a VCO within the L – R decoder. Because there is no L – R carrier at the decoder output, the VCO is usually locked to the 15.734-kHz pilot (or a multiple).

In the circuit of Fig. 3-7, the VCO can be set by adjustment of PLL adjustment VR1 at pin 16 of IC1 and can be monitored at TP31. The VCO signal at TP31 is 15.734 kHz, even though the VCO operates at a different frequency (typically higher).

1. Ground the input to the stereo decoder at pin 2 of the SS connector. This removes all signals to the L – R decoder input, including the 15.734-kHz pilot. If the pilot is present, it is possible that the VCO will lock onto the incoming signal even though the VCO is not exactly on frequency when free-running. Check that stereo indicator D521 is off.
2. Connect a frequency counter (chapter 4) to TP31, and adjust VR1 for a reading of 15.734 kHz. Remove the ground from pin 2 of the SS connector. Check that the TP31 reading remains at 15.734 kHz.

L – R gain adjustment The purpose of this adjustment is to set the level of the stereo L – R signal before dbx NR processing. This is not to be confused with the L – R separation adjustment that sets the L – R level after dbx processing.

1. Apply a modulated L – R signal to pin 2 of the SS connector (using a stereo generator). The stereo indicator D521 should turn on. Use a modulation frequency of 300 Hz (unless otherwise specified in the service literature).
2. Monitor TP37 for the 300-Hz signal, using an audio voltmeter or scope. Adjust VR2 for the correct voltage level at TP37. Always use the values specified in service literature (for both input amplitude at pin 2 of the SS connector and output at the test point).

SAP level adjustment The purpose of this adjustment is to set the level of the SAP signal (before dbx processing) in relation to the stereo L – R signal. If the SAP and L – R signals are not approximately equal at this stage in the audio path, the user audio output or volume control must be reset each time when switching between stereo and SAP.

1. Apply a modulated SAP signal to pin 2 of the SS connector. The SAP indicator D520 should turn on. Use a modulation frequency of 300 Hz, unless otherwise specified.
2. Monitor TP38 for the 300-Hz signal. Adjust VR3 for the correct voltage level at TP38 (using service-literature values). The SAP level adjustment is usually performed after adjustment of the SAP detector (check the service literature).

SAP detector adjustment The purpose of the SAP signal detector (IC3) is to produce a low at output pin 7 when the 78.67-kHz SAP carrier is present at input pin 6 (output pin 7 remains high when the SAP carrier is not present). This is done by comparing the incoming SAP carrier with the IC3 decoder VCO. When both signals are present and locked in frequency and phase, pin 7 goes low. The VCO is set by adjustment of SAP PLL adjust VR4 at pin 1 of IC3.

1. Apply a 78.67-kHz signal to pin 2 of the SS connector. Monitor the voltage across TP35 and TP36. Set VR4 to full counterclockwise. The SAP indicator D520 should be off. The voltmeter should read about –1.0 V.

2. Adjust VR4 clockwise until the voltage at TP35 and TP36 changes to about +1.0 V. This indicates that pin 7 of IC3 is switched to low. You could measure at pin 7 of IC3 or at TP34, but the change in signal level is much more difficult to detect. Also, by checking at TP35/TP36, you also confirm operation of Q13, Q14, and Q15 simultaneously.

SAP detector signal output-level adjustment The purpose of VR5 is to set the level of the SAP detect signal (this signal appears when the SAP carrier is present, and pin 7 of IC3 goes low).

Make certain that no SAP carrier signal is present. On some circuits, it might be necessary to ground the SAP circuit input at pin 2 of the SS connector. Monitor the voltage across TP35 and TP36. Adjust VR5 until the voltage at TP35 and TP36 is – 1.0 V.

Stereo noise-reduction time-constant adjustment The purpose of VR6 is to set the amount of control output from dbx NR IC4 for a given L – R signal amplitude. Connect a digital voltmeter to TP39 and TP40. Adjust VC6 for the correct voltage across R5. Use the service-literature values.

Although you are measuring voltage across R5, VR6 is set for a given current through circuits within IC4 (the more current, the more control). Typically, R6 is set for a current of 15 mA at pin 13 of IC4. Because R5 is 1000 Ω, the voltage reading should be 15 mV.

Stereo-reduction VD adjustment The purpose of VR7 is to set the L – R gain, after spectral processing by IC4, but before wideband processing, to produce the *desired variable deemphasis (VD)*. This is sometimes called the *wideband* or *high-band VD adjustment*.

1. Make certain that no L – R signal is present. On some circuits, this can be done by grounding the input at pin 2 of the SS connector. In other circuits, it is necessary to disable the L – R audio path (such as connecting the emitter to Q18 to +9 or +12 V).
2. Apply a 300-Hz signal to TP37, using the correct level specified in the service literature. A typical input level at TP37 is –24 dB.
3. Monitor the 300-Hz signal (after noise reduction) at TP43. Make certain that the level at TP43 is within the limits specified by the service literature. The typical 300-Hz output level at TP43 should be between –23 and –35 dB. Note the actual level at TP43.
4. Change the frequency of the audio signal applied at TP37 from 300 Hz to 3 kHz. Set the amplitude of the 8-kHz signal, as specified in service literature. A typical input level at TP37 (with the 8-kHz audio) is –17 dB.
5. Adjust VR7 so that the 8-kHz signal (after noise reduction) at TP43 is as specified in the service literature. A typical 8-kHz output level at TP43 is the actual 300-Hz level (as measured in step 3) less –11 dB.

Remember that these noise-reduction adjustments are critical to proper operation of the stereo-TV circuits. Also, the service literature generally recommends that the VD adjustment be performed before the separation adjustment.

SAP noise-reduction time-constant adjustment The purpose of VR8 is to set the amount of control output from dbx NR IC6 for a given SAP signal amplitude.

Connect a digital voltmeter to TP41 and TP42. Adjust VR8 for the correct voltage level across R26 (at TP41 and TP42). Use the service literature values.

Note that although you are measuring across R36, VR8 is set for a given current through the circuits within IC6 (the more current, the more control). Typically, VR8 is set for a current of 15 mA at pin 13 of IC6. Because R26 is 1000 Ω, the voltage reading should be 15 mV.

SAP noise-reduction VD adjustment The purpose of VR9 is to set the SAP gain, after spectral processing by IC6, but before wideband processing, to produce the desired variable deemphasis. The procedure is the same as for adjustment of VR7 (stereo VD adjustment), except that TP38 and TP44 are used instead of TP37 and TP43.

L – R separation adjustment The purpose of this adjustment is to set the level of the stereo L – R signal in relation to the mono L + R signal. Both signals are combined in the Q22 matrix (R59 and R60). If the L – R signal is low in relation to L + R, you will hear only mono. If the L – R is high in relation to L + R, you will hear both signals, but there will be poor separation between the left and right audio (the audio sounds like mono even though stereo is present).

1. Apply a modulated L – R signal to pin 2 of the SS connector. The stereo indicator D521 should turn on. Use a modulation frequency of 300 Hz unless otherwise specified.
2. Monitor TP32 and TP33 for the 300-Hz signal. Adjust VR10 for the correct voltage level at TP32 and TP33. Always use the values specified in service literature (for both input amplitude and output at the test points). Remember that this adjustment can be critical in producing good stereo sound.
3. If you cannot find a separation adjustment procedure in the service literature, use the following as an emergency procedure only.
4. Set VR10 so that L – R is zero (generally, this means setting VR10 full counterclockwise). Apply an L + R signal with 300-Hz modulation at pin 2 of the SS connector. Note the voltmeter reading at TP32 and TP33. This is the mono L + R signal level.
5. Remove the L + R signal and apply L – R with 300-Hz modulation at pin 2 of the SS connector.
6. Adjust VR10 until the readings at TP32 and TP33 are the same as with the L + R (or just below L + R) in step 4.

For a final test, measure stereo separation, as described next. As a practical matter, some technicians recommend adjusting VR10 for a given amount of separation or for maximum separation at the speakers, rather than for a given reading in the decoder circuits. A stereo separation of 60 dB or better is possible on some circuits.

Stereo separation tests Virtually all stereo TV/VCR decoders require a stereo separation test and a stereo indicator (pilot) test. Compare the following procedures with those shown in (or omitted from) the service literature.

1. A pilot signal must be present for all stereo tests. Operate the generator controls as necessary for a pilot signal, and check that the stereo indicator turns on.

2. Before making any stereo tests, make certain that the stereo function is selected on the TV set or VCR.
3. Select the desired audio-modulating frequency of 300 Hz, 1 kHz, or 8 kHz using the appropriate generator controls.
4. Select the desired form of modulation (L, R, L – R, or L + R), using the corresponding generator controls. Many stereo tests are conducted using L (left), followed by a repeat of the test with R (right) modulation. Often, this is followed with a test using L – R modulation.

On some generators, the controls are interlocked, so you can select only one of the four modulation signals at a time. No matter what controls are involved, remember that if you select L + R, there is no stereo operation at the decoder, even though the pilot is present and the stereo indicator is on.

To measure stereo separation, select L (with the pilot on). The stereo indicator should turn on and sound should be heard from the left speaker. Monitor the left-channel speaker (or left-channel audio, whichever is convenient). Measure the left-channel output voltage. Select R (with a pilot). The stereo indicator should remain on, and sound should be heard from the right speaker.

Measure the right-channel output voltage, and compare the right-channel voltage to the left-channel voltage. The ratio of the left-channel output to that of the right-channel is the stereo separation and is usually expressed in dB.

Generally, the actual voltages are not critical. It is the voltage ratio that counts. Of course, both voltages (L and R) must be measured at the same point in the audio path (typically at the L and R speakers or outputs), and both must be measured using the same input voltage level and frequency.

When the pilot is present (stereo on), the channel outputs might (or might not) be equal. However, with the pilot off (stereo off), the channel outputs should be substantially the same, no matter what modulating signal is used. Compare the measured stereo separation to that given in the TV or VCR specifications.

3.4 Color generators

Color generators are essential when testing and servicing any type of color equipment (TV, monitor, computer terminal, VCR, camcorder, videodisc player, CDV, etc.). Color generators produced color signals that are equivalent to the test signals used in TV broadcast studios. At one time, the keyed-rainbow generator was popular for color TV service. Today, the NTSC color-bar generator has all but replaced the rainbow generator.

3.4.1 NTSC color-bar generator

An NTSC generator produces standard EIA colors at the NTSC-prescribed *luma level* (brightness), *chroma phase angle* (hue or tint), and *chroma amplitude* (saturation). Generally, the most useful signal produced by an NTSC instrument is the standard NTSC bar pattern with a –IWQ signal occupying the lower quarter of the pattern as shown in Fig. 3-10.

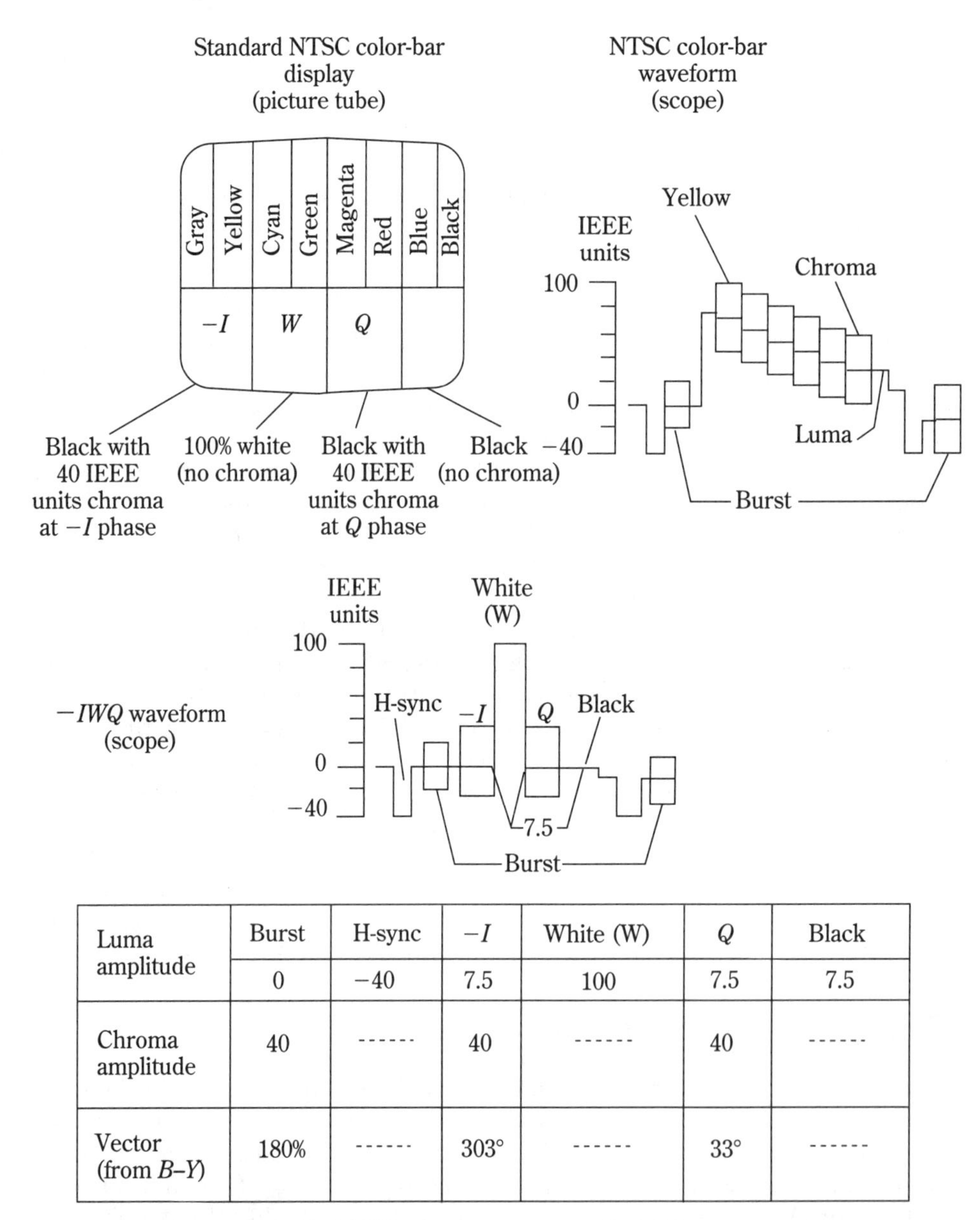

Luma amplitude	Burst	H-sync	−*I*	White (W)	*Q*	Black
	0	−40	7.5	100	7.5	7.5
Chroma amplitude	40	------	40	------	40	------
Vector (from *B–Y*)	180%	------	303°	------	33°	------

3-10 NTSC color-bar pattern with –IWQ.

Figure 3-10 shows both the pattern (as produced on a TV screen when the NTSC signal is applied to the antenna input) and the waveforms (as produced on a scope connected to various points in the TV circuits (for one horizontal line of the pattern). Also, the –IWQ waveform is generated for the bottom 25 percent of the display, and the color-bar waveform is on for the top 75 percent. When using a scope, the two waveforms can be superimposed as described in Sec. 3.4.3.

A typical NTSC generator also produces a number of other patterns, including various convergence, linear staircase, and full-color rasters. Use of these patterns is summarized throughout the remainder of this chapter.

3.4.2 NTSC color-bar pattern

The basic pattern in Fig. 3-10 is the one most often used to test, troubleshoot, and adjust video equipment. Analysis of the pattern shown on the screen of a TV under test, or a pattern produced by playback of a VCR or camcorder, can often localize a problem to a few specific circuits. Here are a few examples.

Overall performance tests An overall performance test of a VCR or camcorder can be conducted by recording the NTSC color-bar pattern, then playing the pattern back on a video monitor or a known good TV set. There should be little difference between the video played back from the VCR or camcorder and an NTSC pattern applied directly to the monitor or TV.

Luma and chroma proportions In a VCR, the luma and chroma signals are separated during the recording process and recombined during playback. If luma and chroma signals are not maintained at the proper proportion when separated, color distortion results, particularly in the vividness of colors (color saturation).

Waveforms can be examined throughout the VCR circuits for proper luma-to-chroma proportions. Figure 3-11A shows the elements of an NTSC color TV signal, and Fig. 3-11B shows the luma-to-chroma proportions. As an example, the magenta bar occurs when the luma is 36 IEEE units and the chroma amplitude is 82 units, at a vector of 61° (using B-Y as a 0° reference). Both amplitudes and vectors can be monitored using a vectorscope, as discussed in Sec. 3.5.

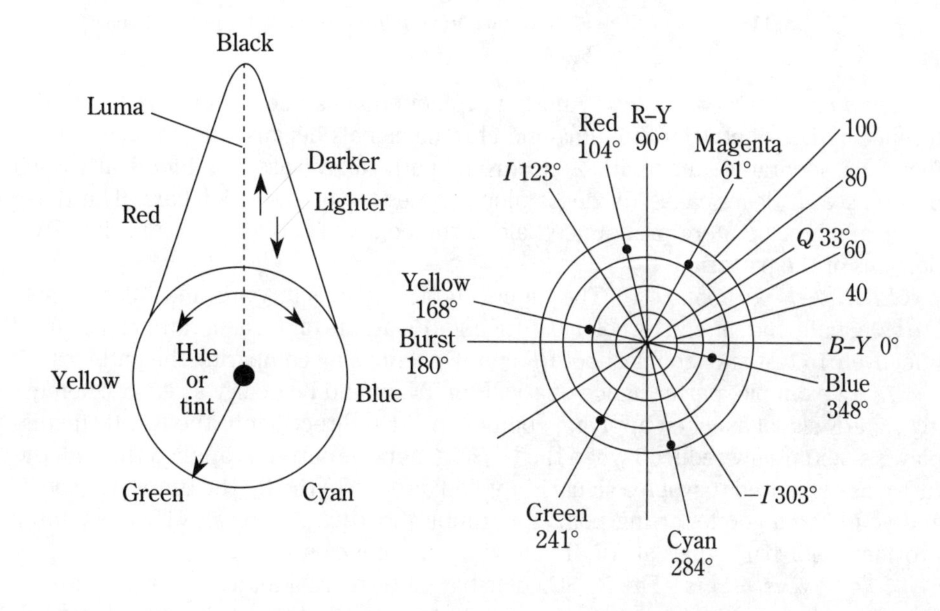

3-11A Elements of an NTSC color TV signal.

	H-sync	Burst	Gray 75% white	Yellow	Cyan	Green	Magenta	Red	Blue	Black
Luma IEEE units	−40	0	77	69	56	48	36	28	15	7.5
Chroma IEEE units	- - -	40	- - -	62	88	82	82	88	62	- - -
Vector (from $B-Y$)	- - -	180°	- - -	167°	284°	241°	61°	104°	347°	- - -

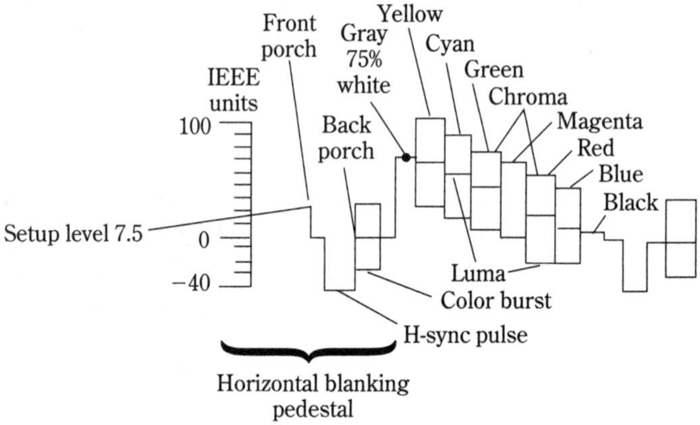

3-11B Composite video signal showing luma-to-chroma proportions.

Luma and chroma delay Another problem that can be found in VCRs is a difference in delay between the luma and chroma signals because of some circuit defect (the separated luma-path and chroma-path delays are not identical). Such differences in delay cause fuzziness along the edges of the color bars. The delay problem might be more pronounced along the edges of the white bar in the –IWQ portions of the pattern.

RF and IF performance The tuners and IF sections of VCRs and color TV sets are essentially the same. The color-bar RF and IF outputs of the generator can be used effectively to test and troubleshoot RF and IF sections by comparing the patterns.

As an example, performance of a VCR or TV should be nearly as good when using the RF signal as when applying composite video directly into the RF. If the display is substantially reduced when the NTSC generator output is applied through the tuner, as compared to when a signal is applied directly to the IF, the tuner is suspect, and you have a good starting point for troubleshooting. A typical NTSC generator provides color-bar signals at RF, IF and video frequencies.

Color adjustments The NTSC color-bar pattern provides a standard reference for color adjustments. The pattern contains bars of the three primary colors: red, blue, and green. These are a good reference for checking 3.58-MHz phase problems.

White, yellow, cyan, and magenta help define problems when the color mixes are not in the correct proportions.

In some troubleshooting applications, it is helpful to know if the problem is chroma or luma related. Many generators are provided with a chroma-off function, which produces the bars and waveforms shown in Fig. 3-12A (normal bar pattern, but without color.)

A TV set can be checked by comparing the displays of Figs. 3-10 and 3-12A. If the display shown in Fig. 3-12A is good, but the one in Fig. 3-10 is not, the problem is in the chroma circuits. In Fig. 3-12A, the chroma-off waveform (scope) is generated for the top 75 percent of the pattern, and the –IWQ waveform (scope) is generated for the bottom 25 percent of the display. The two waveforms are superimposed on the scope display.

Phase lock of the 3.58-MHz oscillator The 3.58-MHz oscillator of a TV or VCR should be locked in phase whenever the color burst is applied. This can be checked using the top-burst-off function available on many NTSC generators. When the top-burst-off function is selected, the color is unsynchronized for the top quarter of the pattern, but there should be no delay or color distortion where the color bars start (middle of the pattern), as shown in Fig. 3-12B.

In Fig. 3-12B, the bars with the burst-off waveform are generated for the top 25 percent of the pattern, the bar waveform is generated for the middle 50 percent, and the –IWQ waveform is generated for the bottom 25 percent. The three waveforms are superimposed on the scope display.

Color-killer function Most TV sets and VCRs have a color-killer circuit that disables the color functions where there is no color burst present. This can be checked using the full-burst-off function available on some NTSC generators. When the full-burst-off function is selected, the bars should appear (in various shades of gray), but there is no color, as shown in Fig. 3-12C. Also, the bars occupy the top 75 percent, and the –IWQ appears on the bottom 25 percent. If there is any color on the picture-tube screen with the full-burst-off function in operation, the color-killer circuit is not operating properly.

Audio or sound functions A 1- or 3-kHz audio signal can be added to the RF output of many generators. In a properly operating TV or VCR, this audio modulation should not affect the normal display. For example, if you select the pattern of Fig. 3-10, there should be no change in the display when the audio modulation is applied or removed. If you note some interference (typically diagonal lines across the display) when audio modulation is applied, sound is leaking into the chroma circuits. Possibly one or more of the sound traps within the TV or VCR are not properly adjusted. The 1- and 3-kHz signals can also be used to test the sound or audio circuit of the TV or VCR.

3.4.3 –IWQ patterns

As shown in Figs. 3-10 and 3-12, the basic –IWQ pattern appears on a split field with the NTSC color-bar pattern (with –IWQ on the lower quarter). When viewing horizontal lines of video (waveforms) on a scope, both the bars and the –IWQ signals are superimposed, which is very handy in many applications. For example, the 7.5 percent black level (right-hand side) and the 100 percent level (second from left) of the –IWQ pattern are the key luma-amplitude references.

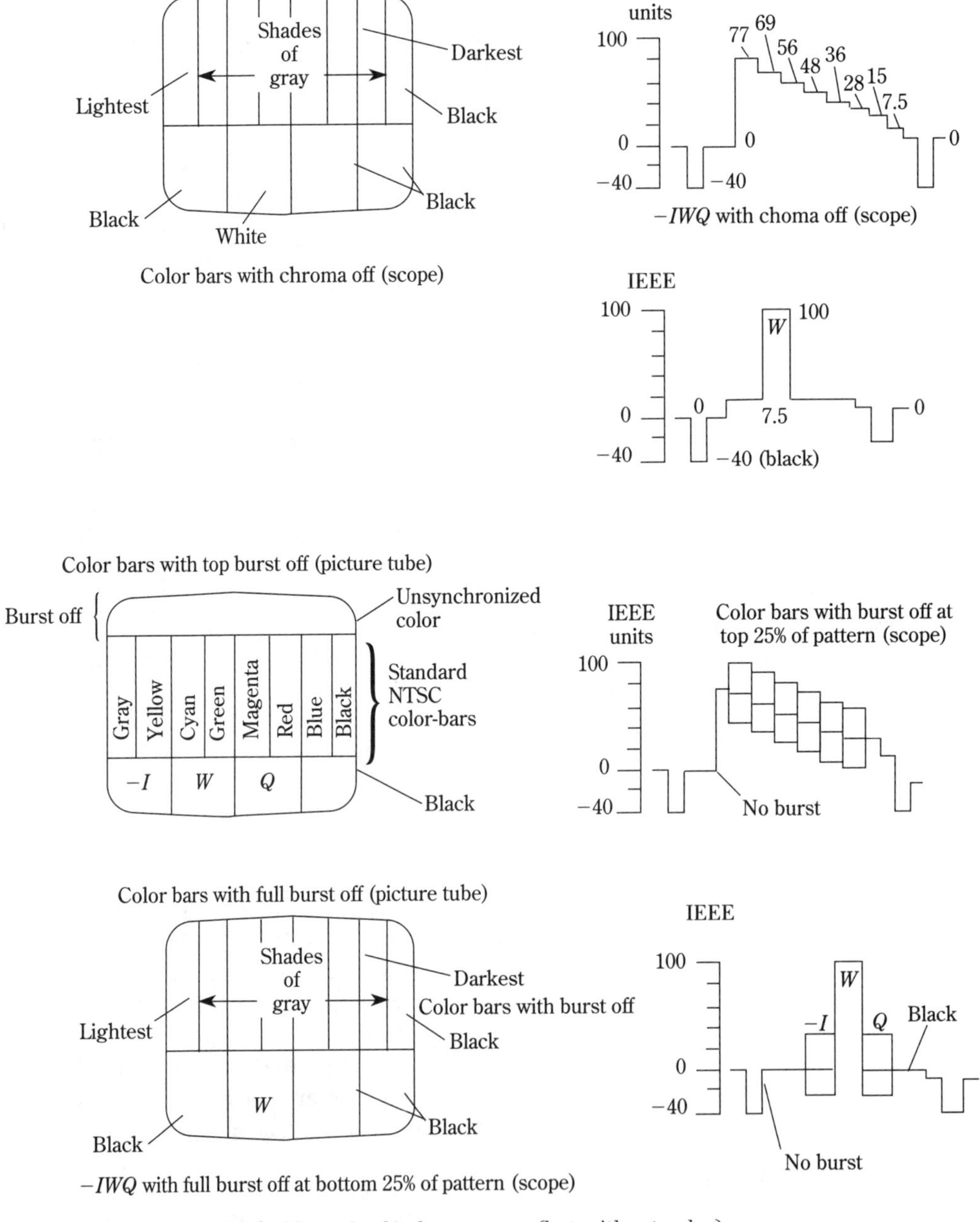

3-12 Normal color-bar pattern (but without color).

These black-and white-level references are used for FM-deviation adjustment and black-clip and white-clip level adjustments in VCRs. The black and white levels are also used whenever luma and chroma ratios are being adjusted or checked during troubleshooting. The –I and Q signals are used primarily in video cameras and studio equipment for setting up the phase and amplitude of the –I and Q signals and maintaining the proper relationship between the two.

3.4.4 Staircase patterns

Figures 3-13 and 3-14 show the various staircase patterns and waveforms available with many generators. When the pattern in Fig. 3-14C is selected, the waveforms in Figs 3-14A and 3-14B are generated for the top 25 percent of the pattern (no burst), and the waveform in Fig. 3-13A is generated for the bottom 75 percent. Both waveforms are superimposed on a scope display.

Amplifier linearity checks The staircase pattern contains five equal steps of increasing luma with a constant chroma amplitude and phase. With the chroma off (Figs. 3-13C and 3-13D), only the luma steps are generated. This luma-only pattern is valuable for checking linearity in TV and VCR amplifier circuits (or any amplifier circuits that will pass the frequencies involved). The amplitude of each step (as measured on a scope) should be equal at the output of an amplifier or other circuit because the amplitude is equal at the input. Nonequal steps monitored at the output of a circuit represents distortion (nonlinear distortion).

Setting white-clip level in VCRs The staircase pattern is desirable for setting the white-clip levels in VCRs. Because the top step is 100 percent white level, it provides the correct reference for white-clip adjustments. If the top step shows less amplitude than other steps, this usually indicates incorrect adjustment of the white-clip level.

Frequency-equalization adjustments in VCRs The staircase pattern is also recommended for frequency-equalization adjustment in the record amplifier of VCRs. The FM signal, which carries luma information in a VCR, is shifted to a different frequency for each step of the staircase signal. However, the record current (the current applied to the tape recording heads in a VCR) should remain constant across the FM frequency band. The frequency-equalization adjustments of a VCR should be set so that the record current is equal for all steps of the staircase input.

Differential gain and differential phase lock The staircase patterns are most effective when checking both differential gain and differential phase. Excessive differential gain or phase can be the cause of color distortion in VCRs, color TVs, video monitors, and computer terminals. Both conditions are checked often in studio equipment that processes video signals.

Theoretically, the chroma amplitude should not change as the luma is varied from 0 to 100 percent. Any interaction is called *differential gain*. To check differential gain, the chroma circuit output is displayed on a precision waveform monitor while the staircase pattern is applied from the generator. As the luma signal steps from 0 to 100 percent, in 20 percent increments, any differential gain causes changes in chroma amplitude (with any or all of the luma steps).

Sometimes, the degree of differential gain can be affected by the peak-to-peak amplitude of the chroma signal. This characteristic can be checked by switching between a low-staircase (20 IEEE units) pattern, such as shown in Fig. 3-13B and a high-staircase (40 IEEE units) pattern (Fig. 3-13A). Differential gain (if any) is usually worse when the high-staircase patterns are used.

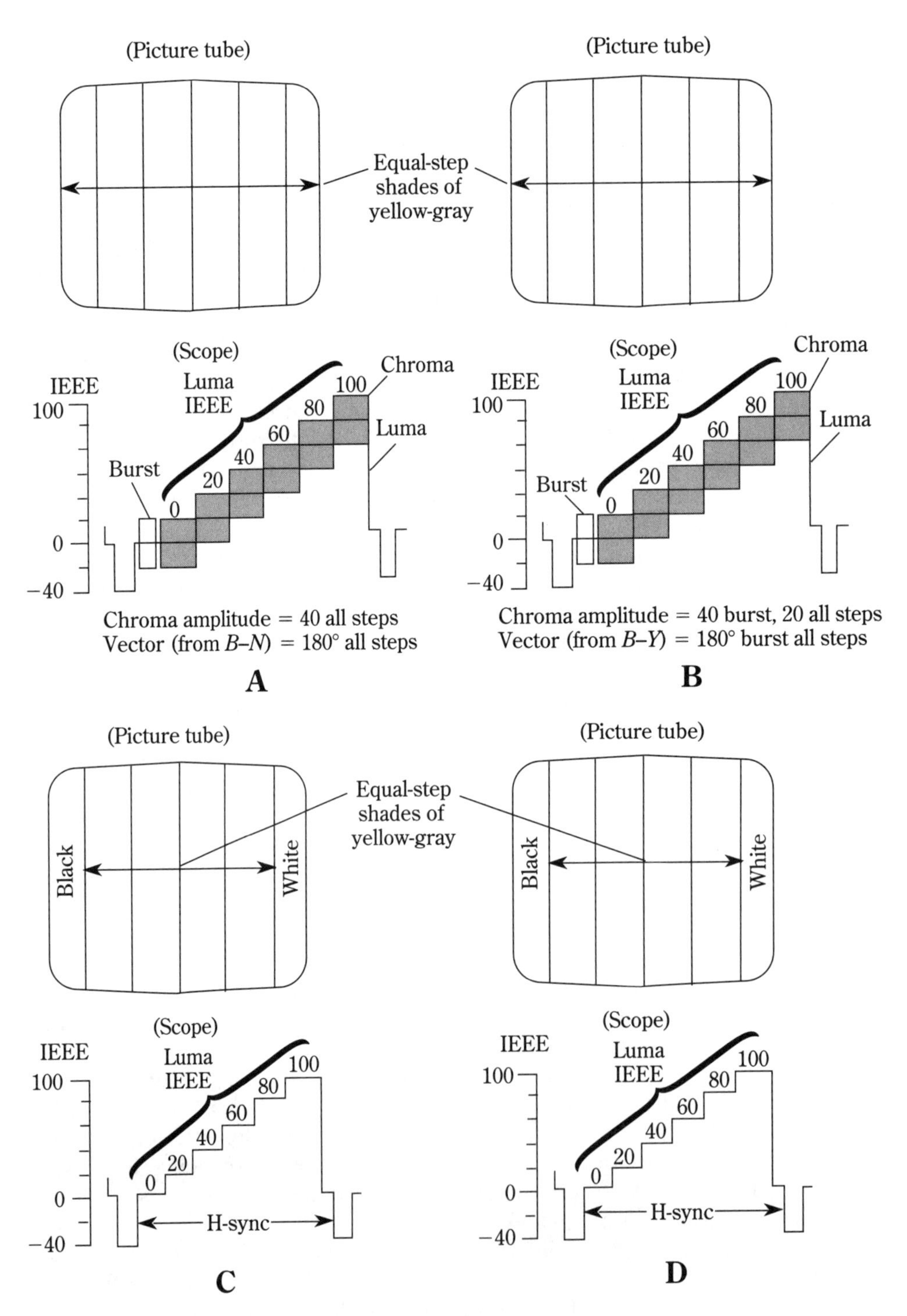

3-13 Staircase patterns.

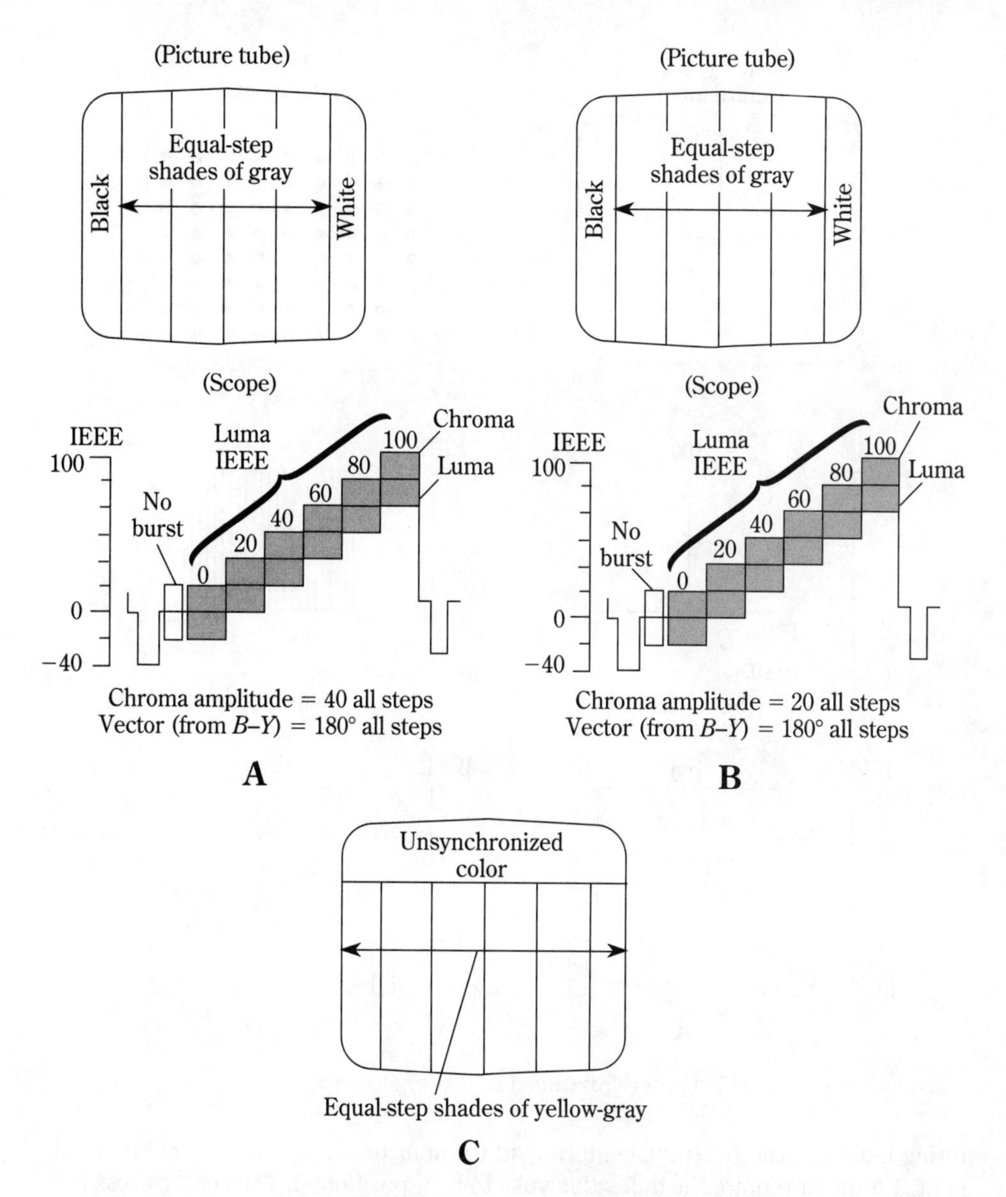

3-14 More staircase patterns.

3.4.5 Convergence patterns

Figures 3-15 and 3-16 show the various convergence patterns and waveforms that are available with many generators. The convergence patterns are used primarily for static and dynamic convergence of color TV, monitors, and terminals.

Center cross The center-cross pattern (Fig. 3-15A) should intersect at the center of the screen, and there should be no tilt of the horizontal line. Improper cen-

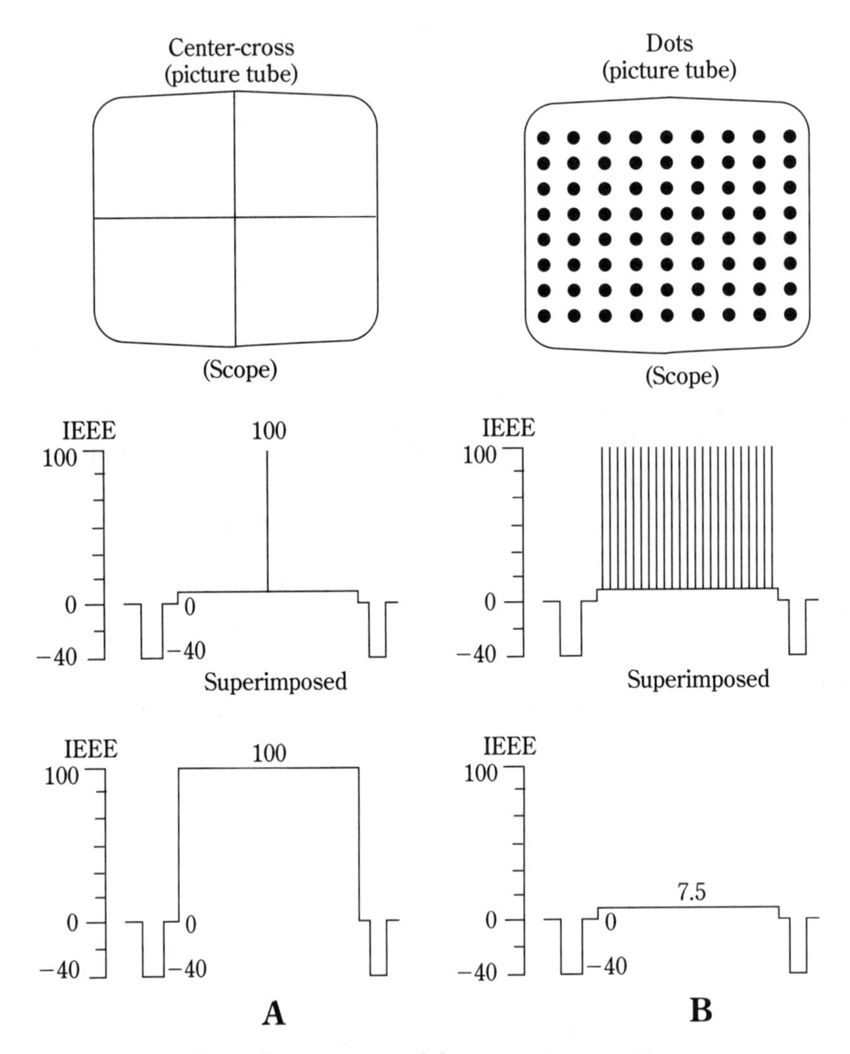

3-15 Center-cross and dots convergence patterns.

tering indicates the need for centering adjustment or a possible deflection-circuit fault. Tilt might require the deflection yoke to be repositioned. The center-cross pattern also provides a good general check of vertical and horizontal sync.

Dots pattern The dots pattern (Fig. 3-15B) is used for static convergence, usually by converging the center dot of the pattern. Most TV sets and some video monitors and terminals have overscan so that all dots are not visible, except possibly under low-voltage conditions. Some TV sets have a tendency toward a greater amount of overscan than other sets. Typically, it is desirable to display at least a 17- by 13-dot pattern.

Crosshatch pattern The crosshatch pattern (Fig. 3-16A) is normally preferred for dynamic convergence, although some technicians prefer the dots pattern for both static and dynamic convergence. The crosshatch pattern is used to

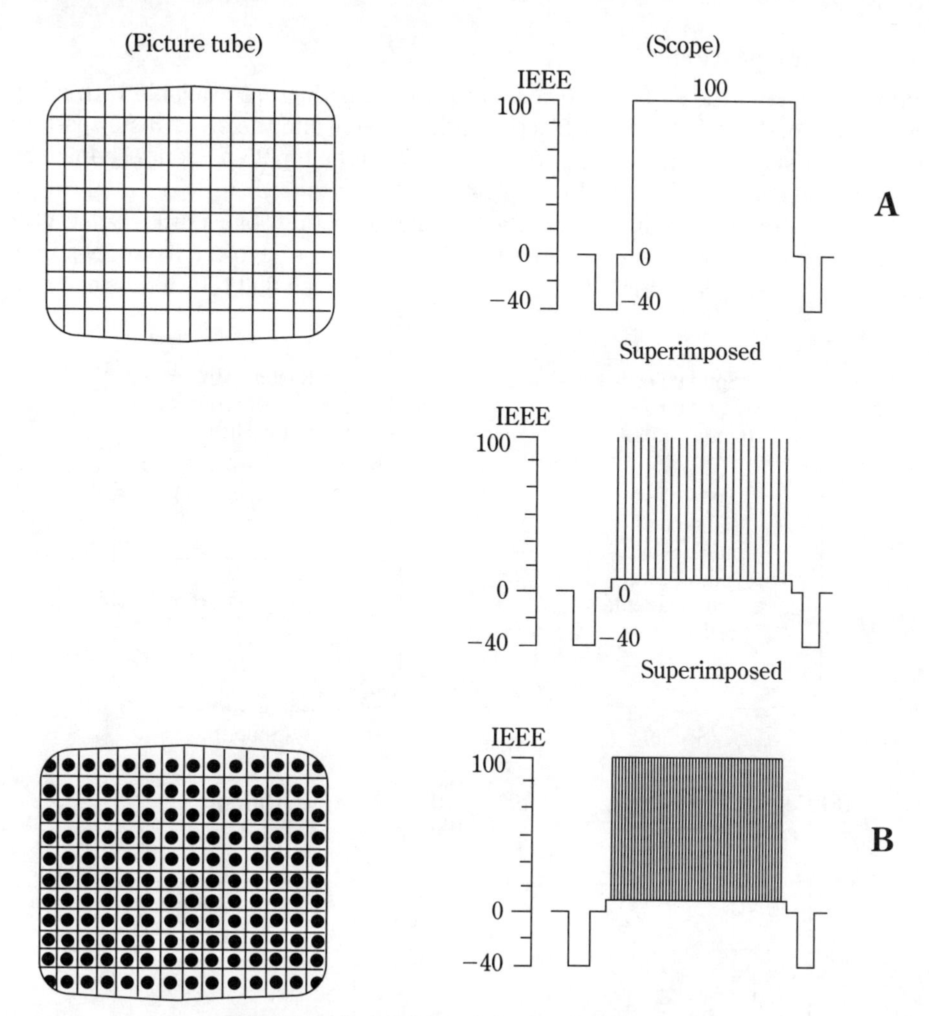

3-16 Crosshatch (a) and dot-hatch (b) patterns.

check both vertical and horizontal linearity. Each square of the crosshatch pattern should be the same size, which is a convenient reference for making linearity adjustments. The crosshatch pattern is also used to check so-called *pincushion distortion*, which sometimes appears at the outside edges of large picture tubes as bends in the lines.

Dot-hatch pattern The dot-hatch pattern (Fig. 3-16B) combines the dots and crosshatch patterns for a quick overall check of static and dynamic convergence, linearity, overscan, and pincushion distortion from a single pattern. There might be some slight amount of vertical jitter when monitoring the convergence patterns with some generators. This jitter is the result of interlaced scan and is normal for an NTSC generator (because the NTSC signal is interlaced). The jitter does not degrade the accuracy of the convergence adjustments and does not indicate a malfunction in vertical sync.

3.4.6 Raster patterns

Figure 3-17 shows the raster patterns and waveforms that are available with many generators. As shown, one raster fills the entire screen, and the other raster pattern (with top burst off) fills the bottom 75 percent (with the top 25 percent occupied by synchronized color).

Raster patterns are valuable for checking and adjusting color purity. Some VCR manufacturers recommend a raster pattern for setting the record current. Not only can the white raster be used in the standard manner (to see that the circuits can pro-

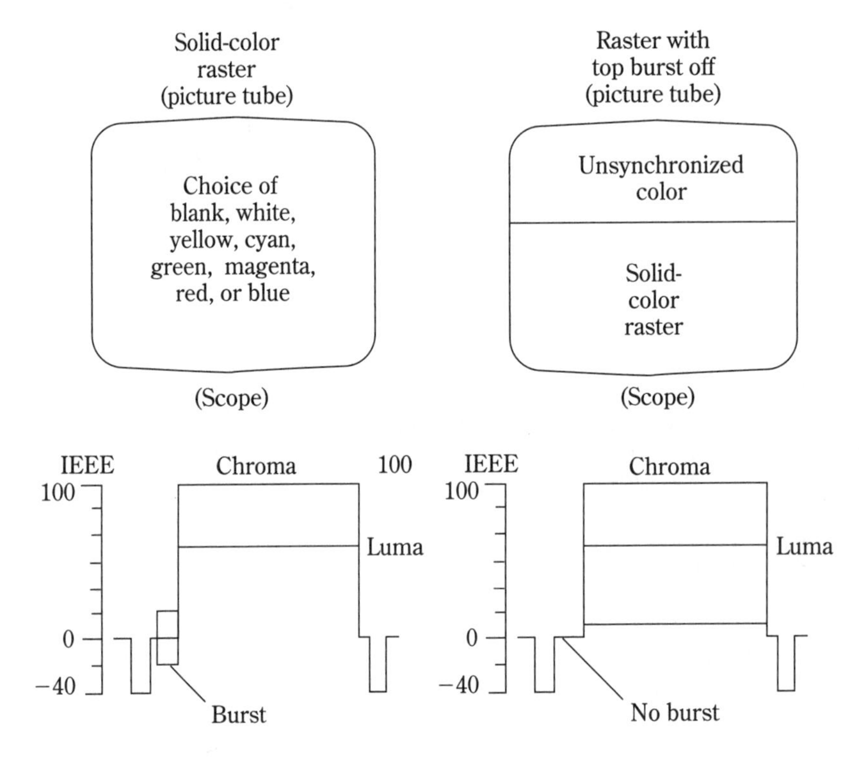

	Blank	White	Yellow	Cyan	Green	Magenta	Red	Blue
Luma (IEEE units)	7.5	77	69	56	48	36	28	15
Chroma (IEEE units)	0	0	60	86	80	80	86	60
Vector (from *B–Y*)	—	—	167°	284°	241°	61°	104°	347°

3-17 Raster patterns.

duce pure white), but the three separate guns of the picture tube can be individually adjusted (in some circuits) using a continuous chroma signal of red, blue, or green.

Analysis of hue and saturation problems can be simplified by analyzing each primary color, or the yellow, cyan, and magenta hues, which are the equal mixture of two primary colors without the third one. On some generators, a black-burst test signal is generated when none of the three primary colors is selected. The luma and chroma components for each raster color are identical to the corresponding bar from the color-bar pattern, but each can be selected individually for analysis.

3.4.7 Sync signals

Every pattern produced by an NTSC generator contains *NTSC sync pulses*. These sync pulses are the same as those produced by a TV broadcast station. The sync amplitude of 40 IEEE units is often the reference against which the remainder of the luma signal is compared. Circuits can be checked for sync clipping by monitoring the staircase pattern on a scope and checking to see if the sync-pulse amplitude remains at 0.4, compared to the 100-percent white step, which has a reference of 1.0.

AGC circuits, which respond to sync-pulse amplitude, can be checked with the composite-video output that is available on many generators. Although the overall amplitude of the video is varied, the sync-pulse amplitude is a constant percentage of the total video amplitude.

The precise timing of the sync pulses allows proper adjustment of the servo and switching circuits in VCRs. The servo circuits control the speed of the videotape in VCRs, while the switching circuits provide for switching between video heads (to allow for a continuous transition from field to field).

3.4.8 Using NTSC generators in color-circuit testing

This chapter does not describe every possible use of an NTSC color-bar generator in color-circuit test and adjustment. To do so is well beyond the scope of the book. If you want a thorough discussion of color-circuit tests, adjustments, and troubleshooting, read the author's best-selling *Lenk's Video Handbook* (McGraw-Hill, 1991). However, the following procedures describe how an NTSC generator can be used to check overall performance of color circuits for any color TV, monitor, or computer terminal.

1. Select a full color-bar display. Switch out any special color controls (such as "accutint"). Adjust the hue or tint and color controls to midrange. Adjust the fine-tuning (if any), brightness, picture, and contrast controls for the best color-bar pattern. The pattern should be similar to that in Fig. 3-10.
2. Pay attention to the magenta and cyan (bluish green) bars. Generally, if you can get good magenta and cyan with the hue or tint control, the color circuits are operating satisfactorily (and all of the remaining colors will be good). This is because magenta (red-blue) and cyan (blue-green) use the three primary colors, and prove that the three picture-tube color guns (red, blue, green) are functioning properly.

   Ideally, the hue or tint control should be near the midrange position when magenta and cyan are good, and should have enough range to shift the color so that both magenta and cyan are off-color. Typically, the magenta becomes red at one extreme of the hue/tint control, while the cyan goes from blue to green (or vice versa) at the other extreme.

3. If it is impossible to get a good magenta and cyan display, suspect that the automatic frequency and phase controls are not working properly.
4. To check accutint or similar functions that enhance flesh tones, switch the function on, and check the color bars. The pattern should change, with several of the bars taking on a more reddish color. The range of the hue/tint control is usually more restricted (only a slight shift of color at both extremes) with accutint on (for most sets). The color control might also be restricted so that the color cannot be taken completely out of the pattern.
5. To check color-sync action, turn the generator chroma-level control slowly to minimum. The color should become pale and finally disappear. Because some sets have an automatic color control (ACC) circuit, the rate of fading depends on the set. Most sets lose color sync just before the color disappears (diagonal lines run through the colors).

These conditions indicate normal operation of the color-sync circuits. However, if a slight reduction of the chroma amplitude causes color to fall out of lock, this indicates that the color-sync ability of the set might be inadequate. For a further check, turn the RF-level control to reduce general signal strength, and note the effect on color sync.

3.5 Vectorscopes

A *vectorscope* is a highly-specialized instrument used in TV broadcast work to test and adjust color demodulators. The vectorscope can also be used to judge the general condition of color TV circuits when used with the NTSC color-bar output of a color generator. A vectorscope measurement is often more precise for both testing and troubleshooting than merely observing the NTSC pattern on the picture tube.

The display of an NTSC vectorscope, with the NTSC color-bar pattern applied, should be a series of interconnected dots within specific areas on a special vectorscope graticule, similar to that shown Fig. 3-18. The pattern or signal to be displayed can be probes from anywhere in the composite video or 3.58-MHz color circuits. Because of the specialized nature of vectorscope, no general operating procedures can be given here. Always follow the vectorscope instruction manual for operating procedures, connections, and color-circuit test/adjustment procedures.

3.5.1 Vectorscope patterns without a vectorscope

If an NTSC vectorscope is not available, a good lab-type scope (chapter 2) can be substituted. The demodulated color signals (directly from the red and blue guns) can be used as X and Y inputs to the scope, as shown in Fig. 3-18A. This connection

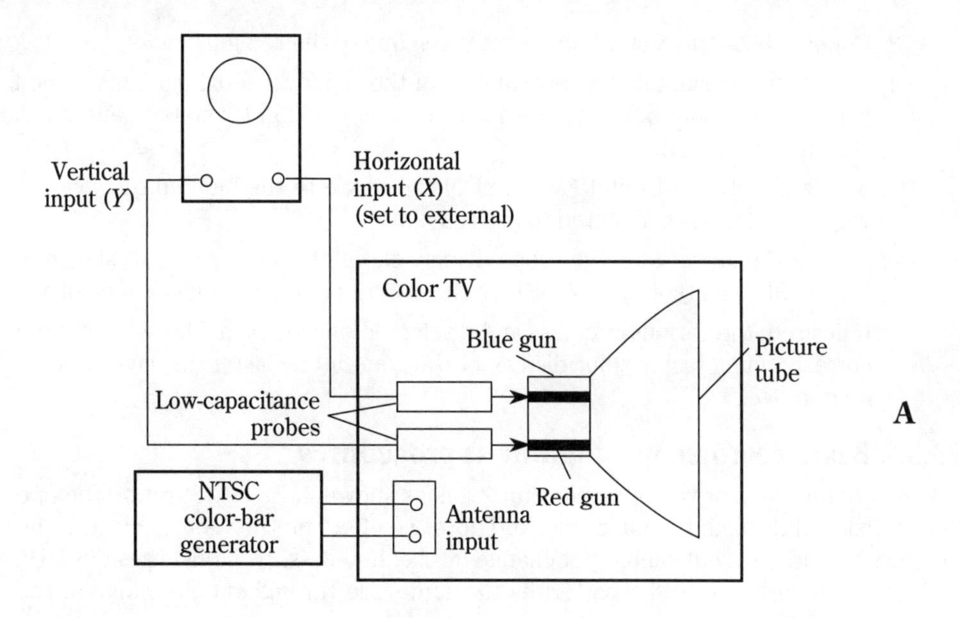

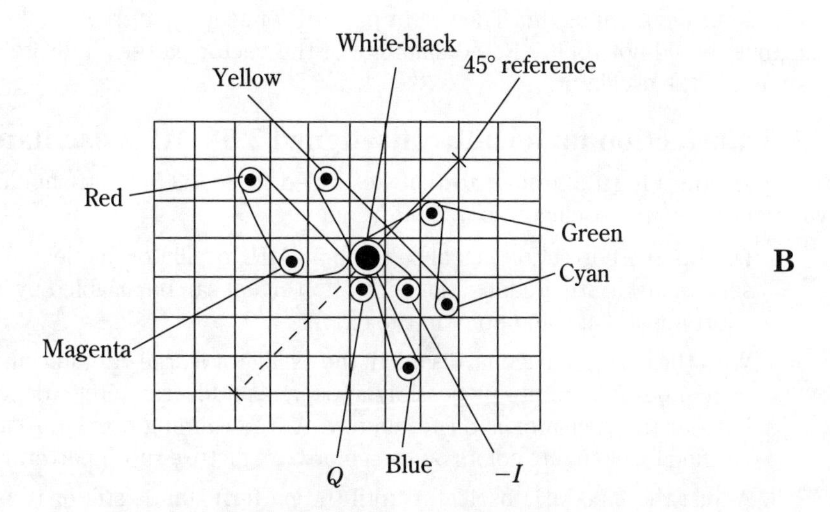

3-18 NTSC vectorscope connections and display.

should produce a scope display, as shown in Fig. 3-18B. The scope can be set up for vectorscope operation, as follows:

1. Select the NTSC color-bar pattern (Fig. 3-10) and apply the generator output to the antenna input of the TV set.
2. Set up the scope for X-Y operation. Adjust the position controls to center the dot on the screen with no signal input to the scope.

3. Connect both the X and Y inputs of the scope to the red gun.
4. Adjust the horizontal and vertical gain of the scope to equal amounts, which will move the spot or dot to the 45° reference position (from the center), as shown in Fig. 3-18B.
5. Now move the horizontal (X) input of the scope to the blue gun. Leave the vertical (Y) input connected to the red gun.
6. For a 90° picture tube, the scope display should be similar to that shown in Fig. 3-18B. The typical 105° picture tube produces a more elliptical display.
7. If desired, the various vectors can be selected one at a time, by selecting the corresponding raster-color display of the generator. Raster displays are covered in Sec. 3.4.6.

3.5.2 Basic vectorscope alignment procedures

With either a scope or vectorscope connected, as shown in Fig. 3-18, rotate the hue or tint control through the full range, and note the effect on the vector pattern. The pattern should turn, but should not change in size. If necessary, adjust the 3.58-MHz reference oscillator until the pattern is the same size throughout the range of the hue or tint control.

As an alternative, set the hue or tint control to midrange and adjust the 3.58-MHz oscillator so that the color signal stays in sync as the chroma-level control on the generator is turned to minimum. The vector pattern (or scope pattern) should reduce in size, but should not rotate. Rapid spinning of the vector pattern indicates a misadjustment of the oscillator.

3.5.3 Correction for a badly misaligned 3.58-MHz oscillator

This procedure is an alternate to that described in Sec. 3.5.2 and should be used when the 3.58-MHz oscillator is way off frequency.

1. Disable the correction signal to the 3.58-MHz oscillator, as described in the service literature. Usually, the correction signal can be disabled by a bias or a short applied at some point in the circuit.
2. With the correction signal disabled, the oscillator is free-running, and the pattern appears to rotate or possibly appears as a blurred circle, depending on how far the oscillator is off frequency. A free-running oscillator can also be verified by changing colors on the color-bar (picture tube) pattern.
3. Adjust the 3.58-MHz oscillator until the pattern stands still or is as close as possible to a motionless condition and the colors change slowly on the color-bar pattern. Then, restore the correction signal to the oscillator.

3.5.4 Troubleshooting with vectorscope patterns

It is possible to check operation of the color circuits by comparing the display in Fig. 3-18B with the correct values shown in Fig. 3-11. That is, you can check each color for correct angle and amplitude. However, this is best done with a vectorscope (using the special vectorscope graticule marked in amplitude and degrees), and is not too practical with a scope (even with a good lab scope).

Of course, any drastic deviations of the vectorscope pattern from that shown in Fig. 3-18B indicate a major problem in the color circuits. Here are some examples:

A loss of R-Y signal causes a loss of vertical deflection, and the vectorscope pattern changes to a straight line. If the B-Y signal is good, the beam is deflected along the horizontal axis. This indicates that the trouble lies in the R-Y demodulator, matrix, or difference amplifier, depending on the circuits. If the R-Y signal is weak, some deflection of the vectorscope pattern occurs, but the pattern is extremely distorted.

A loss of B-Y signal results in no horizontal deflection, and the vectorscope pattern changes to a straight line (a vertical line if the R-Y signal is good). This indicates that the problem is in the B-Y difference amplifier, matrix, or demodulator, depending on circuits. Again, if B-Y is weak, some distortion occurs, but the vectorscope pattern is extremely distorted.

If there is a complete loss of color, the vectorscope pattern usually appears as a center dot, or possibly as a short line in the center of the pattern. Remember that not all circuits produce identical vectorscope patterns. The patterns covered here are for reference only, and must be considered as typical. Always consult the vectorscope instruction manual and all service literature.

4
Electronic counters and frequency standards

There are two basic types of frequency-measuring devices for electronic testing. These include the *heterodyne* (or *zero-beat*) *frequency meter* and the *digital electronic counter*.

4.1 Heterodyne, or zero-beat, frequency meter

In the early days of radio communications, the heterodyne meter (Fig. 4-1A) was the only practical device for frequency measurement of transmitter signals. The signals to be measured are applied to a mixer, together with the signals of a known frequency (usually from a variable-frequency oscillator in the meter). The meter oscillator is adjusted until there is null or "zero beat" on the output device, indicating that the oscillator is at the same frequency as the signals to be measured. The frequency is then read from the oscillator frequency-control dial. Precision frequency meters often include charts or graphs to help interpret frequency-dial readings so that exact frequencies can be pinpointed.

4.2 Electronic digital counter

The electronic counter has all but replaced the heterodyne meter for all types of electronic testing. One reason is that the counter is generally easier to operate and has much greater resolution, or readability. Using the counter, you need only connect the test leads to the circuit or test point, select a time base and attenuator/multiplier range, read the signal frequency on a convenient digital readout.

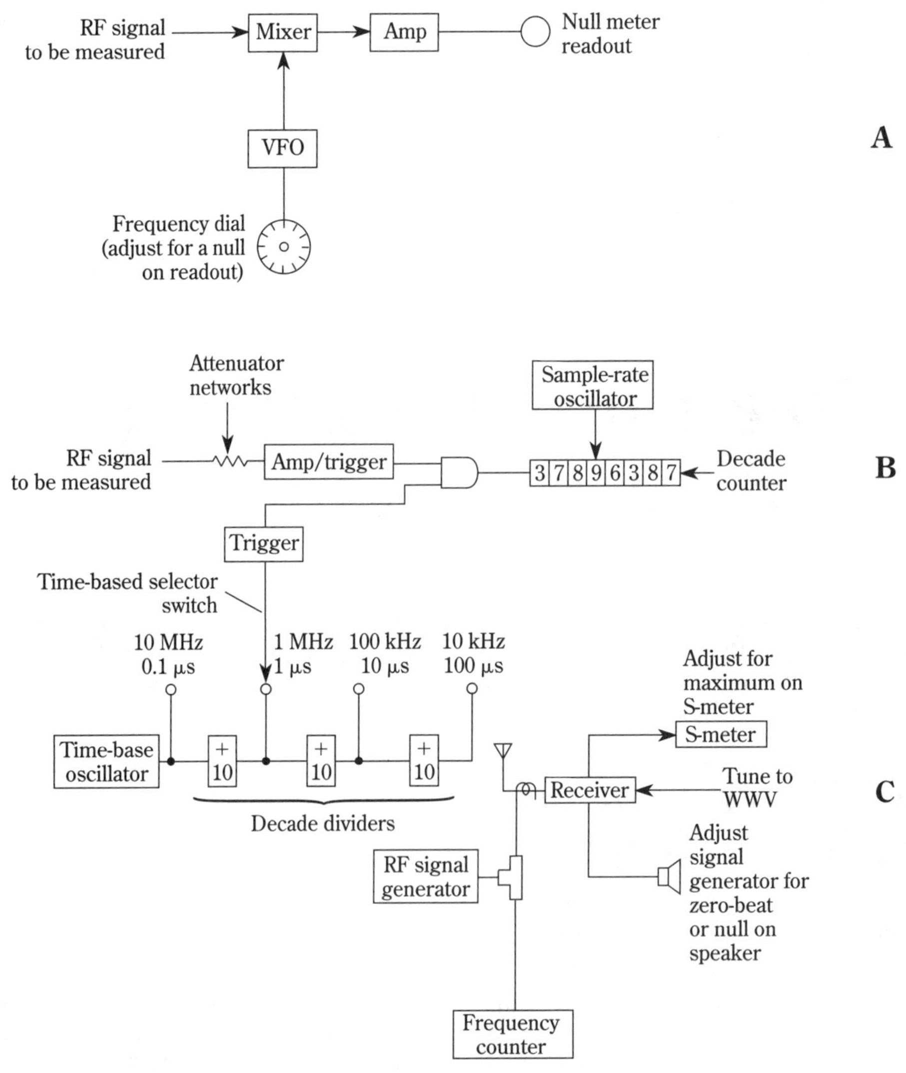

4-1 Frequency meter and counter circuits.

4.2.1 Digital-counter basics

Although there are many types of digital counters, all counters have several basic functional sections in common. These sections are interconnected in a variety of ways to perform the various counter functions. Fig. 4-1B shows the basic counter circuit for frequency measurement (Sec. 4.2.2). A typical digital counter also provides a totalizing operation, period measurement, time-interval measurement, and ratio measurement (Sec. 4.3 through 4.6).

All counters have some form of main gate that controls the count-start and count-stop with respect to time. Counters also have some form of crystal-controlled time base that supplies the precise increment of time to control the gate for a frequency measurement.

The accuracy of any counter depends on the accuracy of the time base, plus or minus one count. For example, if the time-base accuracy is 0.005 percent, the overall accuracy of the counter is 0.005 percent, plus or minus one count. The one-count error occurs because the count might start and stop in the middle of an input pulse, thus omitting the pulse from the total count. Likewise, part of the pulse might pass through the gate before the gate closes, thus adding a pulse to the count (see Sec 4.7 and 4.8).

Most electronic counters have dividers that permit variation of gate time. These dividers convert the fixed-frequency time base to several other frequencies. In addition to the basic sections, most counters have attenuator networks, amplifier and trigger circuits (to shape a variety of input signals to the common form), and logic circuits to control operation of the instrument.

Electronic counters have some form of counter and readout. Early instruments used binary counters and readout tubes that converted the binary count to a decade readout. Such instruments have long since been replaced by decade counters that convert the count to *binary-coded decimal (BCD) form*, decoders that convert the BCD data to decade form (generally BCD-to-seven-segment decoders), and readouts that display the decade information (generally seven-segment LCSs, LEDs, etc.). One readout, or display, is provided for each digit. For example, eight readouts provide for a count up to 99,999,999.

4.2.2 Frequency measurement

The counter circuits are arranged as shown in Fig. 4-1B for frequency measurement. The input signal (for example, from an RF generator) is first converted to uniform pulses by the trigger. The pulses are then routed through the main gate and into the counter and readout circuits, where the pulses are totaled.

The number of pulses totaled during the gate-open inverval is a measure of the average input frequency for that interval. For example, assume that the gate is held open for 1 s and the count is 888. This indicates a frequency of 888 Hz. The count is then displayed (with the correct decimal point) and held until a new sample is ready to be shown. The sample-rate oscillator determines the time between samples (not the internal of gate opening and closing), resets the counter, and initiates the next measurement cycle.

The time-base switch selects the gating interval, thus positioning the decimal point and selecting the appropriate measurement units. The time-base switch selects one of the frequencies from the time-base oscillator. If the 10-MHz signal (directly from the time-base, shown in Fig. 4-1B) is selected, the time interval (gate-open to gate-close) is 0.1 μs. If the 1-MHz signals (from the first decade divider) is chosen, the measurement time interval is 1 μs.

4.2.3 Counter accuracy

The accuracy of a counter is set by the stability of the time base, rather than by the readout. The readout is typically accurate to within plus or minus one count. The time base of the Fig. 4-1B counter is 10 MHz and is stable to within ±0.1 part per million (ppm) or 1 Hz. Counter accuracy is covered further in Sec. 4.7 and 4.8.

4.2.4 Counter resolution

The resolution of a counter is set by the number of digits in the readout. For example, assume that you must use a five-digit counter to test circuits in a CB set. The CB operating frequencies, or channels, are in the 27-MHz range. Now assume that you measure a 27-MHz signal with a five-digit counter. The count can be 26.999 or 27.001, or within 1000 Hz. Because the FCC requires that the operating frequency of a CB set be held within 0.005 percent (or about 1350 Hz in the case of a 27-MHz signal), a counter with five-digit resolution is suitable (that is, the counter has sufficient resolution, but this is no guarantee of accuracy).

4.2.5 Combining accuracy and resolution

To find out if a counter is adequate for a particular test, add the time-base stability (in terms of frequency) to the resolution at the operating frequency. Again using the CB-set example, if the accuracy is 1 Hz and the count can be resolved to 1000 Hz (at the measurement frequency), the maximum possible inaccuracy is 1000 + 1 Hz = 1001 Hz. This is within the approximate 1350 Hz (0.005 percent of 27 MHz) required.

4.2.6 Calibration check of counters

The accuracy of counters should be checked periodically (the major causes of inaccuracy are covered in Sec. 4.8). Always follow the procedures recommended in the counter service literature (some counters have built-in self-test functions). Generally, you can send the instrument to a calibration lab or to the factory, or you can maintain your own frequency standard (the latter is not practical for most service shops!). No matter what procedure or standard is used, remember that the standard must be more accurate and have better resolution than the counter being checked.

4.2.7 WWV signals

In the absence of a frequency standard, or factory calibration, you can use the frequency information broadcast by U.S. government radio station WWV. These WWV signals are broadcast on 2.5, 5, 10, 15, 20, and 25 MHz continuously (night and day), except for silent periods of about 4 minutes beginning 45 minutes after each hour. Broadcast frequencies are held accurate to within 5 parts in 10^{11}. This is far more accurate than precision lab counters.

The hourly broadcast schedules of WWV are subject to change. For full data on WWV broadcasts, refer to National Bureau of Standards (NBS) Standard Frequency and Time Services (Miscellaneous Publication 236), available from the Superintendent of Documents, U.S. Government Printing Office, Washington, DC 20402.

It is the continuous wave (CW) signals broadcast by WWV that provide the most accurate means of calibrating (or checking) counters. It is not practical to use the WWV signal directly (except on some special counters), but the test connections for checking are not complex.

Figure 4-1C shows the basic test connections for checking the accuracy of a counter using WWV. Notice that a receiver and signal generator are required. Their accuracy is not critical, but both instruments must be capable of covering the desired frequency range. The procedure is as follows:

1. Allow the generator, receiver, and counter being tested to warm up for at least 15 minutes.
2. Reduce the generator output (amplitude) to zero. Turn off the generator output if this is possible without turning off the entire generator.
3. Tune the receiver to the desired WWV frequency. It is generally best to use a WWV frequency that is near the operating frequency of the circuits with which the counter will be used. For example, if a 27-MHz CB set is to be tested with the counter, use the 25-MHz WWV signal.
4. Operate the receiver controls until you can hear the WWV signal in the receiver loudspeaker.
5. If the receiver is of the communications type, it will have a beat-frequency oscillator (BFO) and an output signal-strength indicator, or S-meter. Turn on the BFO, if necessary, to locate and identify the WWV signal. Then, tune the receiver for maximum signal strength on the S-meter. The receiver is now exactly on 25 MHz, or whatever WWV frequency is selected.
6. Turn on the generator output and tune the generator until the generator is at zero beat against the WWV signal. As the generator is adjusted so that the generator frequency is close to that of the WWV signal (so that the difference in frequency is within the audio range), a tone, whistle, or "beat note" is heard on the receiver. When the generator is adjusted exactly to the WWV frequency, there is no "difference signal," and the tone can no longer be heard. In effect, the tone drops to zero, and the two signals (generator and WWV) are at zero beat.
7. Read the counter. The readout should be equal to the WWV frequency. For example, with a five-digit counter at 25 MHz, the counter reading should be 24.999 to 25.001.
8. Repeat the procedure at other WWV broadcast frequencies.

Note that a typical shop counter will provide frequency-measurement operation over a frequency range of 5 Hz to 175 MHz (possibly up to 1.2 GHz), with a resolution of 0.1 Hz to 1 kHz, in decade steps (8-digit readout). Accuracy is ±1 count ± time-base error. The display is usually in either kHz or MHz, or both, depending on the control settings.

4.3 Totalizing operation

Counters can be operated in a totalizing mode with the main gate controlled by a manual start/stop switch, as shown in Fig. 4-2A. With the switch in start (gate open), the decade counters totalize the input pulses until the main gate is closed. The counter display or readout then represents the input pulses received during the interval between manual start and manual stop. Some counters can be remotely controlled for totalizing operation.

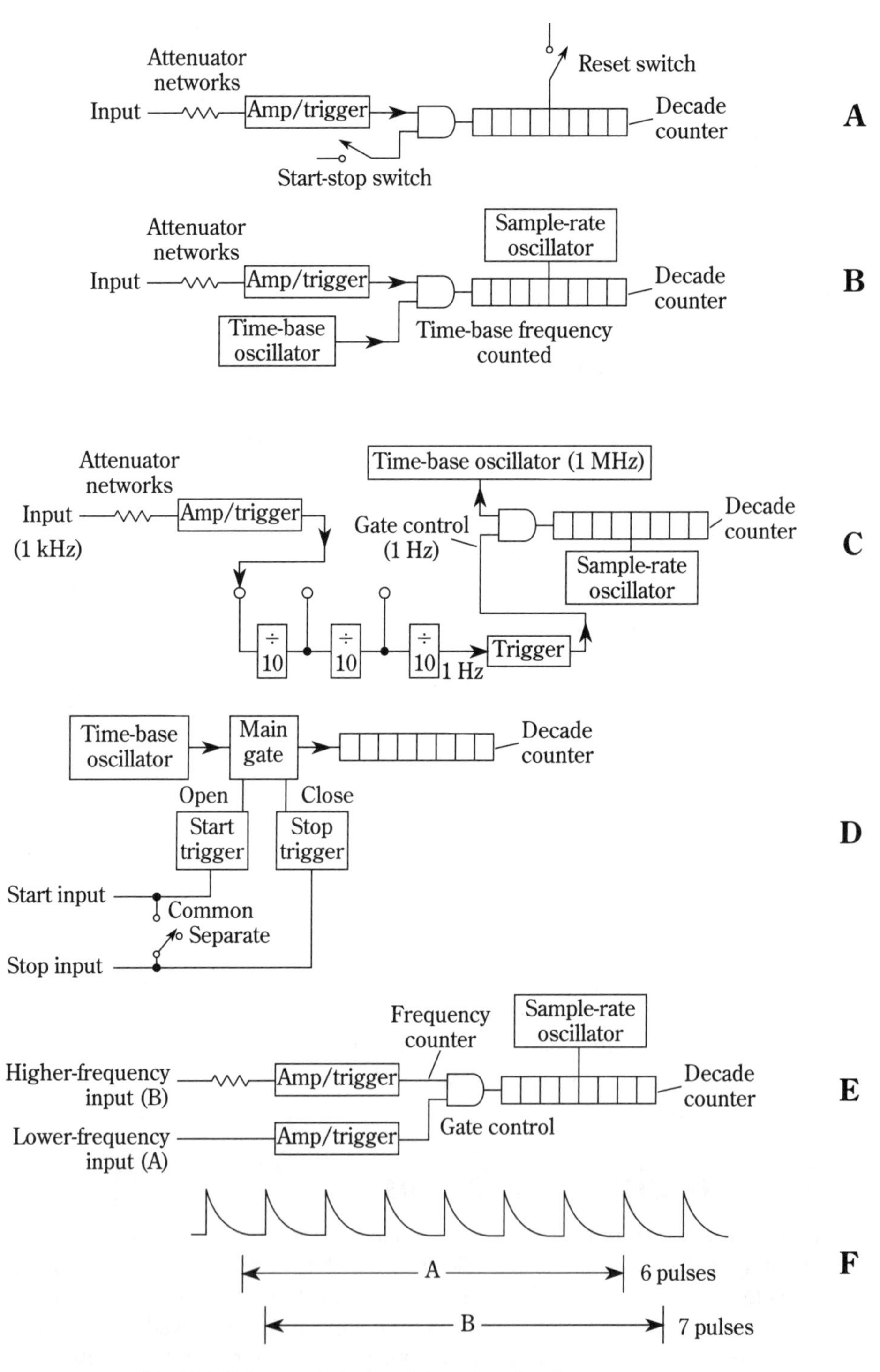

4-2 Totalizing, period, time-interval, and ratio measurements.

Notice that the decade counters must be reset manually in totalizing mode. If not, the new count is added to the previous count each time the gate is opened. For all other modes (frequency, period, time-interval, and ratio) the decade counters are reset by the sample-rate oscillator (which usually operates at 2 or 3 Hz to provide two or three samples each second). The sample-rate oscillator is disabled during totaling mode.

A typical shop counter provides totalizing operation over a frequency range of 5 Hz to 15 MHz, with a capacity of 0 to 99,999,999. Many counters have an overflow LED, which turns on to indicate that the capacity is exceeded.

4.4 Period operation

Period is the inverse of frequency (*period* = 1/*frequency*). Therefore, period measurements are made with the input and time-base conditions reversed from those for frequency measurement. This is shown in Fig. 4-2B. The unknown input signal controls the gate time, and the time-base frequency is counted and read out. For example, if the time-base frequency is 1 MHz, the indicated count is in μs (a count of 70 indicates that the gate is held open for 70 μs). Often, the input-shaping circuit selects the positive-going zero-crossing of successive cycles of the unknown signal at trigger points for opening and closing of the gate.

The accuracy and resolution of period measurements can be increased by period averaging. The connections are shown in Fig. 4-2C. These are the same connections as for regular period measurements, except that the signal to be measured is lowered in frequency by dividers, thus extending the gate-open period. For example, if the input signal is 1 kHz, the period is 1 ms with a regular period measurement. If the time base is 1 MHz, the count is 00001000 on an 8-digit readout. If the period-average method is used and the 1-kHz input frequency is reduced to 1 Hz, as shown in Fig. 4-2C, the period is 1 second, and the count is 01000000 on the same 8-digit readout. Thus, the resolution is increased by 10^3.

A typical shop counter provides period-measurement operation over a frequency range from 5 Hz to 2 MHz (a period range from 0.5 μs to 200,000 μs), with a resolution of 100 ps to 100 ns in decade steps. The accuracy is ±1 count ± time-base error ± trigger error (trigger error is covered in Sec. 4.8). The display is in ms or μs with a decimal point. Minimum pulse width is 250 ns. 1, 10, 100, or 1000 cycles can be averaged, depending on the resolution setting.

4.5 Time-interval operation

Time-interval measurements are essentially the same as period measurements. However, the time-interval mode is more concerned with time between two events, rather than the repetition rate of signals. Counters vary greatly in their time-interval capability. Some counters measure only the duration of an electrical event. Other counters measure the interval between the start of two pulses. The most versatile models, known as *universal counters*, have separate inputs for start and stop commands and have separate controls that permit setting the trigger-level amplitude, polarity, slope, and type of input coupling (ac or dc) for the start and stop control.

The basic time-interval configuration is shown in Fig. 4-2D. The time-base signals are counted and read out when the gate is open. The gate is controlled by two trigger circuits that receive inputs from the signal being measured. With the switch in the separate position, the two triggers receive inputs from separate lines. For example, assume that the start (gate-open) trigger receives an input from a signal applied to an amplifier under test, while the amplifier output is applied to the stop (gate-closed) trigger. The gate is then open for an interval that is equal to the delay through the amplifier. If the time base is 1 MHz, and the count or readout is 33, the delay is 33 μs.

With the switch in the common position, the two triggers receive inputs from the same lines. Each trigger is adjusted so that it responds to a different portion of the same waveform. For example, assume that the start trigger opens the gate when the input signal rises to +10 V and the stop trigger closes the gate when the input signal reaches +15 V. The count then represents the time inverval between the two points.

The characteristics for a typical shop counter operated in time-interval mode are the same as for period mode, except for accuracy. The number of intervals averaged must be taken into consideration. The accuracy factor becomes: ±1 count ± time-base error ± trigger error $\pm N$, where N is the number of intervals averaged.

4.6 Ratio operation

The ratio of two frequencies is determined using one signal (signal A) for the control while the other signal (signal B) is counted, as shown in Fig. 4-2E. With proper transducers (chapter 5), ratio measurements can be applied to any phenomenon that can be represented by pulses or sine waves. Gear ratios and clutch slippage, as well as frequency divider or multiplier operation, are some of the measurements that can be made using ratio operation. Generally, universal counters are required for ratio measurement because you must have access to both gate inputs (as in time-interval operation, Sec. 4.5).

The characteristics for a typical shop counter operated in the ratio mode are: frequency range A = 5 Hz to 10 MHz; frequency range B = 5 Hz to 2 MHz; resolution = Frequency B/(frequency $A + N$), where N = 1, 10, 100, or 1000; accuracy = ± (frequency B + frequency $A \times N$) ± (frequency $B \times$ *trigger error*)/N; the display is a numerical ratio with decimal point.

4.7 Counter display problems

Here are three problems often overlooked by inexperienced counter operators.

4.7.1 Readout capacity

If a long gate time is used when a high frequency is counted, the entire answer is not seen if the readout capacity is exceeded. To find what part of the answer will be visible, you must realize that counting starts with the righthand digit in the readout, progressed to the next digit at the left (after a count of nine is reached), and so on until all digits read nine. Next, account for the effect of gate time. For example, if 0.9 MHz is counted for 1 second, a total of 900,000 counts are gated into the counting circuits. A 6-digit counter will display 900,000, but a 5-digit counter will display

00,000. In a typical 8-digit counter with gate times from 1 μs to 10 s, the entire answer can always be made visible with suitable gate-time selection.

4.7.2 Low-frequency count

The low-frequency end of the count can be another source of problems. Counters can have ac- or dc-coupled inputs, or both. When both inputs are available, the desired input coupling is selected by a front-panel switch, Dc-coupled inputs, or both. When both inputs are available, the desired input coupling is selected by a front-panel switch. Dc-coupled inputs pass all input waveforms, regardless of rise time. Ac-coupled inputs discriminate against slow rise times. Typically, the frequency range specified for a counter defines sine-wave frequencies for which the sensitivity is met. As an example, a typical counter input specification is 20 mVrms 5 Hz to 30 MHz, and 560 mVrms above 30 MHz.

Most ac-coupled counters will count sine waves below the minimum frequency specified, but a higher input amplitude or lower input sensitivity setting is needed. Likewise, most counters will count events of extremely low repetition rate if the input waveshape has a fast rise time. If you must count low-frequency signals with slow rise times, try using the period mode (or period average, if available) and then convert the period to frequency (*frequency* = 1/*period*).

4.7.3 Mechanical contact bounce

When a counter is used to count mechanical contact closure, be alert for spurious counts caused by contact bounce. Any bounce (for example, when measuring relay-contact closure time) has the same effect on the counter as a series of pulses. This is especially true when the counter is set for maximum sensitivity to measure a slow count.

4.8 Counter accuracy

There are three main sources of error in counter measurements: *plus or minus one-count ambiguity*, *time-base stability*, and *trigger error*.

4.8.1 Plus or minus one-count ambiguity

The *plus or minus one-count ambiguity* is inherent in all counter measurements because the input signal and the time base are not normally synchronized. As shown in Fig. 4-2F, the count registered during the gating time can be either six or seven, depending on the moment at which the gating time begins. Notice that gate times *A* and *B* are of equal length. As a result, with any type of operation, the counter display might be incorrect by one count.

The fractional effect of the plus or minus one-count ambiguity is: 1/*total events counted*. Obviously, the more events counted, the smaller the error becomes. This explains why long gate times result in better accuracy in frequency measurement.

4.8.2 Time-base stability

The time base for most counters is a crystal oscillator mounted in a temperature-controlled oven. The frequency and stability of the crystal can be checked against broadcast standards or local standards, as described. Several specifications are used by various manufacturers to describe the stability of the time base. The most important specifications are *crystal aging rate* (also known as *maximum aging rate*), *short-term stability*, and error caused by temperature change and/or line-voltage variation (usually called *temperature stability* and/or *line-voltage stability*).

Crystal or maximum aging rate The *aging rate* (also specified as *long-term stability* or *drift rate*) refers to slow, but predictable variations in average oscillator frequency over time because of changes in the quartz crystal itself. A typical shop counter will have a maximum aging rate of ±1 ppm/year, and a universal counter will have a ratio of ±10 ppm/year.

After an initial period of rapid change when the oscillator is turned on, aging in a good crystal becomes quite slow and assumes a predictable linear characteristic. The slope of this linear characteristic is the overall oscillator aging rate. Because aging is cumulative, it is necessary to check oscillator accuracy periodically (Sec. 4.2.7).

Short-term stability This specification indicates the effects of noise generated internally in the time-base oscillator on the average frequency over a short time (usually 1 second). Short-term effects are so small that the specification is listed for only the most stable time-base oscillators in precision ovens (in lab counters). In the less-stable oscillators (shop counters), other errors make the short-term specification insignificant. In shop counters, make the short-term specifications insignificant. In shop counters, you m*ight* find the term *setability* applied to the time-base oscillator (such as ±0.1 ppm, ±1 Hz). This means that the stability can be considered as ±0.1 ppm over a given measurement time.

When comparing short-term stability specifications, it is important to remember that the averaging time used determines how good the specification appears. A long averaging time hides great frequency variations!

Line-voltage and temperature Counter line-voltage and temperature specifications should be self-explanatory. Typical shop-counter specifications are less than ±0.1 ppm with ±10% line-voltage variation, and less than ±0.0001% or ±1 ppm from 0°C to 50°C ambient. In some counters, an optional user-installed *temperature-compensated crystal oscillator (TCXO)* is available for increased temperature stability. This is especially true for universal counters, which generally have poorer temperature-stability characteristics.

4.8.3 Trigger error

Trigger error arises from noise on the gate-control signal. This noise causes the gate to open or close at incorrect times and results in an erroneous count. Significant trigger error can occur only when an external signal controls the gate (during period, time-interval and ratio modes). Trigger error should not occur in the frequency measurement (Sec. 4.2.2) mode because the trigger is supplied by a (supposedly) noise-free time-base signal.

Absolute trigger error is stated in time units and the fractional effect is given by: error in time/total time gate is open. This explains why long period-averaging is such a good method for reducing measurement error (because the gate time is extended). As more periods are averaged, the effect of both trigger and the plus and minus one-count ambiguity are reduced proportionally.

5
Probes and transducers

Practically all meters and scopes operate with some type of probe. In addition to providing for electrical contact to the circuit under test, probes serve to modify the voltage being measured to a condition suitable for display on a scope or readout on a meter. For example, assume that a very high voltage must be measured and that this voltage is beyond the maximum input limits of the meter or scope. A voltage-divider probe can be used to reduce the voltage to a safe level for measurement. Under these conditions, the voltage is reduced by a fixed amount (known as the *attenuation factor*) usually in the order of 100:1 or 1000:1.

Transducers serve a similar function. However, a transducer operates to convert a physical property to an electrical signal suitable for meter or scope inputs. Although a variety of probes are found in both shop and laboratory, the use of transducers is generally limited to the lab or industrial test and measurement applications.

5.1 Basic probes

In the simplest form, the basic probe is a test probe. Such probes work well on circuits carrying dc and audio signals. However, for higher-frequency signals (even at the low end of the RF range), it might be necessary to use a special low-capacitance probe. The same is true if the meter or scope gain is high. Hand capacitance can cause hum pickup when simple probes are used at higher frequencies. This condition can be offset by shielding low-capacitance probes. More important, use of a low-capacitance probe prevents meter or scope impedance from being connected directly to the circuit being tested (such impedance might disturb circuit conditions).

5.2 Low-capacitance probes

Figure 5-1A shows the basic circuit of a low-capacitance probe. (You can make up such a probe. However, it is far more practical to use the probe designed for the particular

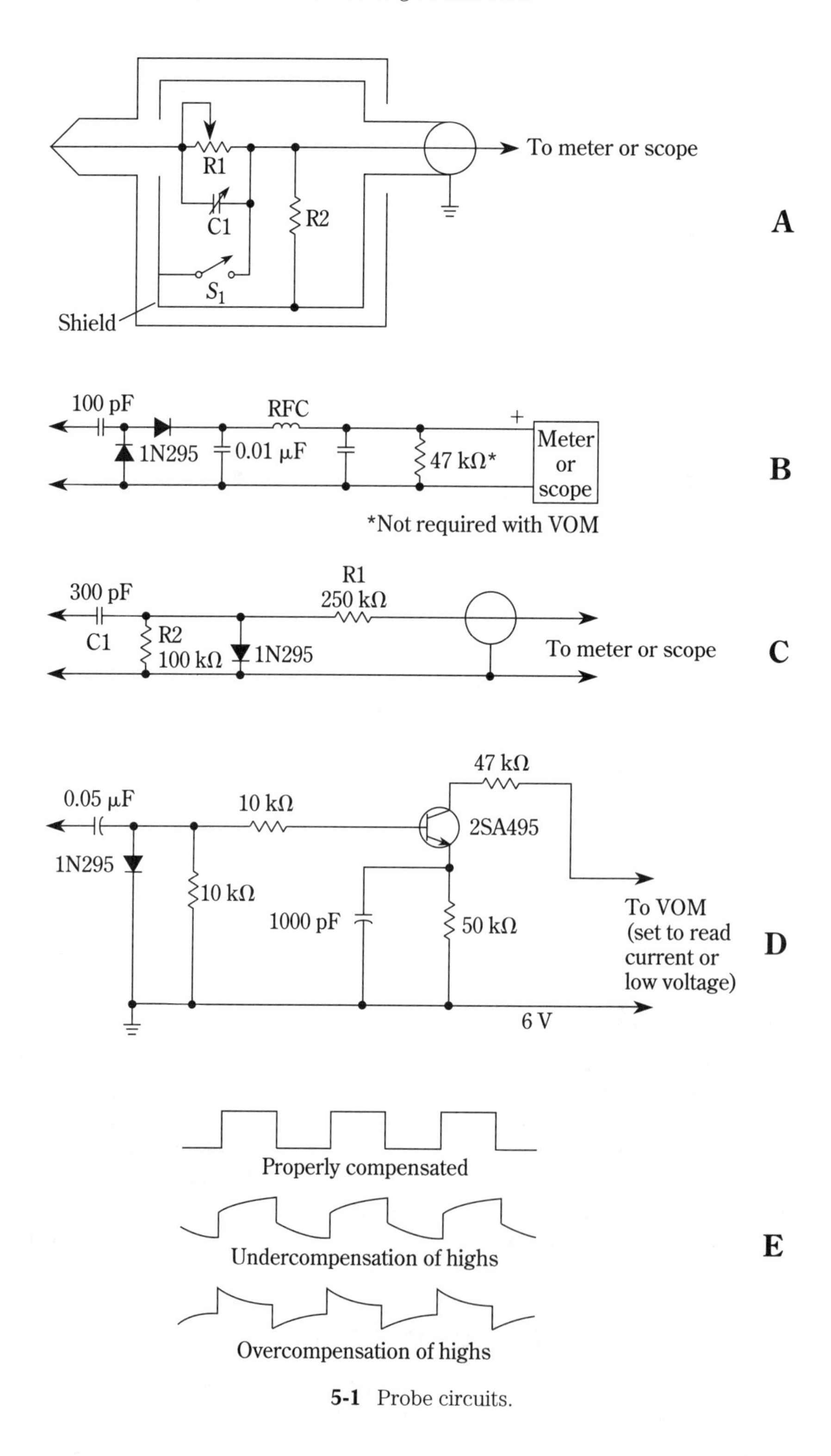

5-1 Probe circuits.

meter or scope.) The series resistance (R_1) and the capacitance (C_1), as well is the parallel or shunt R2, are surrounded by a shielded handle. The values of R_1 and C_1 are preset at the factory and should not be disturbed unless recalibration is required (Sec. 5.7).

In many low-capacitance probes, the values of R_1 and R_2 are selected to form a 10:1 voltage divider between the circuit being tested and the meter or scope input. Such probes serve the dual purpose of capacitance and voltage reduction. Remember that voltage indications are one-tenth (or whatever value of attenuation is used) of the actual value when a voltage-divider probe is used. Capacitance C_1 in combination with R_1 and R_2 also provide a capacitance reduction.

5.3 High-voltage probes

High-voltage probes are not generally needed for solid-state work unless the equipment has a display tube (TV, computer terminals, etc.). An obvious exception is in the high-voltage circuit of some high-power transmitters (such as broadcast transmitters). A typical high-voltage probe provides a voltage reduction of 100:1 or 1000:1.

5.4 RF probes (supplied with meter or scope)

An *RF probe* is required when the signals to be measured are at radio frequencies (chapters 14 and 15) and are beyond the capabilities of the meter or scope. Again, always use the probe supplied with the meter or scope. An RF-probe circuit is covered in Sec. 5.9 and shown in Fig. 5-1B.

5.5 Demodulator probes

The circuit of a demodulator probe is essentially the same as that of an RF probe, but the circuit values and basic functions are somewhat different. When the high-frequency signals contain modulation (which is typical for the modulated RF-carrier signals of most communications equipment, chapter 15), a demodulator probe is more effective for signal tracing.

Figure 5-1C shows a typical demodulator-probe circuit. The circuit is essentially a half-wave rectifier. However, C1 and R2 act as a filter. The demodulator probe produces both an ac and a dc output. The RF signal is converted into a dc voltage approximately equal to the peak value. The modulation voltage on the RF signals appears as ac at the probe output.

In use, the meter is set to dc and the RF signal is measured. Then the meter is set to ac and the modulating voltage is measured. The calibrating resistor R1 is adjusted so that the dc scale (of the meter) reads the correct value. If there is no demodulator probe available for a particular meter, you can use the circuit of Fig. 5-1C on most meters (for troubleshooting and testing but not for precise voltage measurement). The following steps describe the calibration and fabrication procedure:

1. Connect the probe circuit to a signal generator and meter.
2. Set the meter to measure dc voltage. The meter should be high impedance for best results.

3. Adjust the signal generator voltage amplitude to some precise value, such as 10 V, as measured on the generator output meter.
4. Adjust the calibrating resistor R1 until the meter indicates the same value (10 V).
5. As an alternative procedure, adjust the signal generator for a 10-V peak output. Then, adjust R1 for a reading of 7.07 on the meter readout.
6. Remove the power, disconnect the circuit, measure the value of R_1, and replace the variable resistor with a fixed resistor of the same value.
7. Repeat the test with the fixed resistance in place. If the reading is correct, mount the circuit in a suitable package (such as within a test prod). Repeat the test with the circuit in final form. Also repeat the test over the entire frequency range of the probe. Generally, the probe in Fig. 5-1C provides satisfactory response up to about 250 MHz.
8. Remember that the meter must be set to measure direct current because the probe output is dc when there is no modulation.

5.6 Solid-state signal-tracing probe

It is possible to increase the sensitivity of a meter or scope with an amplifier. Such amplifiers are particularly useful with a VOM for measuring small-signal voltages during testing or troubleshooting. An amplifier is usually not required for a digital meter or scope because such instruments contain built-in amplifiers.

Figure 5-1D shows a typical probe and amplifier circuit. Such an arrangement increases sensitivity by at least 10:1 (usually much more) and provides good response up to about 500 MHz. The circuit is not normally calibrated to provide a specific voltage indication. Rather, the circuit is used to increase the sensitivity for signal tracing in RF circuits.

5.7 Probe compensation and calibration

Probes must be calibrated to provide a proper output to the meter or scope with which they are used. Probe compensation and calibration are best done at the factory and require precision test equipment.

The following paragraphs describe the general procedures for compensation and calibration of probes. Never attempt to adjust a probe unless you follow the instruction manual and have the proper test equipment. An improperly adjusted probe produces erroneous readings and might cause undesired circuit loading.

5.7.1 Probe compensation

The capacitors that compensate for excessive attenuation of high-frequency signal components (through the probe resistance dividers) affect the entire frequency range from some midband point upward. Capacitor C1 in Fig. 5-1A is an example of such a compensating capacitor.

Compensating capacitors must be adjusted so that the higher-frequency components are attenuated by the same amount as are low frequency and direct current. It is possible to check the adjustment of the probe-compensating capacitors using a square-wave signal source. This is done by applying the square-wave signal directly to the scope input and then applying the same signals through the probe and noting any change in pattern. In a properly compensating probe, there should be no change (except for a possible reduction of the amplitude).

Figure 5-1E shows typical square-wave displays with the probe properly compensated, undercompensated (high frequencies underemphasized), and overcompensated (high frequencies overemphasized). Proper compensation of probes is often neglected, especially when probes are used interchangeably with meters or scopes having different input characteristics (this is another reason for using the probe supplied with the meter or scope!).

5.7.2 Probe calibration

The main purpose of *probe calibration* is to provide a specific output for a given input. For example, the value of R_1 in Fig. 5-1C is adjusted (or selected) to provide a specific amount of voltage to the meter or scope. During calibration, a voltage of known value and accuracy is applied to the input. The output is monitored, and R1 is adjusted to produce a value or reading (0.707 of RF peak value, for example).

5.8 Probe testing and troubleshooting techniques

Although a probe is a simple instrument, and does not require specific operating procedures, several points should be considered to use a probe effectively in testing and troubleshooting.

5.8.1 Circuit loading

When a probe is used, the probe impedance (rather than the meter or scope impedance) determines the amount of circuit loading. Connecting a meter or scope to a circuit may alter the signal at the point of connection. To prevent this, the impedance of the measuring device must be large in relation to that of the circuit being tested. Thus, a high-impedance probe offers less circuit loading, even though the meter or scope might have a lower impedance.

5.8.2 Measurement error

The ratio of the two impedances (of the probe and the circuit being tested) represents the amount of probable error. For example, a ratio of 100:1 (perhaps a 100-MΩ probe used to measure the voltage across a 1-MΩ circuit) accounts for an error of about 1 percent. A ratio of 10:1 produces an error of about 9 percent.

5.8.3 Effects of frequency

The input impedance of a probe is not the same at all frequencies. Input impedance becomes smaller as frequency increases (capacitive reactance and impedance decrease with an increase in frequency). All probes have some input capacitance. Even an increase at audio frequencies might produce a significant change in impedance.

5.8.4 Shielding capacitance

When using a shielded cable with a probe to minimize pickup of stray signals and hum, the additional capacitance of the cable should be considered. The capacitance effects of a shielded cable can be minimized by terminating the cable at one end in the cable's characteristic impedance. Unfortunately, this is not always possible with the input circuits of some meters and scopes.

5.8.5 Relationship of loading an attenuation factor

The reduction of loading (either capacitive or resistive) caused by a probe might not be the same as the attenuation factor of the probe. (Capacitive loading is almost never reduced by the same amount as the attenuation factor because of the additional capacitance of the probe cable.) For example, a typical 10:1 attenuator probe might reduce capacitive loading by only 2:1.

5.8.6 Checking probe effects

When testing any circuit (but particularly an RF circuit), it is possible to check the effect of a probe on the circuit with the following test. Attach and detach another connection of similar kind (such as another probe) and observe any difference in meter reading or scope display. If there is little or no change when the additional probe is touched to the circuit, it is safe to assume that the probe has little effect on the circuit.

5.8.7 Probe length and connections

Long probes should be restricted to the measurement of dc and low-frequency ac. The same is true of long ground leads. The ground lead should be connected where no hum or high-frequency signal components exist in the ground path between the point and the signal-pickup point. Keep all test connections as short as possible for RF work.

5.8.8 Measuring high voltages

Avoid applying more than the rated voltage to a probe. Fortunately, most commercial probes will handle the highest voltages found in solid-state electronics (with the possible exception of TV picture tube and computer-terminal display-tube high voltages).

5.9 An RF probe for test and troubleshooting

Figure 5-1B shows the circuit diagram of a probe suitable for testing and troubleshooting most RF circuits. The probe is designed specifically for use with a VOM or digital meter and converts both audio-frequency and RF signals to direct current. The probe is full-wave and thus produces a larger signal than the half-wave demodulator probe in Fig. 5-1C.

The probe in Fig. 5-1B operates satisfactorily up to about 250 MHz, and is essentially a signal-tracing device (but does not provide accurate voltage readings). The meter used with the probe must be set to read direct current because probe output is dc. However, if the input signal is amplitude-modulated, the probe output is pulsating direct current.

5.10 Transducers

Although there are many types of *transducers* used as test equipment, transducers can be divided into three broad categories: *resistance-changing*, *self-generating*, and *inductance-* or *capacitance-changing*. Practically any condition can be measured by one or more of these three types of transducers and can be converted to a voltage suitable for readout on a meter or scope. The remaining sections of this chapter summarize the three types of transducers used as test equipment.

5.10.1 Resistance-changing transducers

A *resistance-changing transducer* has the widest application in test equipment. The resistance can be changed by mechanical means or by some external condition, such as temperature. Likewise, the resistance element can be connected as part of a voltage divider or as part of a bridge circuit (bridge circuits are covered further in chapter 6).

Figure 5-2A shows the classic voltage-divider circuit. Here a fixed-voltage source is placed across the resistance element. The element contact is actuated by mechanical force. This force could be a bellows (to measure pressure differences) a spring-loaded weight that slides in a direction parallel to motion (to measure acceleration), or the force could come from movement of an arm coupled to some mechanical device (to measure position). In any event, the output voltage to the meter or scope is proportional to the mechanical force and can be so related.

For example, assume that the mechanical force is pressure that operates against a bellows. In turn, the bellows operates the resistance contact (often called the *wiper*). By applying a known fixed voltage across the resistance element, the output from the transducer is a variable voltage that represents the amount of pressure applied. This relation of pressure to voltage is known as *scale factor* (one foot-pound of pressure equals 1 V, one square-inch of pressure equals 1 mV, and so on).

Another version of the voltage-divider circuit used in test-equipment transducers is shown in Fig. 5-2B. Here, the fixed voltage source is placed across the resistance element and a fixed resistor in series. The resistance element changes resistance value with changes in external condition. An example of this is a *thermistor*, which is a resistor with negative temperature coefficient (resistance decreases with temperature increase). Because the voltage divides across the two resistances, the output voltage is proportional to the value of the resistance element. This, in turn, is in relation to temperature. Thus the output voltage is proportional to the temperature surrounding the thermistor.

Resistance-changing transducers are often used in bridge circuits similar to those covered in chapter 6. Figure 5-2C shows the basic bridge circuit. The bridge is balanced and has zero output when R/R_B is equal to R_C/R_D. When there is a change in one or more of the bridge arms, the bridge becomes unbalanced and there is an output voltage proportional to the change.

One or more of the bridge arms might be active. When only one arm is active (Fig. 5-2D), the transducer bridge is said to be a one-sensitive-arm bridge. Two active arms (Fig. 5-2E) make the bridge a two-sensitive-arm bridge.

There are a number of drawbacks to any resistance-changing type of transducer, as well as some advantages.

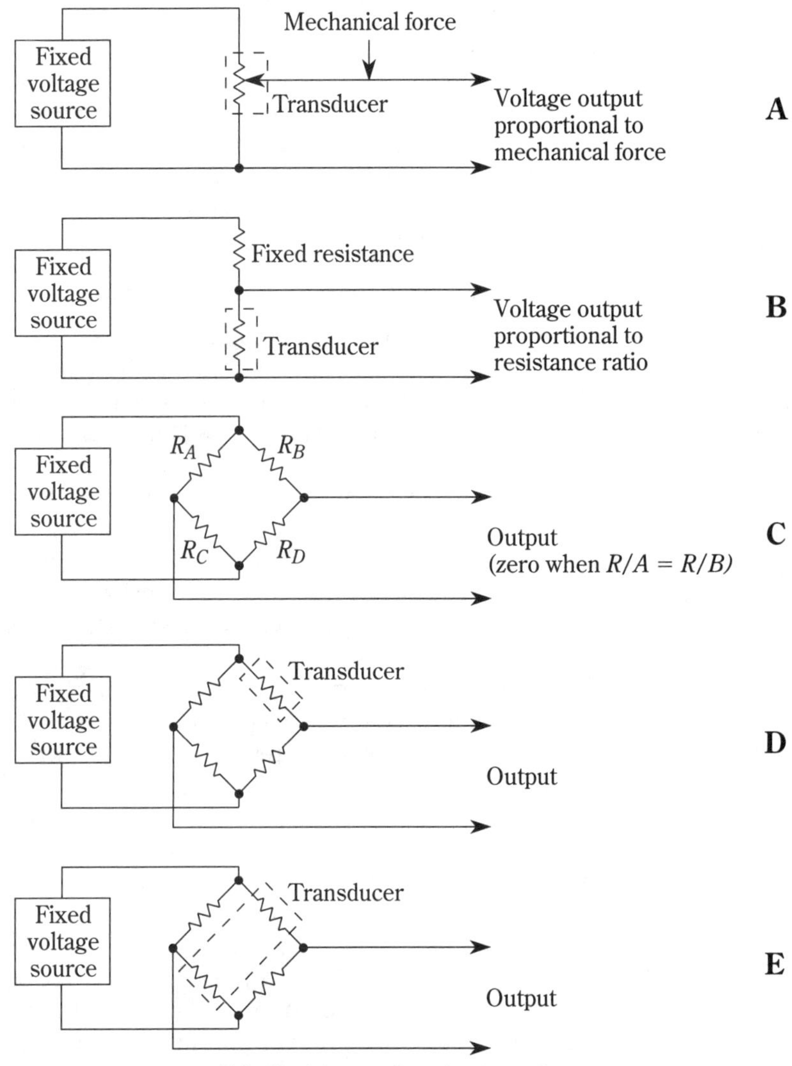

5-2 Resistance-changing transducer.

Advantages The main advantage of resistance-changing is that a minimum of signal conditioning is required, except in very special applications. That is, the transducer output voltage can be applied directly to the meter or scope without modification (amplification, rectification, and so on). Likewise, this high output permits operation, even with severe impedance mismatch (in most cases).

Disadvantages Resistance-changing transducers operated by mechanical movement are subject to error because of shock and vibration because the resistance-element contact (wiper) cannot tell the difference between desired and undesired movement. The thermistor-based transducers also have a problem because current passing through the resistance element creates heat. This heat combines with the ambient heat, and changes the thermistor resistance. For these and various other design considerations, a number of self-generating transducers are used both in scientific and routing lab test-equipment applications.

5.10.2 Self-generating transducers

It is impossible to describe all of the self-generating transducers used in test-equipment applications. In many cases, the various transducers are different versions of the same basic circuit. The following examples are typical of the many commercially available transducer and signal-conditioning systems using the self-generating principles.

Thermocouples Because of their simplicity and ruggedness, thermocouples are often used to measure temperature. Thermocouple transducers are based on the principle that a voltage is produced when two dissimilar metals are joined and subjected to heat. Test-equipment thermocouples often use *chromel* and *alumel* as the dissimilar metals.

One of the drawbacks to a thermocouple transducer is that the voltage output is very low. This is usually resolved by adding a dc amplifier between the thermocouple and the meter or scope. Another drawback is that the thermocouple output varies with changes in ambient temperature, and with temperature changes caused by heating the wires. This is compensated with reference junctions and a bridge circuit.

As shown in Fig. 5-3A, the reference junctions are connected in series with the measuring-junction output. The reference junctions are also placed near a thermistor element that forms one leg of a bridge circuit. The output of the bridge is also in series with the measuring-junction output. Both the reference junctions and the thermistor are subjected to identical temperature conditions. Resistance change in R_A causes changes in bridge balance so that the voltage across R_B is the same value as the change in voltage from the reference junctions, and thus offsets the reference-junction change (but not the desired measuring-junction change). Accuracy of the system depends on the ability of the bridge to track changes in the reference-junction output.

Turbine-type flowmeters These self-generating transducers measure some physical property, such as liquid flow, airflow, and so on. Operation is similar to that of a tachometer, except that turbine flowmeters measure liquid flow, rather than mechanical rotation.

As shown in Fig. 5-3B one type of flowmeter transducer consists essentially of a turbine placed in the flow path, a magnet in one of the turbine blades, and a pickup coil placed near the turbine-blade tips. As the turbine turns in the flow path, the pickup coil produces an alternating current or signal. The frequency of this signal is proportional to flow rate. This signal could be applied directly to a counter (chapter 4), or the ac signal is converted to a dc voltage proportional to frequency. Many ICs can perform such a frequency-to-analog conversion. Refer to the author's best-selling *McGraw-Hill Circuits Encyclopedia* (McGraw-Hill, 1993).

Tachometer-type transducers These transducers measure actual frequency rather than produce an output where frequency is proportional to some other value (as do turbine transducers). Thus, tachometer-type transducers are used with counters (or possibly scopes) but not with meters.

Typically, tachometer-type transducers measure mechanical rotation, as do conventional tachometers, and consist of a pickup coil and magnet similar to that of

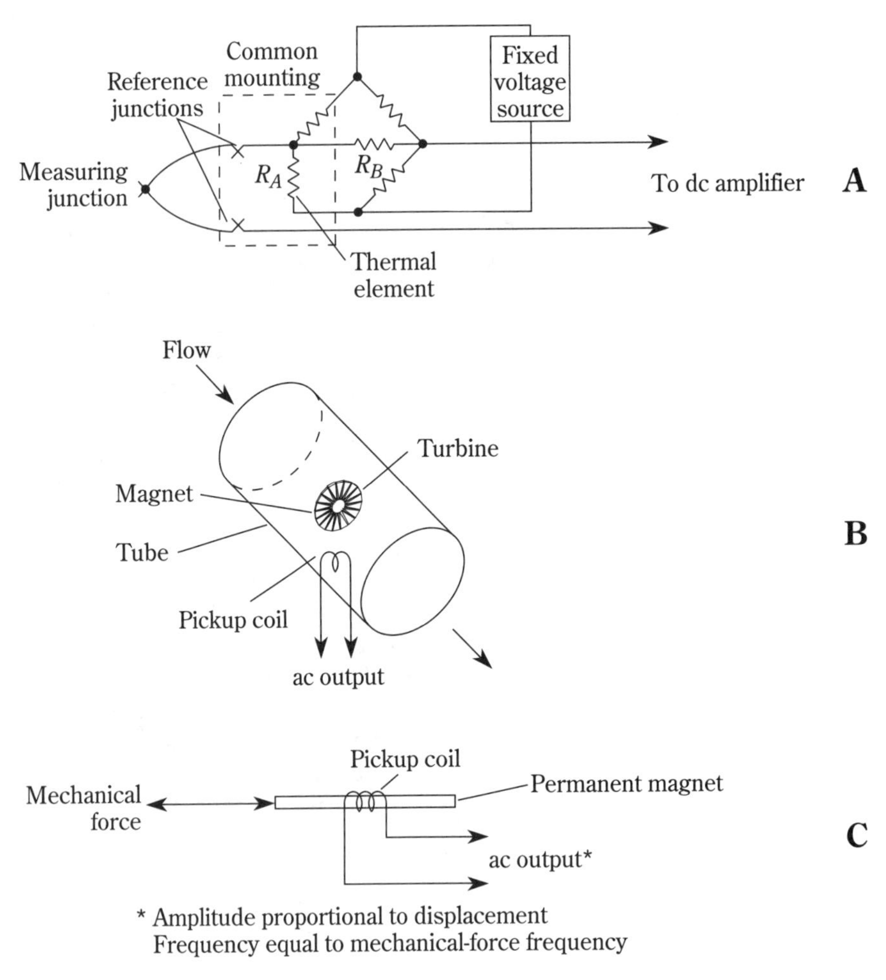

5-3 Self-generating transducers.

Fig. 5-3B. However, the magnet is driven past the coil by mechanical rotation and produces an ac output, usually at the rate of 1 Hz per revolution.

Magnetic-induction transducers The basic operating principle of a magnetic-induction transducer is similar to that of sound-powered telephone, in that both use a high-density permanent magnet with a coil. Any movement of the magnet (in relation to the coil) produces a proportional voltage output from the coil. The magnet is driven by the mechanical force to be measured. One of the main advantages of magnetic-induction transducers is that the magnet can be suspended, thus eliminating the problems of friction and sticking.

Figure 5-3C shows a typical magnetic-induction type of self-generating transducer. The voltage amplitude is proportional to mechanical displacement, and the output frequency is equal to the frequency of the mechanical force.

Piezoelectric-type transducers The piezoelectric principle is based on the fact that certain materials (such as quartz crystal) produce a voltage when subjected to pressure. As shown in Fig. 5-4A, the piezoelectric material is placed between two plates, and the resultant voltage is applied to the meter or scope.

A prime example of this is the piezoelectric accelerometer often used for vibration testing and measurement. Because of the low voltages produced, the output of piezoelectric transducers usually requires considerable amplification to operate with a meter or scope.

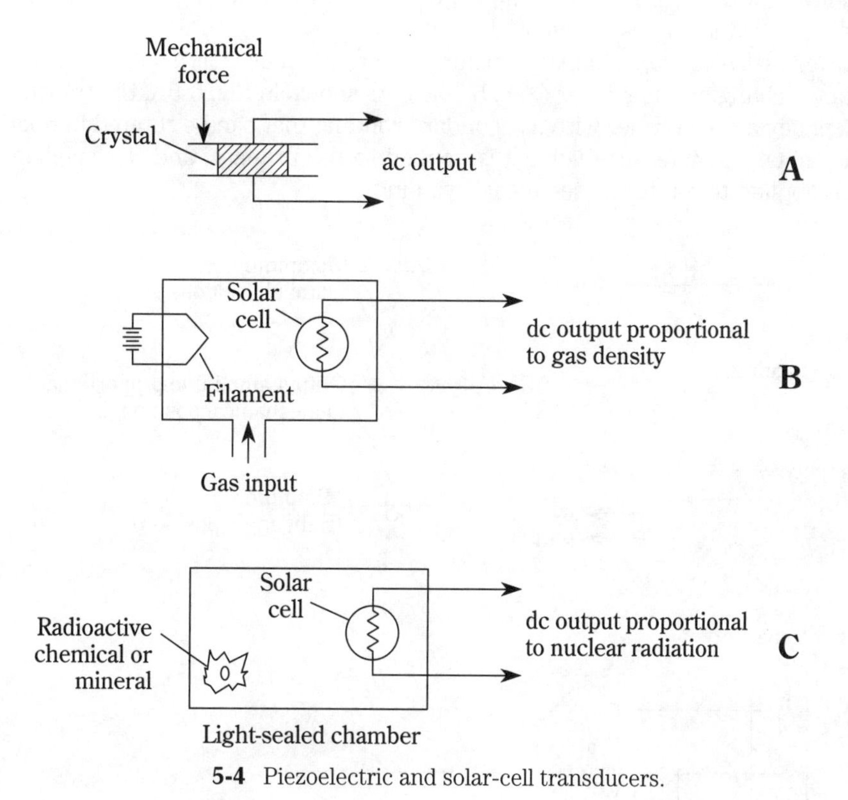

5-4 Piezoelectric and solar-cell transducers.

Solar-cell transducers A solar cell produces a voltage proportional to the amount of light applied to the cell surface. In addition to measuring light, this principle can also be used to measure the presence of nuclear radiation and gas density.

In the case of a gas detector, a metal filament is placed near the solar cell, and a current is passed through the filament, as is shown in Fig. 5-4B. The current causes the filament to heat and glow, producing a voltage output from the solar cell. A metal that glows with greater intensity in the presence of various gases is chosen. One example of this is platinum, which glows with increased brilliance in the presence of hydrocarbons, such as those of gasoline fumes. Such gas-detector transducers are used in industrial test equipment. The output of the solar cell is applied to a meter or scope (sometimes through a dc amplifier).

In the case of a radiation detector, a chemical or mineral is placed near the solar cell, as shown in Fig. 5-4C. The chemical or mineral gives off light in the presence of nuclear radiation. The intensity of the light and solar-cell output are proportional to the amount of radiation.

5.10.3 Inductance- and capacitance-changing transducers

These transducers are usually used in conjunction with a bridge circuit operating on alternating current. There are many such transducers. The linear transformer and liquid-level measure are good examples of inductance-changing and capacitance-changing transducers, respectively.

Linear transformer Linear transformers are often used as a substitute for the resistance-changing transducer (Sec. 5.10.1). As shown in Fig. 5-5A, the transducer is essentially a transformer with a secondary winding that can be rotated in relation to the primary. A reference voltage is applied to the primary, and the mechanical force is applied to rotate the secondary winding.

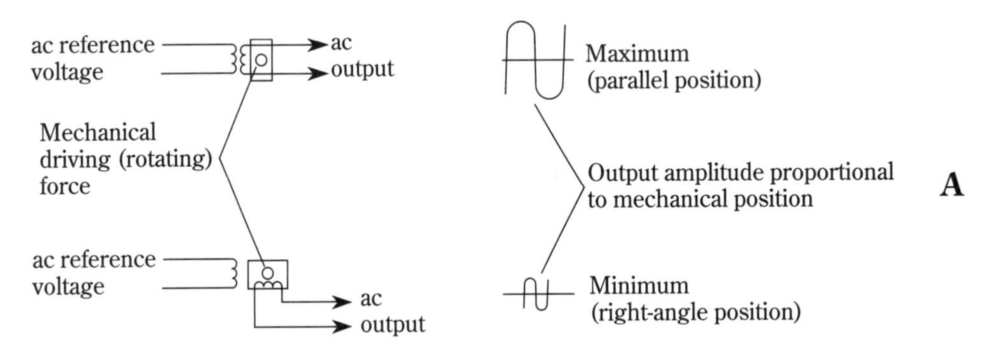

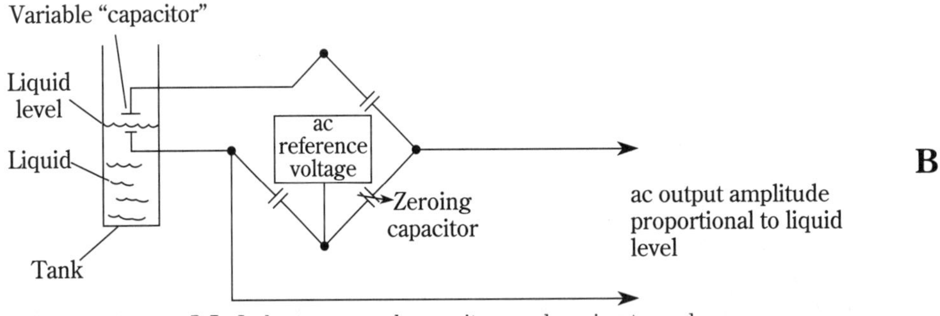

5-5 Inductance- and capacitance-changing transducers.

As the secondary is rotated toward the parallel position, the output voltage (applied to a meter or scope) increases. Conversely, the output decreases when the secondary winding is moved toward the right-angle position. Therefore, the transducer output voltage is proportional to mechanical position.

Liquid-level measure With this transducer, the liquid itself forms one plate of the capacitor, as shown in Fig. 5-5B. Because the value of any capacitor is determined both by the plate size and spacing, any variation in liquid level causes a corresponding variation in capacitance. The capacitor formed by the liquid and the transducer plate is usually connected in a bridge circuit. The matching leg of the bridge is formed by a variable capacitor. This permits the bridge circuit to be zeroed at minimum or maximum capacity. Either way, the bridge output is proportional to liquid level.

6
Special-purpose test equipment

This chapter is devoted to specialized test equipment, such as bridges and time-domain reflectometers. Other specialized test equipment, such as distortion meters, wow/flutter meter, and spectrum analyzers, are covered in related chapters (chapters 12 and 15).

An oscilloscope and pulse generator can be combined to measure the characteristics of transmission lines, including coax cables. Such an arrangement is known as a *time-domain reflectometer (TDR)*, and was originally developed by Hewlett-Packard. The TDR is often described as a "closed-loop radar." A voltage pulse, or step, is propagated down a transmission line, and the incident (outgoing) and reflected voltage waves are monitored by the scope. This "echo" technique reveals at a glance the characteristic impedance of the line, as well as both the position and nature (resistive, inductive, or capacitive) of any discontinuity on the line. Time-domain reflectometers also indicate whether losses in a transmission line are series losses or shunt losses. The first part of this chapter covers TDR.

Many quantities in electronics, such as impedance, admittance, capacitance, inductance, conductance, and so on, are measured by means of bridge circuits and bridge-type test equipment. Some manufacturers produce universal bridges that measure more than one quality. These instruments operate on the balance or null principle, or on the principle of comparison against a standard. The second half of this chapter is devoted to test equipment that operates on null or comparison techniques.

6.1 Basic time-domain reflectometer operation

Figure 6-1A shows the basic time-domain reflectometer circuit. The pulse generator produces a positive-going wave or "step" that is fed into the transmission line. The step wave travels down the transmission line, at a velocity determined by the

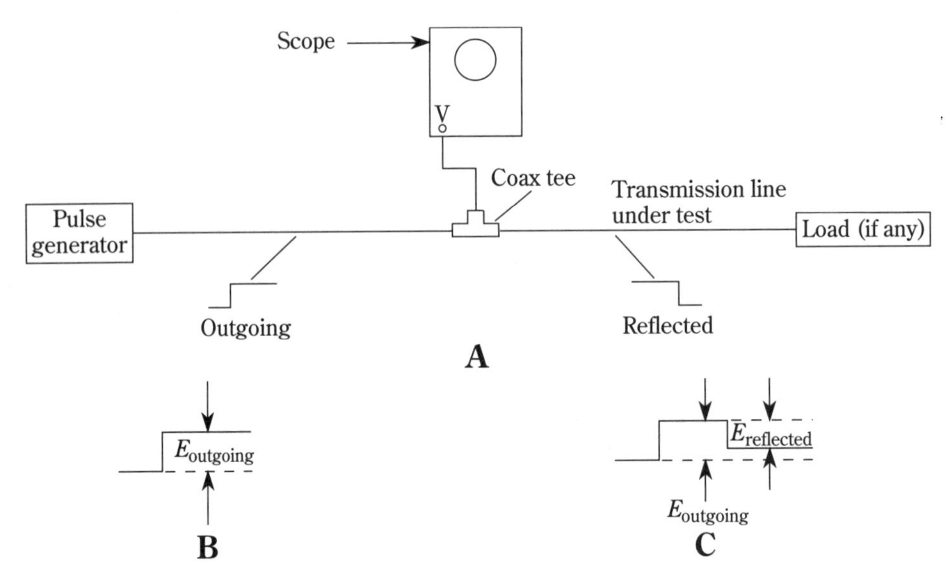

6-1 Basic time-domain reflectometer circuit and scope displays.

type of line. If the load impedance is exactly equal to the characteristic impedance of the line, no wave is reflected, and the scope shows only the outgoing pulse (Fig. 6-1B). If a mismatch exists at the load (the load impedance does not match the line impedance exactly), part of the outgoing wave is reflected. The reflected pulse appears on the scope, algebraically added to the outgoing wave (Fig. 6-1C).

6.2 Analyzing time-domain reflectometer displays

The shape of a waveform on a TDR requires analysis. The shape and amplitude of the reflected wave (in relation to the outgoing wave) reveal the nature and magnitude of mismatches, loads, discontinuities, and so on, along the line.

Figure 6-2 summarizes typical TDR displays and corresponding load or mismatch responsible for the displays. Note that the displays shown in Fig. 6-2 are theoretical, and that actual displays will never be quite as precise!

In Fig. 6-2A (an open-circuit termination where the load impedance is infinity, such as an open-ended coax), the full voltage is reflected back and added to the outgoing voltage.

In Fig. 6-2B (a short-circuit termination where the load impedance is zero, such as a coax with the inner and outer conductors shorted at the free end), the voltage drops to zero and cancels the outgoing wave.

In Fig. 6-2C (a pure resistive load where the load impedance is twice the impedance of the transmission line, such as a 50-Ω coax with a 100-Ω terminating resistor), one-third of the voltage is reflected back and added to the outgoing voltage.

In Fig. 6-2D (a pure resistive load where the load impedance is one-half the impedance of the transmission line, such as a 50-Ω coax with a 25-Ω terminating

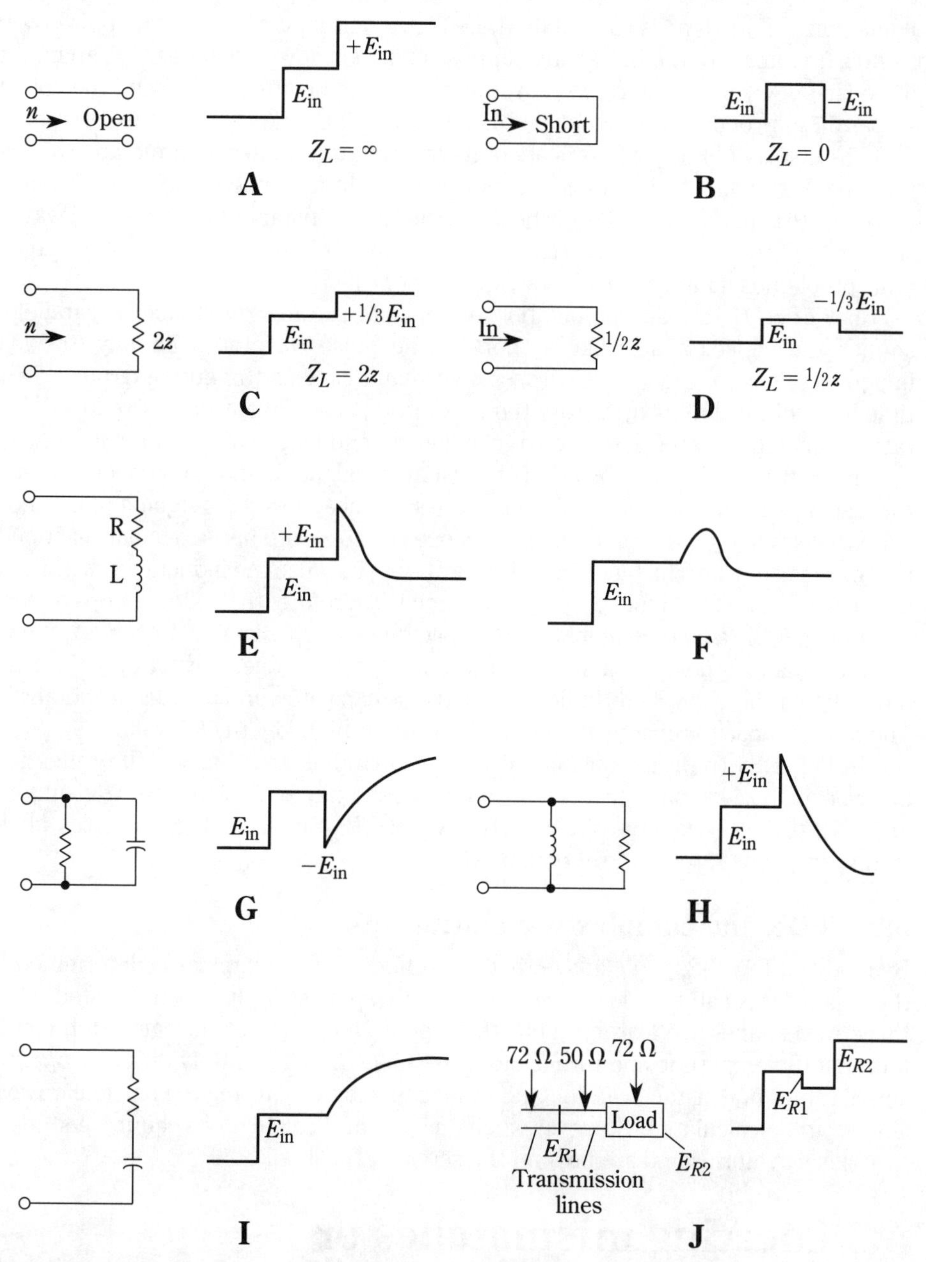

6-2 Typical TDR displays and the corresponding condition that is responsible for displays.

resistor), one-third of the voltage is reflected back and cancels one-third of the outgoing voltage.

In Fig. 6-2E (a load combination of resistance and inductance in series), the leading edge of the reflected wave is the same shape as the outgoing pulse and is added algebraically. With time, the pulse slopes off to a value below that of the outgoing pulse. This slope is caused by the effect of inductance. When the pulse waveform is first applied, the inductance opposes current flow and appears as an infinite impedance. As current starts to flow, the impedance begins to drop, and the voltage slopes off in proportion. Current flow is limited only by the effect of the resistance.

The display of Fig. 6-2E appears only for very large values of inductance and resistance. If the values are small (particularly the inductance value), the display is more like that of Fig. 6-2F. This is because the time constant of the reflected wave is so short that the wave decays to the final value almost before the TDR scope can rise (this problem is minimized with a sampling scope).

In Fig. 6-2G (a load combination of capacitance and resistance in parallel or shunt), the capacitor acts like a short to the pulse waveform. Initially, the load impedance is zero, the voltage drops to zero and cancels the outgoing pulse. With time, the voltage builds up across the capacitors, and current flow is reduced. The rate of capacitor charge and discharge is determined by the *RC* circuit values.

In Fig. 6-2H (a load combination of resistance and inductance in parallel or shunt), the leading edge of the reflected wave is the same shape as the outgoing pulse and is added algebraically. With time, the pulse slopes off to zero. This is a similar reaction to that of a resistance-inductance in series. However, because the inductance is in shunt, the current flow is not limited by the resistance. Therefore, the voltage drops to zero.

In Fig. 6-2I (a load combination of capacitance and resistance in series), the capacitor initially appears as a short to the pulse leading edge, leaving only the resistance. With time, the voltage builds up across the capacitor, and is added algebraically. The rate of capacitor charge and discharge is determined by the *RC* values.

In Fig. 6-2J (multiple mismatches, or mismatches and a load), the reflections must be analyzed separately. The mismatch at the junction of the two transmission lines (72 to 50 Ω) generates a reflected wave, ER_1. Similarly, the mismatch at the load (53 to 72 Ω) creates a reflection, ER_2.

6.2.1 TDR and complex waveforms

By studying Fig. 6-2, you can see that the reflected waveshape at is determined by the type of mismatch or load present on the transmission line being tested. Once these waveshapes are learned, both the type and location of the mismatch can be found. If the amplitude and rate of slope are measured accurately on the scope, the actual values of resistance, inductance, and capacitance causing the particular waveshape can be calculated. However, such analysis and calculations require a study of complex waveforms and are beyond the scope of this book.

6.3 Locating mismatches or discontinuities on lines

In addition to determining the characteristics of mismatches or loads in transmission lines, a TDR can be used to locate mismatches (physically) on lines.

The basic procedure for locating mismatches is to measure the travel time and speed of the step or pulse as it passes down the transmission line and then back to the scope. This is done by measuring the distance (in time) between the outgoing and reflected waves (using the scope horizontal scale in a manner similar to that for measuring pulse delay, Fig. 2-27). The reflected wave is readily identified because it is separated in time from the outgoing wave. Make certain to measure time between the same points on both waves.

By measuring both the time and velocity of propagation, the distance can be calculated by: *distance = velocity of propagation T/2*, where *T* is the transit time from monitoring point to the mismatch and back again, as measured on the scope horizontal scale.

If the velocity of propagation for a particular type of transmission line is not known, it can be found from an experiment on a known length of the same type of line. For example, the time required for the outgoing wave to travel down, and the reflected wave to travel back from an open-circuit termination at the end of a 120-cm length of RG-9A coax is 11.4 ns. This means that the velocity of propagation is 21 cm/ns (11.4/2 = 5.7; 120/5.7 = 21).

6.4 Wheatstone bridges

Null methods have long been used as the most precise and convenient way to measure all types of impedance: resistive and reactive, inductive and capacitive, from low frequencies to ultra-high frequencies. Most null or balance-type instruments are evolved from the basic Wheatstone bridge (Fig. 6-3A).

Resistors R_A and R_B are fixed and of known value. R_S is a variable resistor with the necessary calibration arrangement to read the resistance value for any setting.

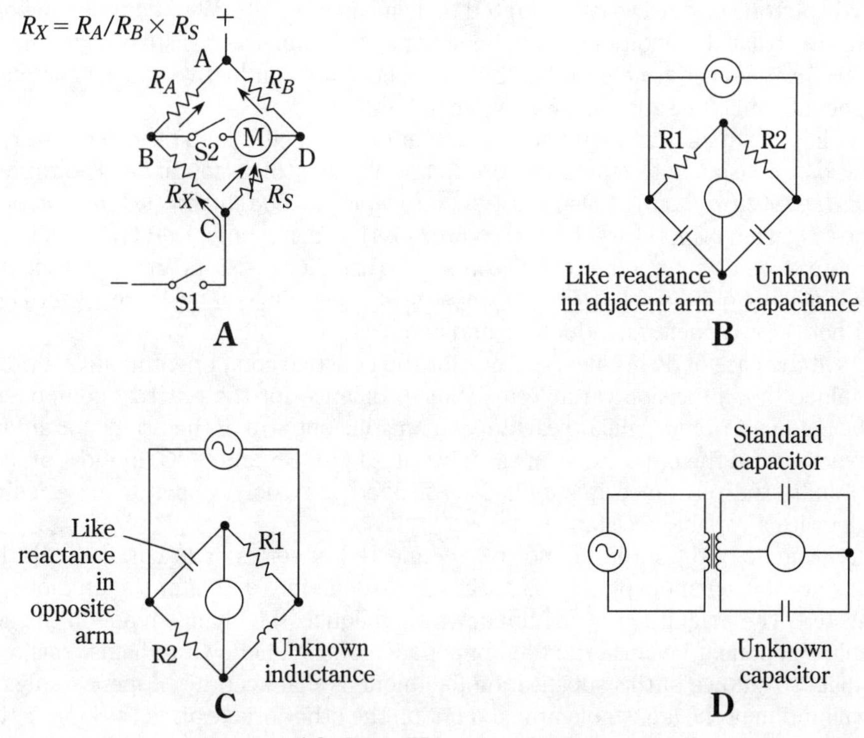

6-3 Basic Wheatstone bridge.

(Most older bridges have a calibrated dial coupled to the variable resistance shaft. Some present-day bridge-type meters have digital readout, such as described in Sec. 6.8.) The unknown resistance value is connected across terminals B and C. A battery or other power source is connected across points A and C.

When S1 is closed, current flows in the direction of the arrows, and there is a voltage drop across all four resistors. The drop across R_A is equal to the drop across R_B (provided that R_A and R_B are of equal resistance value). Variable resistance R_S is adjusted so that the galvanometer M reads zero (center scale) when S1 is closed. At this adjustment, R_S is equal to R_X in resistance. By reading the resistance of R_S (from the calibrated dial), the resistance of R_X is known.

When the variable R_S is equal to R_X, the voltage difference between B and D is zero, and no current flows through the galvanometer, If R_S is not equal to R_X, B and D are not at the same voltage, and current flows through the galvanometer, moving the pointer away from zero (which is at the center on a galvanometer scale).

The equation shown in Fig. 6-3A is used when the values of R_A and R_B are not equal. In some bridges, the value of R_A is 10 times that of R_B. Thus, the actual value of R_X is 10 times the indicated value of R_S. This permits a large value of R_X to be measured with a low-value, R_S. In other bridges, the value of R_B is 10 times that of R_A. Thus, the actual value of R_X is one-tenth the indicated value of R_S.

6.5 Alternating-current bridges

The Wheatstone bridge is easily adapted to ac measurements. With complex impedances, two balance conditions must be satisfied, one for the resistive component, and one for the reactive component. Likewise, both the conductive and susceptance components must be satisfied for complex admittances.

An important characteristic of an inductor or a capacitor, and often a resistor, is the ratio of resistance to reactance, or of conductance to susceptance. The ratio is called *dissipation factor (D)* and the reciprocal is *storage factor (Q)*. In practical bridges, D (which varies directly with power loss), is commonly used for capacitors. Q is more often used for inductors. Q can also be used for resistors, in which case the value is usually quite small. However, most bridge circuits can measure either D or Q (or both) for capacitors, inductors, and resistors.

As in the case of dc bridges, balance for the resistive component in an ac bridge is obtained by a precision variable resistance. Balance for the reactive component can be obtained from a similar reactance in an adjacent arm of the bridge, or an unlike reactance in the opposite arm, as shown in Figs. 6-3B and 6-3C. In most practical circuits, the reactance is supplied by a fixed precision capacitor in series or parallel with a variable resistance.

In some ac bridges, the unknown is connected in series or in parallel with the main adjustable component, and balances are made before and after the unknown is connected. The magnitude of the unknown then equals the change made in the adjustable component because the total impedance of the unknown remains constant. The main advantage of this substitution technique is that accuracy depends only on the calibration of the adjustable arm and not on the other bridge arms (as long as the arms are constant). The substitution principle can also be used to advantage with any bridge if the balances are made with an external, calibrated, adjustable component.

Transformer-ratio arms are used in some ac bridge circuits. The basic circuit is shown in Fig. 6-3D. In practical circuits, the transformer windings are tapped on the standard side in decimal steps (usually from 0.1 to 1), and on the unknown side in decade steps (usually from 1 to 0.001). Several fixed capacitance standards are used, one for each decade step. In a typical transformer-type capacitance bridge, the standard values range from 0.0001 to 1000 pF. Such a combination of internal standards and transformer voltage ratios makes possible a measurement range of 10^9 to 1.

6.6 Universal bridges

Analysis of capacitors, inductors, and resistors for low-frequency applications is commonly made with a universal bridge. Universal bridges have considerable versatility, being able to measure not only resistance, capacitance, conductance, and inductance over wide ranges, but also *D* and *Q*.

Figure 6-4A shows a generalized universal bridge circuit. The bridge is driven by an ac source across D and Q. When the voltage across OP equals the voltage across OS, the output voltage at the detector (across S and P) is zero. With the

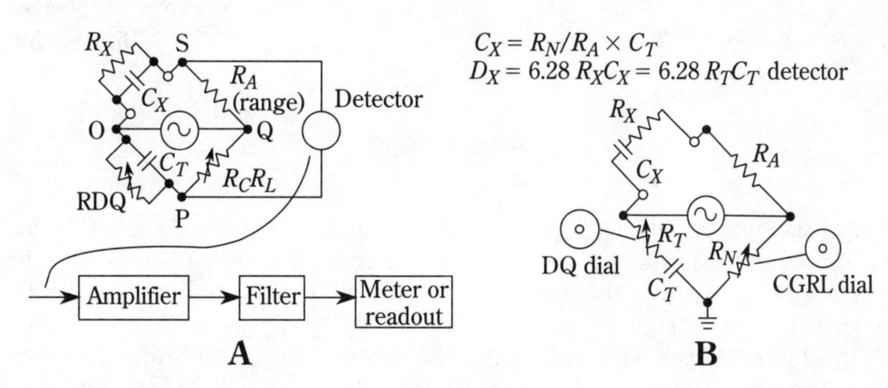

6-4 Generalized universal bridge and typical circuits.

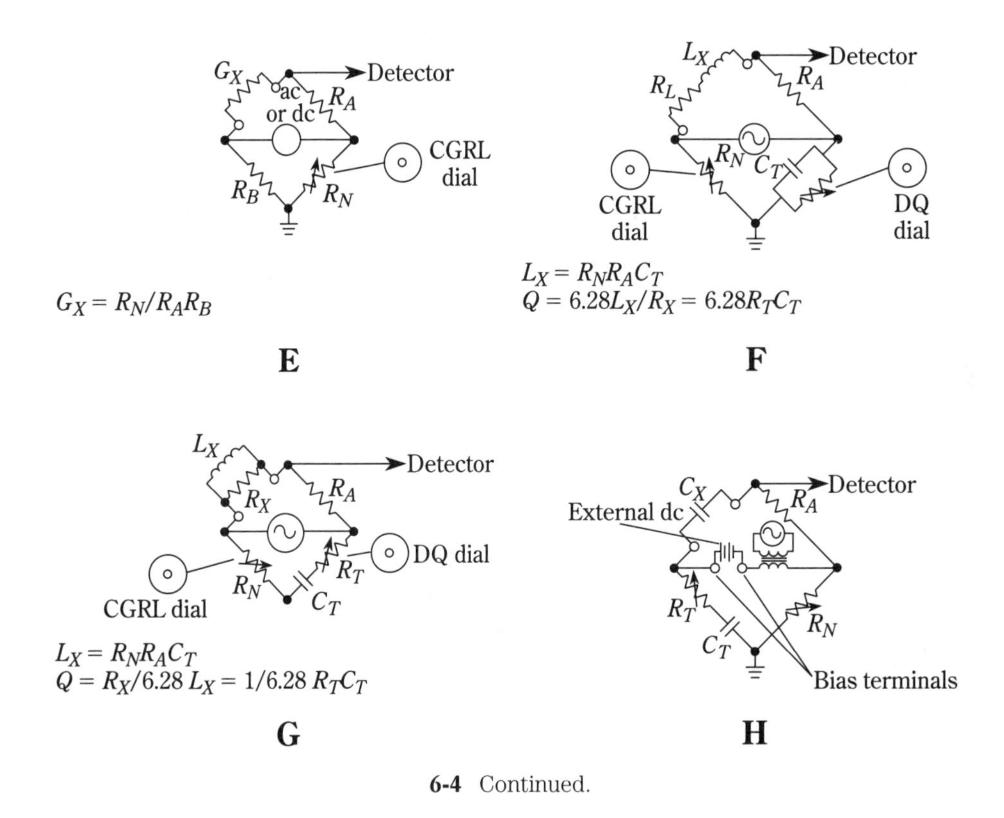

6-4 Continued.

bridge balanced, or nulled, the product of the impedance across OS and that across PQ is equal to the product of the impedance across SQ and that across OP. At balance, the value of any of the four impedances can he calculated if the other three are known.

In addition to the basic bridge circuit, a typical universal bridge includes an audio generator and a null detector. The null detector includes an amplifier (to measure extremely low voltages at or near null), a fixed or tunable filter or frequency-selective circuit (to remove any signals except the audio driving source) and rectifiers (to convert the ac signal to a dc voltage for display on the meter.

Figures 6-4B through 6-4H show typical universal bridge circuits for measuring capacity, inductance, resistance and coincidence, as well as D and Q. These circuits give the user the option of measuring the unknown in terms of either series or parallel equivalents. The choice is a matter of convenience for the problem at hand. For example, if Q is 10 or more (or if D is 0.1 or less), the difference between series and parallel reactance is no more than 1 percent. However, for very low Q (or high D) the difference is substantial (with a Q of 1, parallel reactance is twice that of series reactance).

Operation is essentially the same for each circuit. The DQ and CGRL dials are adjusted for a null (the DQ dial is not used for resistance and conductance measurements). Then, the value is read from the corresponding dial indication. In most cases, the dials are not direct reading. Instead, the dial reading must be multiplied by some factor (a multiplier-control setting in the case of the CGRL dial, and frequency in the case of the DQ dial).

Both capacitance and inductance are measured in terms of impedance presented at a given frequency, rather than by comparison against a standard, even though the readout is in terms of capacity and inductance value (μF, μH, and so on). This is the major difference between the universal bridge and the standard capacitance or inductance bridge described in Sec. 6.7.

Many universal bridges have a connector that permits application of an external bias voltage to the component under test. For example, it might be advantageous to measure a capacitor with the full working voltage applied. A typical example is shown in Fig. 6-4H.

6.7 Standard capacitance and inductance bridges

Capacitance can be measured on a standard capacitance bridge, rather than on a universal bridge. Standard bridges provide greater accuracy (in general) than universal bridges, but they are limited in the variety of measurements that can be made. Standard inductance bridges are also available, but they are not in such common use as capacitance bridges. Standard bridges operate on the principle of comparing an unknown against a standard. Figure 6-5A shows the basic standard bridge for capacitance measurement.

The capacitance of the unknown C_X is balanced by a calibrated, variable, standard capacitor (C_N), or by a fixed standard capacitor and a variable ratio arm, such as R_A. Such bridges have a direct-reading accuracy that seldom exceeds 0.1 percent.

For higher accuracy, resolution, and stability in capacitance measurements at audio frequencies, a bridge with transformer-ratio arms (Fig. 6-3D) has many advantages. Figure 6-5B, 6-5C, and 6-5D show three possible ways of balancing a simple transformer-ratio capacitance bridge, or simplicity; the generator and primary are not shown, but it is assumed that the two secondaries have 100 turns each and are driven so that there is 1 V/turn. The capacitor in the unknown arm is assumed to be 72 pF.

In Fig. 6-5B, the two ratio arms are equal and the bridge is balanced in the conventional way, with a variable standard capacitor adjusted to 72 pF. The detector current can also be adjusted by a variation in the voltage applied to a fixed, standard capacitor. In Fig. 6-5C, the standard capacitor is fixed at 100 pF. This is balanced against the 72-pF unknown connected to the +100-V end of the transformer. The main winding has 100 turns with taps every 10 turns, and the other winding has a tap at each turn. If, as shown, the second winding is connected to the 70-V tap on the main winding, and the capacitor to the 2-V tap on the second winding, the required 72-V is applied to the capacitor.

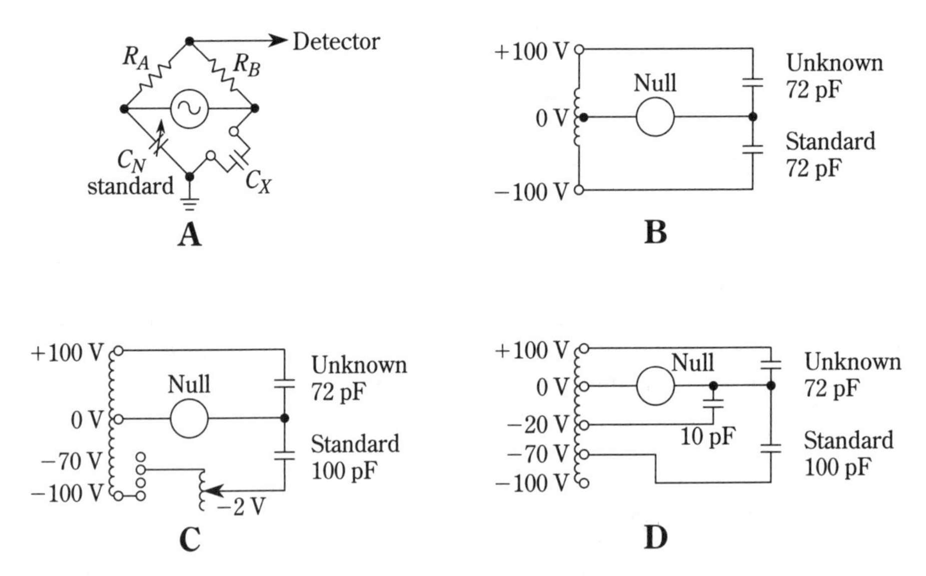

6-5 Basic standard bridge for capacitance measurement.

Another method of balance by voltage variation is shown in Fig. 6-5D. Here, a single decade divider is used in combination with multiple fixed capacitors. The 100-turn secondary is tapped every 10 turns to provide 10-V increments. If a 100-pF capacitor is connected to the 70-V tap and a 10-pF capacitor to the 20-V tap, the resulting detector current balances that of the 72-pF unknown connected to 100 V.

6.8 Digital bridges

The universal and standard bridges covered thus far require that controls (usually calibrated dials) be adjusted for balance as each capacitor (or other device being tested) is connected to the bridge. This means that accuracy and resolution depend on how closely the dial calibrations can be read. As in the case of digital meters, bridges with digital readouts are simpler to use, easier to read, and generally provide greater accuracy (or at least better resolution).

Figure 6-6 shows a typical capacitance bridge with digital readout. (the resolution is to within 0.1 pF). The transformer-ratio type of bridge (Fig. 6-5) is used. The circuit is in balance when the currents through the standard capacitor and the unknown capacitor are equal so that the current in the phase detector is zero. The range is chosen automatically by circuits that select decade taps on the transformer. The phase detector determines whether the current passing through the unknown arm of the bridge is higher or lower than that through the standard arm, and produces an error signal that indicates whether more or less voltage is required on the standard capacitor to reach a balance.

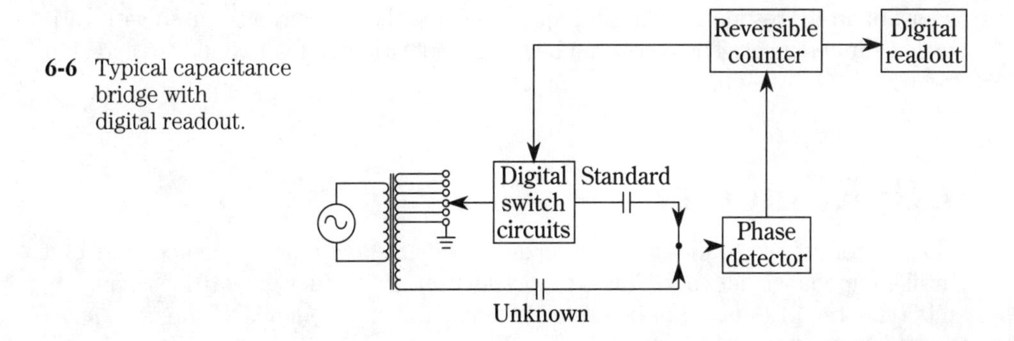

6-6 Typical capacitance bridge with digital readout.

This information is used by a reversible counter that controls, through digital-switch circuits, the standard-capacitor voltage. The counter counts in a direction to minimize the error signal until a balance is reached. At balance, the value of the unknown is displayed on a digital readout (that indicates in capacitance).

6.9 *Q* meters

Although it is possible to measure the Q of a coil with a universal bridge or a standard bridge, a Q meter often provides the most accurate results. Chapter 14 describes additional procedures for measurement of coil and resonant-circuit Q.

Figure 6-7 shows the basic circuit of a Q meter. The resistive voltage E_1 is not actually measured, but is held constant. The reactive voltage E_2 then is measured, but is read out in terms of Q. For example, if E_1 is 1 V, a 15-V E2 indicates a Q of 15, a 20-V E2 indicates a Q of 20 and so on. The meter can be digital (a digital voltmeter).

In many Q meters, provisions are made for inserting low impedances in series with the coil or high impedances in parallel with the capacitor. In this way, parameters of unknown circuits or components can be measured in terms of the effect on circuit Q and the resonant frequency. For example, assume that a coil is to be used

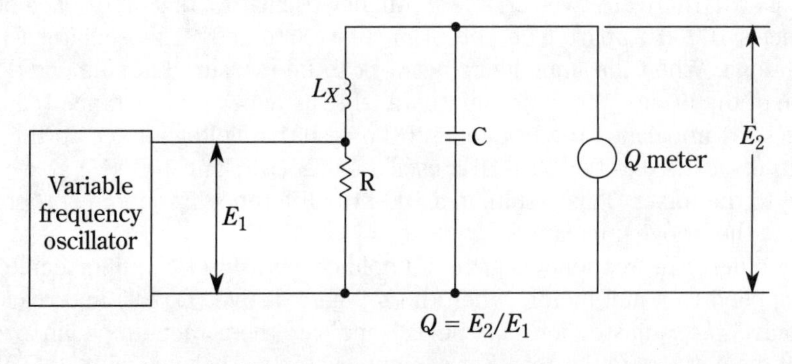

6-7 Basic Q-meter circuit.

as part of an RF resonant circuit (chapter 14), and that the circuit is used with various load impedances. The coil Q can be measured without a load, then with a load to find what effect the change has on coil Q.

6.10 R_X meters

An R_X meter is used to measure the separate resistive and reactive components of a parallel-impedance network. This is especially useful in the design of RF resonant circuits (chapter 14), but it can be used with any parallel impedance. Figure 6-8 shows the basic circuit of a typical R_X meter.

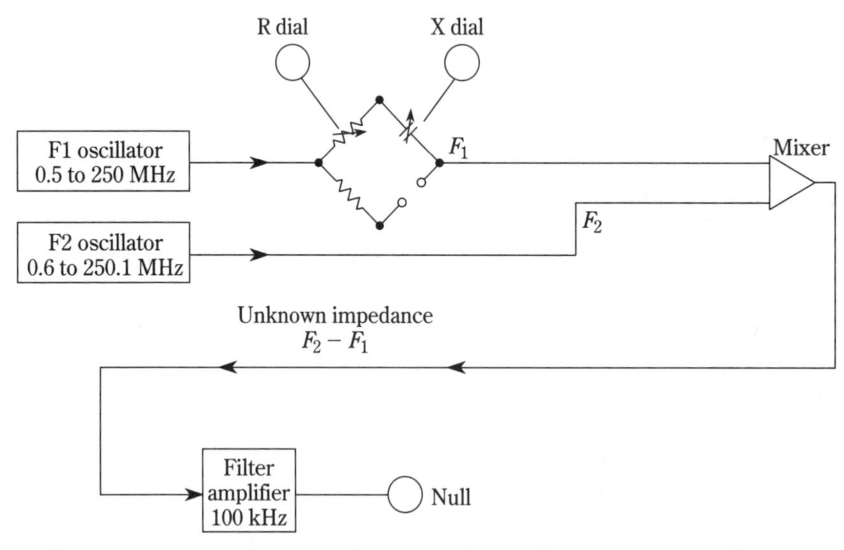

6-8 Basic R_X-meter circuit.

As shown, there are two variable-frequency oscillators that track each other at frequencies 100 kHz apart. The output of the 0.5- to 250-MHz oscillator F1 is fed into a bridge. When the impedance network to be measured is connected across one arm of the bridge, the equivalent parallel resistance and reactance (capacitive or inductive) unbalance the bridge, and the resulting voltage is fed to the mixer. The output of the 0.6- to 250.1-MHz oscillator F2 (tracking 100 kHz above F1) is also fed to the mixer. This results in a 100-kHz difference frequency, proportional in level to the bridge unbalance.

The difference-frequency signal is amplified by a filter/amplifier combination and is applied to a null meter. When the bridge resistive (R dial) and reactive (X dial) controls are adjusted for null, the dials indicate the parallel-impedance components of the network under test. For example, if you get balance with 50 Ω of resistance and 300 Ω of reactance, the network has these same values.

6.11 Admittance meters

In some cases, it is more convenient to measure characteristics in terms of admittance instead of impedance. This is particularly true when characteristics must be measured at high frequencies, such as the measurement of microwave or ultra-high-frequency antennas and transmission lines. Just as an impedance can be broken down into the real (resistive) and imaginary (reactive) components, admittance can be broken into conductance (real) and susceptance (imaginary) by means of an admittance meter.

Figure 6-9 shows the basic circuit of a typical admittance meter. Such meters are normally used at frequencies between 40 MHz and 1.5 GHz, but they can also be used at frequencies down to about 10 MHz. As a null meter, the admittance meter can be used to measure the conductance and susceptance of an unknown circuit directly. The meter can also be used as a comparator to indicate equality of one admittance to another (by measuring the degree of difference of one from another). An admittance meter can also be used to measure voltage standing-wave ratio (VSWR) and reflection coefficient. VSWR and reflection coefficient are covered further in chapter 15.

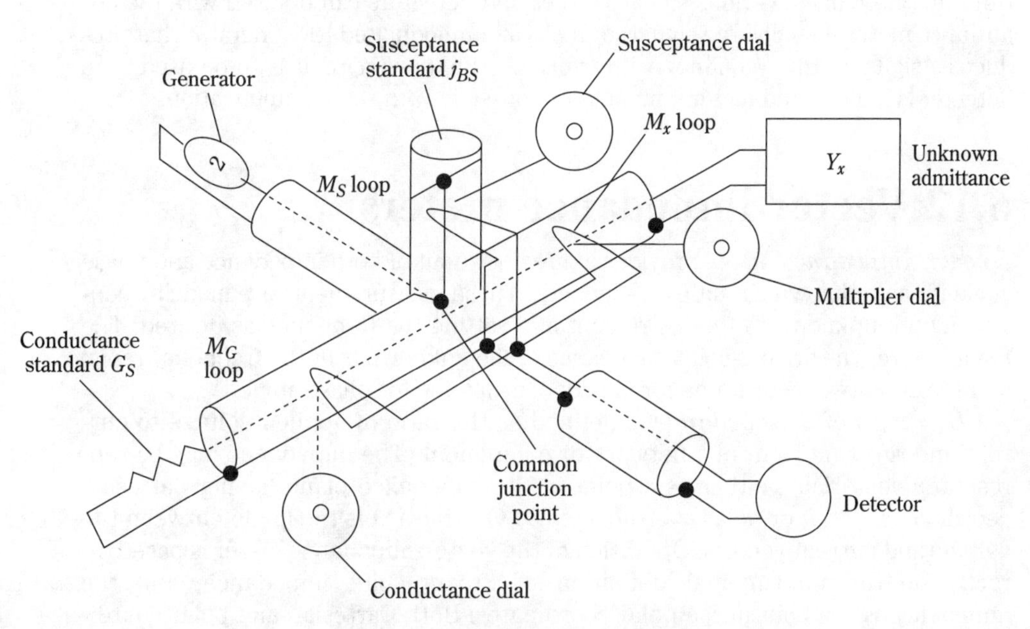

6-9 Basic admittance-meter circuit.

Although the admittance meter makes use of a null indication, it is not a true bridge. In an admittance meter, the currents flowing in three coaxial lines (fed from a common source at a common junction point) are sampled by three adjustable loops that couple to the magnetic field in each line, as shown in Fig. 6-9.

The coupling of each loop can be varied by rotating the loop. One of the coax lines is terminated in a conductance standard (G_S), which is a pure resistance equal to the characteristic impedance of the line. Another line is terminated in a susceptance standard j_{BS}, which is a short-circuited length of coax. The third line is terminated by the unknown circuit or network (Y_X). The outputs of the three loops are combined by connecting all three loops in parallel. When the loops are properly adjusted, the combined output is zero, indicated as a null on the detector.

At balance, the vector sum of the voltages induced in the three loops is proportional to the mutual inductance and to the current flowing in the corresponding line. Because all three lines are fed from a common source, the input voltage is the same for each line, and the current flowing in each line is proportional to the input admittance.

At balance, the sum of the three voltages is zero. Therefore, the amount of coupling required to produce balance is related directly to the amount of susceptance and conductance in the unknown circuit. The amount of coupling is indicated by corresponding dials attached to the three coupling loops (M_G, M_S, and M_X).

As shown in Fig. 6-9, a signal source, a detector, and the unknown circuit are connected to the meter. The three coupling loops are adjusted for a null indication on the detector, and the susceptance, conductance, and multiplying factor are read from the three dials. Various signal sources and detectors can be used with the admittance meter. Usually, the signal source is an unmodulated RF generator that produces a signal at the frequency with which the unknown circuit is to be used. The detector is a diode and meter combination, possibly with some amplification.

6.12 Vector-impedance meters

A *vector impedance meter* provides a direct readout of both impedance and phase angle for an unknown circuit or component. These measurements are made by connecting the unknown to the instrument and setting the frequency as desired. The readout gives the driving-point impedance of the unknown over the frequency range of interest on two meters, one for impedance and one for phase angle.

Driving-point impedance is defined as the ratio of applied voltage to current entering one point of a network or component. The impedance may be represented vectorially, either as a point in the impedance plane having Cartesian coordinates $R \pm j_X$ or polar coordinates $(Z)\ \Theta$, where Θ is the angle between the voltage and current vectors. Operation of the vector-impedance meter is based directly on the fundamental definition of driving-point impedance, and the impedance is read out in the polar coordinates. Both Cartesian and polar coordinates are discussed further in chapter 8.

Figure 6-10 shows a basic vector-impedance meter. In operation, a variable-frequency oscillator applies an unmodulated signal to an amplifier with leveled output. That is, the amplifier output remains constant despite changes in frequency or load. Current from the amplifier passes through the unknown component, mounted across terminals A and B. Current passes from the B terminal through a load to ground. The current through the unknown is used to generate a feedback signal that, in turn, levels the output of the amplifier. The amplifier-feedback arrangement then holds the current through the unknown at a constant level.

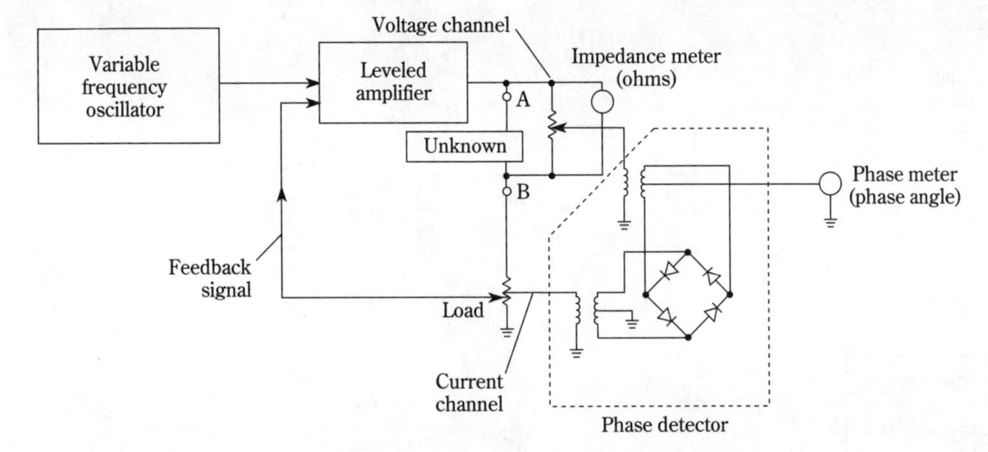

6-10 Basic vector-impedance meter circuit.

Because Z (driving-point impedance) equals E/I, and I is now a constant, Z is directly proportional to the voltage across the unknown. A voltmeter (digital or nondigital) is placed across terminals A and B, and the output is read in terms of impedance.

To measure phase angle, ac outputs from both the voltage and current channels are used to trigger a zero-crossing phase detector. The output of the phase detector is displayed on a meter calibrated directly in phase angle.

6.13 LC meters

Figure 6-11 shows the basic LC meter circuit. An LC meter is actually a reactance meter, but it reads out in terms of capacitance or inductance. LC meters are used primarily to measure the value of very small capacitors and coils. A typical LC meter provides a full-scale reading with capacitances of 0–3 pF and 0–3 μH, but can also read capacitances up to 300 pF and inductances up to 300 μH.

The capacitor or coil to be measured is connected into the resonant circuit of an oscillator. The frequency of this oscillator is compared to that of a reference oscillator in a mixer. A counter measures the difference in frequency and produces a meter reading in μH or pF.

Notice that the variable oscillator operates in a different manner for capacitance and inductance measurements. With capacitance measurements, the unknown capacitor is connected in parallel with the oscillator-circuit capacitor. This increases the capacitance and lowers the oscillator frequency from a nominal 140 kHz. With inductance measurements, the unknown coil is connected in series with the oscillator-circuit coils, also reducing oscillator frequency from 140 kHz.

When capacitance is measured, but the capacitor is not yet connected, the mixer receives two identical 140-kHz signals. The oscillators are identical with equal temperature compensation, so the output-frequency changes when temperatures are essentially the same (after warm-up). With the capacitor connected, the vari-

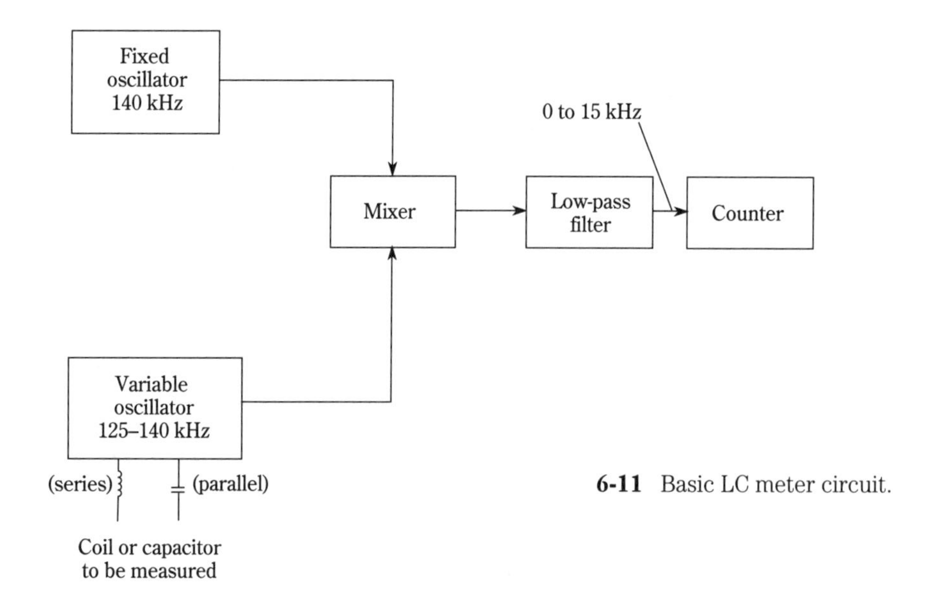

6-11 Basic LC meter circuit.

able-oscillator frequency is reduced to some value between about 125 and 140 kHz, and the mixer produces an output frequency equal to the difference between the two oscillator frequencies.

The filter prevents any 140-kHz signals from passing to the counter, but difference-frequency signals (in the range from 0 to 15 kHz) do pass to the counter. The counter output is actually the difference frequency, but is calibrated in terms of capacitance, typically pF.

When inductance is measured, but the coil is not yet connected, the variable oscillator is not yet operative, and the mixer receives only one 140-kHz signal from the reference oscillator. This signal does not pass to the counter because of the filter, so the counter reads zero. Connecting a coil energizes the variable oscillator, which produces a signal at a frequency between about 125 and 145 kHz. When this occurs, the mixer produces an output frequency equal to the difference, resulting in a corresponding counter readout (in terms of inductance, typically μH).

7
Two-junction (bipolar) transistor tests

This chapter is devoted entirely to test procedures for two-junction or bipolar transistors. The first sections of this chapter describe transistor characteristics and test procedures from the practical standpoint. The information in these sections permits you to test all of the important two-junction transistor characteristics using basic shop equipment. The sections also help you understand the basis for such tests. The remaining sections of the chapter describe how the same tests and additional tests are performed using more sophisticated equipment, such as the oscilloscope curve tracer.

7.1 Basic two-junction transistor tests

Transistors are subjected to a variety of tests during manufacture. It is neither practical nor necessary to duplicate all of these tests in the field. There are four basic tests required for most practical applications: *gain*, *leakage*, *breakdown*, and *switching time*. Unless a transistor is used for pulse or digital work, the switching characteristics are not of special importance.

In the final analysis, the only true test of a transistor is in the circuit within which the transistor is to be used. Except for special circumstances, a transistor will operate properly in-circuit, provided that: (1) the transistor shows proper gain, (2) the transistor does not break down under maximum operating voltage, (3) leakage, if any, is within tolerance, and (4) in the case of pulse circuits, the switching characteristics (delay time, storage time, etc.) are within tolerance.

There are two exceptions to this rule. Transistor characteristics change with variations in operating frequency and temperature. For example, a transistor might be tested at 1 MHz and show more than enough gain to meet circuit requirements. At 10 MHz, the gain of the same transistors might be zero. This can be caused by a number of factors. Any transistor has some capacitance at the input and output. As frequency increases, the capacitive reactance changes and, at some frequency, the transistor becomes unsuitable for the circuit (insufficient gain, unstable oscillation, etc.).

In the case of temperature, the current flow in any junction of the transistor increased with increases in temperature. A transistor might be tested for leakage at a nominal ambient temperature and show leakage well within tolerance. When the same transistor is used in-circuit, the temperature increases, increasing the leakage to an unsuitable level.

It is usually not practical to test transistors over the entire range of operating frequencies and temperature with which the transistor is used. Instead, the transistor is generally tested under the conditions specified in the datasheet. Then, the transistor characteristics are predicted at other frequencies and temperatures using datasheet graphs. In the case of some high-frequency RF applications, the transistor must be tested at the intended operating frequency in a special test circuit that is similar (if not identical) to the operating circuit.

7.2 Two-junction transistor leakage tests

For test purposes, both npn and pnp transistors can be considered as two diodes connected back to back. Thus, the procedures for transistor leakage tests are similar to those of diodes (chapter 10). In theory, there should be no current flow across a diode junction when the junction is reverse-biased. Any current flow under these conditions is the result of leakage. In the case of a transistor, the collector-base junction is reverse-biased and should show no current flow. However, in most practical applications, there is some collector-base current flow, particularly as the collector voltage is operated near the limits and as the operating temperature is increased.

7.2.1 Collector leakage

Collector-leakage current is designated as I_{cbo} or I_{bo} on most datasheets. Collector leakage might be termed *collector cutoff current* on other datasheets, where a nominal and/or maximum current is specified for a given collector-base voltage and ambient temperature. Collector-base leakage is normally measured with the emitter open, but it can also be measured with the emitter shorted to the base or connected to the base through a resistance.

Figure 7-1 shows the basic circuits for the collector-base leakage test. Any of the circuits can be used, but those of Fig. 7-1A and 7-1D are the most popular. The procedure is the same for all of the circuits in Fig. 7-1. The voltage source is adjusted for a given value (thus providing a given reverse bias), and the current (if any) is read on the meter. The current must be below a given maximum for a given reverse bias.

Temperature is often a critical factor in leakage measurements. For example, the maximum collector-leakage for a typical video-range transistor operating at 25°C is 2 μA, with 30 V applied between collector and base. When the collector-base voltage is lowered to 5 V (typical for digital circuits), and the temperature is raised to 150°C, the maximum collector leakage is 50 μA.

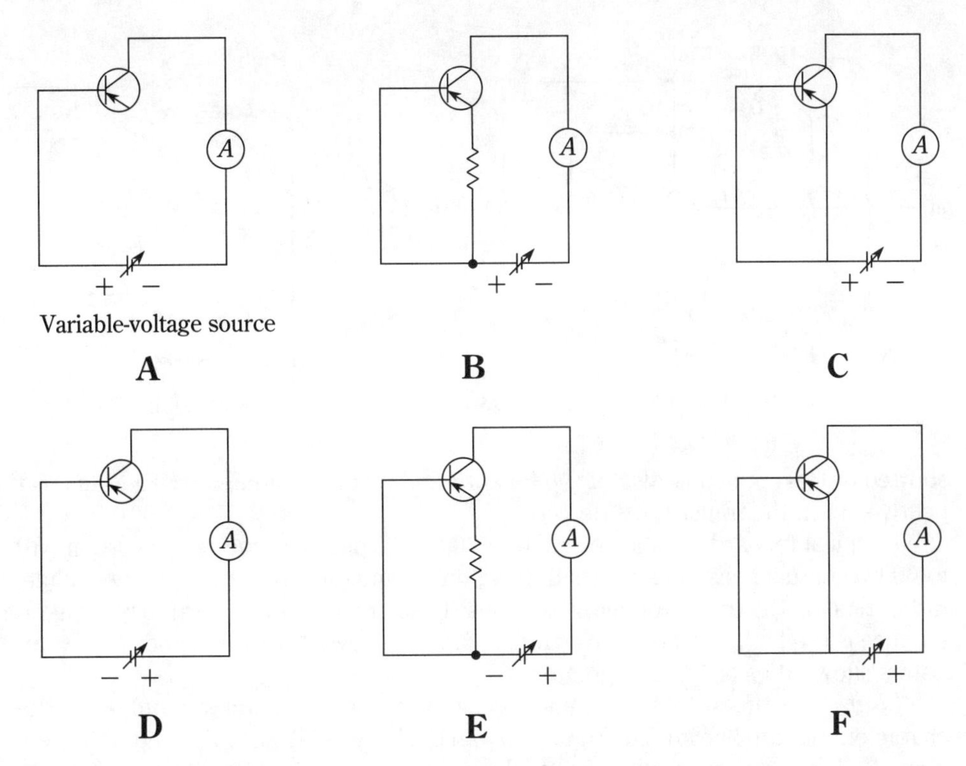

7-1 Basic circuits for collector-base leakage tests.

7.2.2 Emitter leakage

Some datasheets also specify emitter-base current leakage. This is not usually the case because the emitter-base junction is forward-biased for most circuits. If necessary, the circuits of Fig. 7-1 can be used to test emitter-base leakage (I_{eo} or I_{ebo}), except that the collector and emitter connections are interchanged. The emitter-base junction is reverse biased, the collector is left open, and the meter is placed in the emitter-base circuit. The procedures are the same as for collector-base leakage.

7.2.3 Testing two-junction transistor leakage with an ohmmeter

It is possible to quickly check transistor leakage with an ohmmeter. For the purpose of this test, the transistor is considered as two diodes connected back to back. Each diode should show low forward-resistance and high reverse-resistance. These resistances can be measured with an ohmmeter, as shown in Fig. 7-2.

The same ohmmeter range is used for each pair of measurements (base to emitter, base to collector, and collector to emitter). On low-power transistors, there might be a few ohms indicated from collector to emitter. Avoid using the R×1 range ohmmeter of an ohmmeter with a high internal-battery voltage. The voltage can damage a low-power transistor.

If both forward and reverse readings are very high, the transistor is open. Similarly, if any of the readings show a short or very low resistance, the transistor is

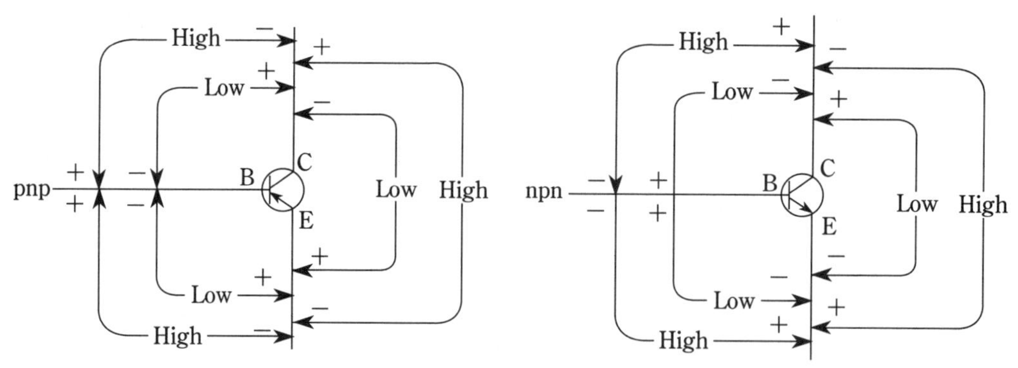

7-2 Transistor leakage tests with an ohmmeter:(a) pnp; (b) npn.

shorted or leaking badly. Also, if the forward and reverse readings are the same (or nearly equal), the transistor is defective.

A typical forward resistance is 300 to 700 Ω. Typical reverse resistances are 10 to 60 kΩ. Actual resistance values depend on ohmmeter range and battery voltage, rather than on the transistor characteristics. Thus, the ratio of forward to reverse resistance is the best indicator. Most transistors will show a 30:1 ratio, and some transistors show ratios of 100:1 or greater.

Caution: Transistors should not be tested for any characteristic unless all the characteristics are known. The transistor can be damaged if this rule is not followed. Even if no damage occurs, the test results can be inaccurate. Never test a transistor with voltages, or currents, higher than the rated values. The maximum current rating is often overlooked. For example, if a transistor is designed to operate with a maximum of 45 V at the collector, it could be assumed that a 9-V battery is safe for all measurements involving the collector. However, assume that the internal (emitter-to-collector) resistance of the transistor is 90 Ω, and that the maximum rated emitter-collector current is 25 mA. With 9 V connected directly between the emitter and collector, the emitter-collector current is 100 mA, four times the maximum rated 25 mA. This can cause the junctions to overheat and damage the transistor.

7.3 Two-junction transistor breakdown tests

The circuits and procedures for *transistor-breakdown tests* are similar to those for leakage tests. The most important breakdown test is to determine the collector-base breakdown voltage. In this test, the collector and base are reverse-biased, with the emitter open, and the voltage source is adjusted to a given value of leakage current. The voltage is then compared with the minimum collector breakdown voltage specified for the transistor.

As an example, using the same transistor described in Sec. 7.2.1, the minimum collector-breakdown voltage is 45 V (with 50 μA flowing and an ambient temperature of 25°C). If 50 μA flows with less than 45 V, the collector-base junction is breaking down.

Another breakdown test specified on some transistor datasheets is the *collector-emitter breakdown voltage.* In this test, the collector and emitter are reverse-biased with the base open. The voltage source is then adjusted for a given value of leakage current through both the emitter-base and collector-base junctions. The collector-emitter breakdown voltage test determines the conditions of both junctions simultaneously.

Breakdown voltage is designated as BV_{cbo} (collector to base emitter open), BV_{ces} (emitter shorted to base), or BV_{ceo} (collector to emitter, base open) on most datasheets. Breakdown is normally measured with the emitter (or base) open, but it can also be measured with the emitter shorted to the base or connected to the base through a resistor or with the emitter and base reverse-biased.

Figure 7-3 shows the basic circuits for breakdown tests. The circuits shown are for pnp transistors. The same circuit can be used for npn transistors when the voltage polarity is reversed. In all cases, the voltage source is adjusted for a given leakage current. Then, the voltage is compared with a minimum specified voltage.

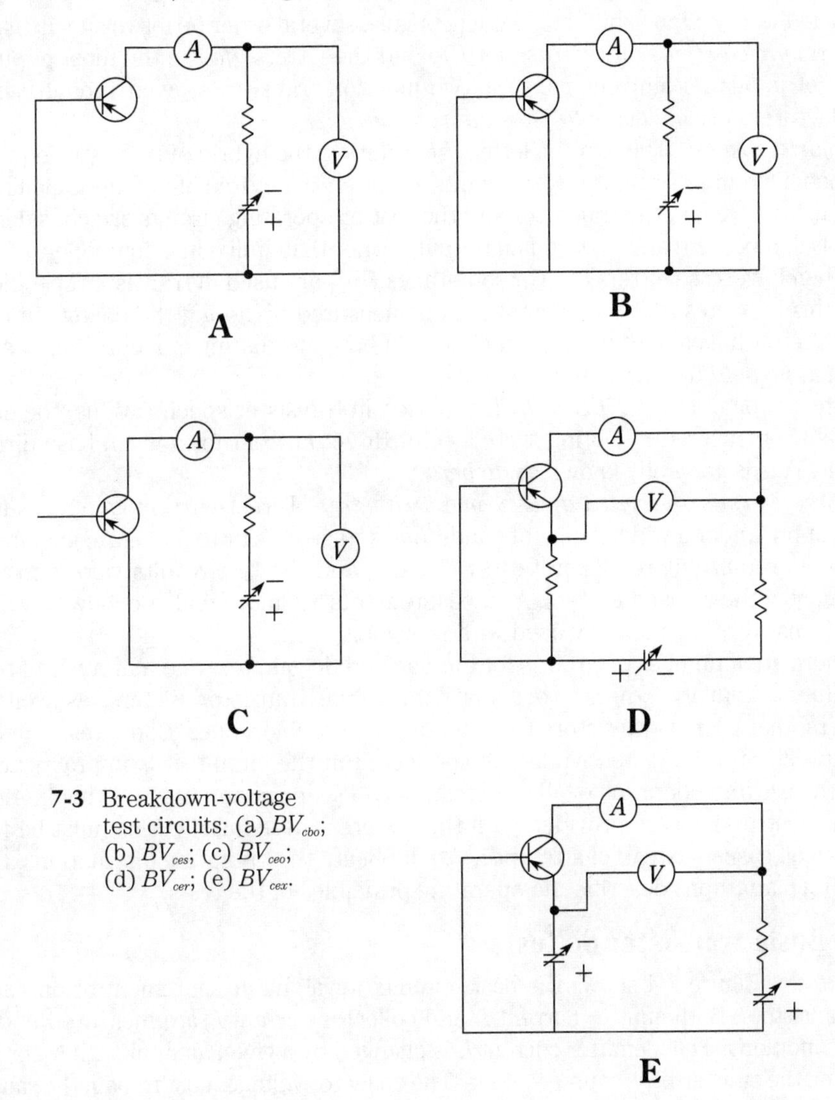

7-3 Breakdown-voltage test circuits: (a) BV_{cbo}; (b) BV_{ces}; (c) BV_{ceo}; (d) BV_{cer}; (e) BV_{cex}.

7.4 Two-junction transistor gain tests

The dynamic gain of a transistor is determined by the amount of change in output for a change in input. The change in output current, for a given change in input current, is measured without changing the output voltage.

When a transistor is connected in a common-base circuit, the collector forms the output circuit, and the emitter forms the output circuit. Common-base current gain is known as *alpha*, indicated by the lowercase Greek letter (α).

Most datasheets specify gain with the transistor connected in a common-emitter circuit, rather than in a common-base circuit. In the common-emitter circuit, the base is the input and the collector is the output. Current gain for a common-emitter circuit is known as *beta*, indicated by the lowercase Greek letter (β).

In addition to alpha and beta, datasheets use several other terms to specify gain. The term *forward current transfer ratio* and the letters *hfe* are the most popular means of indicating current gain for two-junction transistors, even though some manufacturers use *collector-to-base current gain.*

Hybrid system The *h* in the letters *hfe* refers to the hybrid of transistor-equivalent-model circuits. Transistor test circuits are often structured along the same lines. In the hybrid system, the transistor and the test or operating circuits are considered as a "black box" with an input and an output, rather than individual components.

When lowercase letters *hfe* (or sometimes *Hfe*) are used in transistor specifications, this indicates that the current gain is measured by noting the change in collector alternating current for a given change in base alternating current. This is also known as *ac beta* or *dynamic beta*.

When capital letters *HFE* or *hFE* are used in transistor specifications, the current gain is measured by noting the collector direct current for a given base direct current. This is generally known as *dc beta*.

Direct versus alternating gain measurements Direct-current gain measurements apply under a wider range of conditions and are easier to make. Ac gain measurements require more elaborate test circuits, and the test results vary with the frequency of the ac used for test. Ac measurements are more realistic, however, because transistors are normally used with ac signals.

There are a number of circuits for both ac and dc gain test and a number of test procedures. Similarly, there are many commercial transistor testers, as well as adapters, that permit transistors to be tested with oscilloscopes. Some testers permit transistors to be tested while still connected in the circuit. It is impractical to cover the use and operation of all such testers and scope adapters in this book. Also, detailed instructions are provided with the testers. These instructions must be followed in all cases. Instead of attempting to duplicate the operating instructions, the following paragraphs describe the operating principles of the tests.

7.4.1 Basic transistor dc tests

Alpha tests Figure 7-4 shows the basic circuits for alpha measurement of pnp and npn transistors. Both emitter current I_E and collector current I_C are measured under static conditions. Then, emitter current I_E is changed by a given amount with R_1, or by changing the emitter-base source voltage. The collector voltage must remain the same.

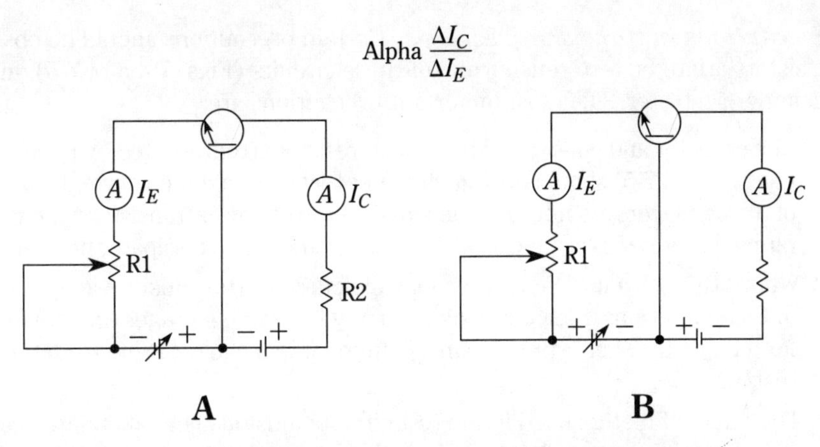

7-4 Basic circuits for alpha measurement of pnp and npn transistors.

The difference in collector current I_C is noted, and the value of alpha is calculated using the equation shown. For example, assume that the emitter current is changed 4 mA, and that this results in a change of 3 mA in collector current IC. This indicates a current gain of 0.75.

Beta tests Figure 7-5 shows the basic circuits for beta measurement of pnp and npn transistors. Both base current I_B and collector current I_C are measured under static conditions. Then, without changing the collector voltage, the base current I_B is changed by a given amount and the difference in collector current I_C is noted.

For example, assume that when the circuit is first connected, base current I_B is 7 mA, and collector current I_C is 43 mA. When base current I_B is increased to 10 mA (a 3-mA increase), collector current I_C is increased to 70 (a 27-mA increase). This represents a 27-mA increase in I_C for a 3-mA increase in I_B, a current gain of 9.

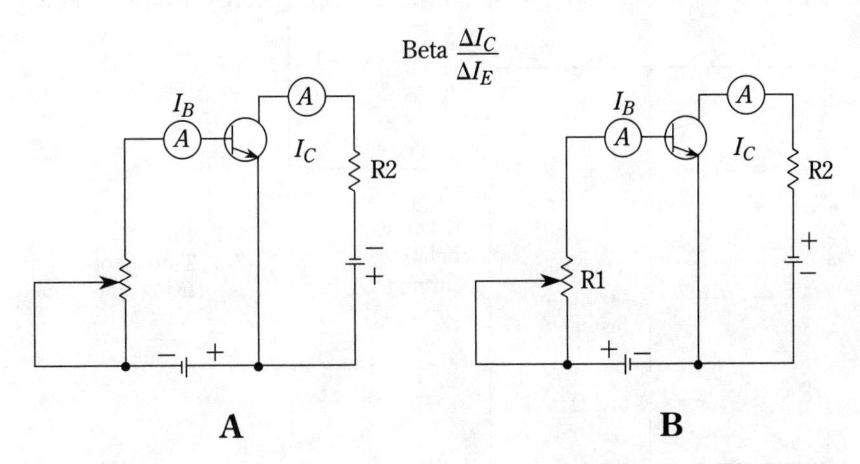

7-5 Basic circuits for beta measurement of pnp and npn transistors.

Precautions in transistor-gain tests Certain precautions should be observed if transistors are to be tested using any of these circuits (Figs. 7-4 and 7-5), instead of a commercial tester. The most important precautions are:

1. The collector and emitter (or base) load resistors (represented by R1 and R2 in Figs. 7-4 and 7-5) should be of such values that the maximum current limitations of the transistors are not exceeded. In the case of power transistors, the wattage rating of the load resistance should be large enough to dissipate the heat.
2. Where there is a large collector-leakage current, this must be accounted for in test conditions. Measure leakage using the same voltages and currents as for gain tests. Then, subtract any significant leakage currents from the gain-test currents.
3. The effect of meters used in the test circuits must also be taken into account. If I_B, I_C, and I_E are small (typically in the microvolt range) any current drawn by the meter might affect the tests.

7.4.2 Basic two-junction transistor ac gain tests

Several types of circuits are used for the ac or dynamic test of transistors. Some of the commercial transistor testers use the same basic circuits shown in Figs. 7-4 and 7-5, except that an ac signal is introduced into the input and gain is measured at the output. Where it is desirable to test transistors at high frequencies, some tester circuits permit an external high-frequency signal to be injected.

7.4.3 Testing two-junction transistor gain with an ohmmeter

It is possible to make a quick check of transistor gain with an ohmmeter. The basic circuit is shown in Fig. 7-6. Normally, there is little or no current flow between emitter and collector until the base-emitter junction is forward-biased. This fact can be used to provide a basic gain test of two-junction transistors.

In this test, the R × 1 range of the ohmmeter should be used. Any internal battery voltage can be used provided that it does not exceed the maximum collector-emitter breakdown voltage. In position A of S1, no voltage is applied to the base and the base-

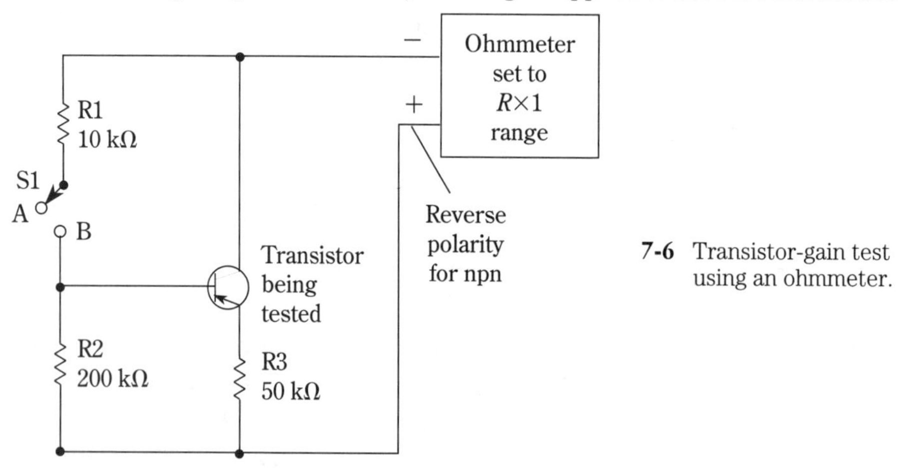

7-6 Transistor-gain test using an ohmmeter.

emitter junction is not forward-biased. Thus, the ohmmeter reads a high resistance. When S1 is set to B, the base-emitter circuit is forward-biased (by the voltage across R1 and R2) and current flows in the emitter-collector circuit. This is indicated by a lower resistance reading on the ohmmeter. A 10:1 (or better) resistance ratio is typical for most modern transistors, and often the ratio is several hundred to one.

7.4.4 Testing two-junction transistor RF gain

The tests described thus far can be used to establish the gain of a two-junction transistor operating at low frequencies. However, the tests do not establish the gain at RF. The only true test of circuit gain (either voltage gain or power gain) is to operate the device in a working circuit and to measure actual gain. The most practical method is to operate the device in the circuit with which the device is to be used. However, it might be convenient to have a standard or universal circuit for gain test.

Very often, the datasheets for transistors to be used in RF work show the circuits used to test gain. Such circuits are shown in Fig. 7-7. The circuits shown are for RF gain tests of a 2N5160 transistor operating at 175 and 400 MHz. The circuit can

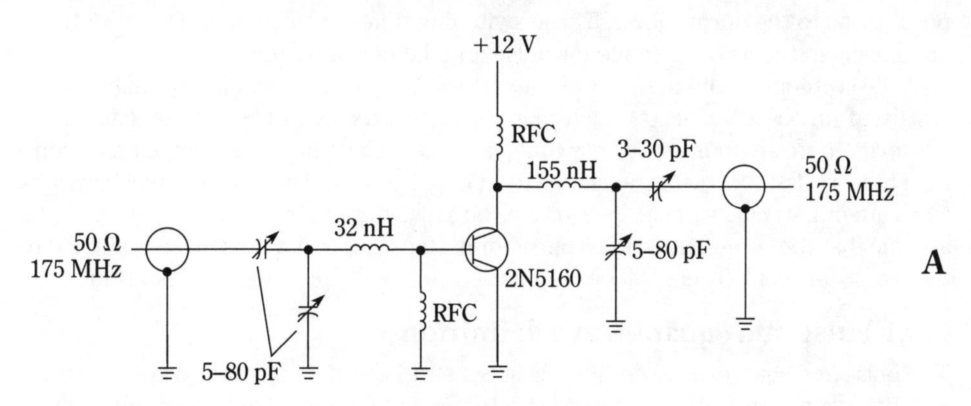

400-MHz test circuit

C1–10 pF
C2,C3,C4,C5–1.0–10 pF variable
C6–2400 pF feed-through
C7–0.004 µF
C8–0.01 µF
L1–30 nH, 1 turn, No. 20 AWG
L2–75 nH, 3 turn, No. 20 AWG
L3–0.33 µH, RFC

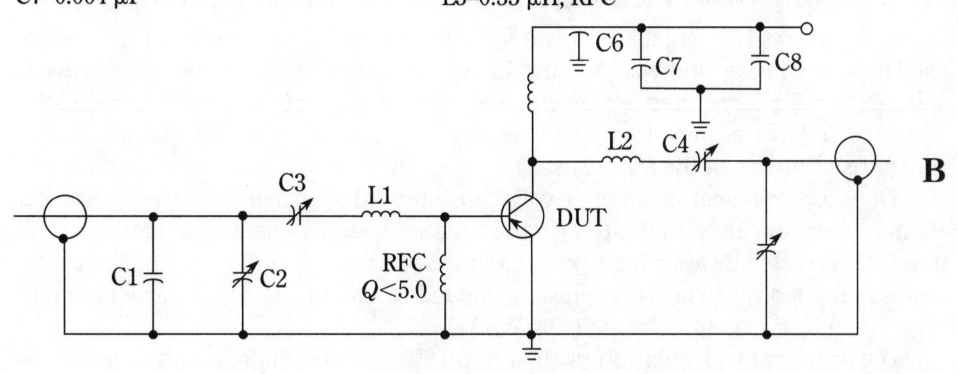

7-7 Typical RF-test amplifier circuits for two-junction transistors.

be modified to test other transistors, at other frequencies, by changing the circuit values. If you are not familiar with the design of such RF circuits, read the author's best-selling *Lenk's RF Handbook* (McGraw-Hill, 1992).

Using either circuit in Fig. 7-7, you simply introduce an RF signal (at the frequency of interest) to the input, and measure both the input and output voltage with an RF voltmeter (or meter with an RF probe). If power-gain measurement is needed, use input and output load resistances (noninductive or composition resistors). Measure the RF voltages across the input and output, then calculate the power gain using $P = E^2/R$. The ratio of output power divided by input power is the power gain.

7.5 Two-junction transistor-switching tests

Transistors used in pulse or digital applications must be tested for switching characteristics. For example, when a pulse is applied to the input of a transistor, there is a measurable time delay before the pulse appears at the output. Similarly, when the pulse is ended at the input, there is additional time delay before the transistor output returns to the normal level. These switching times or turn-on and turn-off times are usually in the μs or ns range for pulse and digital transistors.

The switching characteristics of transistors designed for computer or digital work are listed on the datasheets. Each manufacturer lists its own set of specifications. However, there are four terms (rise time, fall time, delay time, and storage time) common to most datasheets for transistors used in pulse work. These switching characteristics are of particular importance where the pulse durations are short. For example, assume that the turn-on time of a transistor is 10 ns and that a 5-ns pulse is applied to the transistor input. There will be no pulse output or the pulse will be distorted.

7.5.1 Pulse and square-wave definitions

The pulse and square-wave definitions in Sec. 2.14.2 and Fig. 2-25 also apply to transistors. Again, remember that the definitions are for guide purposes only. When pulses are irregular (excessive tilt, droop, sag, overshoot, etc.), the definitions might become ambiguous. However, the relationships for rise time, fall time, delay time, and storage time should remain the same.

7.5.2 Testing two-junction transistors for switching time

Figure 7-8 shows two circuits for testing the switching characteristics of transistors used in pulse or computer work. Figure 7-8A is used when a dual-trace scope is available. Figure 7-8B requires a conventional single-trace scope. In either case, the scope must have a wide frequency response and good transient characteristics (faster rise times than the pulses used).

The scope vertical channel is voltage-calibrated as usual, and the horizontal channel must be time-calibrated (rather than sweep-frequency-calibrated). The transistor is tested by applying a pulse to the base, with a specific bias applied to the base simultaneously. The same pulse is applied to one of the scope vertical inputs (Fig. 7-8A) or the monitor scope (Fig. 7-8B).

The transistor collector output (inverted 180° by the common-emitter circuit) is applied to the other scope vertical input (Fig. 7-8A) or to the monitor scope (Fig.

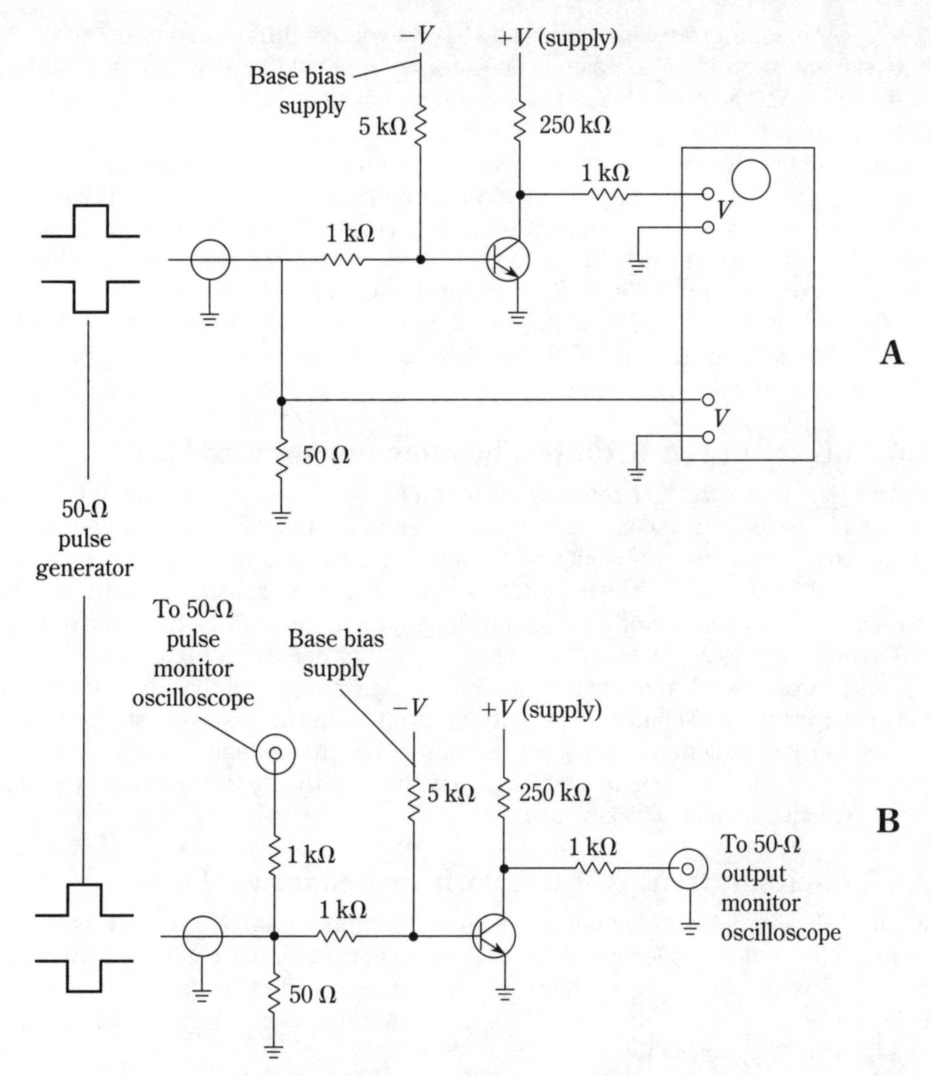

7-8 Switching-test circuits for two-junction transistors.

7-8B). In the Fig. 7-8B circuit, the same scope can be moved between the two monitor points. The two pulses (input and output) are then compared with rise time, fall time, delay time, storage time, and so on. The transistor output pulse characteristics can then be compared with transistor specifications. Remember the rules of thumb for pulse testing, which are described in Sec. 2.14.2.

7.6 Testing transistors in circuit

The normal forward-bias characteristics of transistors can be used to test transistor circuits without removing the transistor, and without using an in-circuit tester. Germanium transistors normally have a voltage differential of 0.2 to 0.4 V between emitter and base. Silicon transistors normally have a voltage differential of 0.4 to 0.8 V. The polarities at the emitter and base depend on the type of transistor (npn or pnp).

The voltage differential between the emitter and base acts as a forward bias for the transistor. That is, a sufficient differential or forward bias turns the transistor on, resulting in a corresponding amount of emitter-collector flow. Removal of the voltage differential, or an insufficient differential, produces the opposite results. That is, the transistor is cut off (no emitter-collector flow or very little flow).

The following sections describe two methods of testing transistors in-circuit. One method involves removal of the forward bias. The other method introduces an external forward bias to the circuit.

7.6.1 In-circuit transistor test by removing forward bias

Figure 7-9A shows the test connections for an in-circuit test by removal of forward bias. First, measure the emitter-collector differential voltage under normal circuit conditions. Then, short the emitter-base junction and note any change in emitter-collector differential. If the transistor is operating, the removal of forward bias causes the emitter-collector current flow to stop, and the emitter-collector voltage differential increases (the collector voltage rises to or near the supply value).

As an example, assume that the supply voltage is 12 V and that the differential between collector and emitter is 6 V when Q1 is operating normally (no short). When the emitter-base junction is shorted, the emitter-collector voltage differential should rise to (or near) 12 V. If there is no change with (or without) the short, the transistor is defective, probably leaking badly.

7.6.2 In-circuit transistor test with applied forward bias

Figure 7-9B shows the test connections for test by application of forward bias. First, measure the emitter-collector differential under normal circuit conditions (or measure the voltage across R4, as shown). Next, connect a 10-kΩ resistor between the collector and base, and note any change in emitter-collector differential (or any change in voltage across R4).

If the transistor is operating, the application of forward bias causes the emitter-collector current flow to start (or increase), and the emitter-collector voltage differential decreases (or the voltage across R4 increases). Again, if there is no change with (or without) the forward bias, the transistor is defective.

7.6.3 Go/no-go tests

The tests in Fig. 7-9 show that the transistor is operating on a go/no-go basis, which is usually sufficient for most troubleshooting applications. However, the tests do not show transistor gain or leakage and do not establish operation of the transistor at high frequencies. For these reasons, some troubleshooters reason that the only satisfactory test of a transistor is in-circuit operation. If the transistor does not perform the intended function in a given circuit, the transistor must be replaced.

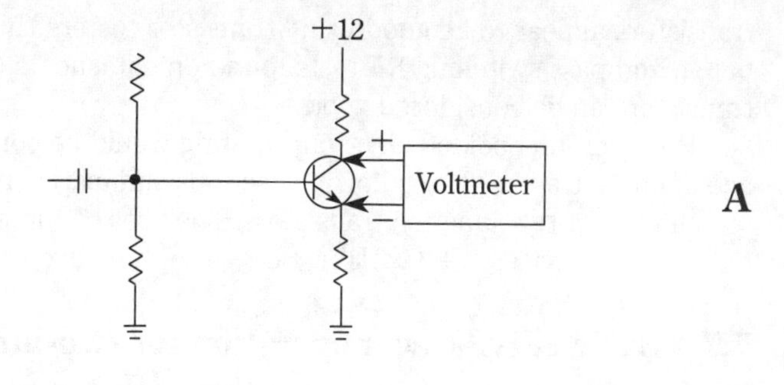

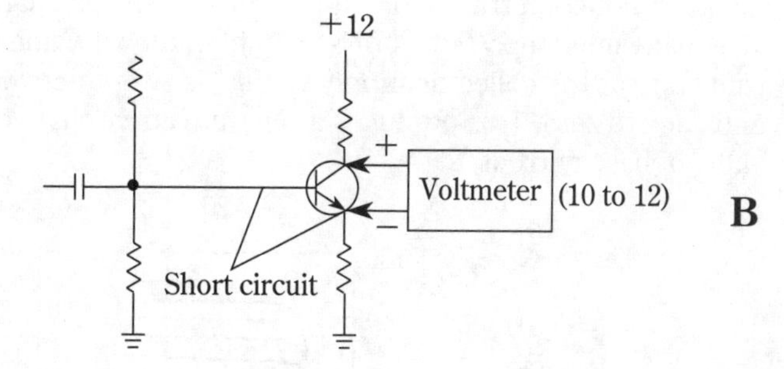

7-9 In-circuit two-junction transistor tests.

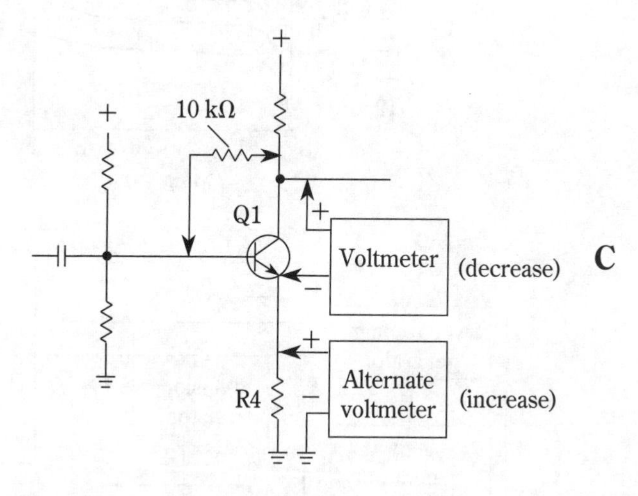

7.7 Two-junction transistor tests using a curve tracer

At one time, there were a number of oscilloscopes (or *oscilloscope adapters*, called *curve tracers*) manufactured specifically to display transistor characteristic curves. Curve tracers have all but disappeared from typical shop use, and now appear only in the lab. For this reason, and because curve-tracer instruction manuals are quite detailed as to operation, curve tracers are not described to the full extent here. If transistors appear to be good using transistor testers (in circuit or out of circuit, or both), and pass switching/RF tests found on datasheets (or in Sec. 7.4 and 7.5), the transistor can be considered good.

However, no book on electronic testing would be complete without some coverage of curve tracers. The following sections summarize basic curve-tracer operating techniques for two-junction transistors. The uses of curve tracers to test characteristics of other devices (FETs, UJTs, SCRs, etc.) are given in the appropriate chapters.

7.7.1 Basic curve-tracer operation for two-junction transistor tests

Figure 7-10 shows the functional block diagram of a typical curve tracer. When testing a two-junction transistor, the tracer introduces changes in base current in form of equal-value steps (steps of selectable, known value). These steps occur at the same rate as the collector supply voltage is swept between 0 V and some peak value and back to zero. This produces a separate curve that corresponds to each different value of base current.

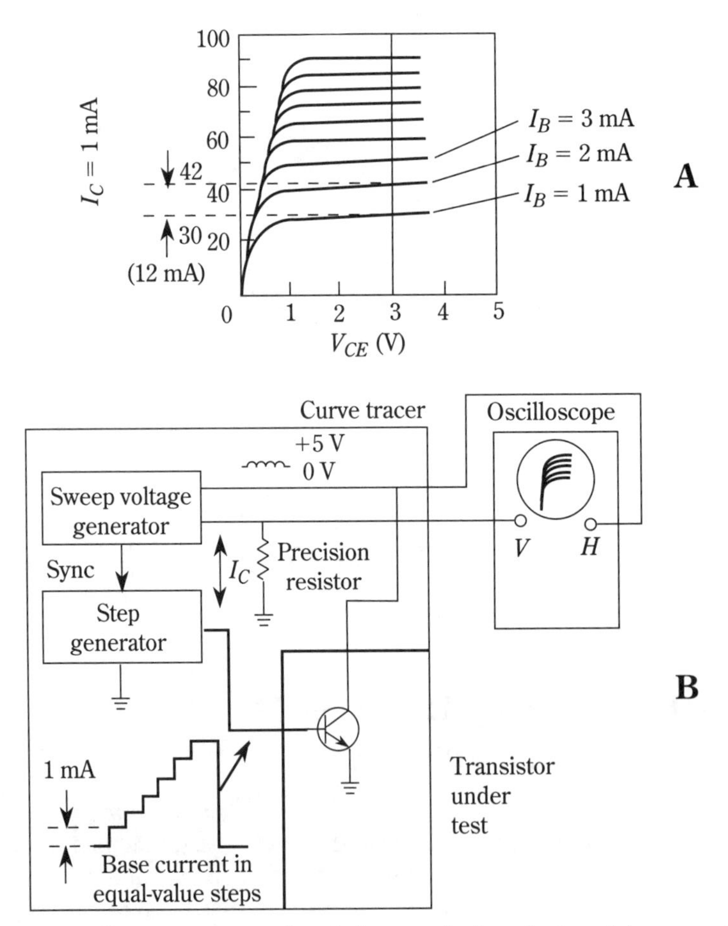

7-10 Basic test connections and typical scope displays for transistor curve tracer.

When curves show collector-current versus collector-voltage (for different values of base current), the change in collector current induced by one step of the base current is proportional to the vertical distance between adjacent curves. This change can be read directly from the screen scale. Which vertical line is chosen for the scale depends on what collector voltage is specific (because each vertical line corresponds to a particular collector voltage).

As shown in Fig. 7-10, the transistor is connected into a grounded-emitter circuit. Precise 1-mA current steps are applied to the base. Voltage sweeps from 0 to about 5 V are applied to the collector. The scope vertical deflection is obtained from the resulting collector current. All of this is done by setting switches on the curve tracer.

On some curve tracers, each sweep must be initiated individually, whereas other instruments produce a series of 10 (or more) curves in sequence automatically. Once the curves are made, they are interpreted as follows:

Choose the vertical line corresponding to the specified collector voltage. For example, the 3-V vertical line is chosen in Fig. 7-10. Note (on that line) the distance between the two curves that appear above and below the specified base current (or specified collector current).

As an example, assume that the transistor of Fig. 7-10 is to be operated with a collector current of 30 to 40 mA, with 3 V at the collector. The two most logical curves are then the 1- and 2-mA base-current curves. The distance between these two curves represents a difference of 12 mA in collector current. The 1-mA curve intersects the 3-V line at 30 mA, and the 2-mA curve intersects the 3-V line at 42 mA.

Divide the collector-current difference by the base-current difference that caused it. Because the base current per step is 1 mA and the difference in collector current is 12 mA, the gain (beta) is 12. If the base current per step is 0.1 mA (as it is for some curve tracers), and all other conditions are the same, the beta is 120 (12/0.1 = 120).

7.7.2 ac beta tests with a curve tracer

Figure 7-11 shows the technique of ac beta measurement using a curve tracer (this is one of the major uses for curve tracers). As stated, the ac or dynamic current gain of a two-junction transistor can be defined as the ratio of change in collector current to the change in base current (at a specified collector voltage).

Measure the difference in collector current (ΔI_C) between the two curves on the scope display. (The settings of the curve tracer and/or scope controls determine the amount of collector current, represented by each vertical division of the display scale. In Fig. 7-11, each vertical division represents 2 mA of collector current.) Be sure that both curve readings are taken at the same collector voltage. In Fig. 7-11, both readings are taken at 5 V.

Note the change in base current (ΔI_B) that produces each curve. In Fig. 7-11, each curve is produced by a change of 10 µA. Thus, ΔI_B is 10 µuA for any of the curves. Calculate beta by dividing ΔI_C by ΔI_B. For example, if ΔI_C is 2 mA and ΔI_B is 10 µA, as shown in Fig. 7-11A, beta is 200 (2/0.01 = 200).

It is generally easier to use the two centermost curves of the display to measure ac beta. However, this is not always practical. If so, the ΔI_C measurement can be made between two nonadjacent curves. For example, the difference between the collector current of the second and fourth curves can be made for measurement of

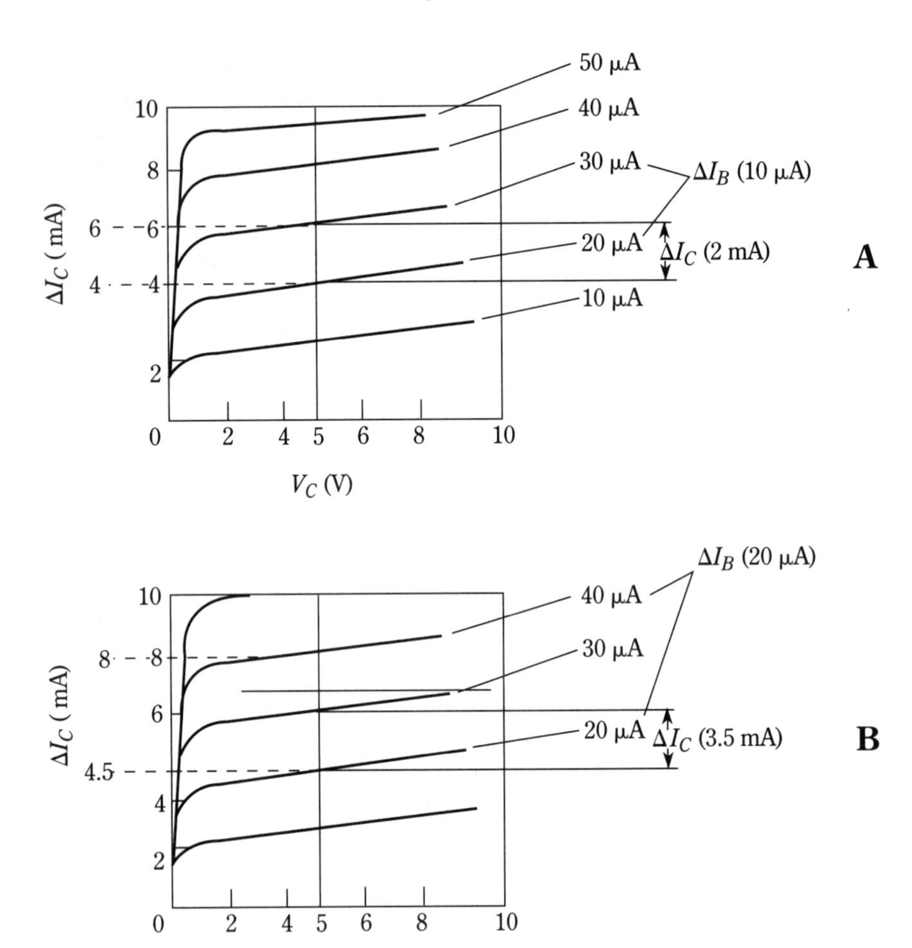

7-11 Ac beta measurement using a curve tracer.

ΔI_C, as shown in Fig. 7-11B. If this method is used, make certain to use two steps of base current for determining ΔI_B when calculating beta. Using the values of Fig. 7-11B, ac beta 3.5 mA/20 μA = 175 (at a V_C of 5 V).

If the transistor datasheet is available, measure the ac beta at the approximate collector current and voltage specified. If no values are specified, adjust the curve-tracer controls for a display of the most evenly and widely spaced curves.

8
Field-effect transistor (FET) tests

This chapter is devoted entirely to test procedures for *field-effect transistors (FETs)*. The first sections of the chapter describe FET characteristics and test procedures from a practical standpoint. The information in these sections permits you to test all important FET characteristics using basic shop equipment, and also helps you understand the basis for such tests. The remaining sections describe how the same tests, and additional tests, are performed using more sophisticated equipment.

8.1 FET operating modes

As in the case of two-junction transistors, FETs are tested on both a static and dynamic basis. Static tests indicate the FET response to dc variations. Dynamic tests show the response to ac or signals. FETs operate in three modes, as shown in Fig. 8-1. Also, although both JFETs and MOSFETs operate on the principle of a "channel" current controlled by an electric field, the control mechanisms of the two are different, resulting in considerably different characteristics. The main difference between the JFET (junction FET) and the MOSFET (metal-oxide-semiconductor FET) is in the gate characteristics. The input of the JFET acts like a reverse-biased diode, whereas the input of a MOSFET is similar to a capacitor.

8.1.1 Depletion mode

The *depletion-only FET* (Fig. 8-1A) is classified as type A and has considerable drain-current (I_D) flow for zero gate voltage. No forward gate voltage is used. Maximum drain-current flow occurs when the gate-source voltage (V_{GS}) is zero. Drain current is reduced by applying a reverse voltage to the gate terminal. That is, I_D decreases when reverse V_{GS} is increased.

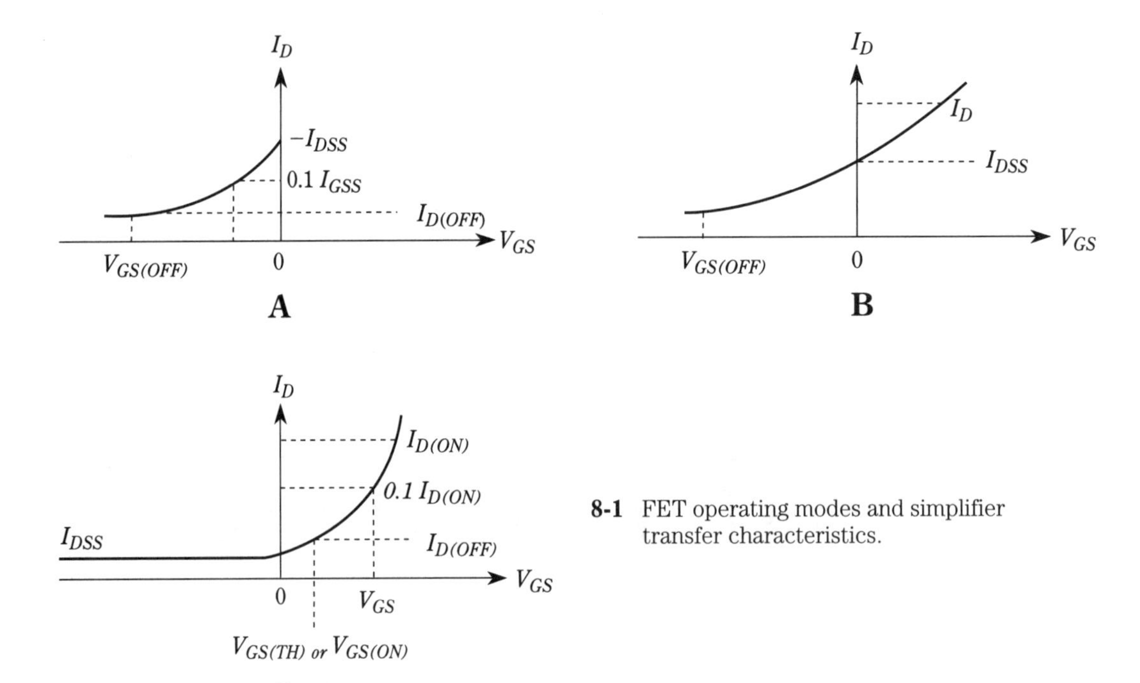

8-1 FET operating modes and simplifier transfer characteristics.

8.1.2 Depletion/enhancement mode

The *depletion/enhancement FET* (Fig. 8-1B) is classified as type B and also has considerable drain-current flow for zero V_{GS} (but not as much as type A). Drain current is increased by application of a forward gate voltage and reduced by application of a reverse gate voltage. For a JET used as type B, drain current can be increased by gate voltage only until the gate-source PN junction becomes forward-biased. At this point, a further increase in forward gate voltage does not produce an increase in drain current.

8.1.3 Enhancement mode

The *enhancement-only FET* (Fig. 8-1C) is classified as type C and has little or no current flow for zero gate voltage. Drain current does not occur until a forward gate voltage is applied. This voltage is known as the *threshold voltage*, and is indicated as $V_{GS(TH)}$. Once the threshold is reached, the transfer characteristics for a type-C FET are similar to those of the type B.

8.2 Handling MOSFETS

It is assumed that you are familiar with handling procedures for MOSFETs during test. If not, here is a quick review.

To prevent possible damage from static charges when removing or handling a MOSFET, first turn the power off. If the MOSFET is to be removed, your body should be at the same potential as the unit from which the device is removed. This can be done by placing one hand on the chassis before removing the MOSFET. If the MOS-

FET is to be connected to an external tester, put the hand holding the MOSFET against the tester panel and connect the leads from the tester to the MOSFET leads (or plug in the MOSFET). *Caution*: Make certain that the external unit is at ground potential before touching it. In some obsolete or defective equipment, the panel or boards are above ground (typically by one half of the line voltage).

When handling a MOSFET, the leads should be shorted together. Generally, this is done in shipment by a shorting ring or spring. When testing MOSFETs, connect the tester lead to the MOSFET (preferably the source lead first). Then, remove the shorting ring.

When soldering or unsoldering a MOSFET, the soldering-tool tip should be at ground potential (no static charge). Connect a clip lead from the barrel of the soldering tool to the tester or board. The use of a soldering gun is not recommended for MOSFETs.

Remove power to the circuit before inserting or removing a MOSFET (or a plug-in module containing a MOSFET). The voltage transients developed when terminals are separated might damage the MOSFETs. The same condition applies to circuits with conventional transistors. However, the chances of damage are greater with MOSFETs.

8.3 MOSFET protection circuits

Because of the static-discharge problem, MOSFETs are often provided with some form of protection circuit. Generally, this takes the form of a diode or diodes incorporated as part of the MOSFET substrate material. These protection circuits can be ignored, except when breakdown-voltage tests are performed, as described in Sec. 8.7.

In testing a MOSFET for breakdown, the main point to remember is that the breakdown voltage is that of the protective device (diode), not the MOSFET. Also, datasheets sometimes list the *protective-diode clamp voltage* (often referred to as the *knee voltage* (V_{KNEE}). In any event, gate or input voltages cannot exceed the knee value, even though the drain and source voltages might be higher.

8.4 FET control-voltage tests

The *gate-source voltage* (V_{GS}) is considered to be the control voltage (or signal) for a FET. That is, the amount of V_{GS} controls the amount of source-drain current. $V_{GS(OFF)}$ is the gate voltage that is necessary to reduce the drain-source current (I_D) to zero or to some specified value near zero. $V_{GS(TH)}$ is more descriptive for enhancement-only (type C) mode, and is the gate voltage where I_D just starts to flow. Some datasheets specify $V_{GS(OFF)}$ as the gate voltage that produces some given value of I_D, such as 1 pA or 1 nA.

8.4.1 $V_{GS(OFF)}$ test

Figure 8-2 shows the basic circuit for the $V_{GS(OFF)}$ test. This test applies to depletion-only and depletion/enhancement-mode FETs. As shown, VDS is set at some fixed value, and reverse-bias V_{GS} is adjusted until I_D is at some specific negligible value. This is essentially a cutoff value test. As an example, the N-channel 3N128 specifies a V_{DS} of 15 V and an I_D of 50 μA. The V_{GS} should be between –0.5 and –8 V for

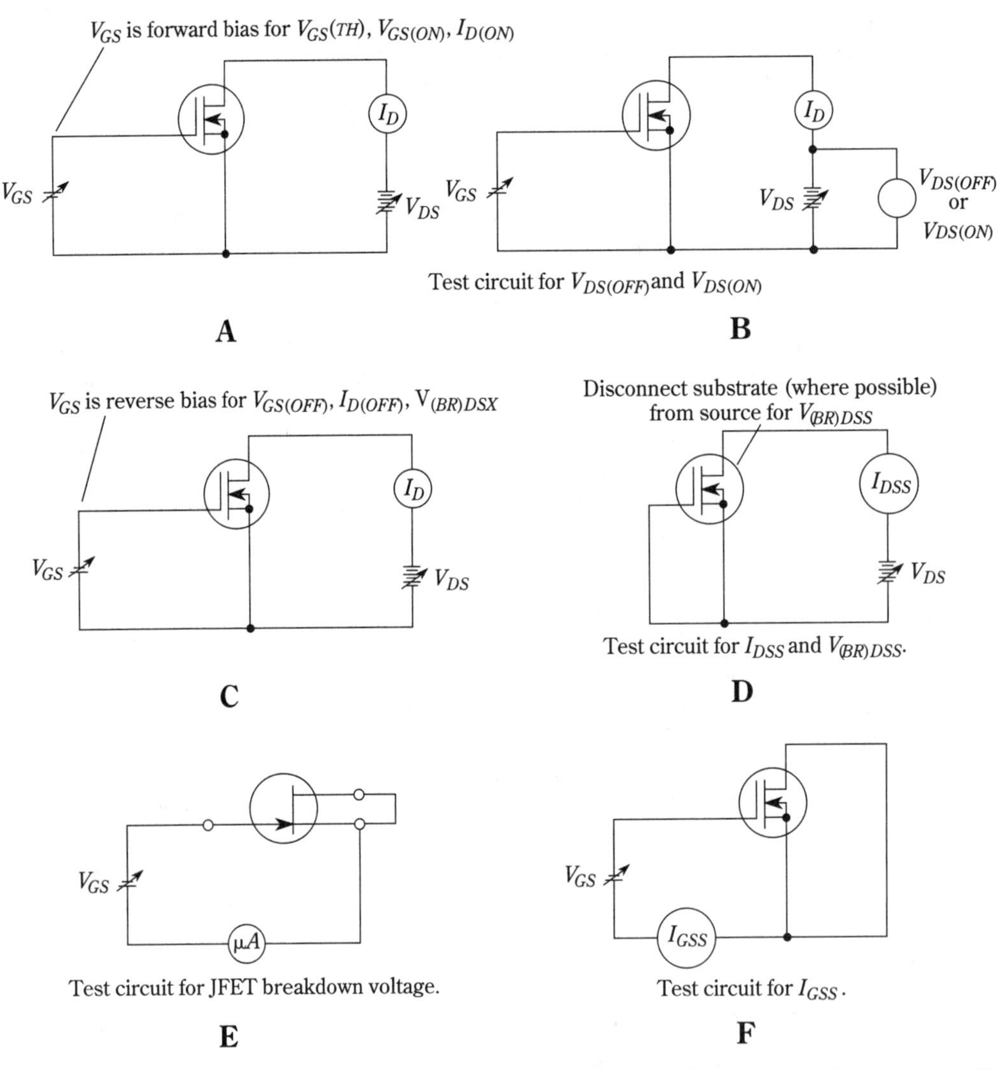

8-2 FET control-voltage, operating-voltage, operating-current, and breakdown-voltage test circuits.

the 50 μA of I_D. As a guideline, VDS should be a minimum of 1.5 times $V_{GS(OFF)}$ to provide proper circuit operation.

8.4.2 $V_{GS(TH)}$ or $V_{GS(ON)}$ test

$V_{GS(TH)}$ or $V_{GS(ON)}$ are essentially the same specification as $V_{GS(OFF)}$, except that $V_{GS(ON)}$ and $V_{GS(TH)}$ are generally applied to enhancement-only devices. The basic circuit of Fig. 8-2 is used, except that V_{GS} is connected for forward bias (gate positive for N channel). A forward bias is required because I_D flows in an enhancement-mode device only when the gate is forward-biased. During test, V_{GS} is increased from zero volts until the I_D is at a specific value. For example, an MRF137 has a typical $V_{GS(TH)}$ of 3 V with an I_D of 25 mA.

8.5 FET operating-voltage tests

The *drain-source voltage* (V_{DS}) is considered to be the operating voltage for a FET. V_{DS} is the equivalent of collector voltage in a two-junction transistor. Generally, the only concern is that the maximum datasheet value for V_{DS} must not be exceeded. In most applications, however, V_{DS} must be a minimum of 1.5 times V_{GS}.

When FETs are used as switches and choppers, the terms $V_{DS(ON)}$ and $V_{DS(OFF)}$ sometimes appear on datasheets. $V_{DS(ON)}$ is similar to saturation voltage in two-junction transistors. $V_{DS(OFF)}$ is the drain-source voltage at which the I_D increases very little for an increase in train-source voltage, with V_{GS} held at zero. Sometimes, the term V_P, the pinch-off voltage, is used instead. However, V_P generally applies to JFETs, rather than to MOSFETs.

The various values of V_{DS} depend on source resistance R_S, drain resistance R_D, or drain-to-source resistance R_{DS}. All of these are static dc values and rarely appear on datasheets. Dynamic values are generally far more important. Although FETs do not usually have an operating voltage between drain and gate, the term V_{DG} usually appears on most FET datasheets. Generally, this is a maximum voltage and is usually the same as V_{DG} maximum.

8.5.1 $V_{DS(OFF)}$ and $V_{DS(ON)}$ saturation tests

The basic circuit for saturation tests is shown in Fig. 8-2. Saturation tests are generally of concern only when the FET is used for switch or chopper applications. During either test, V_{GS} is held at some fixed value, and V_{DS} is increased until I_D is at maximum, or increases very little for further increases in V_{DS}. Take care not to exceed the maximum V_{DS} when making this test. Generally, $V_{DS(ON)}$ is made with some specific value of V_{GS}, whereas $V_{DS(OFF)}$ is made with V_{GS} at zero (gate shorted to source).

8.6 FET operating-current tests

The drain-source current (I_D) is considered as the operating current for a FET. I_D is the equivalent of collector current in a two-junction transistor. Generally, the only concern is that the maximum datasheet value for I_D must not be exceeded.

In addition to I_D, the terms $I_{D(ON)}$ and $I_{D(OFF)}$ are used on some datasheets. $I_{D(ON)}$ is an arbitrary current value (usually near the maximum rated current) that locates a point in the enhancement operating mode (Fig. 8-1). $I_{D(OFF)}$ is the current that flows when $V_{GS(OFF)}$ is applied. On some FETs operating in the enhancement-only mode, the datasheet value of $I_{D(OFF)}$ is the current when $V_{GS(TH)}$ is applied.

8.6.1 I_{DSS} test

The basic circuit for the I_{DSS} test is shown in Fig. 8-2. This test is applied to depletion-only and depletion/enhancement-mode devices. As shown, V_{DS} is set at some fixed value, and V_{GS} is zero (gate-shorted to source). I_{DSS} is a zero-bias current-flow test. For example, a 3N128 specifies a V_{DS} of 15 V and V_{GS} of zero. Under these conditions, I_D should be between 5 and 25 mA.

8.6.2 $I_{D(ON)}$ test

$I_{D(ON)}$ is essentially the same specification as I_{DSS}, except that $I_{D(ON)}$ is generally applied to enhancement-mode devices (type B or C). The basic circuit of Fig. 8-2 is used, except that V_{GS} is connected to forward bias (gate positive for N channel). Forward bias is required because I_D flows in an enhancement-mode device only when the gate is forward-biased. Some I_D flows in a class-B device without forward bias. However, $I_{D(ON)}$ is the current near the maximum (or saturation current). During test, V_{GS} and V_{DS} are set to some fixed value, and the resulting value of $I_{D(ON)}$ is measured.

8.6.3 $I_{D(OFF)}$ test

The test circuit for $I_{D(OFF)}$ is the same as for $V_{GS(OFF)}$, as shown in Fig. 8-2. This is because $I_{D(OFF)}$ is the current that flows when $V_{GS(OFF)}$ is applied. V_{DS} and V_{GS} are set to some fixed value, and the resulting value of $I_{D(OFF)}$ is measured. For example, a 3N128 specifies a V_{DS} of 20 V and a V_{GS} of –8 V. Under these conditions, the I_D should be 50 μA maximum. On enhancement-only devices (type C), the value of $I_{D(OFF)}$ is the I_D that flows when $V_{GS(TH)}$ is applied. Thus, the fixed value of V_{GS} is always some forward-bias voltage.

8.7 FET breakdown-voltage tests

Voltage breakdown is a particularly important characteristic of FETs. There are a number of specifications to indicate the maximum voltage that can be applied to the various elements. These include:

$V_{(BR)GSS}$ gate-to-source breakdown voltage, which is the breakdown voltage with the drain and source shorted. Under these conditions, the gate-channel junction (PN diode for a JET; oxide layer for a MOSFET) also meets the breakdown specification because the drain and source are the connections to the channel. Notice that for a MOSFET, $V_{(BR)GSS}$ will physically puncture the oxide layer.

$V_{(BR)DGO}$ drain-to-gate breakdown voltage is essentially the same specification as $V_{(BR)GSS}$, except that $V_{(BR)DGO}$ represents breakdown from gate to drain. $V_{(BR)DGO}$ is more properly applied to JFETs, but they might appear on a MOSFET datasheet.

8.7.1 JFET voltage-breakdown tests

For JFETs, $V_{(BR)GSS}$ is the maximum voltage that can be applied between any two terminals and is the lowest voltage that leads to breakdown of the gate junction. As shown in Fig. 8-2, an increasingly high reverse voltage is applied between the common gates and source. Junction breakdown can be determined by the gate current (beyond normal I_{GSS}), which indicates the beginning of an avalanche (V_A) condition.

8.7.2 MOSFET voltage-breakdown test

For MOSFETS, $V_{(BR)GSS}$ is the breakdown voltage from gate to source, and it is a maximum voltage, not a test voltage. Such tests are usually made only during manufacture (as part of destructive testing to establish breakdown for a particular type of device). Instead, some datasheets specify $V_{(BR)DSS}$ (for type-C devices). As shown in Fig. 8-2, $V_{(BR)DSS}$ is made with the gate and source shorted, because no reverse-bias is necessary for cutoff with a type-C device (this is the same circuit as for I_{DSS}). With V_{GS} at zero, V_{DS}

is increased until there is some heavy flow of I_D. This is nondestructive. For example, an MFR137 specifies a $V_{(BR)DSS}$ of 65 V minimum with V_{GS} at zero, and an I_D of 10 mA.

8.8 FET gate-leakage tests

Gate leakage is a very important characteristic of a FET because such leakage is directly related to input resistance. When gate leakage is high, input resistance is low, and vice versa. Gate leakage is usually specified at I_{GSS} (reverse-biased gate-to-source current with drain shorted to source) and is a measure of the static short-circuit input impedance.

8.8.1 I_{GSS} test

As shown in Fig. 8-2, V_{DS} is zero (drain shorted to source), V_{GS} is set to some given (reverse-biased) values, and resultant leakage current I_{GSS} is measured. For example, an MRF137 specifies a V_{DS} of zero and a V_{GS} of 20 V. Under these conditions, I_{GSS} should not exceed 1 μA at 25°C.

8.9 Dual-gate FET tests

In addition to the static characteristics described thus far, dual-gate FETs have certain other characteristics that might require testing. For example, with dual-gate MOSFETs, both gates control I_D. As a result, there are a number of datasheet specifications that describe how the voltage on one gate affects I_D, with the other gate held at some specific voltage, or at 0 V. Similarly, there are specifications that describe how gate leakage or gate current flow is affected by gate voltage. For practical applications, the most important dual-gate characteristics include cutoff voltage, breakdown voltage, and gate current.

8.9.1 $V_{G1S(OFF)}$ and $V_{G2S(OFF)}$ tests

These tests apply to dual-gate devices, operating in the depletion and depletion/enhancement modes. Dual-gate devices can also be tested for $V_{GS(OFF)}$, as described in Sec. 8.4.1, when both gates are tied together.

Figure 8-3 shows the basic circuit for $V_{G1S(OFF)}$ and $V_{G2S(OFF)}$ tests. In both cases, V_{DS} is set at some specific value, one gate is forward-biased to a specific value, and the other gate is reverse-biased. The reverse-gated V_{G1S} or V_{G2S} is adjusted until I_D is at some negligible value, indicating a cutoff condition.

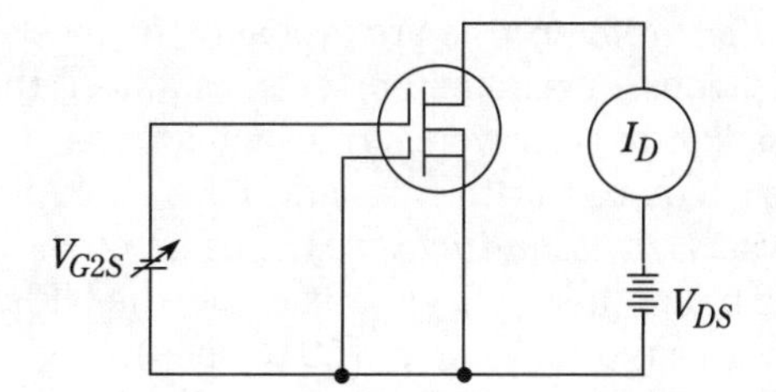

8-3 Dual-gate FET test circuits.

8.9.2 Dual-gate voltage-breakdown tests

Dual-gate MOSFETs are often tested for forward and reverse gate-to-source breakdown voltage. Figure 8-3 shows the basic test circuit. V_{DS} and one gate are at 0 V (both shorted to source). A variable voltage is applied to the opposite gate, and the gate current is measured. For example, to measure $V_{(BR)G1SSF}$, a 40841 MOSFET specifies a V_{DS} and V_{G2S} of 0 mV, and an I_{G1SSF} of 100 µA, with a typical 9 V of V_{G1} applied. $V_{(BR)G1SSR}$ is measured in the same way, except that the gate is reverse-biased. The results should be the same (100 µA of I_{G1SSR} for the 40841). $V_{(BR)G2SSR}$ is measured in the same way, except that gate 1 is connected to the source and the voltage is applied to gate 2.

8.9.3 Dual-gate current tests

Dual-gate current tests use the same basic test circuit as for gate-voltage breakdown tests, as shown in Fig. 8-3. The difference in procedure is that the gate voltage is set to a specific value and the resultant current is measured. For example, to measure I_{G1SSF}, a 40841 MOSFET specifies a V_{DS} and V_{G2S} of 0 V and a V_{G1} of 6 V. A maximum I_{G1SSF} of 60 nA should flow under these conditions. I_{G1SSR} is measured in the same way, except that V_{G1S} is –6 V. I_{G2SSF} and I_{G2SSR} are measured in the same way, except that gate 1 is connected to the source and the voltage is applied to gate 2.

8.9.4 I_{DS} test

This test applies to dual-gate FETs, operating in the depletion and depletion/enhancement modes. Dual-gate devices can also be tested for I_{DSS}, as described in Sec. 8.6, when both gates are tied together.

The circuit for I_{DS} is shown in Fig. 8-3, V_{DS} is set at some specific value, gate 1 is shorted to the source, and gate 2 is set to a specific value; I_{DS} is considered as a zero-bias current-flow test, even though one gate has a forward bias. For example, the 40841 specifies a V_{DS} of 15 V, a V_{G1S} of 0 V, and a V_{G2S} of +4 V. Under these conditions, the I_D should be a typical 10 mA.

8.10 FET dynamic characteristics

Unlike the static characteristics described thus far, the characteristics (ac or signal) of FETs apply equally to types A, B, and C. However, conditions and presentations of the dynamic characteristics depend mostly on the intended application. This section does not cover FET applications, but it concentrates on the test procedures whats, whys, and hows. It starts with y-parameters.

8.10.1 y parameters

The *y-parameter tests* are probably the most important dynamic tests for any FET applications. Y-parameter tests establish the four basic admittances (forward transadmittance, reverse transadmittance, input admittance, and output admittance) required in the design of FET circuits. *Forward transadmittance* (also known as *transconductance*) appears on all FET datasheets, although it might be called by another name. For example, the MFR137 lists forward transconductance as gfs (with a typical value of 750 mmhos).

In simple terms, *admittance* is the reciprocal of impedance. In turn, impedance

(Z) is a combination of resistance (R, the real part) and reactance (X, the imaginary part). Admittance (y) is composed of conductance (g, the real part) and susceptance (j_b, the imaginary part). Thus, g is the reciprocal of R, and j_b is the reciprocal of X.

To find g, divide R into 1; to find R, divide g into 1. Z is expressed in ohms. y, being a reciprocal, is expressed in mhos or millimhos (mmhos). For example, an impedance Z of 50 ohms equals 20 mmhos (1/50 = 0.02 mho = 20 mmhos).

A y-parameter is an expression for admittance in the form:

$$y_i = g_i + j_{bi}$$

Where:

g_i = real (conductive) part of input admittance
j_{bi} = imaginary (susceptive) part of input admittance
y_i = input admittance (the reciprocal of Z_i).

The term $y_i = g_i + j_{bi}$, expresses the y-parameter in rectangular form. Some manufacturers describe the y-parameter in polar form. For example, they give the magnitude of the input as $|y_i|$ and the angle of the input admittance as $< y_i$. Quite often, manufacturers mix the two systems of vector algebra on datasheets.

8.10.2 Conversion of vector-algebra forms

In case you are not familiar with the basics of vector algebra, the following notes summarize the steps necessary to manipulate vector-algebra terms. With this background, you should be able to perform all calculations for y-parameter tests.

To convert from rectangular to polar form:

1. Find the magnitude from the square root of the sum of the squares of the components:

$$\textit{Polar magnitude} = g^2> + j_b^2$$

2. Find the angle from the ratio of the component values:

$$\textit{Polar angle} = \arctan j_b/g$$

The polar angle is leading if the j_b term is positive and is lagging if the b term is negative. For example, assume that y_{fs} is given as $j_{gfs} = 30$ and $j_{bfs} = 70$.

$$|y_{fs}|\ \textit{polar magnitude} = 30^2 + 70^2 = 76$$

$$<y_{fs}\ \textit{polar angle} - \arctan 70/30 = 67°$$

Converting from polar to rectangular form:

1. Find the real (conductive, or g) part when polar magnitude is multiplied by the cosine of the polar angle.
2. Find the imaginary (susceptance, or j_b part) when polar magnitude is multiplied by the sine of the polar angle.

If the angle is positive, the j_b component is also positive. When the angle is negative, the j_b component is also negative. For example, assume that the y_{fs} is given as $|y_{fs}| = 20$ and $<y_{fs} = -30°$. This is converted to rectangular form by:

$$20 \times \cos 33° = g_{fs} = 16.8$$

$$20 \times \sin 33° = j_{bfs} = 11$$

8.10.3 The four basic y parameters

Figure 8-4 shows the y-equivalent circuit for a FET. A similar circuit can be drawn for a two-junction transistor.

Notice that y parameters can be expressed with number or letter subscripts. The number subscripts are universal and can apply to two-junction transistors, FETs, and ICs. When used, the letter subscripts are most popular on FET datasheets.

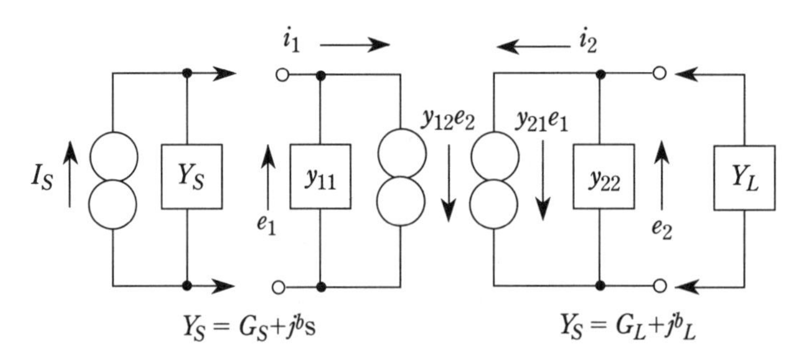

8-4 The y-equivalent circuit for a FET.

The following notes can be used to standardize y-parameter terms. Note that the letter s in the letter subscripts refers to common-source operation of a FET, and is equivalent to a common-emitter two-junction circuit.

y_{11} is input admittance and can be expressed as y_{os}

y_{12} is reverse transadmittance and can be expressed as y_{rs}

y_{21} is forward transadmittance and can be expressed in y_{fs}

y_{22} is output admittance and can be expressed as y_{os}

Input admittance, with Y_L = infinity (a short circuit of the load), is expressed as:

$$y_{11} = g_{11} + g_{b11} = d_{i1}/d_{e1} \text{ (with } e_2 = 0)$$

This means that y_{11} is equal to the difference in current i_1, divided by the difference in voltage e_1, with voltage e_2 at 0. The voltages and currents involved are shown in Fig. 8-4.

Some datasheets do not show y_{11} at any frequency but give input capacitance C_{iss} instead. For example, the MRF137 lists C_{iss} as 48 pF at a frequency of 1 MHz. Input capacitance is covered in Sec. 8.12. If you assume that the input admittance is entirely (or mostly) capacitive, the input impedance can be found when the input capacitance is multiplied by $6.28f$ (f = frequency in Hz), and the reciprocal is taken.

Because admittance is the reciprocal of impedance, admittance is found when input capacitance is multiplied by $6.26f$ (where admittance is capacitive). For example, if the frequency is 100 MHz, and the input capacitance C_{iss} is 9 pF, the input admittance is: $6.28 \times (100 \times 10^6) \times (9 \times 10^{12})$, about 5 or 6 mmhos.

This assumption is accurate only if the real part of y_{11} (or g_{11}) is negligible. Such an assumption is reasonable for most FETs, but not necessarily for all two-junction transistors. The real part of two-junction transistor input admittance can be quite large in relation to the imaginary j_{b11} part.

Forward transadmittance (or forward transconductance), with Y_L = infinity (a short circuit of the load) is expressed as:

$$y_{21} \text{ (or } g_{fs}) = g_{21} + j^{b}{}_{21} = d_{i2}/d_{i1} \text{ (with } e_2 = 0)$$

This means that y_{21} is equal to the difference in output current i_2 divided by the difference in input voltage e_i, with voltage 32 at 0. In other words, y_{21} represents the difference in output current for a difference in input voltage.

Two-junction transistor datasheets often do not give any value for y_{21}. Instead, forward transadmittance is shown by a hybrid notation using h_{fe} or h_{21} (which means hybrid forward transadmittance with common emitter). No matter what system is used, it is essential that the values of forward transadmittance be considered at the frequency of interest.

Output admittance, with Y_S = infinity (a short circuit of the source or input) is expressed as:

$$y_{22} = g_{22} + j^{b}{}_{22} = d_{is}/d_{e2} \text{ (with } e_1 = 0)$$

Reverse transadmittance, with Y_S = infinity (a short circuit of the source of input) is expressed as:

$$y_{12} = g_{12} + j_{b12} = d_{i1}/d_{i2} \text{ (with } e_1 = 0)$$

y_{12} is usually not considered an important two-junction transistor parameter. However, y_{12} might appear in equations related to RF design.

8.10.4 y-parameter measurement

It is obvious that y-parameter information is not always available or in a convenient form. In practical design, it might be necessary to measure the y-parameter using test equipment. The main concern in measuring y-parameter is that the measurements are made under conditions simulating those of the final circuit. For example, if supply voltages, bias voltages, and operating frequency are not identical (or close) to the final circuit, the tests might be misleading.

Although the datasheets for transistors to be used in RF circuits usually contain input and output admittance data, it might be helpful to know how this information is obtained. There are two basic methods for measuring the y-parameters of transistors.

One method involves direct measurement of the parameter (such as measuring changes in outputs for corresponding changes in input). The other uses tuning substitution (where the transistor is tuned for maximum transfer of power, and the admittances of the tuning circuits are measured). Both methods are summarized in the following paragraphs.

8.10.5 Direct measurement of forward transadmittance (y_{fs}, g_{fs}, y_{21})

Figure 8-5 shows a typical test circuit for direct measurement of forward transadmittance (forward transconductance). Although an FET is shown, the same circuit can apply to any single-input device, such as a two-junction transistor.

The value of R_L must be such that the drop produced by the FET current drain is negligible, and the operating-voltage point (V_{DS} in the case of a FET) is correct for

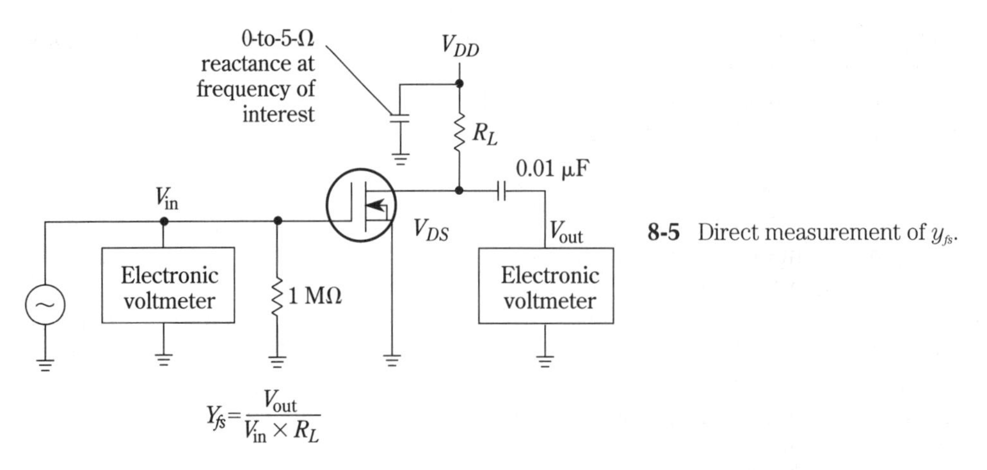

8-5 Direct measurement of y_{fs}.

a given power-supply voltage (V_{DD}) and operating current (I_D). For example, if I_D is 10 mA, V_{DD} is 20 V, and V_{DS} is 15 V, R_L must drop 5 V at 10 mA. Thus, the value of R_L is 5 V/0.01 A = 500 Ω.

During test, the signal source is adjusted to the frequency of interest. The amplitude of the signal source V_{in} is set to some convenient number, such as 1 V or 100 mV. The value of y_{fs} (g_{fs} or y_{21}) is calculated from the equation of Fig. 8-5, and is expressed in mhos (or mmhos or μmhos). For example, assume that the value of R_L is 1000 ohms, V_{in} is 1 V, and V_{out} is 8 V. The value of y_{fs} is 8/(1 × 1000) = 0.008 mho = 8 mmho = 8000 μmho.

8.10.6 Direct measurement of output admittance (y_{os}, y_{22})

Figure 8-6 shows a typical test circuit for direct measurement of output admittance. The value of R_S must be such as to cause a negligible drop (so that V_{DS} can be maintained at the desired level, with given V_{DD} and I_D). During testing, the signal source is

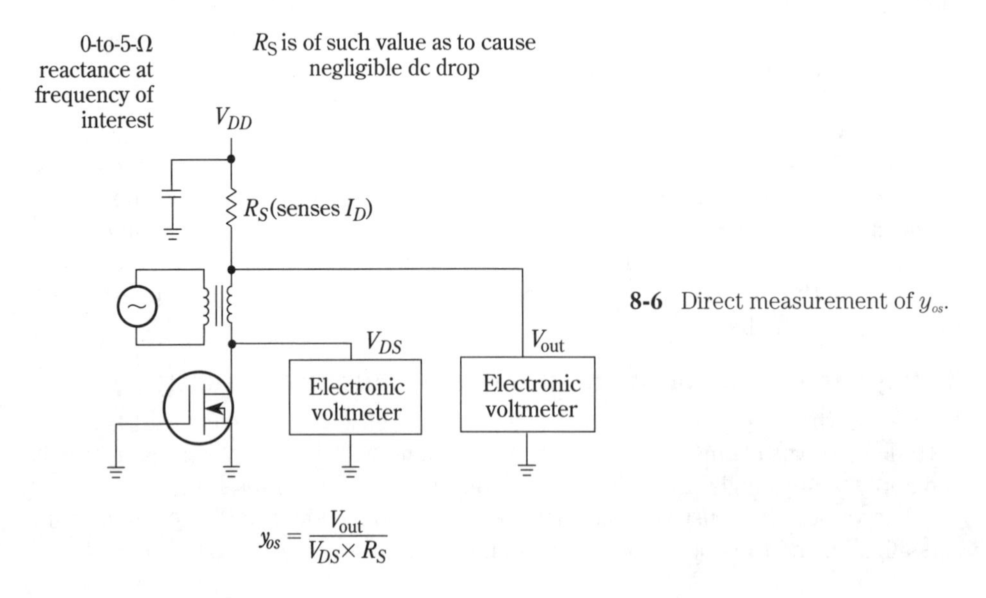

8-6 Direct measurement of y_{os}.

adjusted to the frequency of interest. Both V_{out} and V_{DS} are measured, and the value of y_{os} or y_{22} is calculated from the equation of Fig. 8-6.

Some datasheets do not show y_{os} or y_{22}, but give output capacitance C_{oss} instead (output capacitance is covered in Sec. 8.13). For example, the MRF137 lists C_{oss} as 54 pF. Output admittance is found when output capacitance is multiplied by 6.28*f* (where the admittance is capacitive).

8.10.7 Direct measurement of input admittance (y_{is}, y_{11})

Although y_is or y_{11} is generally not a critical factor for FETs, it is necessary to know the values of y_is to calculate impedance-matching networks for FET RF amplifiers (with some design procedures). If necessary to establish the imaginary part ($j^b{}_i$s), use an admittance meter or R_X meter (chapter 6).

Most FET datasheets give input capacitance C_{iss} (input capacitance is covered in Sec. 8.12). For example, the MRF137 lists C_{iss} as 48 pF. Input admittance is found when input capacitance is multiplied by 6.28*f* (where admittance is capacitive).

8.10.8 Direct measurement of reverse transmittance (y_{rs}, y_{12})

Again, although y_{rs} is not generally a critical factor for FETs, it is necessary to know the values of y_{rs} to calculate impedance-matching networks (with some design procedures). Although the real part of g_{rs} remains at zero for all conditions and at all frequencies, the imaginary part ($j^b{}_{rs}$) does vary with voltage, current and frequency. The reverse susceptance varies and, under the right conditions, can produce undesired feedback from output to input. This condition must be accounted for in the design of RF amplifiers to prevent feedback from causing oscillation. If it is necessary to establish the imaginary part (j_{brs}), use an admittance meter or R_X meter (chapter 6).

Most FET datasheets give reverse transfer capacitance C_{rss}. Reverse transfer capacitance is covered in Sec. 8.14. For example, the MRF137 lists C_{rss} as 11 pF. Reverse transadmittance is found when reverse transfer capacitance is multiplied by 6.28*f* (where admittance is capacitive).

8.10.9 Tuning-substitution measurement of y parameters

Figure 8-7 is a typical test circuit for measurement of y parameters using the tuning-substitution method. Although a two-junction transistor is shown, the circuit can be adapted for use with FETs.

During testing, the transistor is placed in a test circuit designed with variable components to provide wide tuning capabilities. This is necessary for correct matching at various power levels. The circuit is tuned for maximum power gain at each power level for which admittance information is desired.

After the test amplifier is tuned for maximum power gain, the dc power, signal source, circuit load, and test transistors are disconnected from the circuit. For total circuit impedance to remain the same, the signal-source and output-load circuit connections are terminated at the characteristic impedances (typically 50 Ω).

After the substitutions are complete, admittances are measured at the base- and connector-circuit connections of the test transistor (points A and B, respectively, in Fig. 8-7), using a precision admittance meter (chapter 6).

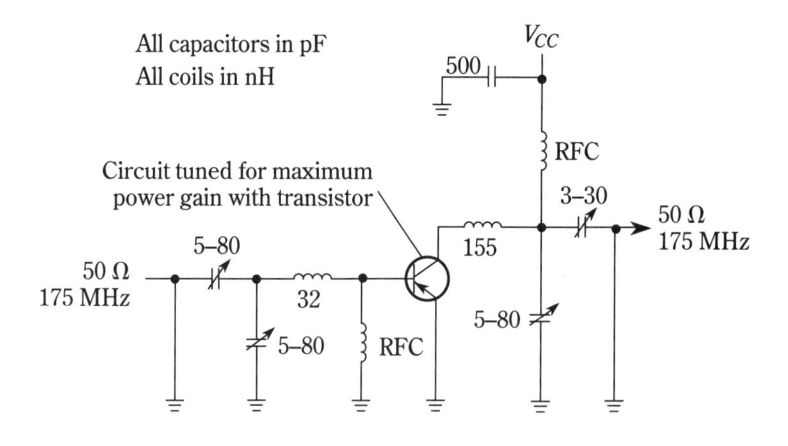

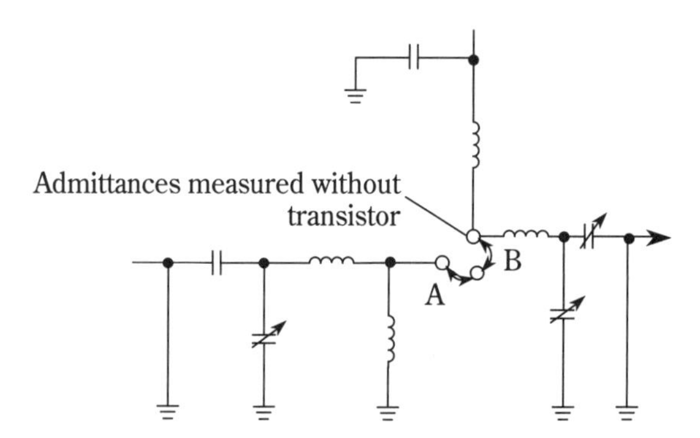

8-7 Tuning-substitution measurement of y-parameters.

The transistor input and output admittances are the conjugates of the base-circuit connection and the collector-circuit admittances, respectively. For example, if the base-circuit connection (point A) admittance is $8 + j_3$ (as shown on the admittance meter), the input admittance of the transistor is $8 - j_3$.

In some systems of circuit design, the networks are calculated on the basis of input-output resistance and capacitance, instead of admittance (although admittances are often used to determine stability before going into design of the networks). With this approach, usually called *large-signal design*, the admittances measured in the circuit of Fig. 8-7 are converted to resistance and capacitance.

Resistance is found by dividing the real part of the admittance into 1. Capacitance is found by dividing the imaginary part of the admittance into 1 (to find the capacitive reactance). Then, the reactance is used in the equation $C = 1/(6.28f \times Reactance)$ to find the actual capacitance.

8.11 FET amplification factor

The amplification factor does not usually appear on most FET datasheets. This is because amplification does not usually have a great significance in most small-signal applications for FETs (common-source power gain or G_{ps} is used more often). However, amplification factor is sometimes used as a figure of merit in isolated cases. Amplification factor defines the relationship between output-signal voltage and input-signal voltage, with the output current held constant, or amplification factor = $\Delta V_{DS}/V_{GS}$, with I_D held constant. Amplification factor can also be calculated by y_{fs}/y_{os}.

8.12 FET input-capacitance tests

Input capacitance (C_{iss}) is the common-source input capacitance with the output shorted and is used as a low-frequency substitute for y_{is}. This is because y_{is} is entirely capacitive at low frequencies. To find an approximate value for y_{is} at low frequencies (below about 1 MHz), multiply C_{is} by 6.28f. The result is j_{bis}, or the imaginary part of y_{is}. At these low frequencies, g_{is} can be considered as zero. C_{iss} is an important characteristic for FETs used in switching or chopper applications. This is because a large voltage switch at the gate must appear across the input capacitance C_{iss}.

8.12.1 C_{iss} tests

Figures 8-8 and 8-9 show the basic circuits for C_{iss} tests. Figure 8-8 is for JFETs and Fig. 8-9 is for MOSFETs. For dual-gate MOSFETs, the capacitance measurement is between gate 1 and all other terminals. For single-gate MOSFETs, the capacitance measurement is between the gate and all other terminals.

The circuit of Fig. 8-9A is for MOSFETs where a specific V_{GS} must be applied, but V_{DS} is of no concern. The circuit of Fig. 8-9B is used where the capacitance is mea-

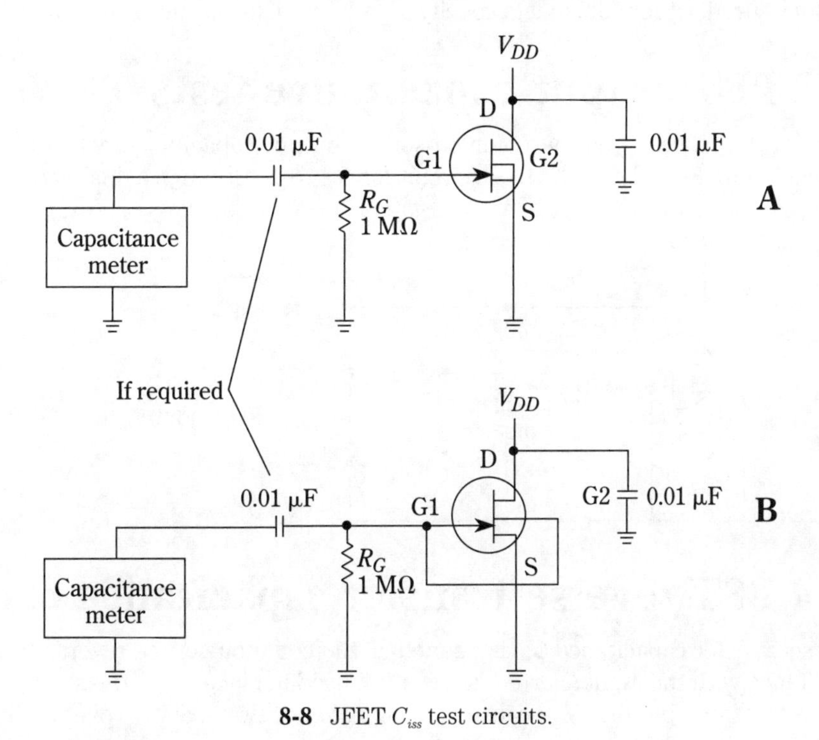

8-8 JFET C_{iss} test circuits.

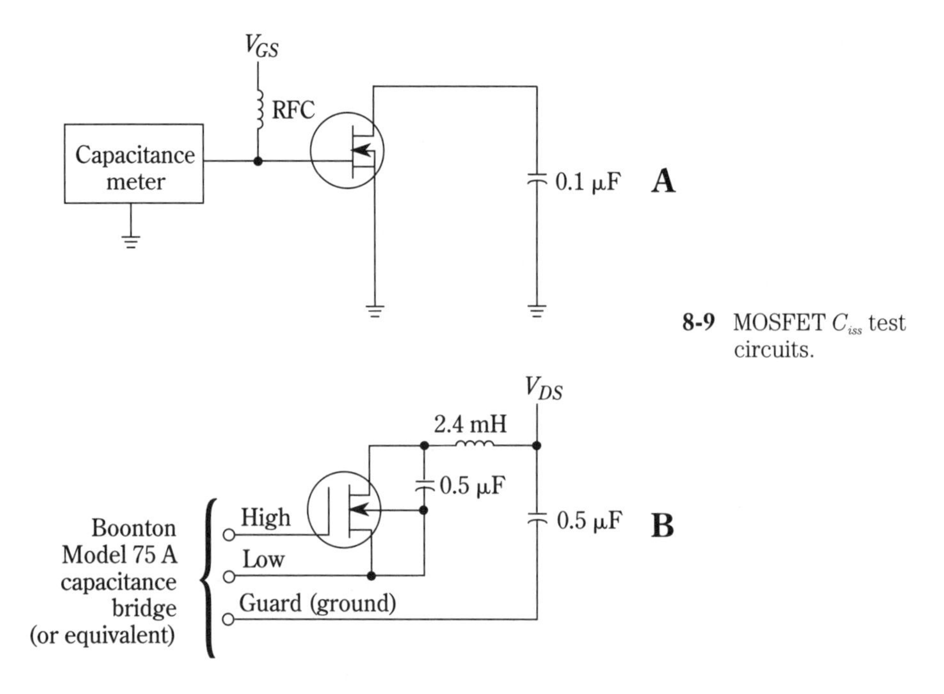

8-9 MOSFET C_{iss} test circuits.

sured with a three-terminal bridge (high, low, and guard or ground). The reason for the two circuit is that some MOSFET datasheets specify a given V_{DS}, I_D, and V_{GS} when C_{iss} is measured. Other datasheets specify zero V_{DS} and sometimes zero V_{GS}.

8.13 FET output-capacitance tests

Output capacitance C_{oss} is the common-source output capacitance with the input shorted. Figure 8-10 shows the basic circuit for C_{oss} test. Although a dual-gate FET is shown, the circuit also applies to a single gate. However, with single gates, V_{GS} is zero (gate shorted to source).

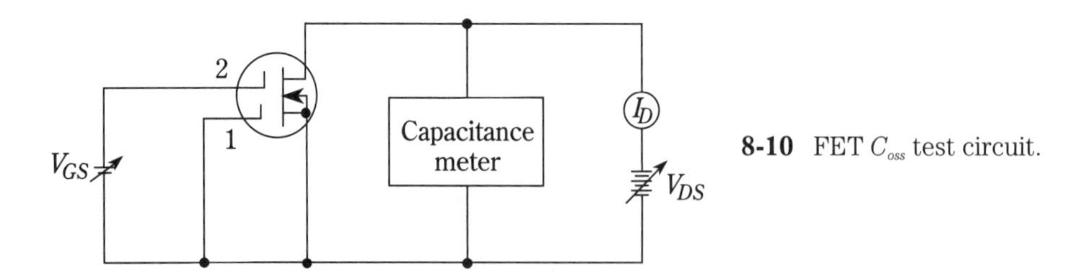

8-10 FET C_{oss} test circuit.

8.14 FET reverse-transfer capacitance tests

Reverse-transfer capacitance C_{rss} is defined as the common-source reverse-transfer capacitance with the input shorted. C_{rss} is often used in place of y_{rs}, the short-circuit reverse-transfer admittance, because y_{rs} is almost entirely capacitive over the useful frequency range of most FETs and is of relatively constant capacity. Consequently, the low-frequency C_{rss} is an adequate specification.

C_{rss} is also of major importance in FETs used as switches. Similar to the C_{os} of two-junction transistors, C_{rss} must be charged and discharged during the switching interval. For chopper applications, C_{rss} is the feed-through capacitance for the chopper drive. C_{rss} is also known as the Miller-effect capacitance because the reverse capacitance can produce a condition similar to the Miller effect in vacuum tubes (constantly changing frequency-response curves).

8.14.1 C_{rss} tests

Figures 8-11 and 8-12 show the basic tests for C_{rss}. Figure 8-11 is for JFETs and Fig. 8-12 is for MOSFETs. Three-terminal measurements are used in all cases. For dual-gate MOSFETs, gate 2 and the source are returned to the guard terminal and the capacitance measurement is made between gate 1 and the drain.

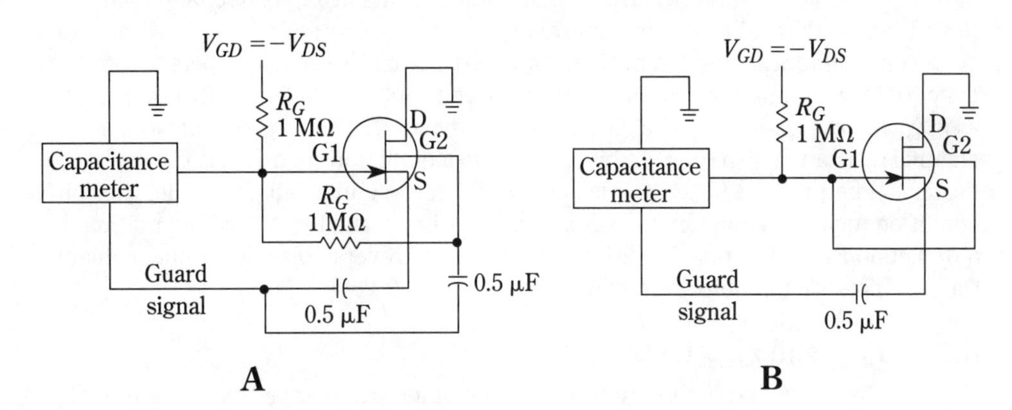

8-11 JFET C_{rss} test circuits.

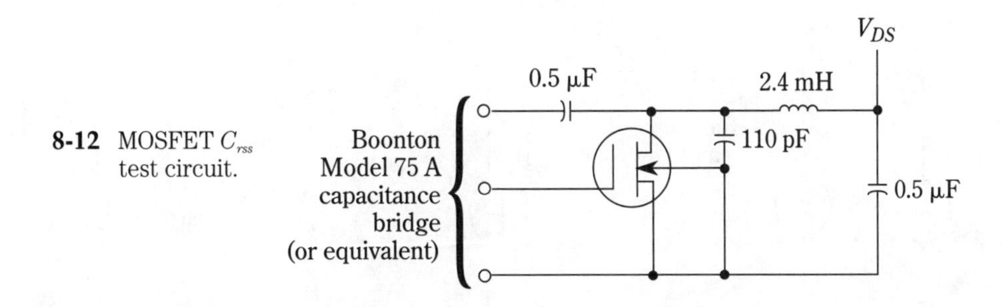

8-12 MOSFET C_{rss} test circuit.

8.15 FET element-capacitance tests

The elements of a FET have some capacitance between them, as do two-junction transistors. Some of these capacitances have an effect on the dynamic characteristics of the FET. Drain-substrate junction capacitance, $C_{d(sub)}$, is the most important FET element capacitance. $C_{d(sub)}$ is often found on datasheets for FETs used in switching. This is because $C_{d(sub)}$ appears in parallel with the load in a switching circuit, and must be charged and discharged between the two logic levels during the switching intervals. C_{ds} drain-to-source capacitance is another specification found on some switching FET datasheets. C_{ds} also appears in parallel with the load in switching and logic applications. The capacitance between elements of a FET can be measured with a capacitance meter. No special test connections are required. Some datasheets specify certain test connections or conditions, such as all remaining elements connected to source, or gate connected to source.

8.16 FET channel-resistance tests

The channel resistance is an important characteristic for FETs used in switching. Channel resistance describes the bulk resistance of the channel in series with the drain and source, and is listed as $r_{d(on)}$, r_{DS}, R_{DS}, R_{ds}, $r_{d(off)}$, etc., depending on the datasheet. Some of these descriptions are static, some are dynamic, and some are mixed (again, depending on the datasheet).

From a practical standpoint, there are two channel-resistance specifications of concern in switching applications. These are the "on" and "off" specifications, such as $r_{ds(on)}$ and $r_{ds(off)}$, which can be either static or dynamic. The "on" specification is the channel resistance with the FET biased "on." In a depletion-mode FET, the "on" condition can be produced by zero bias ($V_{GS} = 0$). In the enhancement-mode, the "on" condition requires some forward bias. The opposite is true for the "off" condition. In a depletion FET, the "off" condition requires some reverse bias. For enhancement, the "off" condition require zero bias.

8.16.1 $r_{ds(on)}$ and $r_{ds(off)}$ tests

Figure 8-13 shows the basic circuit for both channel-resistance tests. For a depletion/enhancement FET, $r_{ds(on)}$ is measured by adjusting V_{GS} to 0 V (or simply connecting gate to source).

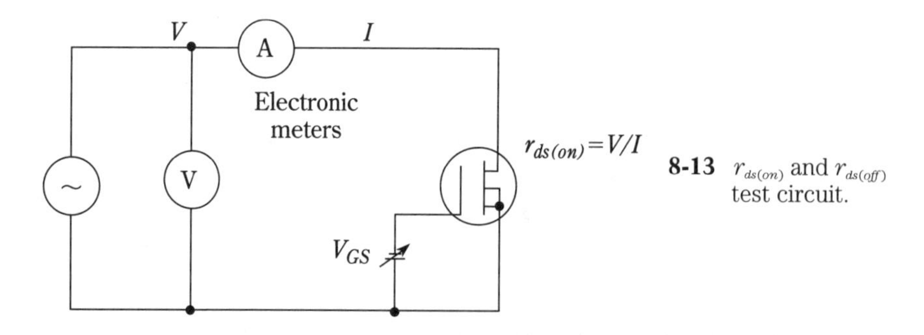

8-13 $r_{ds(on)}$ and $r_{ds(off)}$ test circuit.

During testing, the ac voltage (V) is adjusted to some convenient value (1 V, 10 V, etc.), and the channel current (I) is measured. The value of r_{ds} is found when V is divided by I. For example, if V is adjusted to 10 V and 50-mA channel current is measured, the "on" channel resistance, $r_{ds(on)}$, is 200 Ω (10/0.05 = 200). If the FET operates in the enhancement-only mode (type C), it is necessary to forward-bias the gate by adjusting V_{GS} to some specific value.

Figure 8-13 can also be used to measure the "off" channel resistance $r_{ds(off)}$. However, the bias conditions are opposite those for $r_{ds(on)}$. In the depletion and depletion/enhancement modes, the gate must be reverse-biased by adjusting V_{GS} to some specific value. In the enhancement-only mode, the gate can be connected directly to the source. Figure 8-13 can also be used to measure channel resistance of dual-gate FETs. Generally, the simplest way is to connect both gates together. However, some datasheets specify a fixed bias on one or both gates. As a point of reference, a typical MOSFET "on" resistance is about 200 Ω, whereas the "off" resistance is greater than 10^{10} Ω.

8.17 FET switching tests

The switching tests for two-junction transistors (Sec. 7.5) also apply to FETs, as do the pulse and square-wave definitions covered in Sec. 2.14.2 and shown in Fig. 2-25. Figure 8-14 shows a circuit for testing FET switching characteristics (rise time, fall time, delay time, and storage time). The scope must have a wide frequency response and good transient characteristics (faster rise times than the pulse used). The scope vertical channels are voltage-calibrated, with the horizontal channel time-calibrated. The FET is tested by applying a pulse to the gate. The same pulse is applied to one of the vertical inputs of the scope. In some cases, a bias is also applied to the FET gate. The FET output is applied to the other vertical input.

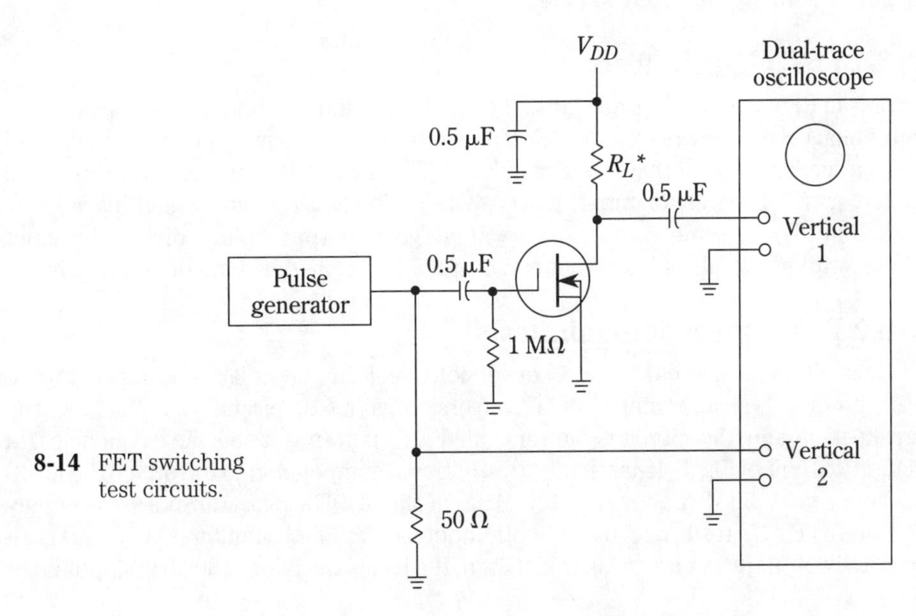

8-14 FET switching test circuits.

The two pulses (input and output) are then compared as to rise time, fall time, delay time, storage time, shape, etc. The FET output pulse characteristics can then be compared with the FET specifications (remember the rules of thumb for pulse testing, described in Sec. 2.14.2.)

8.18 FET gain tests

The datasheets for FETs used as amplifiers usually list some gain characteristics. Typically, these are power-gain figures at some specified frequency and under specified test conditions (such as V_{GS}, V_{DD}, and I_D). The gain figures are expressed in decibels, sometimes as maximum or minimum values. For example, the MRF137 shows a typical 16-dB common-source power gain (G_{ps}) at 150 MHz with a V_{DD} of 28 V, an I_D of 25 mA, and a power out (P_{out}) of 30 W. A minimum G_{ps} of 13 dB is also listed for the same test conditions. At 400 MHz, the typical G_{ps} drops to 7.7 dB.

The Y_{fs} tests described in Sec. 8.10 can be used to establish gain of a FET. However, the tests do not establish the gain in a working circuit. The only true test of circuit gain is to operate the FET in a working circuit and to measure actual gain. The most practical method is to operate the FET in the circuit in which the FET is to be used. However, it might be convenient to have a standard or universal circuit for FET gain tests.

Figure 8-15 shows two circuits, typical of the power-gain test circuits found on FET datasheets. These same circuits can be used for the noise-figure tests described next. Remember that these test circuits are for a specific FET (the MRF137), operating with specific loads (50 Ω) and at specific frequencies (150 and 400 MHz). However, similar test circuits can be fabricated for other loads and frequencies. Once the test circuit is assembled, the following procedures for measuring power gain are relatively simple.

8.18.1 FET G_{ps}-gain tests

In brief, an RF voltage is applied across the input load, and the input power is calculated from voltage and load resistance (E^2/R). The amplified output voltage is measured across a similar load and the power calculated (RF power measurements are described in chapter 15). The ratio of output power to input power is the power gain (usually expressed in dB). In some cases, simple voltage gain (output voltage divided by input voltage, with identical loads at input and output) is specified instead of power gain.

8.18.2 FET conversion-gain tests

For those FETs used as converters, conversion gain is of particular importance. Figure 8-16 shows a typical example of a conversion-gain test circuit (supplied on the datasheet). Again, this circuit is for a specific FET operating at specific frequencies. In brief, the circuit of Fig. 8-15 is used, with additional components to introduce the local-oscillator signal (30 MHz above the 150-MHz RF signal). The procedures are essentially the same as for G_{ps} tests, except that both inputs are applied simultaneously, and both input and output power are measured. Often, the local-oscillator (180 MHz) input is set

150-MHz test circuit

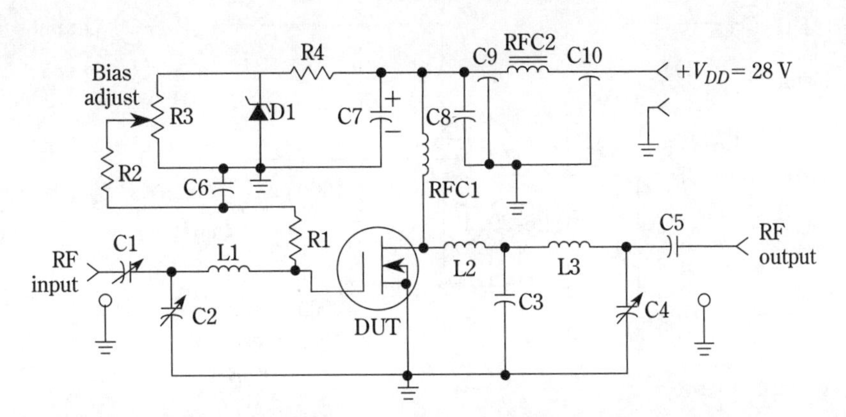

C1— Arco 406, 15–115 pF, or equivalent
C2— Arco 403, 3.0–35 pF, or equivalent
C3—56 pF Mini-Unleco, or equivalent
C4—Arco 404, 8.0–60 pF, or equivalent
C5—680 pF, 100 mils chip
C6—0.01 μF, 100 V, disc ceramic
C7—100 μF, 40 V
C8—0.1 μF, 50 V, disc ceramic
C9, C10—680 pF feedthru
D1—1N5925A Motorola Zener

L1—2 turn 0.29" ID, #18 AWG enamel, closewound
L2—1¼ turns, 0.2" ID, #18 AWG enamel, closewound
L3—2 turns, 0.2" ID, #18 AWG enamel, closewound
RFC1—20 turns, 0.30" ID, #20 AWG enamel, closewound
RFC2—Ferroxcube VK-200—19/4B
R1—10 kΩ, ½ W thin film
R2—10 kΩ, ¼ W
R3—10 turns, 10 kΩ
R4—1.80 kΩ, ½ W
Board—G10, 62 mils

400-MHz test circuit

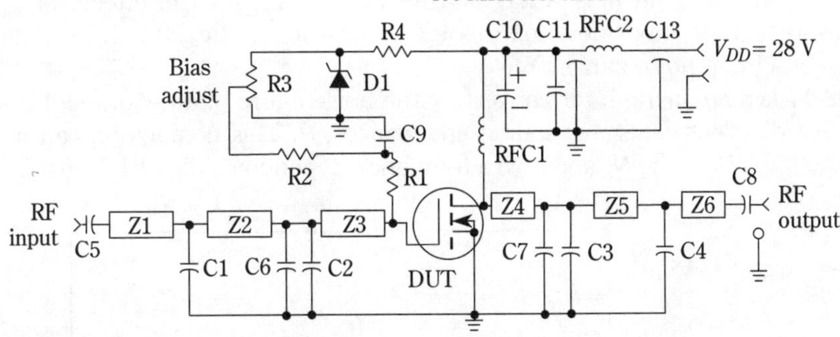

C1,C2,C3,C4—0–20 pF Johanson, or equivalent
C5,C8—270 pF, 100 mil chip
C6,C7—24 pF Mini-Unleco, or equivalent
C11-0.1 μF, 50 V, disc cera mic
C9—0.01 μF, 100V, disc ceramic
C12,C13—680 pF feedthru
D1—1N59215A Motorola Zener
R1,R2—10 kΩ, ¼ W
R3—10 turns, 10 kΩ

R4—1.8 kΩ, ½ W
Z1— 2.9" ×0.166" microstrip
Z2,Z4—0.35"×0.166" microstrip
Z3—0.40"×0.166" microstrip
Z5—1.05"×0.166" microstrip
Z6—1.9"×0.166" microstrip
RFC1—6 turns, 0.300" ID. #20 AWG enamel, closewound
RFC2—Ferroxcube VK-200 — 19/4B
Board—glass Teflon, 62 mils

8-15 FET power-gain and N_F test circuits.

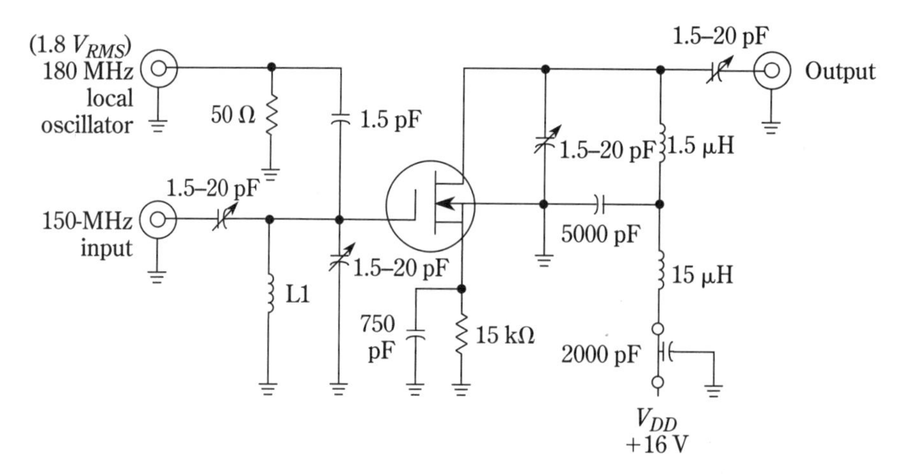

8-16 FET conversion-gain test circuit.

at some specific voltage level, and the 150-MHz RF power is measured. Then, the input power is compared to the 30-MHz output power to find conversion gain.

8.19 FET noise figure tests

The noise figure for FETs is usually listed as N_F on the datasheets and represents a common-source figure. N_F represents a ratio between input signal-to-noise (S/N) ratio and output S/N ratio. In most FET datasheets, N_F includes the effects of e_n (equivalent short-circuit input noise expressed in volts-per-root cycle) and i_n (the equivalent open-circuit noise current).

Figure 8-17 is a nomograph for converting the noise figure to an equivalent input voltage for different generator source impedances, R. This nomograph can be used with any FET. Because N_F and e_n are frequency-dependent, Fig. 8-17 must be

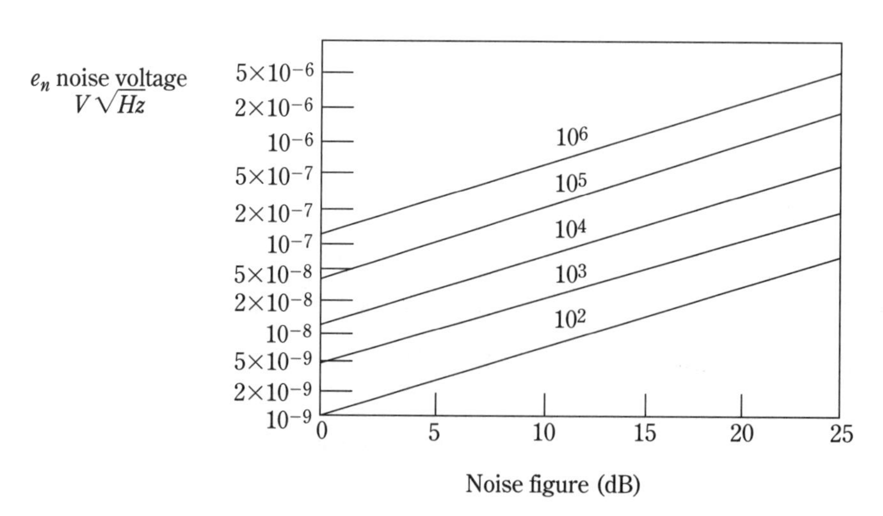

8-17 Noise-figure conversion chart.

used with a datasheet graph or table to determine e_n. For example, assume that the datasheet lists N_F as 5 dB at 200 MHz. Also assume that the circuit with which the FET is to be used has 1-MΩ input. Using these values, Fig. 8-17 shows e_n to be about 2×10^{-7}, about 200 nV.

8.19.1 N_F tests

The circuit of Fig. 8-15 can also be used to measure the noise figure. Such noise-figure tests are best done using specialized test equipment, such as a VHF noise source and VHS noise meter. The instructions supplied with the noise-measurement test equipment are generally complete and provide full descriptions of noise tests. In any event, the test procedures are not duplicated here. However, the nomograph of Fig. 8-17 can be used to find a noise figure for a given e_n and circuit input impedance. For example, if e_n is 200 nV (2×10^{-7}) for a 100-kΩ (10^5) input impedance, the N_F should be about 12 dB.

8.20 FET cross-modulation tests

Cross-modulation values might be listed on a FET datasheet. Figure 8-18 shows the basic block diagram and some typical circuits for cross-modulation tests. These circuits and values are for a FET being tested at frequencies of 200 MHz (desired operating frequency) and 150 MHz (interfering frequency). The circuit provides for unneutralized, neutralized, and cascode amplifier operation.

During testing, signals at both frequencies are applied to the input. The test-circuit output is monitored by the receiver. The amount of attenuation produced by the interfering signal is measured at various signal levels, and compared with the output when no interfering signal is present.

8.21 FET intermodulation tests

Intermodulation values might be listed on a FET datasheet. Figure 8-19 shows the basic circuit for intermodulation tests. This circuit and values are for an FET being tested at frequencies of 175 MHz (f_1) and 200 MHz (f_2). If intermodulation is present, there will be signals of various frequencies at the output.

During testing, f_1 and f_2 are set to zero (amplitude) and the background-noise level is measured (on the RF indicator of a receiver tuned to the intermodulation frequency, and connected at the output of the test circuit). The amplitudes of f_1 and f_2 are increased until the reading on the indicator is 1 mV above the move level. The voltage levels required to produce the output indications are measured on the voltmeter at the test-circuit input.

8.22 FET tests using a curver tracer

In many respects, testing a FET with a curve tracer is similar to testing two-junction transistors (chapter 7). A family of curves is displayed on the scope, and the curves have a similar appearance. As shown in Fig. 8-20, N-channel FETs have a family of curves similar to npn transistors, and P-channel FET curves are a similar to pnp transistors. Two-junction transistor curves are a graph of collector-current versus

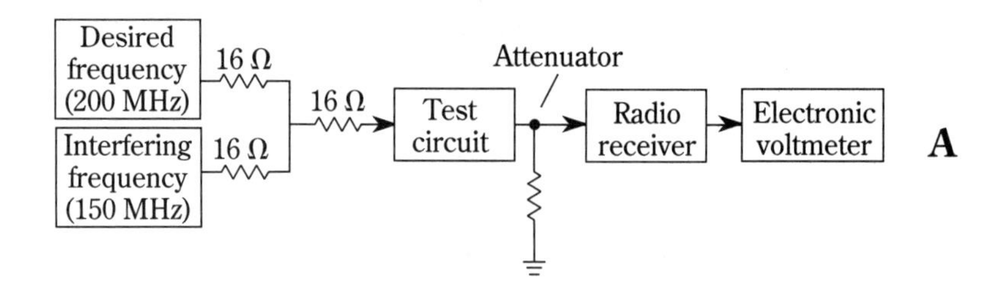

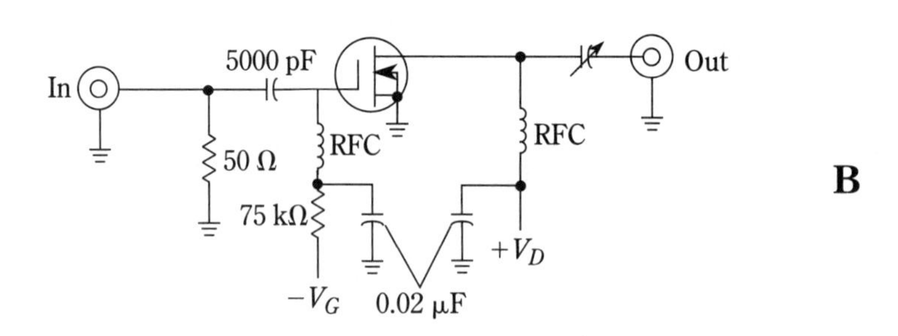

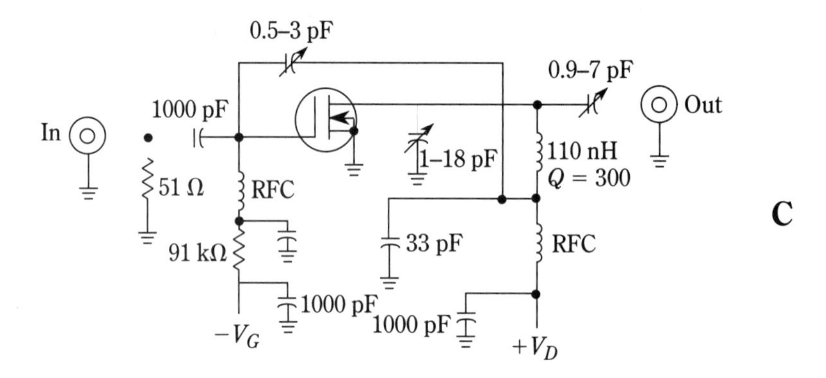

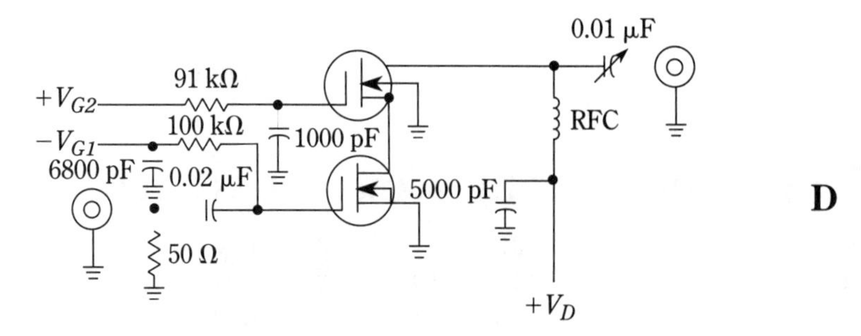

8-18 FET cross-modulation test circuits.

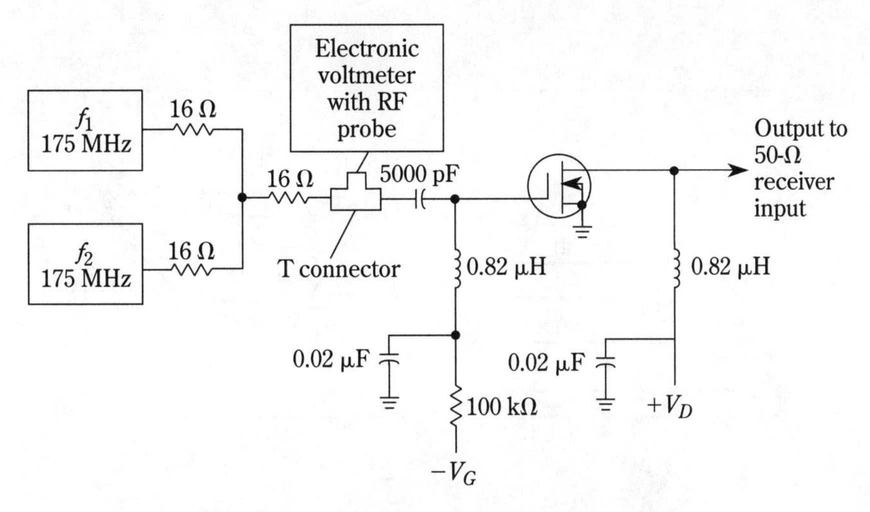

8-19 FET intermodulation test circuits.

collector-voltage at various base currents. FET curves are a graph of drain-current versus drain-voltage at various gate voltages. Similarly, FET breakdown voltages can be observed and measured by the same method used for transistors.

In other respects, testing FETs is different from testing two-junction transistors. For FETs, the step-selector switch (or whatever the curve-tracer control might be called) is placed in a "volts per step" position. That is, the curve tracer supplies constant-voltage steps to the FET, rather than constant-current steps for two-junction transistors (Fig. 7-10). Also, the polarity of the step voltage is reversed in relation to the sweep voltage. Although the zero-base-current step of a two-junction transistor usually provides no collector current, the 0-V step at the gate of most FETs produces the highest drain current. Each reverse-bias voltage step results in less drain current, and when the gate voltage is sufficiently high, drain current is pinched off. The point of pinch-off (sometimes listed as V_P on datasheets) can be measured with the curve tracer.

8.22.1 FET transconductance (gain) measure

The most useful and common measurement to be made for a FET is the gain measurement. *Dynamic gain* or *gate-to-drain forward transconductance* (g_{fs}) in the common-source configuration is the ratio of change in drain current to the change in gate voltage at a given drain voltage. Transconductance is measured in mhos. The MRF137 has a typical g_{fs} of 750 mmhos, and a minimum g_{fs} of 500 mmhos, with a V_{DS} of 10 V and an I_D of 500 mA.

Figure 8-20C shows how typical FET curves can be used to find gain. As shown, the change in drain current (ΔI_D) is 1.5 mA (from 7 mA to 5.5 mA), for a change in gate voltage (ΔV_G) of 0.1 V (from 0.1 V to 0.2 V), at a V_{DS} of 6 V. This indicates a gain (transconductance or g_{fs}) of 15 mmhos (0.0015 A/0.1 V = 0.015 mho = 15 mmhos).

As with a two-junction transistor, FET gain is not constant over the entire voltage and current range. The gain is normally calculated in the typical operating range.

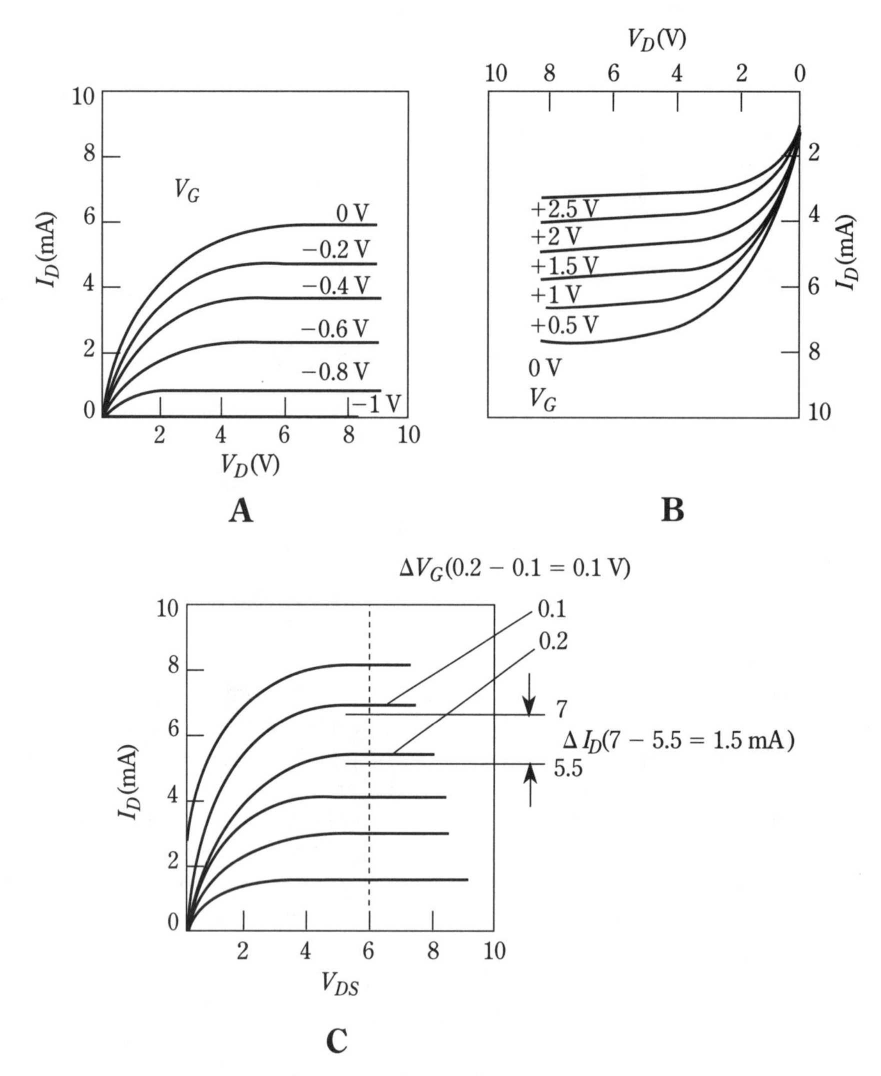

8-20 FET characteristic curves (N-channel, P-channel and gain).

Distortion and linearity can be determined by the same method as described for two-junction transistors. That is, if the spacing between curves is equal, the FET forward transconductance is linear.

9
Unijunction and programmable UJT (PUT) tests

This chapter is devoted entirely to test procedures for *unijunction transistors (UJTs)* and *programmable unijunction transistors (PUTs)*. The first sections of this chapter describe UJT and PUT characteristics and test procedures from the practical standpoint. The information in these sections permits you to test all the important UJT and PUT characteristics using basic shop equipment. The sections also help you understand the basis for such tests. The remaining sections of the chapter describe how the same tests, and additional tests, are performed using more sophisticated equipment.

9.1 Basic UJT and PUT functions

The UJT and PUT operate on entirely different principles from those of the two-junction transistor and FET (chapters 7 and 8). Likewise, the test procedures for UJTs and PUTs are quite different from other transistors. A UJT is a negative-resistance device where, under proper conditions, the input voltage or signal can be decreased, yet the output or load current increases. Once the UJT is turned on, it does not turn off until the circuit is broken or the input voltage is removed.

For these reasons, the UJT makes an excellent trigger source. The UJT can be biased just below the "firing" point. When a small trigger voltage (either intermittent or constant) is applied, the UJT fires and produces a large output voltage pulse or signal that remains on until the circuit is broken (by switching off the base voltage).

The PUT is a four-layer device similar to an SCR (chapter 11), except that the anode gate (rather than the cathode gate) is brought out. The PUT is normally used in conventional UJT circuits. The characteristics of both UJT and PUT devices are similar, but the triggering voltage of the PUT is programmable, and can be set by an external voltage-divider network. The PUT is faster and more sensitive than the UJT. The PUT finds limited application as a phase-control device, and is more often used in long-duration circuits. In general, the PUT is more versatile and economical than the UJT, and can replace the UJT in many applications.

Although the basic function of the UJT is that of a switch, a relaxation oscillator is the primary building block of most UJT circuits. The basic relaxation oscillator provides phase control of thyristors, such as SCRs, triacs, and so on. For that reason, UJT and PUT test procedures are often based on the relationship to a relaxation oscillator.

9.1.1 UJT symbol and static characteristics

As shown in Fig. 9-1, the UJT is a three-terminal device: emitter (E), base-1 (B1) and base-2 (B2). Figure 9-1 also shows the static UJT emitter characteristic curves for a single value of V_{B2B1}. The emitter curve is not drawn to scale to show the different operating regions in more detail. The region to the left of the peak point is called the *cutoff region.* The emitter junction is reverse-biased in most of the cutoff region, but it is slightly forward-biased at the peak point. The region between the peak point and valley point, where the emitter junction is forward-biased, is called the *negative-resistance region.* The region to the right of the valley point is called the *saturation region.*

As shown in Fig. 9-2, the PUT has three terminals: an anode (A), a gate (G), and a cathode (K). As seen from the equivalent circuit, the PUT is actually an anode-gated SCR (chapter 11). This means that if the gate is made negative with respect to the anode, the PUT switches from a blocking state to an "on" state. Because the PUT

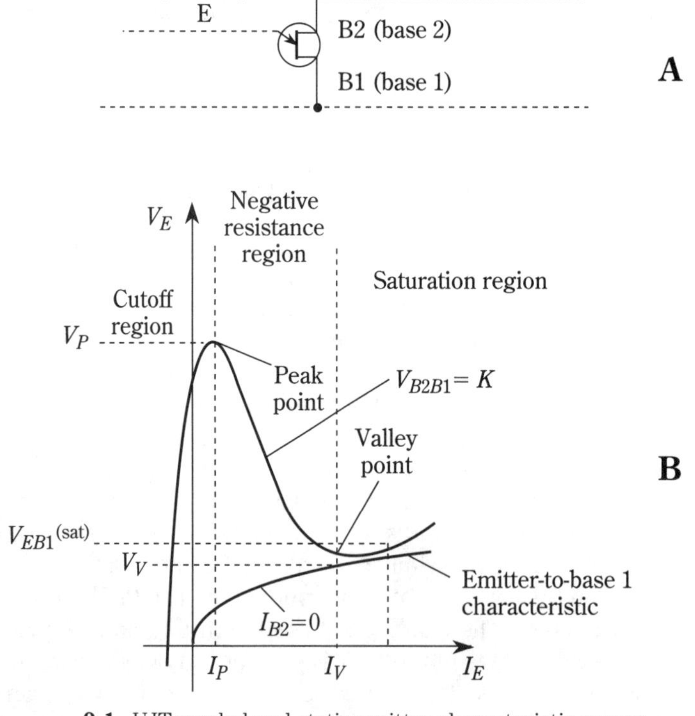

9-1 UJT symbol and static emitter characteristic curves.

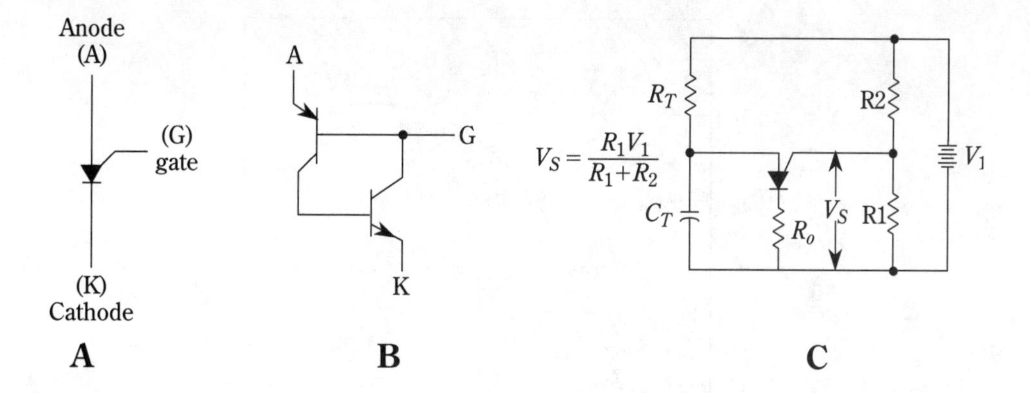

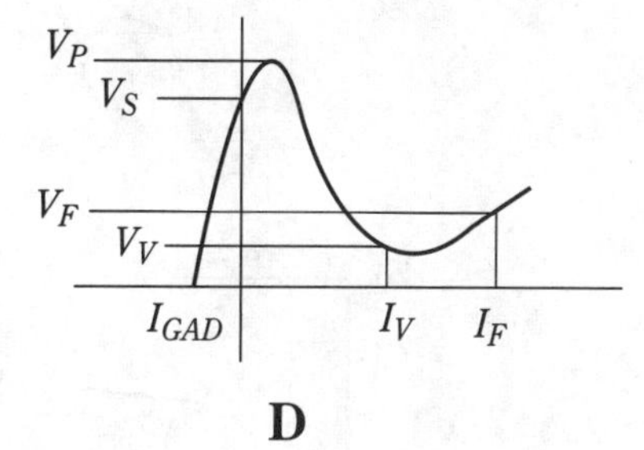

9-2 PUT symbol, equivalent circuit, relaxation oscillator circuit, and static characteristic curve.

is normally used as a UJT, the UJT terminology is used to describe PUT parameters. To operate a PUT as a UJT, an external reference voltage must be maintained at the gate terminal.

Figure 9-2 also shows a typical PUT relaxation oscillator together with the characteristic curve (looking into the anode-cathode terminals). As in the case of a UJT, the peak and valley points of the PUT curve are stable operating points at either end of a negative-resistance region. The peak-point voltage (V_P) is essentially the same as the external gate reference (taken from the voltage divider R_1/R_2), the only difference being the gate diode drop (V_S). Because V_S is circuit-dependent (values of R_1, R_2 and V_1), rather than device-dependent (as is the case with a UJT), the V_S reference can be varied. As a result, the V_P of a PUT is made programmable by varying the reference. This feature is the most significant difference between the UJT and PUT.

9.2 UJT characteristics

Figure 9-3 shows a typical UJT emitter curve for a V_{B1B2} of 20 V, drawn to scale (unlike Fig. 9-1). Figure 9-3A shows part of the cutoff region plotted on a linear scale. When the emitter voltage is zero, the emitter current is negative. Peak-point voltage is reached at a forward-emitter current of about 10 μA.

As shown in Fig. 9-3B, the voltage remains constant until the emitter current reaches about 100 nA, at which point the voltage starts to decrease. Peak current (I_P) for the UJT graphed in Fig. 9-3 thus equals 100 nA, while peak voltage V_P equals 16 V. Valley voltage (V_V) is about 1.6 V and valley current I_V is about 8 mA.

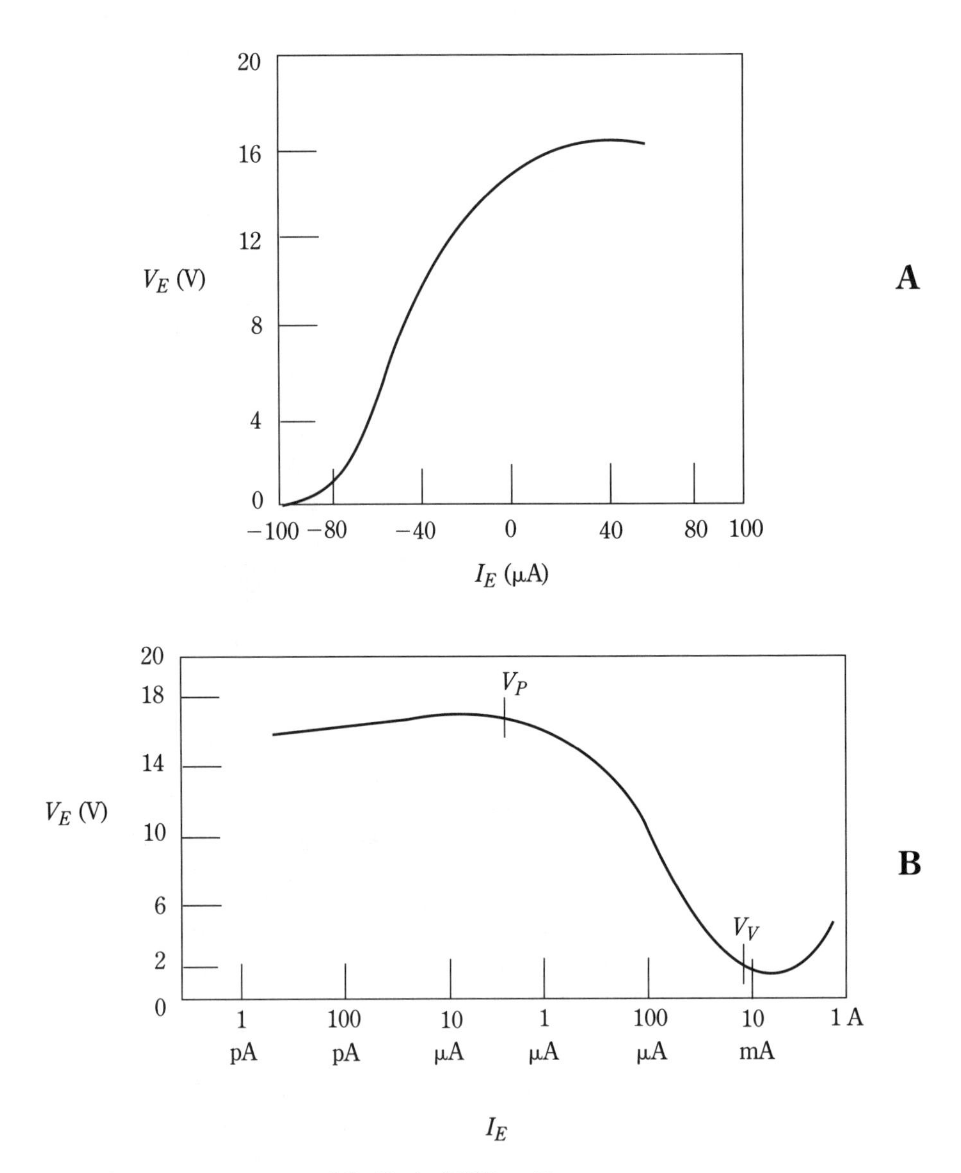

9-3 Typical UJT emitter curves.

9.2.1 Saturation resistance r_s

The *saturation resistance* (r_s) can be found from the slope of the emitter characteristic in the saturation region (above about 8 mA) and is about 5 Ω.

9.2.2 Diode voltage drop V_D

Diode voltage drop (V_D)is defined as the forward voltage drop of the emitter junction. Because V_D is essentially equivalent to the forward voltage drop of a silicon diode (chapter 10), the value of V_D depends on both forward current and temperature. V_D is particularly important because it is one of the major factors in peak voltage V_P. The equation for V_P is: $V_P = V_D + NV_{B2B1}$, where N is the intrinsic standoff ratio.

In applications such as timers and oscillators, any changes in V_P result in inaccuracy because oscillator and timer accuracy depend on the repeatability of V_P. It is important to hold the emitter current near peak when measuring V_P.

9.2.3 Intrinsic standoff ratio

Intrinsic standoff ratio is defined by $N = (V_P - V_D)/V_{B2B1} = r_{B1}/r_{BB}$. The intrinsic standoff ratio is somewhat temperature-dependent, and is also slightly dependent on V_{B2B1}. The procedure for measuring intrinsic standoff ratio is described in the following sections of this chapter. It is also possible to calculate the ratio if the values of V_{B2B1}, V_P and V_D are known.

The ratio can also be measured if you have a very sensitive ohmmeter. First, measure the base-1 to emitter (r_{B1}) resistance. Then, measure the base-1 to base-2 (r_{BB} or interbase) resistance. Divide r_{B1} by r_{BB} to find the intrinsic standoff ratio.

Notice that r_{BB} is highly temperature-dependent and can vary with interbase voltage V_{B2B1}. Also note that the peak-point characteristics V_P and I_P, as well as the valley-point characteristics V_V and I_V, decrease as ambient temperature increases. When V_{B2B1} increases, both V_V and I_V increase.

9.2.4 Interbase characteristics

Interbase characteristics are usually indicated by measurement of base-2 current I_{B2} as a function of interbase voltage V_{B2B1} and emitter current I_E. Usually, interbase characteristics are measured on a sweep basis, rather than with constant voltages and currents, to avoid heating effect because of power dissipation.

Figure 9-4 shows a circuit for sweep test of UJT pulse interbase characteristics. A constant-current pulse is applied to the emitter from time zero (t_0) to time (t_1). Simultaneously, a voltage ramp going from 0 to 30 V is applied to base-2. Base-1 is grounded to complete the circuit. The pulse and ramp can be supplied by function generators, chapter 3.

The current (I_{B2}) is measured with a current probe (chapter 5) and applied to the vertical input of a scope. The voltage ramp, applied to base-2, is also applied to the scope horizontal input. Figure 9-4B shows typical interbase characteristics at ambient temperatures of –55, +25, and +125°C. As shown, the percentage increase in I_{B2} decreases with increasing emitter current and temperature.

9.2.5 Transient characteristics

The *transient characteristics* of a UJT are usually not specified in the same way as for a two-junction transistor or FET. For example, switching times are usually not specified on a UJT datasheet. Instead, a parameter of f_{max} is given for most UJTs. The f_{max} indicates the maximum frequency of oscillation that can be obtained using the UJT in a specified relaxation-oscillator circuit (the capacitor and resistor values are specified).

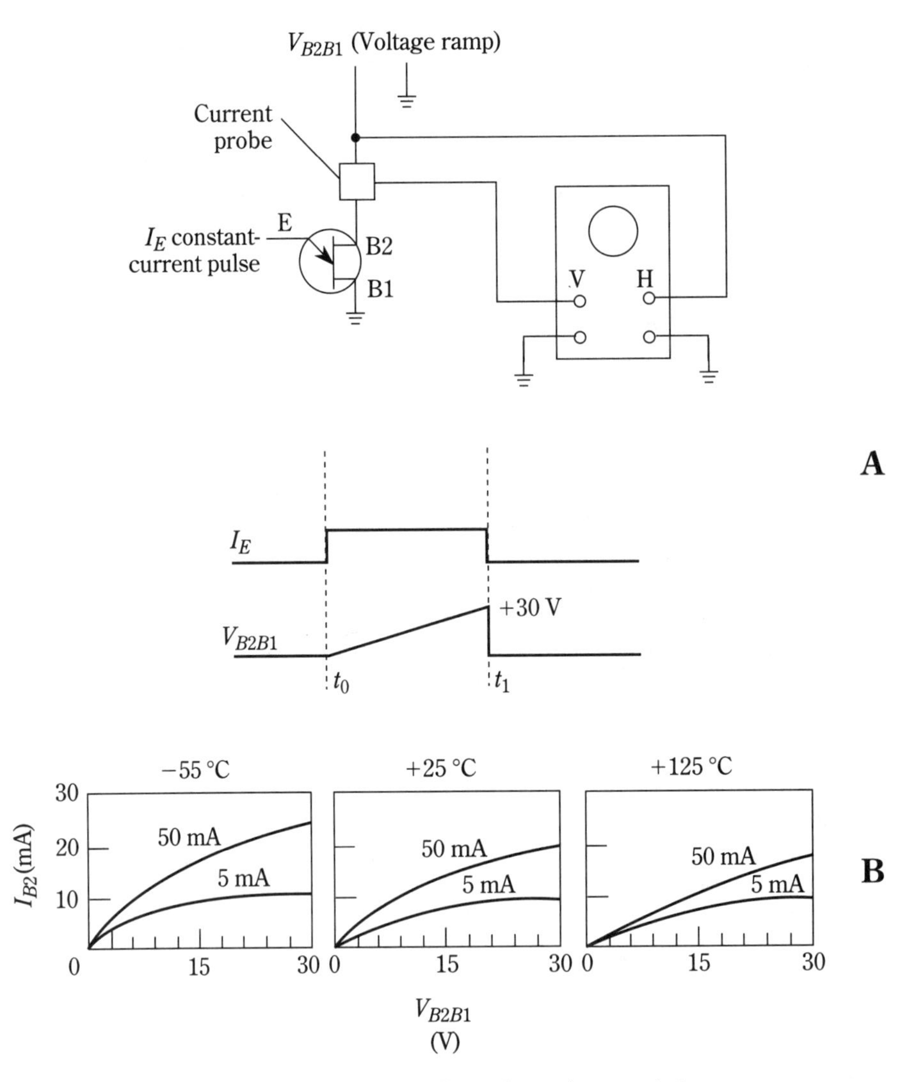

9-4 Sweep test of UJT interbase characteristics.

In some applications, such as critical timers, it might be of interest to determine turn-on and turn-off times associated with the UJT. The following paragraphs describe basic procedures for these measurements.

Turn-on and turn-off for a purely-resistive circuit The circuit of Fig. 9-5A can be used to measure t_{on} and t_{off} for the case where the UJT emitter circuit is purely resistive. Typical switching-time values are t_{on} = 1 μs, and t_{off} = 2.5 μs. The waveform observed at the base-1 terminal when the UJT turns off is shown in Fig. 9-5B. Operation of the test circuit is as follows.

When the emitter is returned to ground through the relay contacts, stored charge in the junction causes a current to flow out of the emitter, and the output voltage across R1 is smaller than the steady-state off-value. Immediately following

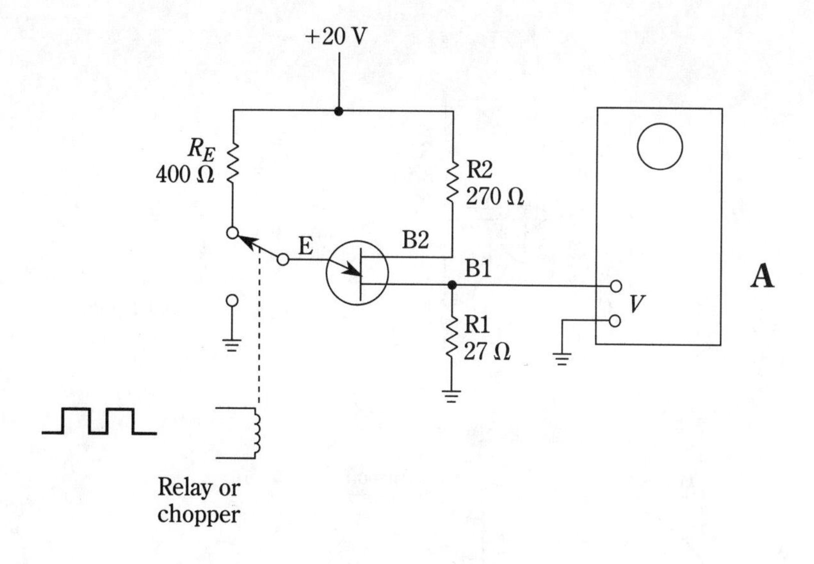

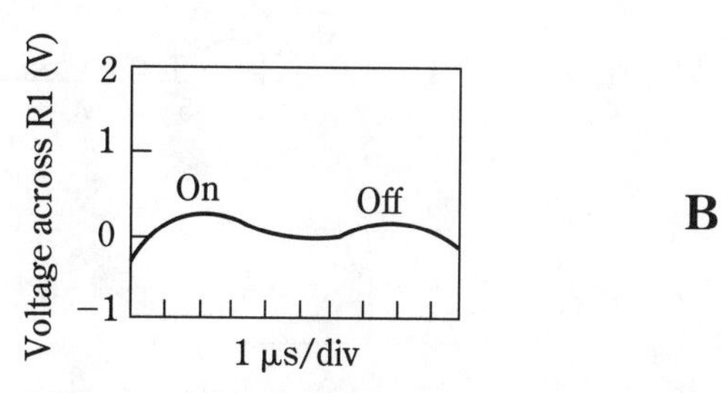

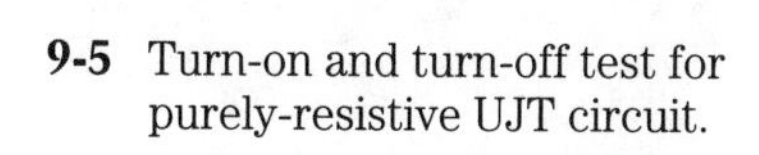

9-5 Turn-on and turn-off test for purely-resistive UJT circuit.

the removal of the excess charge, the voltage across R1 goes higher than the steady-state off-value because r_{B1} has still not returned to normal, and I_{B2} is larger than the steady-state off-value.

Turn-on and turn-off for capacitive/resistive circuit. The circuit of Fig. 9-6A can be used to measure t_{on} and t_{off} for the case where the UJT emitter circuit is both resistive and capacitive (which is the usual case). The test circuit of Fig. 9-6A is a relaxation oscillator, with turn-on and turn-off time being measured at the base-1 terminal. The turn-on and turn-off waveforms are shown in Fig. 9-6B.

Turn-on time is measured from the start of the turn-on to the 90% point. A typical turn-on time is 0.5 μs with the capacitance C_E (0.1 μF) shown. This is shown in Fig. 9-7, which gives turn-on and turn-off times versus capacitance in relaxation-oscillator test circuits.

Turn-off time is measured from the start of turn-off to the 90% point, and is about 12 μs, (because of the long discharge time of the capacitor). Turn-off time also increased with an increase in C_E capacitance, as shown in Fig. 9-7C.

Go/no-go UJT tests The circuits of Figs. 9-6 and 9-7 can be used to provide a quick check of UJT operation. If the UJT oscillates and produces output pulses when connected, as shown in Figs. 9-6 and 9-7, it is reasonable to assume that the UJT is good. The frequency of oscillation depends primarily on the values of C_E and R_E. The output voltage depends on the ratio of R_1 and R_2 (and the supply voltage).

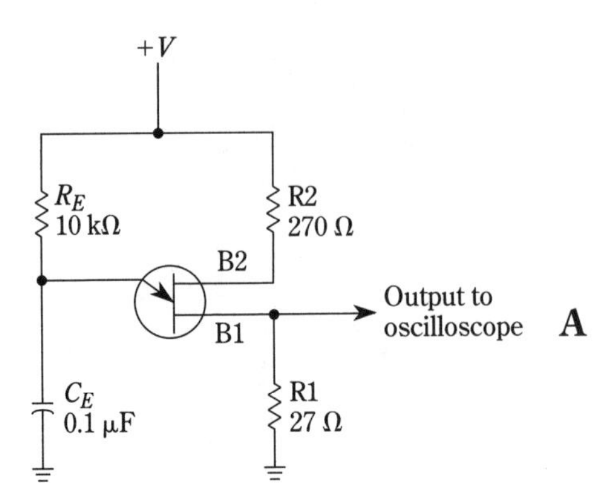

9-6 Turn-on and turn-off test for capacitive/resistive UJT circuit.

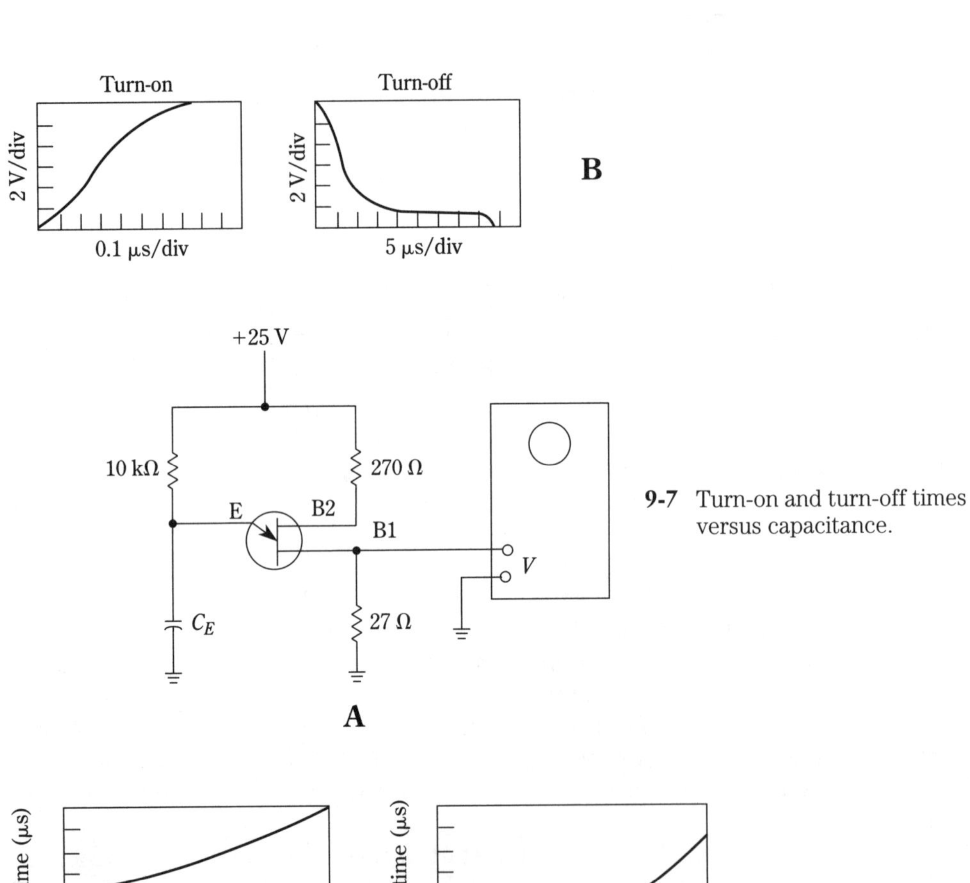

9-7 Turn-on and turn-off times versus capacitance.

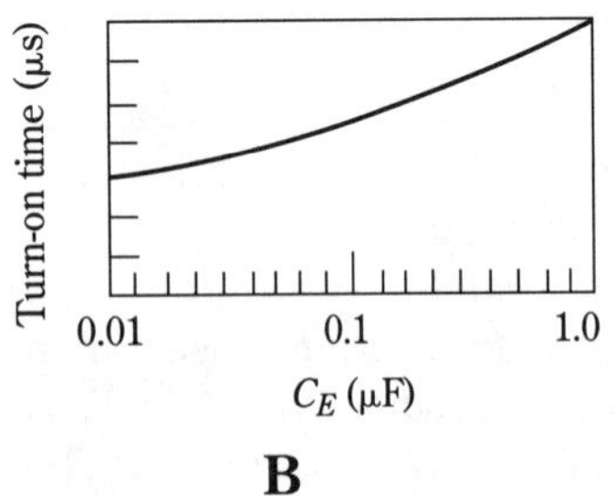

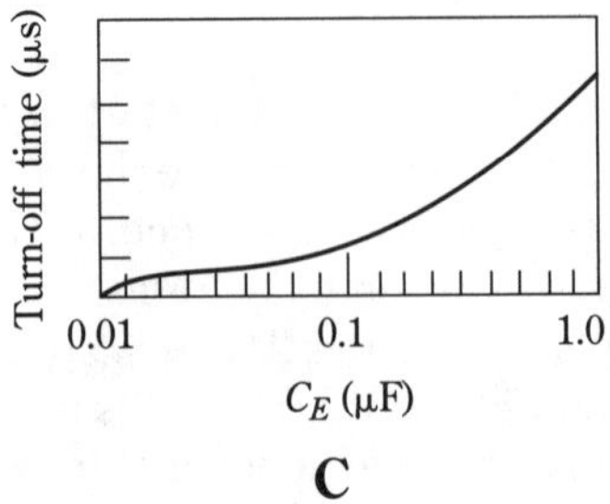

9.3 PUT characteristics

The following paragraphs describe test procedures for the most important PUT parameters.

9.3.1 Peak-point current I_P

The *peak point* is indicated graphically by the static curve in Fig. 9-2. Reverse anode current flows with anode voltage less than the gate voltage (V_S) because of leakage from the bias network to the charging network. With currents less than I_P, the PUT is in a blocking state. With currents above I_P, the PUT goes through the negative-resistance region to the on-state.

The charging current, or the current through a timing resistor, must be greater than I_P at V_P to ensure that the PUT switches from a blocking to on-state in an oscillator circuit. For this reason, maximum values of I_P are given on most PUT datasheets. These values depend on V_S, temperature, and gate resistance. Typical curves on the datasheet indicate this dependence and must be consulted for most applications.

Figure 9-8 shows a peak-point current I_P test circuit for a typical PUT. This circuit is a sawtooth oscillator that uses a 0.01-μF timing capacitor, a 20-V supply (to provide V_S), an adjustable charging current, and equal biasing resistors R. The 2N5270 FET circuit is used as an adjustable charging current source. A variable gate-supply voltage V_G is used to control the FET current source, and thus control operation of the PUT. The PUT can be adjusted to operate throughout the entire curve by adjustment of V_G.

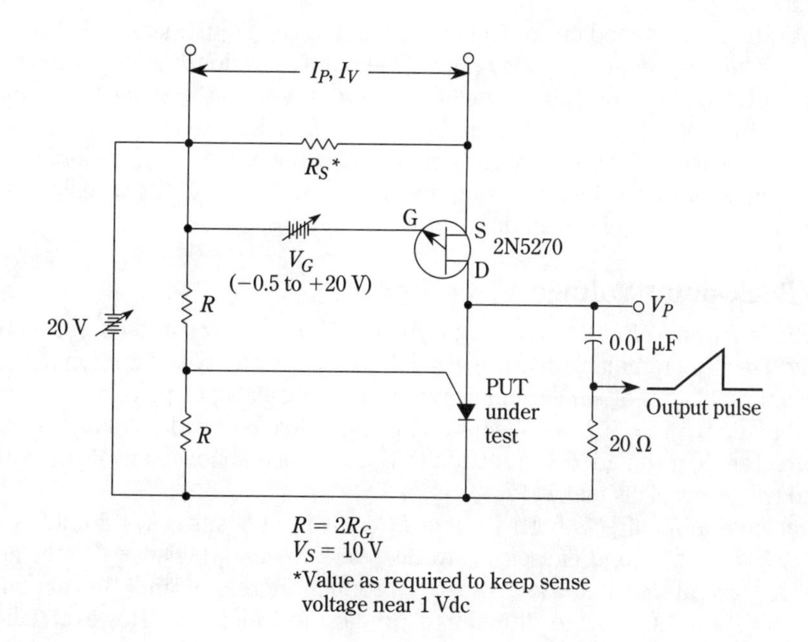

9-8 Test circuit for I_P, V_P, and I_V.

The output pulse (sawtooth sweep) from the circuit is displayed on a scope. The peak-point current is measured across sense resistor R_S by means of a voltmeter ($I = E/R$). The value of R_S is not critical, but it should be some precision value that keeps the sense voltage in the 1-V range.

To measure I_P, the PUT is set to off (just prior to oscillation) by the adjustment of V_G. This condition is indicated by the absence of an output pulse to the scope. Then, the voltage across R_S is measured, and the value of I_P is calculated. For example, if the indicated voltage is 1 V at I_P, and R_S is 1 MΩ, I_P is 1 μA.

9.3.2 Valley-point current I_V

The *valley point* is indicated graphically in Fig. 9-2. With currents slightly less than I_P, the PUT is in an unstable negative-resistance state. A voltage minimum occurs at I_P, and with higher currents the PUT is in a stable on-state. When the PUT is used as an oscillator, the charging current or current through a timing resistor must be less than I_V, at the valley-point voltage V_V. For this reason, minimum values of I_V are given on the datasheet. When the PUT is used in the latching mode, the anode current must be greater than I_V.

Valley-point current (I_V) is tested using the same circuit of Fig. 9-8. Again, the circuit operates as a sawtooth oscillator, and operation of the circuit is controlled by V_G. Circuit output is monitored on the scope, and I_V is measured across sensing resistor R_S in the form of a voltage (as is the case for I_P).

It is sometimes difficult to differentiate among N valley-point current I_V, latching current I_L, and holding current I_H. The following summarizes how the three values can be measured.

With the PUT latched on, reducing the current (by adjustment of V_G) produces a minimum indication at the valley point (the point at which I_V is measured). However, the PUT remains on (producing a sawtooth sweep to the scope) at lower currents until the holding-current I_H point is reached. I_H is detected by the absence of an output pulse (just before oscillation occurs). Latching current I_L is generally higher than I_H, and is measured by increasing the current while the PUT is oscillating, then noting the value at which oscillation stops.

9.3.3 Peak-point voltage V_P

The unique feature of a PUT is that the *peak-point voltage* can be determined externally. This programmable feature gives the PUT the ability to function in voltage-controlled oscillators or similar applications. The triggering or peak-point voltage is approximated by: $V_P = V_{offset} + V_S$, where V_S is the unloaded divider voltage. V_{offset} is always greater than the anode-gate voltage V_{AG}, because I_P flows out of the gate just prior to triggering. This makes $V_{offset} = V_{AG} + I_{PRG}$.

A change in R_G affects both V_{AG} and I_{PRG}, but in opposite ways. First, as R_G increases, I_P decreases and causes V_{AG} to decrease. Second, because I_P does not decrease as fast at R_G increases, the I_{PRG} product increases and the actual V_{offset} increases. These effects are difficult to predict and measure. However, allowing V_{offset} to be 0.5 V is usually accurate for most applications.

Actual peak-point voltage V_P is also tested using the circuit of Fig. 9-8. During testing, V_G is adjusted until the PUT is at peak (as described for the I_P test). Then, V_P

is measured across the PUT, as shown. V_P can be measured with a high-impedance (10 MΩ or higher) scope.

9.3.4 Peak output voltage V_o

The *peak output voltage* (V_o) of a PUT used in a relaxation-oscillator circuit depends on many factors, including dynamic impedance, switching speed, and V_P. When the timing capacitor used in the circuit is small (less than about 0.01 μF), the effect of switching speed on V_o is increased. This is because small-value capacitors lose part of their charge during the turn-on interval.

Peak output voltage V_o is tested using the circuit of Fig. 9-9. This test circuit is a relaxation oscillator similar to that shown in Fig. 9-2. The Fig. 9-9 circuit uses a relatively large timing capacitor (0.2 μF) to minimize the effect of losing part of the charge during turn-on. The output of the relaxation-oscillator circuit is displayed on a scope across the 20-Ω resistor (in series with the cathode K lead).

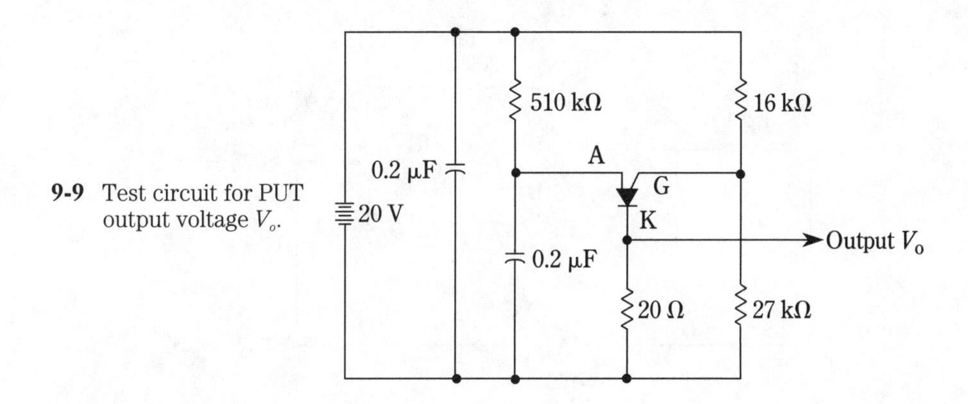

9-9 Test circuit for PUT output voltage V_o.

9.3.5 Rise time t_r

Rise time (t_r) is a useful parameter in pulse circuits that use capacitive coupling, and can be used to predict the amount of current that flows between such circuits. The rise time for a typical PUT is about 40 ns. This is measured in the circuit of Fig. 9-10 (which requires a sampling scope). In this circuit, t_r is considered the time between the 0.6-V point and the 6-V point on the leading edge of the output pulse. These values represent the 10% and 90% points, respectively, on the leading edge, with the circuit values shown in Fig. 9-10.

9.4 Additional UJT test circuits

The following paragraphs describe some additional UJT test circuits.

9.4.1 Simplified test for UJT firing point

It is possible to measure the firing point of a UJT using the circuit shown in Fig. 9-11. The test also shows the amount of emitter-to-base-1 current flow after the UJT has

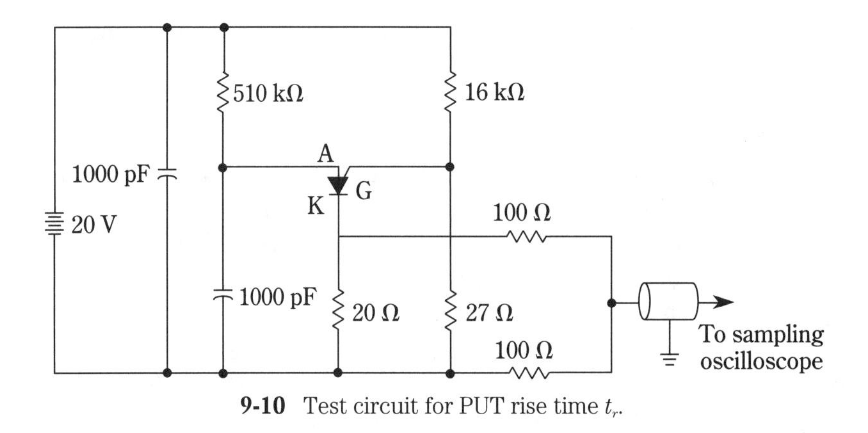

9-10 Test circuit for PUT rise time t_r.

9-11 Simplified test circuit for UJT firing point.

fired. If a UJT fires with the correct voltage applied and draws the rated amount of current, the UJT can be considered satisfactory for operation in most circuits. This test can be combined with the basic relaxation-oscillator test shown in Figs. 9-6 and 9-7, to show UJT operation on a go/no-go basis.

The base-2 voltage is shown in Fig. 9-11 as +20 V. However, any value of base-2 voltage can be used to match a particular UJT. Initially, R_1 is set to 0 V (at the ground end). The setting of R_1 is gradually increased until the UJT fires. The firing voltage is indicated on the voltmeter. When the UJT fires, the ammeter indication suddenly increases. The amount of emitter-to-base-1 current is read on the ammeter. Typically, the firing point is in the range of 0 to 20 V, whereas emitter-to-base-1 current is less than 50 μA.

9.4.2 Intrinsic standoff ratio test

In the circuit of Fig. 9-12, the interbase voltage is provided by a rectified 10-Hz source (in effect, the interbase circuit is swept at a 10-Hz rate). With the values

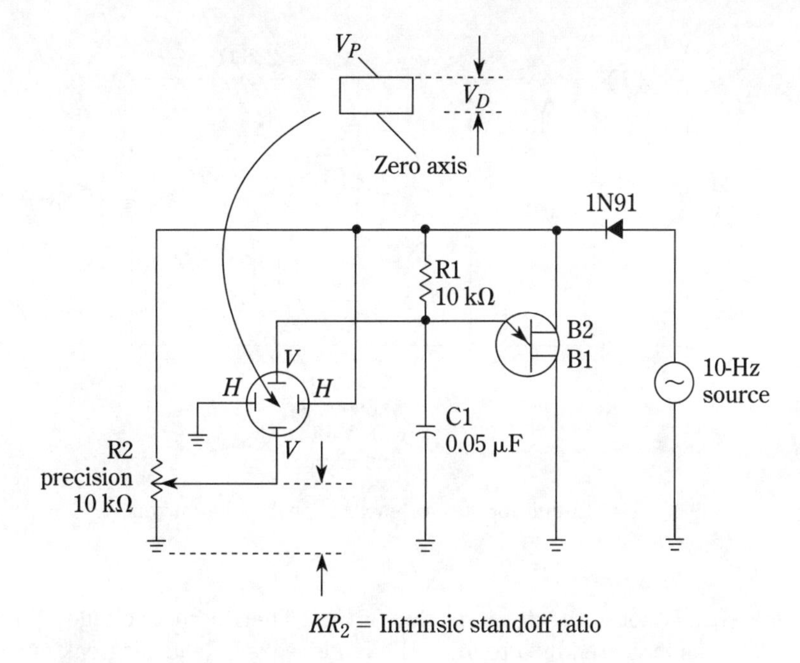

9-12 Test circuit for measurement of intrinsic standoff ratio and V_D.

shown, the UJT oscillates at about 2 kHz in the basic relaxation-oscillator configuration. The interbase voltage is also applied to the horizontal axis of the scope and to R2. The UJT emitter voltage is applied to one input of the scope vertical-deflection amplifier. The R2 voltage is applied to the other input of the vertical-deflection amplifier. With these connections, the scope pattern consists of a plot of V_{B2B1} (on the horizontal axis) against $V_P - KV_{B2B1}$ (on the vertical axis). K is equal to the fractional setting of R_2. V_P is the upper envelope of the displayed emitter voltage.

During test, K (potentiometer R2) is adjusted until the upper envelope of the display is horizontal (parallel to the zero axis). When the upper envelope is horizontal, K (or the fractional setting of R2) is equal to the intrinsic standoff ratio. For example, if R2 is 10 kΩ and the resistance from the contact arm of R2 to ground is 7 kΩ, the ratio is 0.7. If a precision potentiometer is used for R2, the intrinsic standoff ratio can be measured with an accuracy of better than 0.05%.

With R2 adjusted to produce a horizontal envelope on the scope display, note the displacement of the upper envelope from the zero axis of the scope (in volts). This displacement is equal to V_D. If the scope vertical screen can be read accurately, V_D can be measured within about 20 mV.

9.4.3 Peak-point current test

The peak-point current I_P can be measured with the circuit of Fig. 9-13. This circuit actually measures the minimum emitter current required for oscillation in a relaxation-oscillator circuit. This minimum emitter current is a good approximation of the peak-point emitter current.

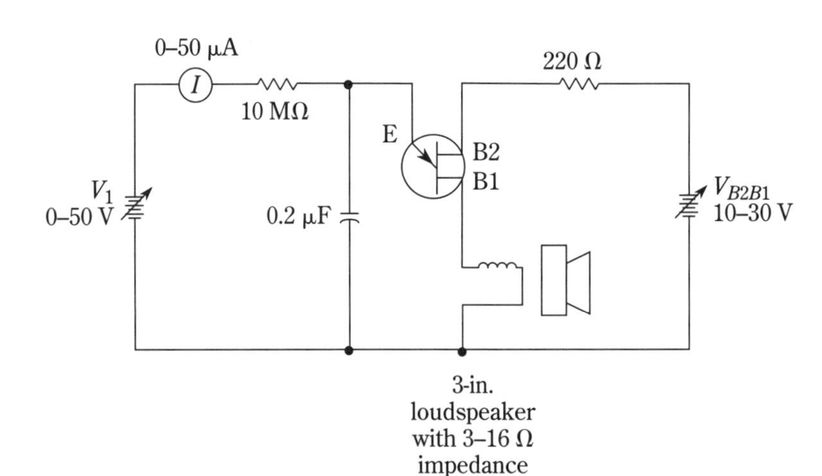

9-13 Test circuit for measurement of peak-point current I_P.

To measure I_P, set in the desired value of V_{B2B1}. Then, adjust voltage V_1 until the UJT just fires, as indicated by a tone on the loudspeaker. Read the peak current on the microammeter. Take care to avoid any ripple on the V_{B2B1} supply (if a battery is not used) because ripple can reduce the apparent peak-point current considerably.

9.4.4 Valley-point current and voltage test

Because of the slight change of emitter voltage at or near the valley point of a UJT, it is difficult to locate the exact position of the valley point by inspection of the emitter-characteristic curve. To overcome this problem, the circuit of Fig. 9-14 can be used to measure V_V and I_V by the null method.

With the circuit of Fig. 9-14, supply voltage V_2 is adjusted for a null on the digital voltmeter. I_V and V_V can then be measured directly on the corresponding meters. The values given in Fig. 9-14 are typical. Other values of interbase voltage and R_2 series resistance can be used as desired for special tests.

9.4.5 Frequency response (emitter input resistance)

The UJT small-signal frequency-response characteristic of most value is the emitter input resistance, which can be measured with the circuit of Fig. 9-15. In this circuit, the UJT is biased in the negative-resistance region, and a null is obtained on the electronic voltmeter by adjusting R1 and C1. At null, the emitter input impedance is approximately equal to $R_1 + 1/(6.28C_1)$.

9.5 UJT tests using a curve tracer

Some curve tracers provide special circuits and connections for measurement of UJT characteristics. Most curve tracers can be converted by simple external circuits to test UJTs. Figure 9-16 shows the connections for conversion of a curve tracer to

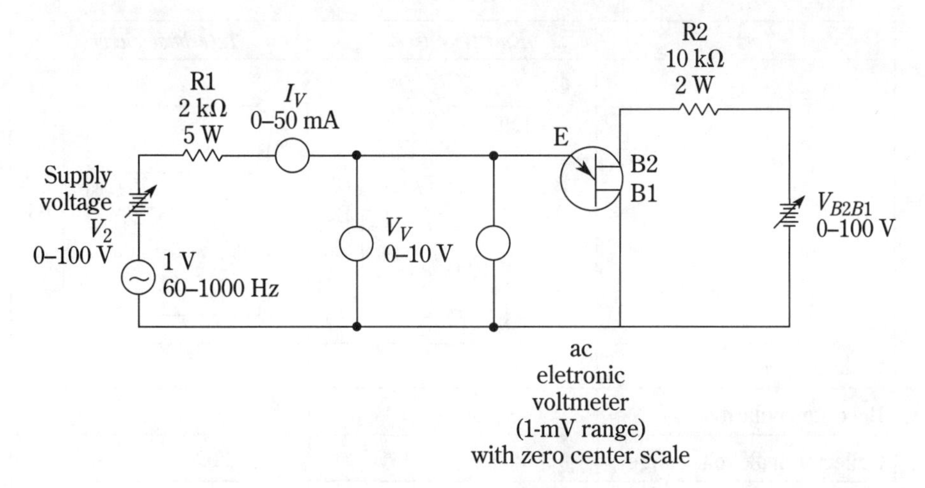

9-14 Test circuit for measurement of valley voltage V_V and valley current T_V.

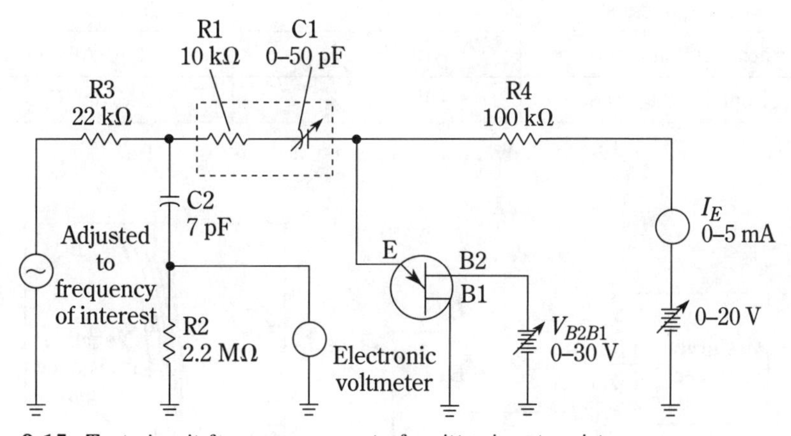

9-15 Test circuit for measurement of emitter input resistance.

test UJTs. For testing of UJT interbase characteristics, the curve-tracer terminals normally used for a two-junction transistor base are connected to the UJT emitter. For testing of emitter curves, the UJT emitter is connected to the collector terminal of the tracer. When making these connections, make certain to observe any limits placed by the curve tracer. For example, when displaying emitter-characteristic curves, the interbase voltage should not exceed 12 V (for a particular tracer) because of the voltage limitation of the base-current step generator.

9.5.1 Displaying UJT curves on a curve tracer

To display UJT curves on the curve tracer, connect the UJT to the tracer, as shown in Fig. 9-17. As shown, base-1 is connected to the emitter terminal or jack, base-2 is

Test	*Emitter curves*	*Interbase curves*
Circuit	C, 1N91, B, 100, E, B2, B1	C, B, E, B2, B1
Collector sweep polarity	+	+
Base step polarity	+	+
Collector peak volts range	200 V	20 V
Collector limiting resistor	5 kΩ	500 Ω
Base current steps selector	20 mA/step	10 mA/step
Number of current steps	5	5
Vertical current range	2 mA/div. I_E	2 mA/div. I_{B2}
Horizontal voltage range	1 V/div. V_E	2 V/div. V_{BB}

9-16 Conversion of a curve tracer to test UJTs.

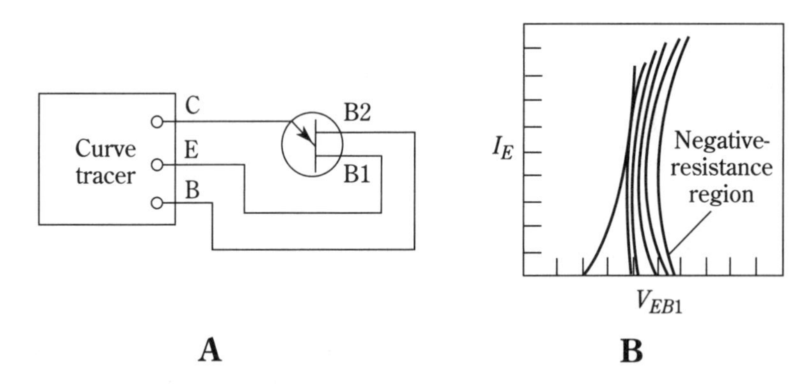

9-17 Displaying UJT curves on a curve tracer.

connected to the base jack, and the (UJT) emitter is connected to the controller jack. (Note that base-1 and base-2 are interchangeable.)

Set the curve-tracer polarity switch to npn. Increase the sweep voltage from zero until the trigger point is exceeded. This should produce the high emitter-current spike on the scope. Set the step selector to the "current per step" position that produces the most curves on the display.

The curves might appear quite close together (compared to the display of a two-junction transistor or FET). This means that you must observe the display carefully to distinguish the curves. It might be helpful to spread out the display by increasing the horizontal sensitivity of the scope, or to use the expanded sweep magnification (if available) at the area of interest.

With the test connection, as shown in Fig. 9-17, the step current of the curve tracer is applied from base-1 to base-2, and the sweep voltage is applied to the emitter. Thus, the sweep voltage is the UJT trigger voltage. As the sweep voltage is slowly increased from the trigger threshold producing the first current spike, the other curves are added one by one. Thus, for each base-current step, the emitter trigger voltage can be measured.

9.5.2 Measuring UJT interbase resistance with a curve tracer

Interbase resistance (shown as r_{BB} or R_{BB} on UJT datasheets) can be displayed using the connections shown in Fig. 9-18. As shown, base-1 and base-2 are connected to the curve-tracer collector and emitter jacks, respectively. The UJT emitter is left open-circuited (no connection). The display should be a linear trace, as shown. The vertical scale of the display represents forward current (I_F), and the horizontal scale represents interbase voltage (V_{BB}). Interbase resistance equals interbase voltage divided by forward current, as shown in Fig. 9-18.

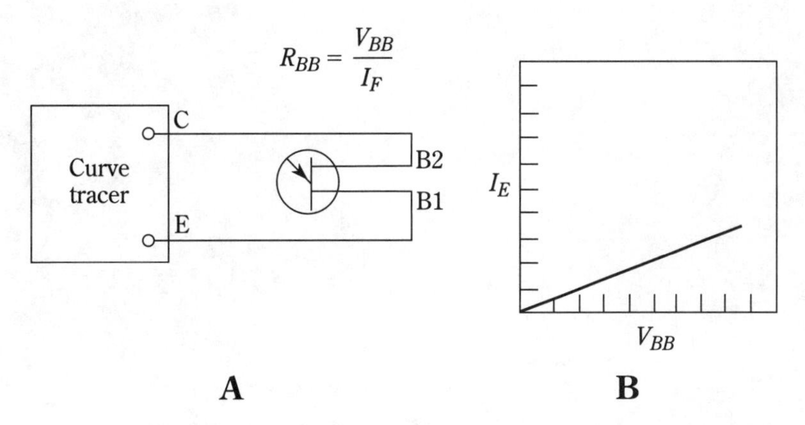

9-18 Measuring UJT interbase resistance with a curve tracer.

10
Solid-state diode tests

This chapter is devoted entirely to test procedures for solid-state diodes. These diodes include signal and rectifying diodes, zener diodes, and tunnel diodes. The first sections of this chapter describe diode characteristics and test procedures from the practical standpoint. The information in these sections permits you to test all the important diode characteristics using basic shop equipment. The sections also help you understand the basis for such tests. The remaining sections of the chapter describe how the same tests, and additional tests, are performed using more sophisticated equipment, such as the curve tracer.

10.1 Basic diode tests

Three basic tests are required for power-rectifier and small-signal diodes. First, any diode must have the ability to pass current in one direction (forward current) and prevent or limit current flow (reverse current) in the opposite direction. Second, for a given reverse voltage, the reverse current should not exceed a given value. Third, for a given forward current, the voltage drop across the diode should not exceed a given value.

All of these tests can be made with a multimeter. If the diode is to be used in pulse or switching circuits, the switching time must also be tested. This requires a scope and pulse generator. In addition to the basic tests, a zener diode must also be tested for the correct zener-voltage point. Similarly, a tunnel diode must be tested for negative-resistance characteristics.

10.2 Diode continuity tests

The elementary purpose of a diode (both power rectifier and small signal) is to prevent current flow in one direction while passing current in the opposite direction.

The simplest test of a diode is to measure current flow in the forward direction with a given voltage, then reverse the voltage polarity and measure current flow, if any. If the diode prevents current flow in the reverse direction, but passes current in the forward direction, the diode meets most basic circuit requirements. Further, if there is no excessive leakage-current flow in the reverse direction, it is quite possible that the diode will operate properly in all but the most critical circuits.

A simple resistance measurement or continuity check can often be used to test a diode's ability to pass current in one direction only. A basic ohmmeter can be used to measure the forward and reverse resistance of the diode. Figure 10-1 shows the basic circuit. A good diode shows high resistance in the reverse direction and low resistance in the forward direction.

If the resistance is high in both directions, with the ohmmeter connected as shown in Fig. 10-1, the diode is probably open. A low resistance in both directions usually indicates a shorted diode. If the resistance is close to the same in both directions (high or low), the diode is probably leaking.

It is possible for a defective diode to show a difference in forward and reverse resistance. The important factor in making a diode-resistance test is the ratio of *forward-to-reverse resistance* (sometimes called the *front-to-back ratio*). The actual ratio depends on the type of diode. However, as a guideline, a signal diode typically has a ratio of several hundred to one, whereas a power diode can operate satisfactorily with a ratio of about 10 or 20 to one.

Diodes used in power circuits are usually not required to operate at high frequencies. Such diodes can be tested effectively with dc or low-frequency ac. Diodes used in other circuits, even audio equipment, must be capable of operation at higher frequencies and should be so tested.

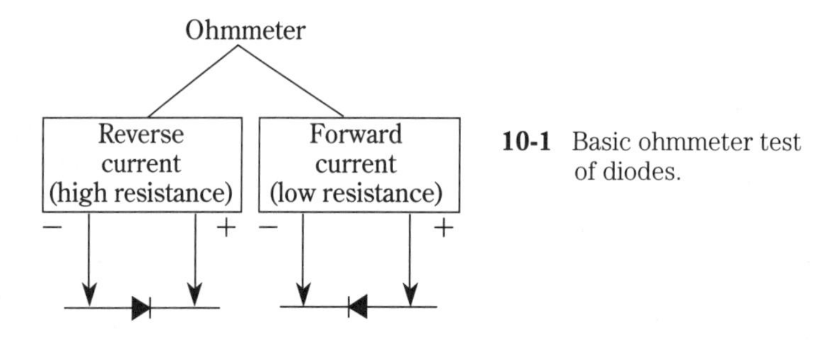

10-1 Basic ohmmeter test of diodes.

10.2.1 Diode test circuit

There are many commercial diode testers available for use in the shop or laboratory. Most simple diode testers operate on the continuity-test principle. This is similar to testing a diode by measuring resistance, except that the actual resistance value is of no concern. Instead, arbitrary circuit values are used, and the diode condition is read out on a good-bad meter scale.

Figure 10-2 shows the basic circuit of a simple diode tester using the continuity principle. The condition of the diode under test is indicated by lamps. With an open diode, or no connection across the test terminals, lamp I1 turns on because of cur-

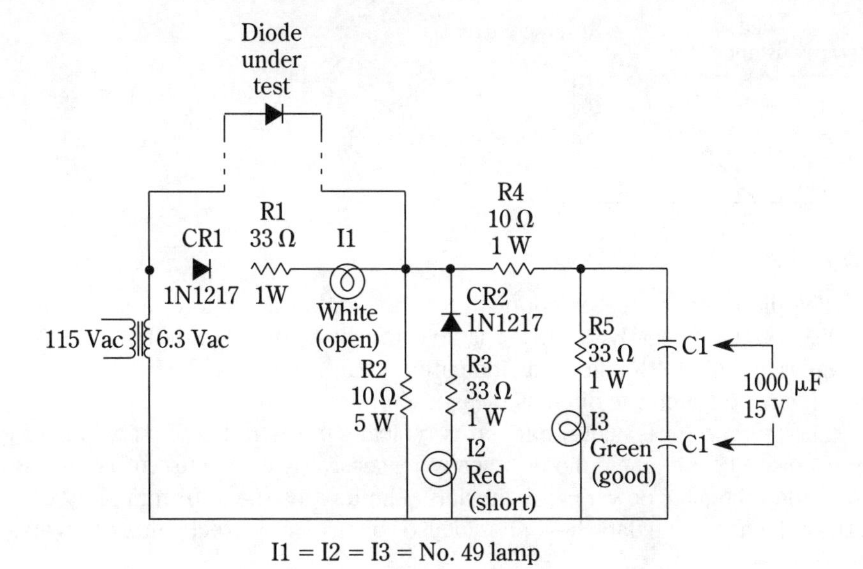

10-2 Simple diode tester.

rent flow through CR1, R1, and R2. The voltage drop across R2 is too low to turn on lamp I3. Because CR2 is reverse-biased, lamp I2 does not turn on.

If a good diode is connected across the test terminals with the polarity, as shown in Fig. 10-2, lamp I1 is shorted out and does not turn on. Also, lamp I2 does not turn on because CR2 is reverse-biased by the voltage developed from the diode under test. Capacitors C1 and C2 charge, permitting lamp I3 to light, thus indicating that the diode is good.

A shorted diode also shorts out lamp I1 and allows both half-cycles of the ac to be applied across R2. CR2 conducts on the negative half-cycles (when the junction of CR2 and R4 is negative), causing lamp I2 to turn on. Although current is also applied across R4, R3, I3, C1, and C2. Capacitors C1 and C2 appear as a short across R5 and I3. Thus, lamp I3 does not turn on. This leaves only lamp I2 to light and indicate a short.

If a good diode is connected, but with the polarity reversed from that shown in Fig. 10-2, all lamps turn on. Under these conditions, CR1 conducts on the positive half-cycles, causing lamp I1 to turn on. The test diode conducts on negative half-cycles and develops a dc voltage across R2 (with the junction of R2 and I1 negative). CR2 then conducts, and I2 turns on. Capacitors C1 and C2 charge, permitting I3 to turn on.

10.3 Diode reverse-leakage tests

Reverse leakage is the current flow through a diode when a reverse voltage (anode negative with respect to cathode) is applied. The basic circuit for measurement of reverse leakage is shown in Fig. 10-3. Similar circuits are incorporated in some commercial diode testers, or can be duplicated with the basic test equipment shown.

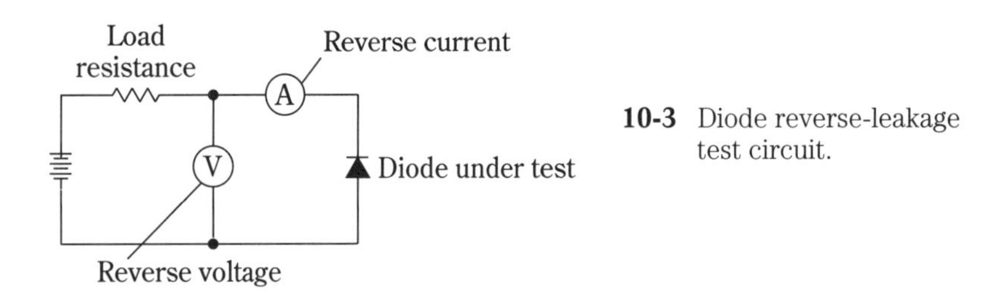

10-3 Diode reverse-leakage test circuit.

The diode under test is connected to a variable dc source in the reverse-bias condition (anode negative). The variable source is adjusted until the desired reverse voltage is applied to the diode, as indicated by the voltmeter. Then, the reverse current (if any) through the diode is measured on the ammeter.

The reverse (or leakage) current is typically in the range of a few μa (or possibly nA or pA) for a single diode. Usually, excessive leakage current is undesired in any diode (signal or power), but the actual limits must be determined by reference to the datasheet. The datasheet should also specify the correct value of reverse voltage for the test.

10.4 Diode forward voltage drop tests

Forward voltage drop is the voltage that appears across a diode when a given forward current is band-passed. The basic circuit for measurement of forward voltage is shown in Fig. 10-4. Similar circuits are incorporated in some commercial diode testers, or it can be duplicated with the basic test equipment shown.

The diode under test is connected to a variable dc source in the forward-bias condition (anode positive, cathode negative). The variable source is adjusted until the desired amount of forward current is passing through the diode, as indicated by ammeter. Then current is passing through the diode as indicated by ammeter. Then the voltage drop across the diode is measured on the voltmeter. This is the forward voltage drop. Usually, a large forward voltage drop is not desired in any diode. The maximum limits of forward voltage drop (for a corresponding forward current) should be found on the datasheet.

Typically, the forward voltage drop for a germanium diode is about 0.2 V, whereas a silicon diode has a forward voltage drop of about 0.5 V, at nominal currents. Increased forward currents produce increased voltage drops.

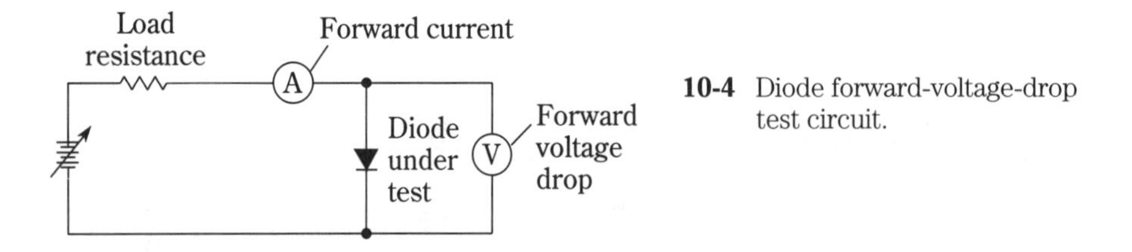

10-4 Diode forward-voltage-drop test circuit.

10.5 Diode dynamic tests

The circuits and methods covered in preceding sections of this chapter provide static test of diodes, meaning that the diode is subjected to constant direct current when the leakage and voltage drop are measured. Diodes do not usually operate this way in circuits. Instead, diodes are operated with alternating current, which tends to heat the diode junctions and change the characteristics. It is more realistic to test a diode under dynamic conditions. The following paragraphs describe two typical dynamic tests for diodes using a scope.

10.5.1 Dynamic tests for power-rectifier diodes

Power-rectifier diodes can be tested with a scope to display dynamic current and voltage characteristics. A dc scope is required and both the vertical and horizontal channels must be voltage-calibrated. As discussed in chapter 2, the horizontal channel is usually time-calibrated. However, the horizontal channel can be voltage-calibrated using the same procedures as for the vertical channel.

In brief, the scope sweep circuit is disconnected from the input to the horizontal amplifier (the horizontal input is set to EXTERNAL), a reference voltage is applied to the horizontal-amplifier input, and the horizontal gain or width control is set for some specific deflection or width (1 cm on the horizontal scale for 1 V, etc.). For best results, the horizontal and vertical channels must be identical, or nearly identical, to eliminate any phase difference.

As shown in Fig. 10-5, a power diode is tested by applying a controlled ac voltage across the anode and cathode through resistor R1. The ac voltage (set to or just below the maximum rated peak inverse voltage, or PIV, of the diode) alternately biases the anode positive and negative, causing both forward and reverse current to flow through R1. The voltage drop across R1 is applied to the vertical channel and causes the scope trace spot to move up and down. Vertical deflection is proportional to current through the diode under test. The vertical-scale divisions can be converted directly to current when *R1* is Ω. For example, a 3-V vertical deflection indicates a 3-A current. If R1 is 1000 Ω, the readout is in mA.

The same voltage applied across the diode is applied to the horizontal channel (which has been voltage-calibrated), and causes the spot to move right or left. Horizontal deflection is proportional to voltage across the diode (neglecting the small voltage drop across R1).

The combination of the horizontal (voltage) deflection and vertical (current) deflection causes the spot to trace out the complete current and voltage characteristics. The basic test procedure is as follows:

1. Connect the equipment, as shown in Fig. 10-5. Place the scope in operation. Voltage-calibrate both the vertical and horizontal channels, as necessary. The spot should be at the vertical and horizontal center with no signal applied to either channel.

2. Switch off the internal sweep of the scope. Set the sweep selector and sync selector (if any) of the scope to EXTERNAL. Leave the horizontal and vertical gain controls set at the (voltage) calibrate position, as established in step 1.
3. Adjust the variac so that the voltage applied across the power diode under test is at (or just below) the maximum rated value (as determined from the diode datasheet).
4. Check the oscilloscope pattern against the typical curves of Fig. 10-5 and/or against the diode specifications. The curve of Fig. 10-5 is a typical response pattern. That is, the forward current (deflection above the horizontal centerline) increases as forward voltage (deflection to the right of the vertical centerline) increases. Reverse current (deflection below the horizontal centerline) increases only slightly as reverse voltage (deflection to the left of the vertical centerline) is applied, unless the breakdown or "avalanche" point is reached. In conventional (non-zener) diodes, it is desirable (if not mandatory) to operate considerably below the breakdown point. Some diodes will break down if operated in the reverse condition for any period of time.
5. Compare the current and voltage values against the values specified in the diode datasheet. For example, assume that a current of 3 A should flow with 7 V applied. This can be checked by measuring along the horizontal scale to the 7-V point, then measuring from that point up (or down) to the trace. The 7-V (horizontal) point should intersect the trace at the 3-A (vertical) point, as shown.

10.5.2 Dynamic tests for small-signal diodes

The procedures for checking the current-voltage characteristics of a signal diode are the same as for power-rectifier diodes. However, there is one major difference. In a signal diode, the ratio of forward voltage to reverse voltage is usually quite large. A test of forward voltage at the same amplitude as the rated voltage will probably damage the diodes. On the other hand, if the test voltage is lowered for both forward and reverse directions, the voltage is not a realistic value in the reverse direction.

Under ideal conditions, a signal diode should be tested with a low-value forward voltage and a high-value reverse voltage. This can be done using a circuit shown in Fig. 10-6. The circuit of Fig. 10-6 is essentially the same as that of Fig. 10-5 (for power diodes), except that diodes CR1 and CR2 (Fig. 10-6) are included to conduct on alternate half-cycles of the voltage across transformer T1. Rectifiers CR1 and CR2 are chosen for a linear amount of conduction near zero.

The variac is adjusted for maximum-rated reverse voltage across the diode under test, as applied through CR2, when the upper secondary terminal of T1 goes negative. This applies the full reverse voltage. Resistor R1 is adjusted for maximum-rated forward voltage across the diode, as applied through CR1, when the upper secondary terminal of T1 goes positive. This applies a forward voltage limited by R1. With R1 properly adjusted, perform the current-voltage check as described for power diodes (Sec. 10.5.1).

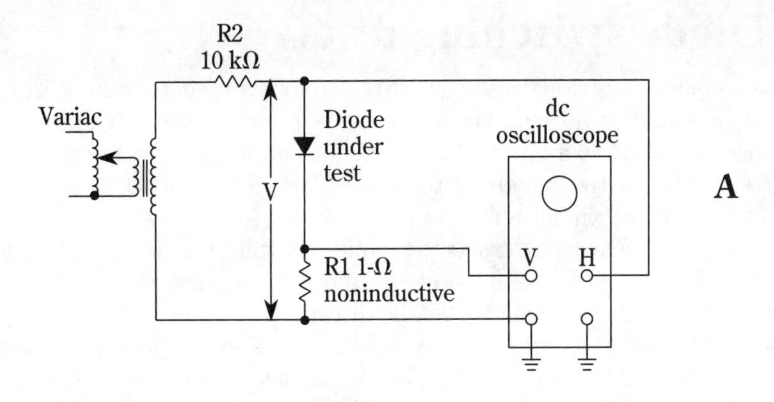

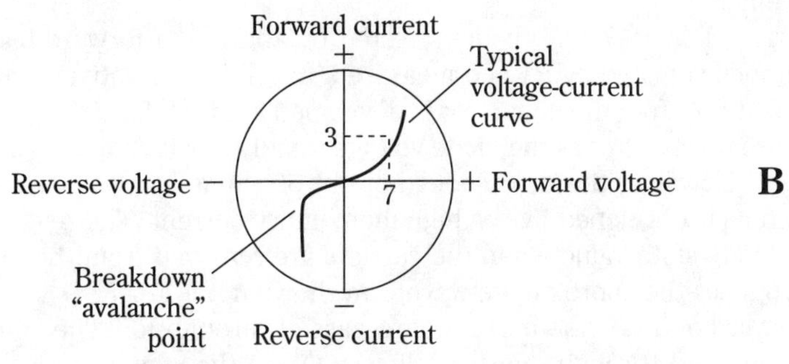

10-5 Power diode test connections and display.

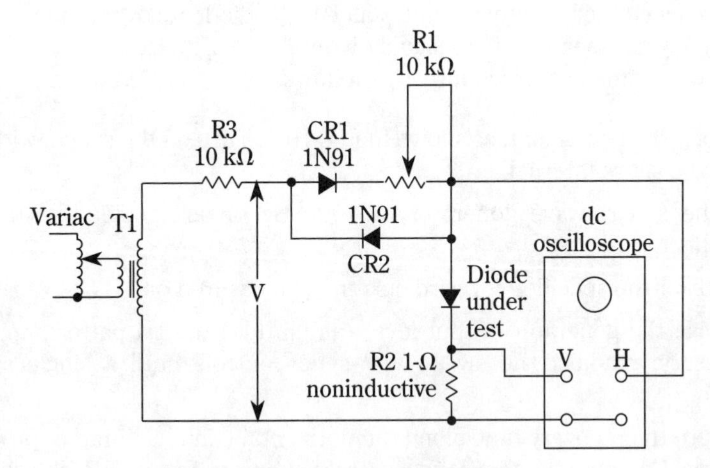

10-6 Signal diode test connections.

10.6 Diode switching tests

Diodes used in pulse or digital work must be tested for switching characteristics. The single most important switching characteristic is recovery time. When a reverse-voltage pulse is applied to a diode, there is a measurable time delay before the reverse current reaches the steady-state value. This delay period is listed as the recovery time (or some similar term) on the diode datasheet.

The duration of recovery time sets the minimum width for pulses with which the diode can be used. For example, if a 5-ns reverse-voltage pulse is applied to a diode with a 10-ns recovery time, the pulse will be distorted.

A scope with wide frequency response and good transient characteristics can be used to check the switch and recovery time of diodes. The scope vertical channel must be voltage-calibrated and the horizontal channel must be time-calibrated, in the usual manner.

As shown in Fig. 10-7, the diode is tested by applying a forward-bias current from a dc supply, adjusted by R1 and measured on M1. The negative portion of the square-wave output from the generator is developed across R3.

The square wave switches the diode voltage rapidly to a high negative value (reverse voltage). However, the diode does not cut off immediately. Instead, a steep transient voltage is developed by the high momentary current. The reverse current falls to the steady-state value when the carriers are removed from the diode junction. This produces the approximate waveform shown in Fig. 10-7.

Both forward and reverse currents are passed through R3. The voltage drop across R3 is applied through emitter-follower Q1 to the scope vertical channel. The coax provides some delay so that the complete waveform is displayed. CR1 functions as a clamping diode to keep the R4 voltage at a level that is safe for the scope.

The time interval between the negative peak and the point at which the reverse current has reached the low, steady-state value is the diode recovery time. Typically, this time is a few nanoseconds for a signal diode.

The recovery-time test procedure is as follows:

1. Connect the equipment, as shown in Fig. 10-7. Turn on the scope, with sweep and sync set to internal.
2. Set the square-wave generator to 100 kHz or as specified in the diode datasheet.
3. Set R1 for the specified forward current as measured on M1.
4. Increase the generator output level (amplitude) until a pattern appears. If necessary, readjust the sweep and sync controls until a single sweep is shown.
5. Measure the recovery time along the horizontal (time-calibrated, probably in ns) axis. Compare the theoretical pattern shown in Fig. 10-7B with the actual recovery-time pattern in Fig. 10-7C (which shows a recovery time of about 4 or 5 ns).

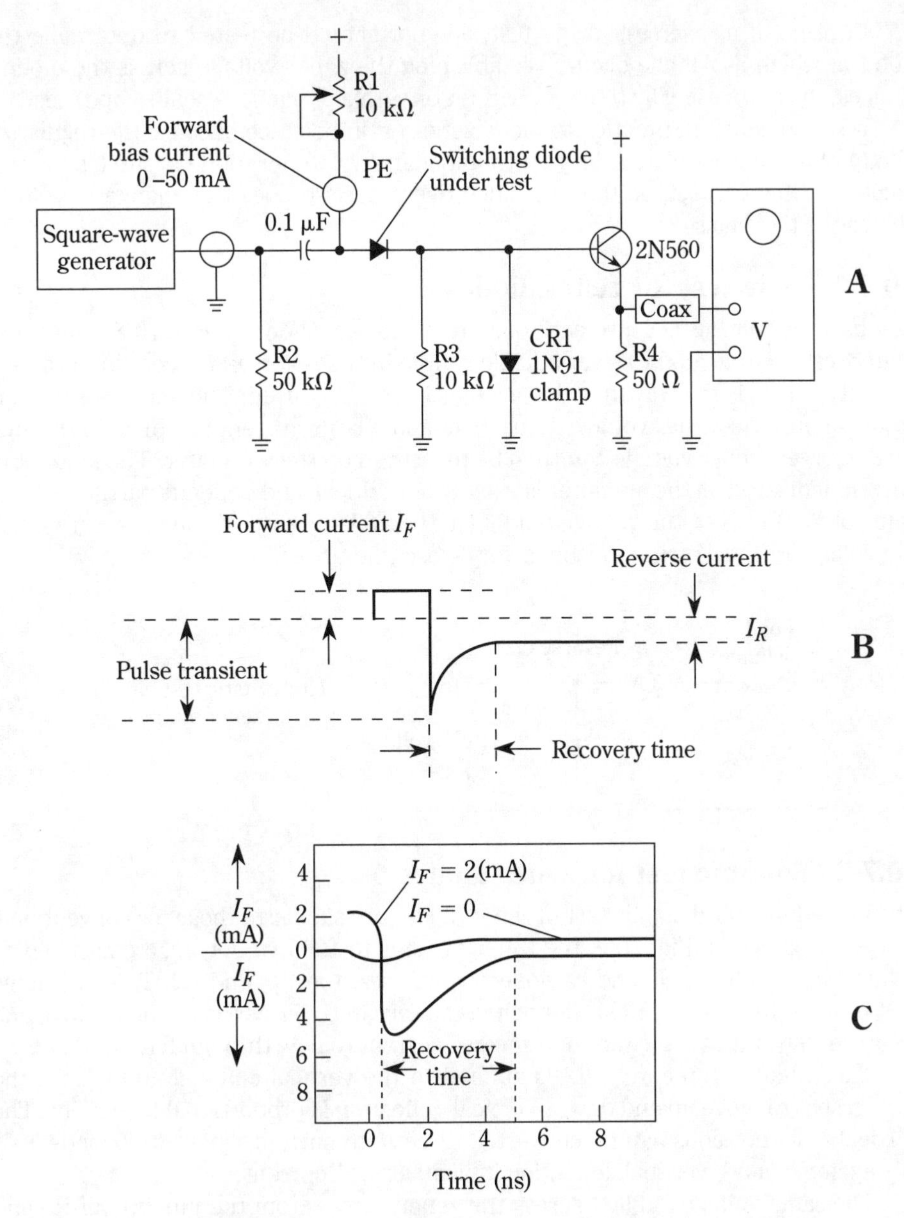

10-7 Diode switching test circuit and displays.

10.7 Zener diode tests

The test of a zener diode is similar to that of a power-rectifier or signal diode. The forward voltage drop test for a zener is identical to that of a conventional diode's (Sec. 10.4). A reverse-leakage test is usually not required because a zener goes into the avalanche condition when sufficient reverse voltage is applied.

In place of a reverse-leakage test, a zener should be tested to determine the point at which avalanche occurs (establishing the zener voltage across the diode). This can be done using a static-test circuit or with a dynamic (oscilloscope) test.

It is also common practice to test a zener for impedance because the regulating ability of a zener is related directly to impedance. A zener diode is similar to a capacitor in this respect; as the reactance decreases, so does the change in voltage across the terminals.

10.7.1 Static test for zener diodes

The basic circuit for measurement of zener voltage is shown in Fig. 10-8. The zener under test is connected to a variable dc source in the reverse-bias condition (anode negative). This is the way in which a zener is normally used. The variable source is adjusted until the zener voltage is reached, and a large current is indicated through the ammeter. Zener voltage can then be measured on the voltmeter. The amount of current indicated on the ammeter is usually not critical, and depends partially on the value of R_1. However, the voltage reading in the avalanche condition must agree with the voltage rating of zener (within a very close tolerance).

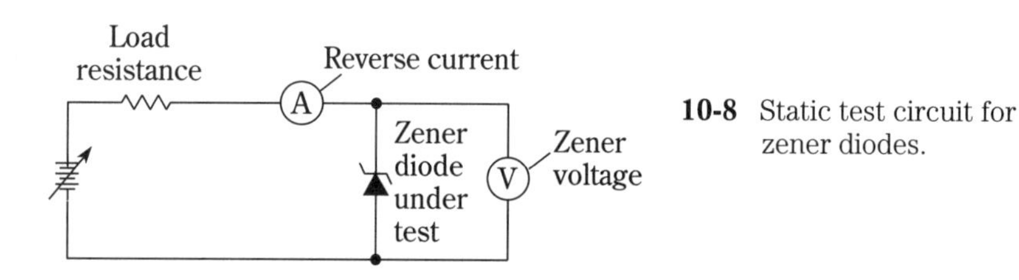

10-8 Static test circuit for zener diodes.

10.7.2 Dynamic test for zener diodes

The procedures for dynamic test of zener diodes are similar to those for conventional diodes. As shown in Fig. 10-9, the zener diode is tested by applying a controlled ac voltage across the anode and cathode through resistors R1 and R2. The ac voltage (set to some value above the zener voltage) alternately biases the anode positive and negative, causing both forward and reverse current to flow through R1 and R2.

The voltage drop across R2 is applied to the vertical channel and causes the screen spot to move up and down. Vertical deflection is proportional to current. The vertical scale divisions can be converted directly to current when R2 is made 1 Ω. For example, a 3-V vertical deflection indicates a 3-A current.

The same voltage applied across the zener (taken from the junction of R1 and the zener) is applied to the horizontal channel (which has been voltage-calibrated, Sec. 10.5.1) and causes the spot to move right or left. Horizontal deflection is proportional to voltage. The combination of the horizontal (voltage) deflection and vertical (current) deflection causes the spot to trace out the complete current and voltage characteristics.

The zener-avalanche test procedure is as follows:

1. Connect the equipment as shown in Fig. 10-9. Turn on the scope. Voltage-calibrate both the vertical and horizontal channels. The spot should be at the vertical and horizontal center with no signal applied to either channel.

2. Set the sweep and sync controls to external.
3. Adjust the variac so that the voltage applied across the zener (and resistors R1 and R2 in series) is greater than the rated zener voltage.
4. Check the scope pattern against the typical curves of Fig. 10-9 and/or against the diode specifications. The curve of Fig. 10-9 is a typical response pattern. That is, forward current increases as forward voltage increases. Reverse (or leakage) current increases only slightly as reverse voltage is applied until the avalanche or zener is reached. Then, the current increases rapidly (thus the term avalanche).

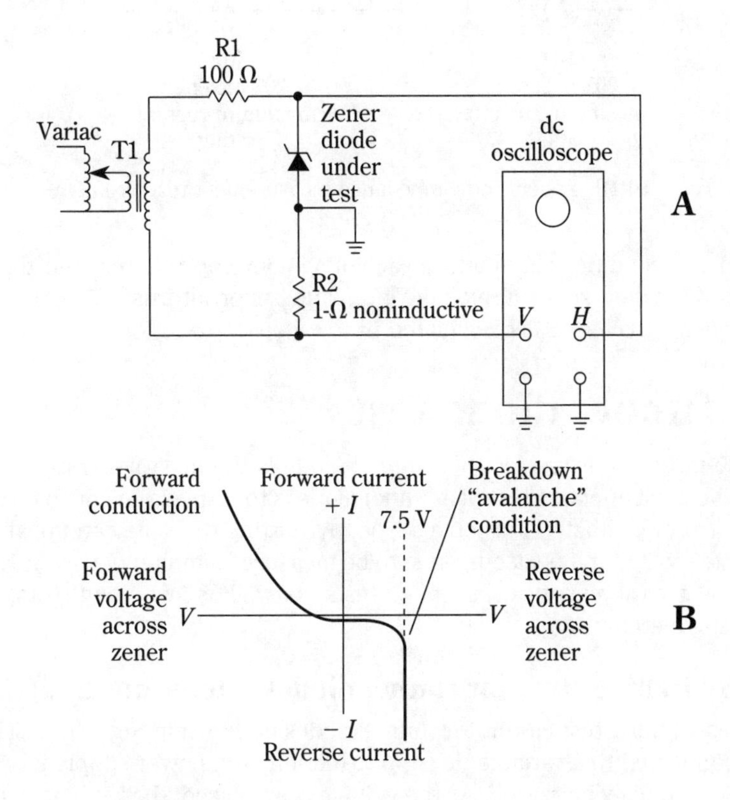

10-9 Zener-diode test connections and display.

10.7.3 Impedance test for zener diodes

As discussed, the regulating ability of a zener diode is directly related to the diode impedance. Similarly, zener-diode impedance varies with current and diode size. To properly test a zener for impedance, you must make the measurements with a specific set of conditions. This can be done using the circuit of Fig. 10-10.

With the circuit of Fig. 10-10, the zener dc is set to about 20% of the zener maximum current by R1. The zener dc is indicated on M1. Alternating current is also applied to the zener, and is set by R2 to about 10% of the maximum zener current rating. The zener ac is indicated on M3.

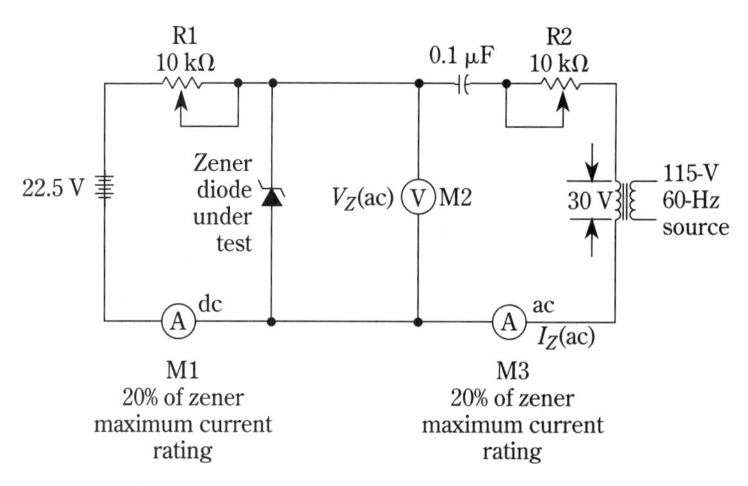

10-10 Zener-diode impedance-measurement test circuit.

When test conditions are met, the ac voltage developed across the zener is read on meter M2. When zener ac voltage $V_{Z(ac)}$ and zener alternating current $I_{Z(ac)}$ are known, the impedance (Z) is calculated by $Z = V_{Z(ac)}/I_{Z(ac)}$.

10.8 Tunnel diode tests

The single most important test of a tunnel diode is the negative-resistance characteristic. The most effective test of a tunnel diode is to display the entire forward voltage and current characteristics on a scope, permitting the valley and peak voltages, as well as the valley and peak currents, to be measured simultaneously. It is also possible to make basic negative-resistance tests, as well as switching tests, of tunnel diodes using meters.

10.8.1 Switching time for tunnel diodes (meter method)

The basic switching-test circuit for tunnel diodes is shown in Fig. 10-11. The tunnel diode is connected to a variable dc supply. Initially, the power supply is set to zero, and then is gradually increased. As the voltage is increased, there is some voltage indication across the tunnel diode. When the critical voltage is reached, the voltage indication "jumps" or suddenly increases. This indicates that the diode has switched and is operating normally. Typically, the voltage indication is about 0.25 to 1 V. The power supply is then decreased gradually. With a normal tunnel diode, the voltage indication gradually decreases until a critical voltage is reached. Then, the voltage indication again "jumps" and suddenly decreases.

10.8.2 Negative-resistance test for tunnel diodes (meter method)

Although the negative-resistance characteristics of a tunnel diode are best tested with a scope, it is possible to get fairly accurate results using meters connected as shown in Fig. 10-12. The diode under test is connected in the reverse-bias condition (anode negative). Thus, any current indication on the ammeter is reverse current.

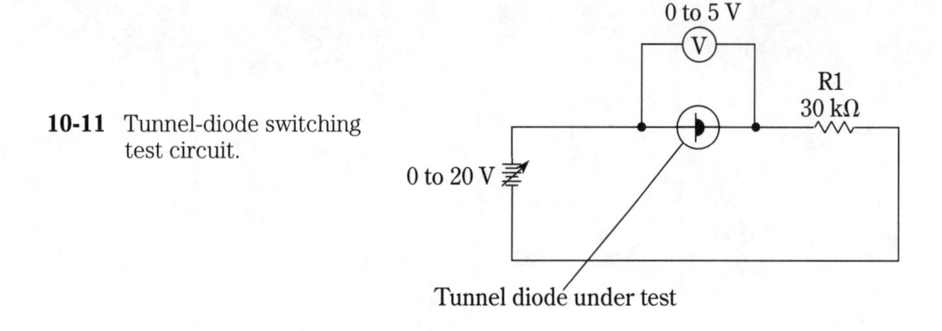

10-11 Tunnel-diode switching test circuit.

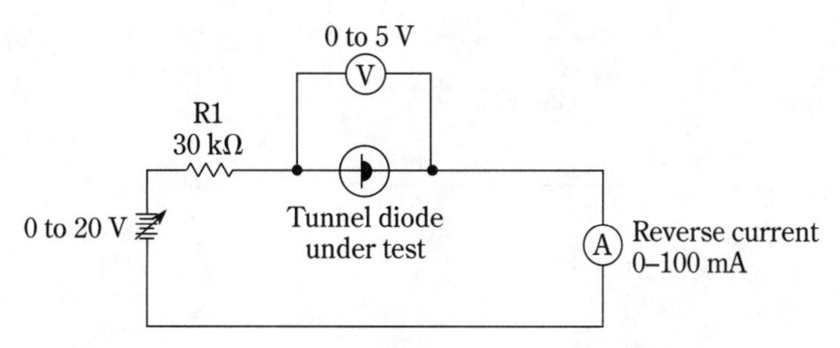

10-12 Tunnel-diode negative-resistance test circuit (meter method).

Initially, the power supply is set to zero and is gradually increased until the voltage reading starts to drop (indicating that reverse current is flowing and the diode is in the negative-resistance region). The negative-resistance region should not be confused with leakage. True negative resistance is indicated when further increases in supply voltage cause an increase in current reading, but a decrease in voltage across the diode. The amount of negative resistance can be calculated using the equation: *Negative resistance (ohms) = Decrease (volts) across diode/Increase (amperes) through diode.*

Do not subject a conventional diode to a negative-resistance test unless there is a special need for the test (even though conventional diodes might show some tunnel-diode characteristics). Also, do not operate a conventional diode in the negative-resistance region for any longer than is necessary. Considerable heat is generated and the diode can be damaged.

10.8.3 Negative-resistance test for tunnel diode (scope method)

A dc scope is required for test of a tunnel diode. Both the vertical and horizontal channels must be voltage calibrated. Also, the horizontal and vertical channels must be identical, or nearly identical, to eliminate any phase difference.

As shown in Fig. 10-13, the tunnel diode is tested by applying a controlled dc voltage across the diode through R1. This dc voltage is developed by rectifier CR1 and is controlled by the variac. Current through the tunnel diode also flows through R3. The voltage drop across R3 is applied to the vertical channel and causes the spot to move up and down. Thus, vertical deflection is proportional to current. Vertical

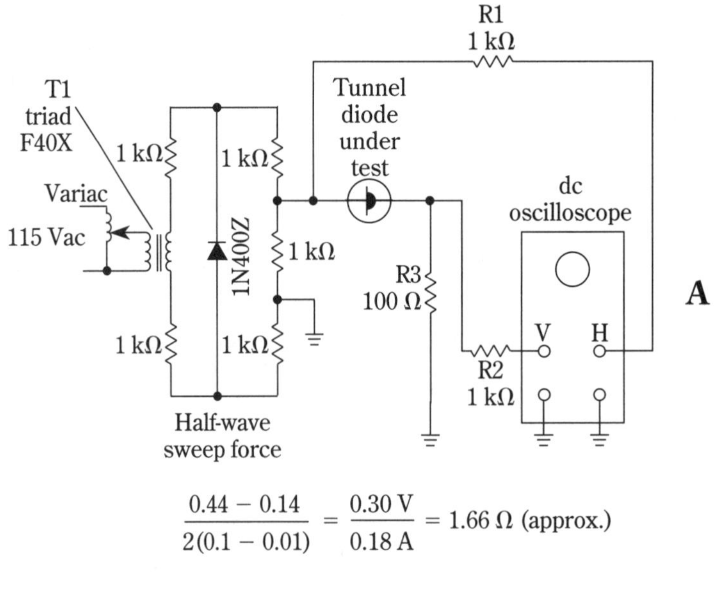

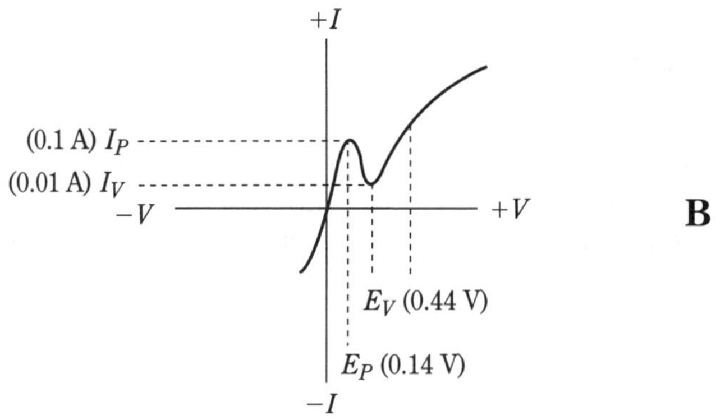

10-13 Tunnel-diode negative-resistance test circuit (scope method).

scale divisions can be converted directly to a realistic value of current when R1 is made 100 Ω. For example, a 3-V vertical deflection indicates 30 mA.

The same voltage applied across the tunnel diode is applied to the horizontal channel (which has been voltage-calibrated) and causes the spot to move from left to right. For a tunnel diode, the horizontal and vertical zero-reference point (no-signal spot position) should be at the lower left of the screen, rather than in the center. The horizontal deflection is proportional to voltage. The combination of the horizontal (voltage) deflection and vertical (current) deflection causes the spot to trace out the complete negative-resistance characteristic.

The negative-resistance test procedure is as follows:

1. Connect the equipment, as shown in Fig. 10-13. Turn on the scope. Voltage calibrate both the vertical and horizontal channels, as necessary. The spot should be at the lower left-hand side of center with no signal applied to either channel.
2. Set the sweep and sync controls to external.
3. Adjust the variac so that the voltage applied across the tunnel diode is the maximum rated forward voltage (or slightly below). This can be read across the voltage-calibrated horizontal axis.
4. Check the scope pattern against the curve of Fig. 10-13, and/or against the tunnel-diode datasheet.
5. The following equation can be used to get a rough approximation of negative resistance in tunnel diodes: negative resistance = $(E_V - E_P)/2(I_P - I_V)$, where E_V = valley voltage, I_V = valley current, E_P = peak voltage and I_P = peak current.

Figure 10-13 shows some typical tunnel-diode voltage-current values, and the calculation for the approximate negative resistance.

10.9 Diode tests using a curve tracer

Signal and power diodes conduct easily in one direction and are nonconducting in the opposite direction. These properties can be tested and observed with a curve tracer and scope. For diode tests, the pulsating dc sweep voltage is applied across the diode, and the diode current/voltage are plotted on the scope screen. The step-current or step-voltage used for transistor tests are not used in diode tests, and the curver-tracer STEP SELECTOR has no effect on test results.

10.9.1 Diode forward-bias display on a curve tracer

To display forward bias, connect the diode and operate the curve tracer as shown in Fig. 10-14. For realistic values, the scope horizontal channel should be voltage-calibrated with a low voltage. For a typical diode, a horizontal sensitivity of 0.5 V/division (or preferably less) is necessary to get any degree of accuracy in the voltage reading. Most scopes used with curve tracers can be calibrated for a sensitivity of 0.25 V/division in the horizontal axis.

When testing diodes as described here, only one curve is displayed, not a family of curves, as displayed for transistors. Both forward voltage drop and dynamic resistance can be measured using the connections of Fig. 10-14. No current flows until the applied voltage exceeds the forward voltage drop (typically 0.2 to 0.4 V for germanium and 0.5 to 0.7 V for silicon). Above the forward-voltage point, current increases rapidly with an increase in forward voltage. As shown, the current increases more rapidly and the "elbow" has a sharper bend for silicon diodes than for germanium diodes.

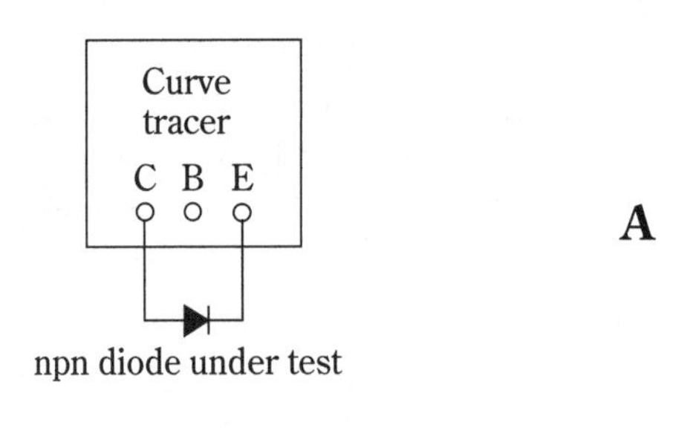

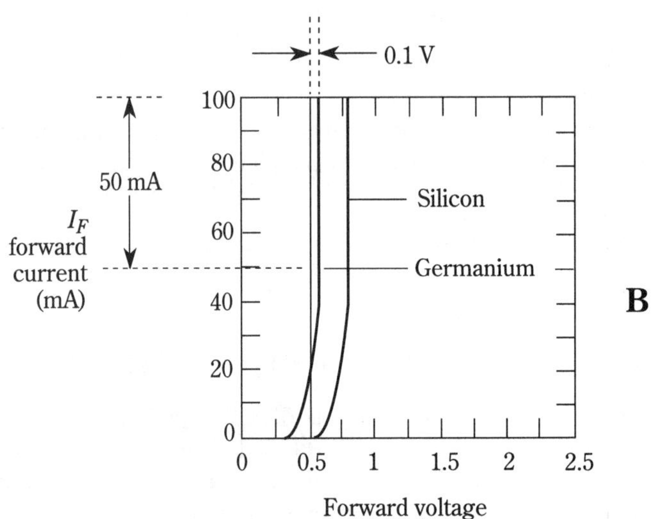

10-14 Diode forward-bias display on a curve tracer.

The dynamic resistance of a diode equals the change in forward voltage divided by the change in forward current. For example, the germanium diode curve of Fig. 10-14 shows an increase of about 50 mA in forward current for an increase of about 0.1 V of forward voltage. This indicates a dynamic resistance of about 2 Ω. Germanium diodes, with more slope to their curves, have higher dynamic resistance than do silicon diodes.

There is no need to increase the curve-tracer SWEEP VOLTAGE control setting beyond that which gives a full-scale vertical presentation, although there is generally very little danger that a higher setting will do harm. Typically, the scope vertical-sensitivity control can be set to the 1-mA/division position for examining the low-current characteristics, or to a lower sensitivity position (such as 10 mA/division) for observing a wider range of forward current conduction.

11 Thyristor and control-rectifier (SCR) tests

This chapter is devoted entirely to test procedures for control rectifiers and thyristors. The terms *control rectifier* and *thyristor* are alternately applied to many devices used in electronic-control applications. The most common such devices are the SCR, SCS, PNPN switch, triac, diac, SUS, SBS, and SIDAC. The first sections of this chapter describe control-rectifier or thyristor characteristics and test procedures from the practical standpoint. The information in these sections permits you to test all the important control-rectifier and thyristor characteristics using basic shop equipment. The sections also help you understand the basis for such tests. The remaining sections of the chapter describe how the same tests, and additional tests, are performed using more sophisticated equipment

11.1 Thyristor and control-rectifier basics

Before going into specific test procedures, review some thyristor and control-rectifier basics. The control rectifier (also called a *controlled rectifier*) is similar to the basic diode (chapter 10), with one exception. The control rectifier must be triggered or turned on by an external voltage source.

Control rectifiers have a high forward and reverse resistance (no current flow) without the trigger. When the trigger is applied, the forward resistance drops to zero (or very low), and a high forward current flows, as with the basic diode. The reverse current remains high, and no reverse current flows, so the control rectifier rectifies ac power in the normal manner. As long as the forward voltage is applied, the forward current continues to flow. The forward current stops, and the control rectifier turns off, if the forward voltage is removed.

Of the numerous control rectifiers in use, many are actually the same type (or slightly modified versions), but manufactured under different trade names or designations. The term *thyristor* is applied to many control rectifiers. Technically, a thrysitor is defined as any semiconductor switch, where feedback occurs in normal operation. Thyristors can be two-, three- or four-terminal devices, and are capable of both unidirectional and bidirectional operation. The following paragraphs describe the most common control-rectifier or thyristor devices.

11.1.1 Common control-rectifier devices

Figure 11-1 shows the most common control-rectifier symbols. With some manufacturers, the letters *SCR* refer to semiconductor control rectifier and can mean any type of solid-state control rectifier. However, SCR usually refers to silicon control (or controlled) rectifiers. SCRs are normally used to control ac, but can be used to control dc. Either ac or dc voltage can be used as the gate signal, provided that the gate voltage is large enough to trigger the SCR into the on condition.

The *silicon-controlled switch (SCS)* is similar to an SCR. However, the SCS can be both turned on and turned off by gate signals (where it is necessary to remove the forward voltage to turn off an SCR). Often, the SCS is used as an SCR, with the extra gate terminal not connected. The SCS is sometimes called a *PNPN switch*.

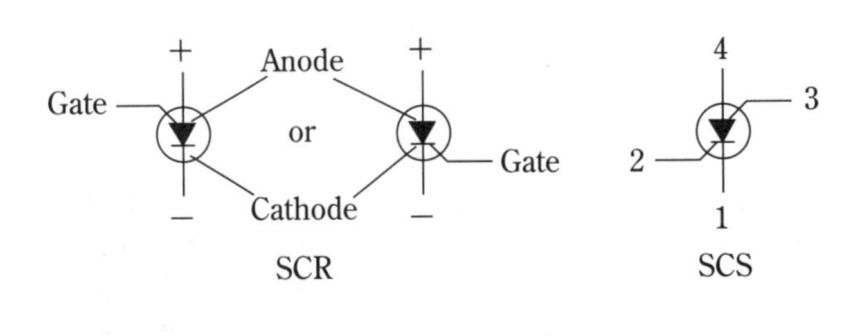

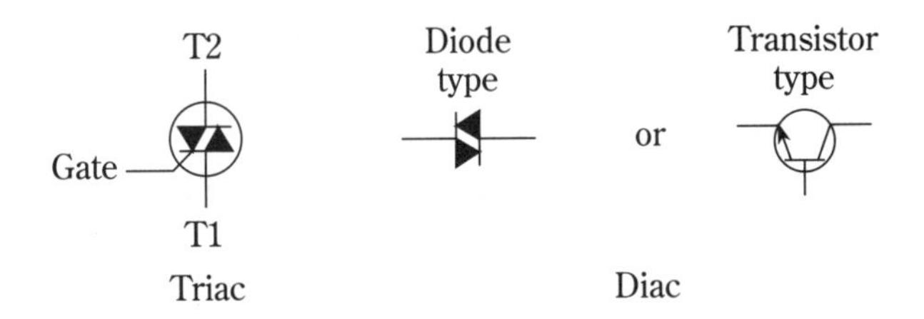

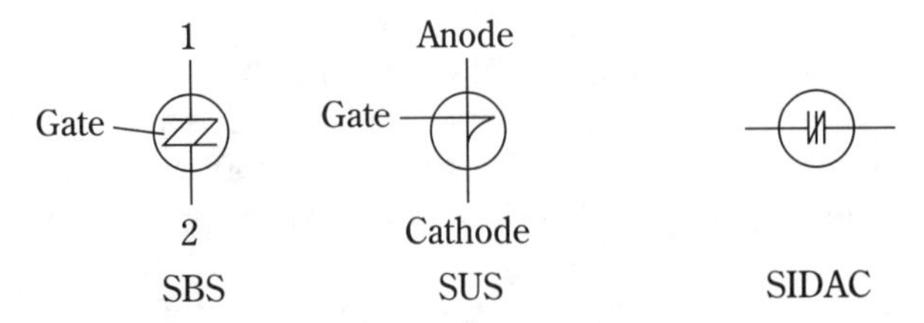

11-1 Common control-rectifier symbols.

Like the SCR and SCS, the triac is triggered by a gate signal. Unlike either the SCR or SCS, the triac conducts in both directions, and is most useful for controlling devices operated by ac power (such as ac motors). Triacs can be triggered from many sources, as can SCRs and SCSs.

Common trigger sources for control rectifiers are the diac and the SIDAC. These devices do not have a trigger, and do not conduct in either direction until a certain breakover voltage is reached. Typically, these devices are connected to the gate of an SCR or triac. When the trigger voltage reaches the breakover point, the device passes the trigger signal to the SCR or triac. The SIDAC can also be used as a relaxation oscillator.

The SUS is essentially a miniature SCR with an anode gate and a built-in low-voltage zener between the gate and cathode. The SUS operates in a manner similar to the UJT (chapter 9). However, the SUS switches at a fixed voltage, determined by the zener, rather than by a fraction or percentage of the supply voltage. From a test standpoint, the SUS can be tested as a UJT or an SCR.

The SBS is essentially the identical SUS structures arranged in an inverse-parallel circuit. Because the SBS operates as a switch with both polarities of applied voltage, the SBS is useful for triggering triacs with alternate positive and negative gate pulses (for both turn-on and turn-off).

11.1.2 Basic SCR operation

An SCR is used to best advantage when both the load and trigger are ac. With ac power, control of the power applied to the load is determined by the relative phase of the trigger signal versus the load voltage. This is shown in Fig. 11-2. Because the trigger is lost once the SCR is conducting, an ac voltage at the load permits the trigger to regain control. Each alternation of ac through the load causes conduction to be interrupted (when the ac voltage drops to zero between cycles), regardless of the polarity of the trigger signal.

As shown in Fig. 11-2, the SCR conducts for each successive positive alternation at the anode, if the trigger voltage is in phase with the ac power input signal. When the trigger is positive-going at the same time as the load or anode voltage, load current starts to flow as soon as the load voltage reaches a value that causes conduction. When the trigger is negative-going, the load voltage is also negative-going, and conduction stops. Thus, the SCR acts as a half-wave rectifier.

If there is a 90° phase difference between the trigger voltage and load voltage (for example, the load voltage lags the trigger voltage by 90°), as shown, the SCR does not start conducting until the trigger voltage signals positive, even though the load voltage is initially positive. When the load voltage drops to zero, conduction stops, even though the trigger voltage is still positive.

If the phase shift is increased between trigger and load voltages (as shown), conduction time is even shorter, and less power is applied to the load circuit. By shifting the phase of the trigger voltage in relation to the load voltage, it is possible to vary the power output, even though the voltages are not changed in strength.

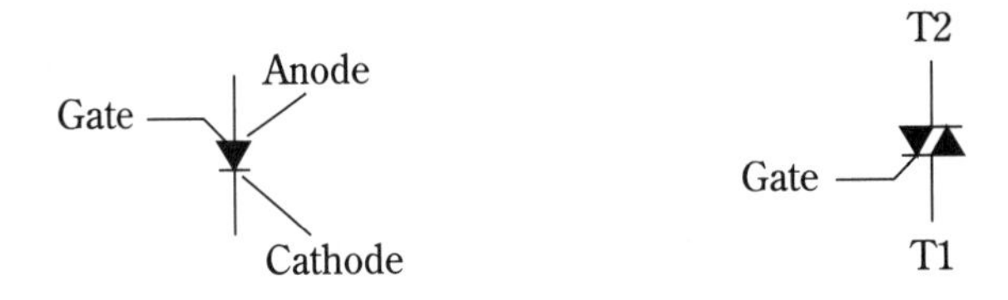

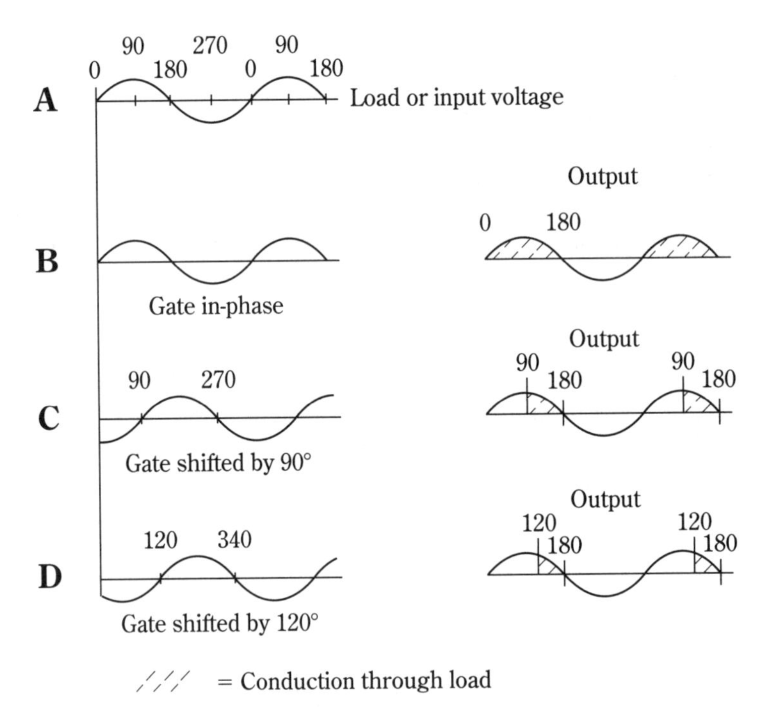

11-2 Symbols and operating characteristics for SCRs and triacs.

11.2 Control rectifier test parameters

Manufacturers have their own set of symbols, letters, and terms to identify the parameters of control rectifiers and thyristors. Many of the symbols and terms are duplicates used by different manufacturers. In a few cases, special terms and letter symbols are used by a manufacturer to identify the parameters of their own particular type of control rectifier. No attempt is made to duplicate all of this information here. However, the following paragraphs cover the most important parameters. These parameters can be compared with those found on the datasheets of a particular control rectifier or thyristor.

Forward voltage is the voltage drop between the anode and cathode at any specified forward anode current, when the device is in the on condition. Forward anode current is any value of positive current that flows through the device in the on condition.

Forward blocking voltage is the maximum anode-cathode voltage in the forward direction that the device can withstand before conduction, at zero gate current. Forward breakover voltage is the value of forward anode voltage at which the device switches to the on state, with a shunt resistance between gate and cathode. The basic test connections for forward-on conditions are shown in Fig. 11-3.

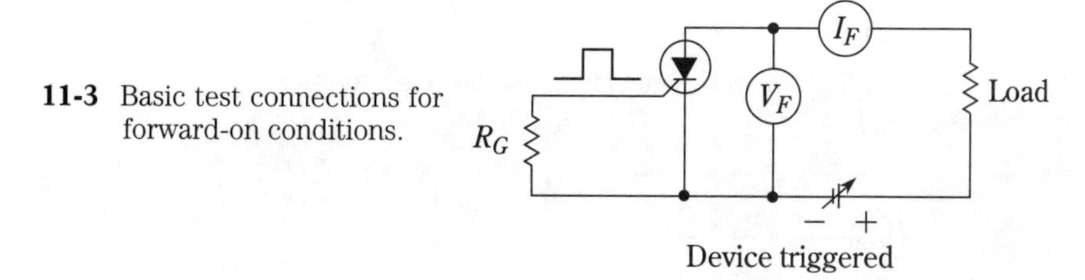

11-3 Basic test connections for forward-on conditions.

Forward off current is the anode current that flows when the device is in the off condition (with a positive voltage applied) and is sometimes listed as forward leakage current. Figure 11-4 shows the basic test connections for forward-off measurements.

Reverse anode voltage is any value of negative voltage that can be applied to the anode. The rated reverse anode voltage is less than the reverse avalanche voltage, and is the maximum peak inverse voltage (PIV) of the device.

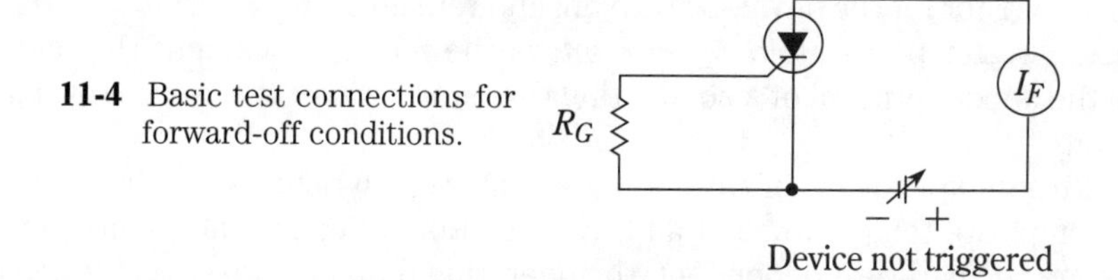

11-4 Basic test connections for forward-off conditions.

Reverse blocking voltage is the maximum reverse anode-cathode voltage (at zero gate current) that the device can withstand before voltage breakdown, and is similar to the peak inverse voltage of a diode. *Reverse current* is the negative current that flows through the device at any specified condition of reverse anode voltage and temperature. The basic test connections for reverse conditions are shown in Fig. 11-5.

Forward gate voltage is the voltage drop across the gate-cathode junction at any specified forward-gate current. *Forward gate current* is any value of positive current that flows into the gate, with a stunt resistance between gate and cathode. Figure 11-6 shows the basic test connections for both forward gate voltage and current measurements.

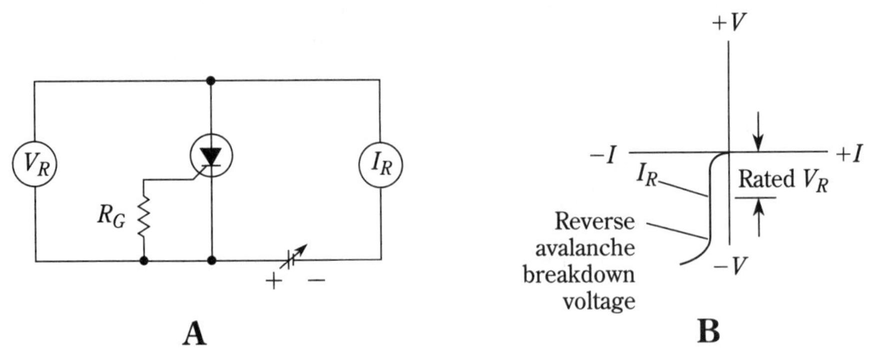

11-5 Basic test connections for reverse conditions.

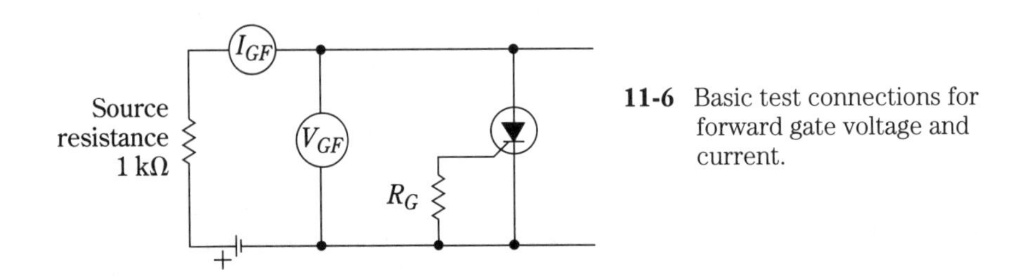

11-6 Basic test connections for forward gate voltage and current.

Latching current is the minimum current that must flow through the anode terminal of a thyristor for the device to turn on, and remain in the on state after removal of the gate trigger pulse. *Holding current* is the minimum current that can flow through the anode terminal of a conducting thyristor without the device reverting to the off state.

Junction temperature in any control rectifier or thyristor is usually considered to be a composite temperature of all three junctions. Because all parameters of a thyristor are temperature-dependent, the operating temperature of the device must be considered when making any tests. This can be especially critical in holding-current and delay-time measurements.

Reverse gate voltage is the voltage drop across the gate-cathode junction at any specified reverse gate current. In turn, *reverse gate current* is any value of negative current that flows into the gate, with a shunt resistance between gate and cathode. Figure 11-7 shows the basic test connections for both reverse gate voltage and current measurements.

Delay time is the interval between the start of the gate pulse and the instant at which the output has cleansed to 10% of the maximum amplitude. Figure 11-8 shows the basic test connections for delay measurements, together with a typical graph or display. As shown, delay time is the time (following initiation of the gate pulse) required for the anode voltage to drop to 90% of the initial value. Delay time decreases with increased gate current, but it eventually reaches a minimum.

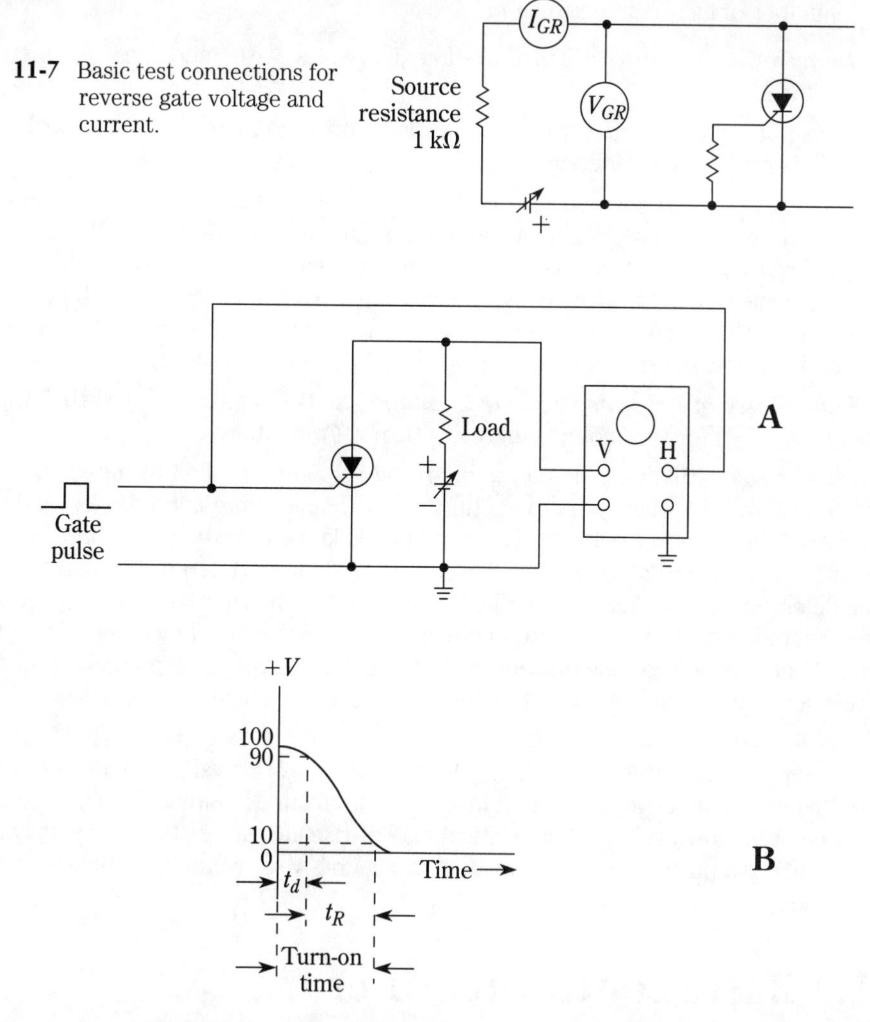

11-7 Basic test connections for reverse gate voltage and current.

11-8 Basic test connections for delay measurements.

Rise time is the interval during which the output pulse changes from 10% to 90% of the maximum amplitude. In Fig. 11-8, rise time is the time required for the anode-cathode voltage to drop from 90% to 10% of the initial value.

Turn-on time is usually expressed as a combination of delay time plus rise time. Turn-off time is the interval between the start of the turn-off and the instant at which the anode voltage can be reapplied, without turning on the device. When a thyristor is triggered, and the anode voltage is positive, the device conducts. When the anode voltage swings negative, conduction stops. However, if the anode swings positive immediately, the device can conduct even though the trigger is not present. There must be some delay between the instant that conduction stops and the instant the anode can be made positive. This delay is the turn-off time.

A number of factors affect turn-off time:

Junction temperature Turn-off time increases with increases in junction temperature.

Forward current and rate of decay Turn-off time increases as forward current and its rate of decay increase.

Reverse recovery current If the device is subjected to a reverse bias (anode made negative) immediately after a condition of forward conduction (such as occurs when alternating current is used, and the anode swings from positive to negative each half-cycle), a reverse (or recovery) current flows from anode to cathode. This is essentially the same as recovery current of a conventional diode (chapter 10). Turn-off time decreases as reverse (or recovery) current increases.

Rate of rise of reapplied forward voltage As the rate of rise and the amplitude of reapplied forward voltage increase, turn-off time increases.

Rate of rise (dV/dT). When a rapidly-rising voltage is applied to the anode of a thyristor, the anode might start to conduct, even though there is no trigger, and the breakdown voltage is not reached. This condition is known as *rate effect*, or *dV/dT effect*, or (sometimes) *dI/dT effect.* The letters *dV/dT* signify a difference in voltage for a given difference in time. The letters *dI/dT* signify a difference in current for a given difference in time. Either way, the letter combinations indicate how much the voltage (or current) changes or a specific time interval. Usually, the terms are expressed in volts of change per microsecond, or amperes of current change per microsecond.

Critical rate of rise Every thyristor has some critical rate of rise. That is, if the voltage (or current) rises faster than the critical rate-of-rise value, the device turns on (with or without a trigger) even though the actual anode voltage does not exceed the rated breakdown voltage. This critical rate-of-rise characteristic is especially important where a pulse-type signal, rather than a sine-wave voltage, is applied across the anode.

11.3 Basic control-rectifier and thyristor tests

As in the case of diodes and transistors, control rectifiers are subjected to many tests during manufacture. Few of these tests need be duplicated in the field. One of the simplest and most comprehensive tests for a control rectifier is to operate the device in a circuit that simulates actual circuit conditions (typically, ac and an appropriate load at the anode, ac or a pulse signal at the gate), then measures the resulting conduction angle on a dual-trace scope.

With such a test, the trigger and anode voltages, as well as the load current, can be adjusted to normal (or abnormal) dynamic operating conditions and the results noted. For example, the trigger voltage can be adjusted over the supposed minimum and maximum trigger levels. Or the trigger can be removed and the anode voltage raised to the actual breakover. The conduction-angle method should test all important characteristics of a control rectifier, except for turn-on, turn-off, and rate of rise (which are also covered in this section).

11.3.1 Conduction-angle test

A dual-trace scope can be used to measure the conduction angle of a control rectifier or thyristor. As shown in Fig. 11-9, one trace of the scope displays the anode current, while the other trace displays the trigger voltage. Both traces must be voltage-calibrated. The anode load current is measured through a 1-Ω noninductive resistor. The voltage developed across the resistor is equal to the current. For example, if a 3-V indication is obtained on the scope trace, a 3-V current is flowing in the anode circuit.

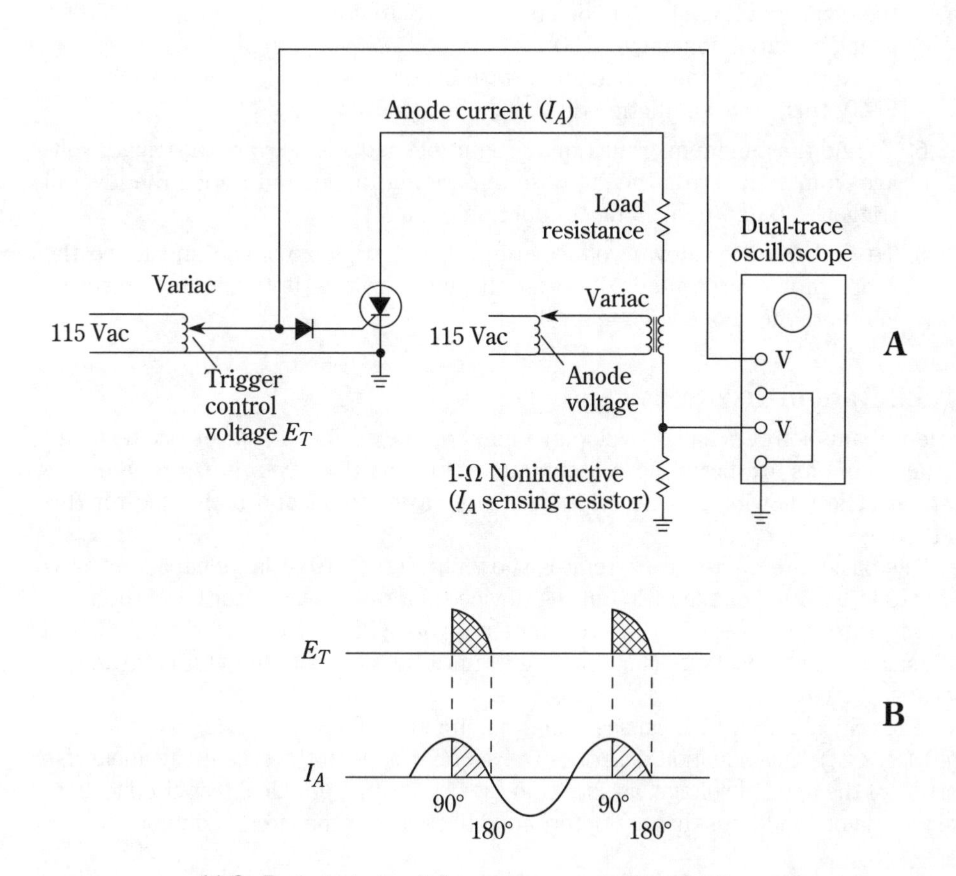

11-9 Basic test circuit for measurement of conduction angle.

The trigger voltage is read out directly on the other scope trace. Notice that a diode is shown in the trigger circuit to provide a pulsating dc trigger. This diode can be removed if an ac trigger is preferred. Because the trigger is synchronized with anode current (both are obtained from the same source), the portion of the trigger cycle in which anode current flows is the conduction angle.

The conduction-angle test procedure is as follows:

1. Connect the equipment, as shown in Fig. 11-9. Turn on the scope. Set the sweep and sync to internal.

2. Apply power to the control rectifier. Adjust the trigger voltage, anode voltage, and anode current to the desired levels. Anode voltage can be measured by temporarily moving the scope probe (normally connected to measure gate voltage) to the anode.
3. Adjust the scope sweep and sync controls to produce two or three stationary cycles of each wave on the screen.
4. On the basis of one conduction pulse equalling 180°, determine the angle of anode current flow, by reference to the trigger-voltage trace. For example, in the display of Fig. 11-9, anode current starts to flow at 90° and stops at 180°, giving a conduction angle of 90°.

   Notice that if the device under test is a triac (or similar device such as an SBS), there is a conduction display on both half-cycles.
5. To find the minimum or maximum required trigger level, vary the trigger voltage from zero across the supposed operating range, and notice the level of trigger voltage when anode conduction starts.
6. To find the breakdown voltage, remove the trigger voltage and move the scope probe to the anode. Increase the anode voltage until conduction starts, and note the anode voltage level.

11.3.2 Rate-of-rise tests

Various manufacturers have developed a number of circuits for rate-of-rise tests. All of these circuits are based on a technique known as the *exponential waveform method.* The following is a description of the basic circuit and technique for this method.

The basic rate-of-rise test circuit is shown in Fig. 11-10. A large capacitor C1 is charged to the full voltage rating of the device under test. Capacitor C1 is then discharged through a variable time-constant network (R2 and C2). This is repeated with smaller tune constants (higher dV/dT) until the device under test is turned on by the fast dV/dT.

The critical rate, which causes firing, is defined as $dV/dT = (0.632 \times \textit{Anode voltage})/(R_2 \times C_2)$. This equation describes the average slope of the essentially linear rise portion of the applied voltage, as shown in Fig. 11-11. In a practical test circuit, there are two major conditions that are determined the value of the circuit components.

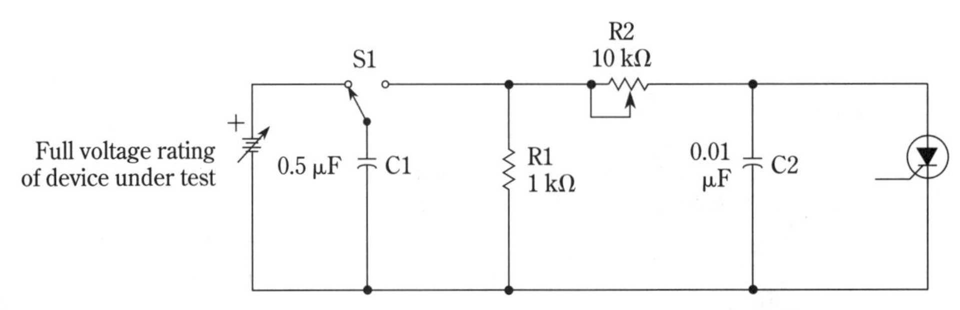

11-10 Basic test circuit for measurement of rate-of-rise (dV/dT).

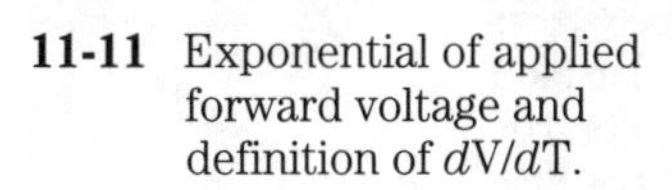
11-11 Exponential of applied forward voltage and definition of dV/dT.

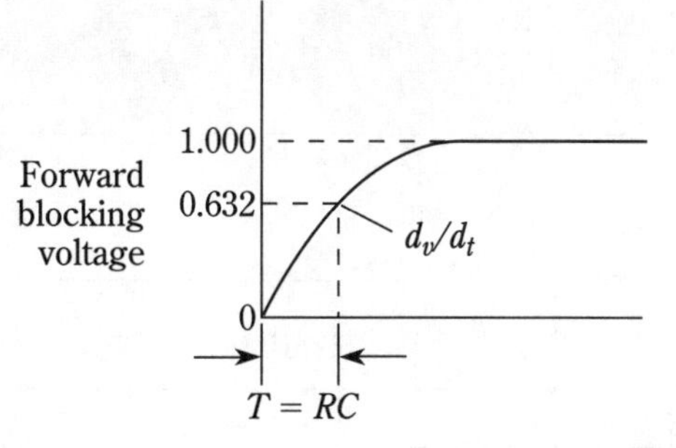

First, C1 should be large enough to serve as a constant-voltage source during the discharge of C1 and the charging of C2. Second, C2 should be much larger than the intrinsic cathode-to-anode capacitance of the device under test plus any stray device and test wiring capacitances (a typical control rectifier junction capacitance is 800 pF for zero applied voltage). The values shown for C1 and C2 in Fig. 11-10 are practical for the test of most control rectifiers. Notice that any stray inductance and capacitance in the test circuit should be minimized. This is especially true for measurement of high dV/dT values.

11.3.3 Turn-on and turn-off (recovery) tests

Figure 11-12 shows a circuit capable of measuring both turn-on and turn-off (recovery) time. Circuit inductances must be kept to a minimum with short connections, thick wires, and closely spaced return loops or wiring on a grounded PC board. External pulse sources must be provided for the circuit. These pulses are applied to transformers T1 and T2, and serve to turn the device under test on and off (the transformers can be Sprague 11Z12, or equivalent). The pulses can come from any source, but should be of the amplitude, duration, and repetition rate that correspond to the normal operating conditions of the device under test.

When a suitable gate pulse is applied to T1, the device under test is turned on. Load current is set by R_L. A predetermined time later, the turn-off control rectifier is turned on by a pulse applied to T2. This places C2 across the device under test, applying a reverse bias, and turning the device under test off.

Any scope capable of a 10-μs sweep can be used for viewing both the turn-on and turn-off action of typical control rectifiers. The scope is connected with the vertical input across the device under test. Turn-on time is displayed when the scope is triggered with the gate pulse applied to time device. Turn-off time is displayed when the score is triggered with the gate pulse applied to the turn-off control rectifier.

The actual spacing between the turn-on and turn-off pulses is usually not critical. However, a greater spacing causes increased conduction and heats the junction. Because the operation of control rectifiers is temperature-dependent, the rise in junction temperature must be taken into account for accurate test results. (Both turn-on and turn-off tunes increase with an increase in junction temperature.)

Figure 11-13 shows turn-on action. Turn-on time is equal to delay time (t_d) plus rise time (t_r). After the beginning of the gate pulse, there is a short delay before appreciable load current flows. Delay time is the time from the leading edge of the gate-current pulse (beginning of scope sweep) to the point of 10% load current flow (delay time can be decreased by over-driving the gate).

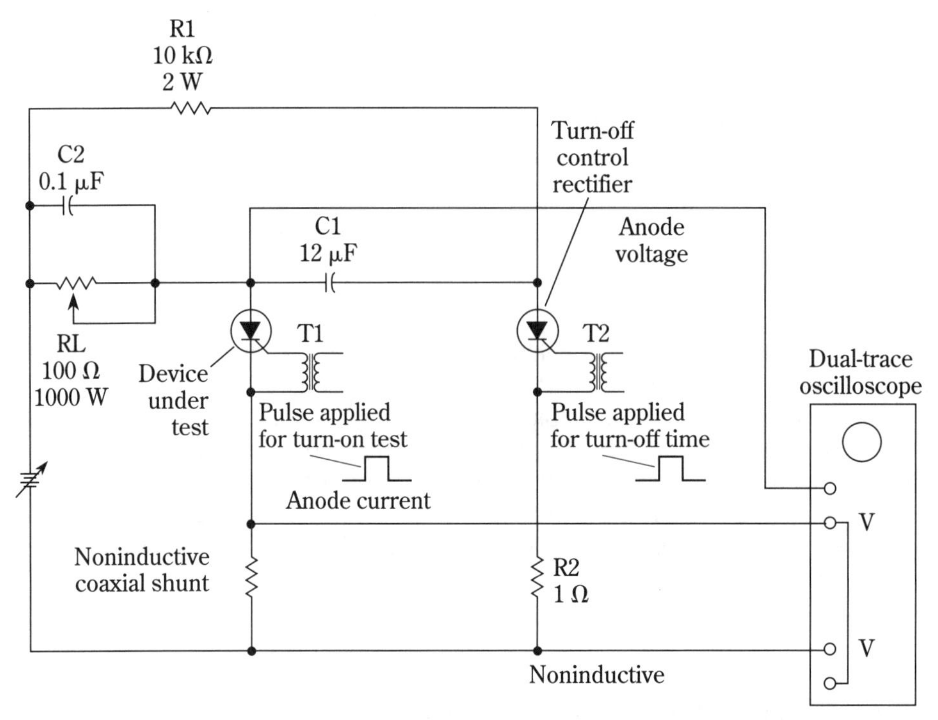

11-12 Test circuit for measurement of turn-on and turn-off (recovery) time.

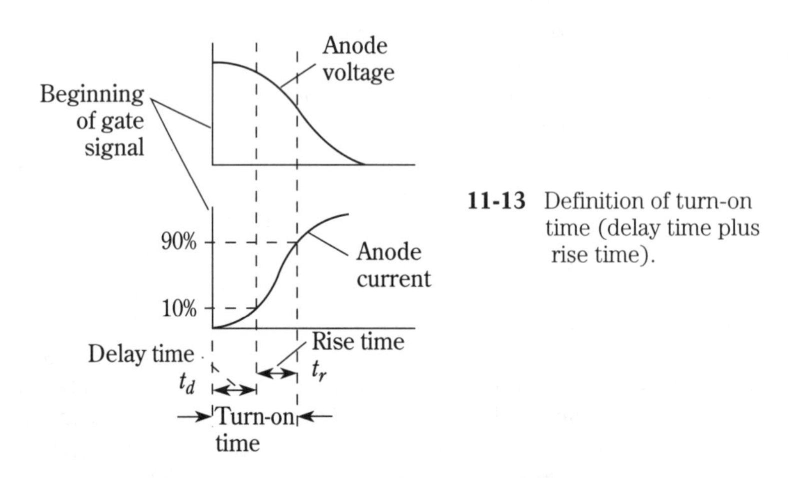

11-13 Definition of turn-on time (delay time plus rise time).

Rise time (t_r) is the time required for the load current to increase from 10 to 90% of the total value. Rise time depends on load inductance, load current amplitude, junction temperature, and (to a lesser degree) on anode voltage. The higher the load current and inductance, the longer the rise time. An increase in anode voltage tends to decrease the rise time. Capacitor C2 (Fig. 11-12) tends to counter the load inductance, thus lessening the rise time.

By triggering the scope with the gate pulse applied to the device under test, the sweep starts at the gate-pulse leading edge. Thus, the scope presentation shows the anode voltage from this point on. In noninductive circuits, when the anode voltage decreases to 90% of initial value, this time is equal to 10% of the load current, and is thus equal to the delay time.

With the scope set at a 10-μs horizontal sweep, each division represents 1 μs. The delay time is read directly by counting the number of divisions on the scope screen. If the circuit is noninductive, the decrease from 90 to 10% of the anode voltage is about equal to the load current increase from 10 to 90%. The time this requires is equal to the rise time.

The total time from zero time (start of scope sweep) to the 10% of anode voltage is equal to the turn-on time. Therefore, turn-on time is determined by counting the number of divisions from the start of the scope sweep to the 90% anode-load current.

Figure 11-14 shows the reverse current and reverse recovery (turn-off) action of the device under test. Turn-off time is the time necessary for the device under test to turn off and recover the forward-blocking ability. The *reverse recovery time* (t_h) is the length of the interval between the time the forward current falls to zero when going reverse, and the time the current returns back to zero from the reverse direction.

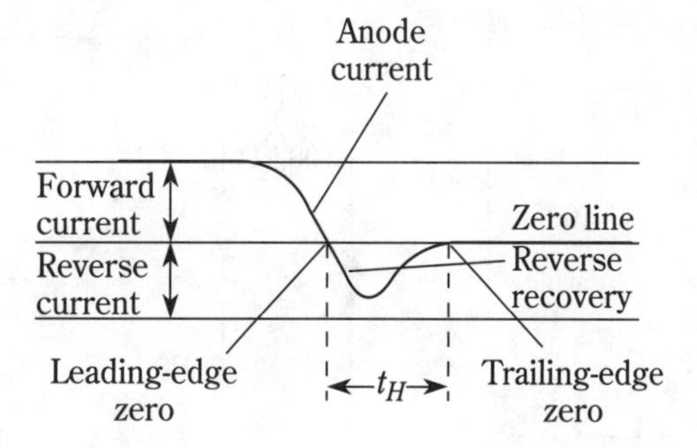

11-14 Definition of reverse recovery (turn-off) time.

In Fig. 11-12, the time available for turn-off is determined by the values of C_1 and R_2. Decreasing C_1 decreases the time the device under test is reverse-biased. Resistor R2 limits the magnitude of the reverse current. The current shape of the reverse voltage and pulses are determined by capacitor-resistor discharge. At the end of the reverse pulse, forward voltage is reapplied. Having turned off, the device under test blocks forward voltage, and no current can flow.

Figure 11-14 shows the reverse-current pulse. With scope set for a 20-μs sweep, the value of reverse recovery time (turn-off time) can be measured by counting off the divisions from the zero point on the leading edge of the reverse-current pulse, to time zero point on the trailing edge, as shown.

11.4 Blocking-voltage and leakage-current tests

This section describes test circuits and procedures for both forward and reverse blocking-voltage, as well as for leakage-current measurements. In these circuits, the voltages and currents are measured on meters, rather than on a scope. However, one circuit provides for a simultaneous scope display of the characteristics.

11.4.1 Test circuit for thyristors above 2 A

Figure 11-15 shows a circuit for test of the forward and reverse blocking characteristics of SCRs, triacs, etc. with operating currents above 2 A (although there will be no damage to devices operating below 2 A, the test results might be ambiguous). The circuit of Fig. 11-15 consists of a variable-voltage current-limited power supply that develops a half-wave voltage across the thyristor under test to minimize junction heating. Instrumentation consists of suitable meters to indicate blocking voltage and leakage currents. If desired, a scope can be connected to provide a visual display of the forward and reverse voltage-current characteristics of the device under test.

To measure forward breakdown voltage and forward leakage current, set S1 to FWD, and raise the voltage with T1 until the meter V reads the rated forward breakdown voltage. Read the full-cycle average leakage current from meter I, or the peak leakage current from the voltage-current trace on the scope.

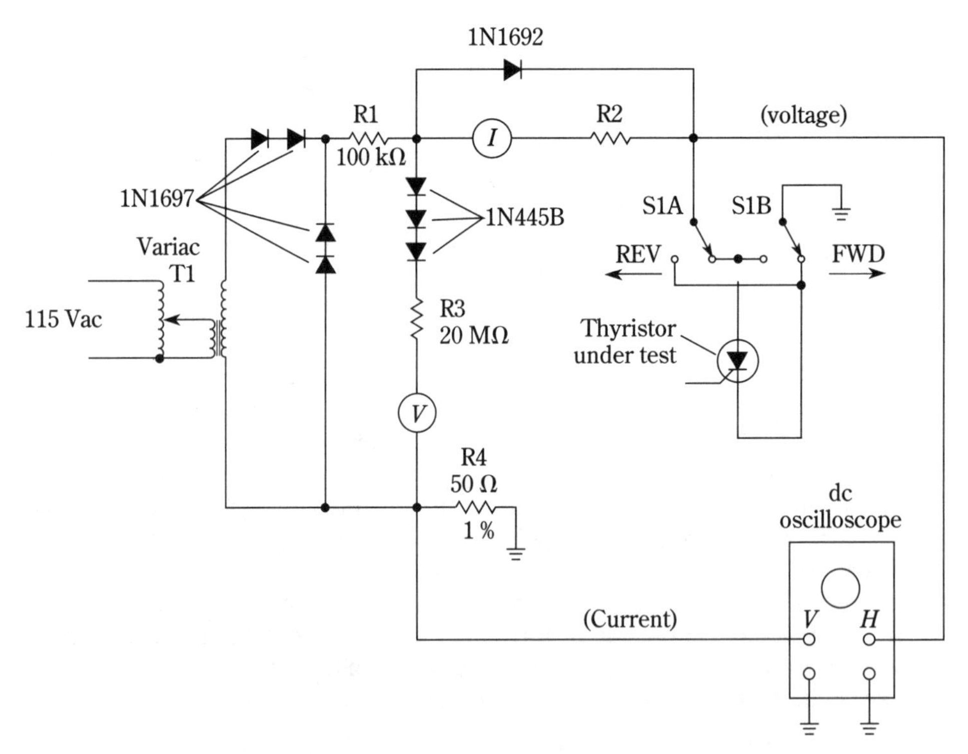

11-15 Test circuit for blocking characteristics (operating currents above 2 A).

To measure the actual forward breakover voltage, increase the applied voltage with F1 until the scope trace indicates that the device under test is breaking over (the current will rise sharply just prior to breakdown). Be careful that the applied voltage does not exceed the peak forward voltage rating (PFV) of the device.

Because a triac conducts in both directions (as does an SBS) it is often desirable to measure breakdown voltage and leakage currents in both forward and reverse directions. A triac can be tested in the reverse direction (sometimes called the *third quadrant direction*) by setting S1 to REV, and repeating the test (read voltage on meter V and current on meter I).

To measure reverse voltage and current of an SCR, set S1 to REV, and raise the voltage with T1 until meter V reads the rated reverse voltage. Read the full-cycle average reverse leakage current on meter I, or read peak leakage current on the scope display.

11.4.2 Test circuit for thyristors below 2 A

The circuit of Fig. 11-16 provides a simple and inexpensive means for checking the instantaneous-leakage characteristics and blocking-voltage capabilities of low-current thyristors. Push-button switch S1 minimizes junction heating and should not be omitted from the circuit. If required, the tests can be conducted at elevated temperatures by placing the thyristor in an oven.

To use the circuit, set S2 to FWD, press S1 and increase the input with R1 so that meter V reads the rated blocking voltage. Read leakage current, if any, on meter I.

To measure actual forward breakdown voltage, adjust R1 until the current reading on I increases sharply and the voltage reading on V decreases. The reading on meter V just prior to this point is the forward breakover voltage of the device under test. Reverse blocking-voltage and leakage-current measurements are made with S2 in the REV position. Again, do not keep any device in the breakdown condition for any length of time.

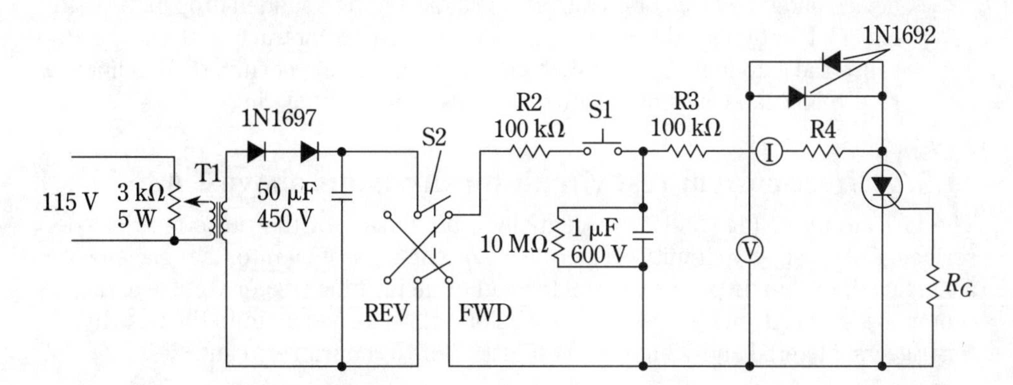

V = 500 V–0–500 V, center zero
I = 100 μA–0–100 μA, center zero
R4 = 2.5 kΩ, minus resistance of meter I
R_G = gate shunt resistance (if required by specification)
T1 = triac R29A or equivalent

11-16 Test circuit for blocking characteristics (operating currents below 2 A).

11.5 Gate-trigger voltage and current tests

This section describes test circuits and procedures to determine, under stated conditions, the magnitude of the gate-trigger voltage and current necessary to switch a thyristor from forward-blocking to the on-state. One circuit provides a pulse trigger source for SCRs and triacs, while a second circuit provides a direct-current trigger. A third circuit provides for gate test of low-current SCRs.

11.5.1 Pulse test circuit for thyristors above 2 A

The circuit of Fig. 11-17 provides for pulse testing of thyristors with operating currents greater than 2 A. The blocking voltage waveform applied to the device under test consists of a clipped half wave with peak magnitude of 6 or 12 V, depending on the setting of S1. The correct setting of S1 is determined from the thyristor datasheet, as is the value of anode load resistor R_S.

The gate-source voltage is a square-wave pulse, the magnitude of which can be varied from 0 to 6 V. The pulse width can also be adjusted from about 5 μs to greater than 100 μs. Gate voltage can be switched either positive or negative for testing triacs.

A typical test procedure using the circuit of Fig. 11-17 is as follows:

1. Adjust R1 so that a gate pulse occurs only once during each half-cycle of applied anode blocking voltage. Time the pulse (by adjustment of R1) to occur approximately 4 ms after the start of the half-cycle.
2. Adjust R2 to give the desired width of the pulse. Note that gate pulse widths in excess of about 100 μs result in measured values of trigger voltage and current that are equivalent to continuous dc measurements.
3. Initially, set R3 for 0 V output. Then gradually increase the setting of R3 until the device under test triggers. Triggering is indicated by a sudden drop in the reading of V1, or by a sudden step in the gate voltage-current trace on the scope. Because the gate impedance can change when triggering occurs, the readings of gate voltage and current must be made just prior to triggering!

11.5.2 Direct-current test circuit for thyristors above 2 A

The test circuit of Fig. 11-18 is essentially a dc version of the pulse test just described. Anode-supply circuitry is identical, and the pulse generator is replaced with a simple adjustable dc power source. Instead of monitoring trigger voltage and current with a scope, dc meters are used. As before, R3 is turned up until the test thyristor triggers. Meters I and V2 are read off just prior to the trigger point.

11.5.3 Gate-trigger test circuit for thyristors below 2 A

The measurement of low-current SCR triggering voltage and current is complicated by the fact that the gate impedance changes drastically when the SCR triggers. Also, the gate-trigger voltage and current values depend on the source impedance of the test circuit. Source impedance must be specified when making tests.

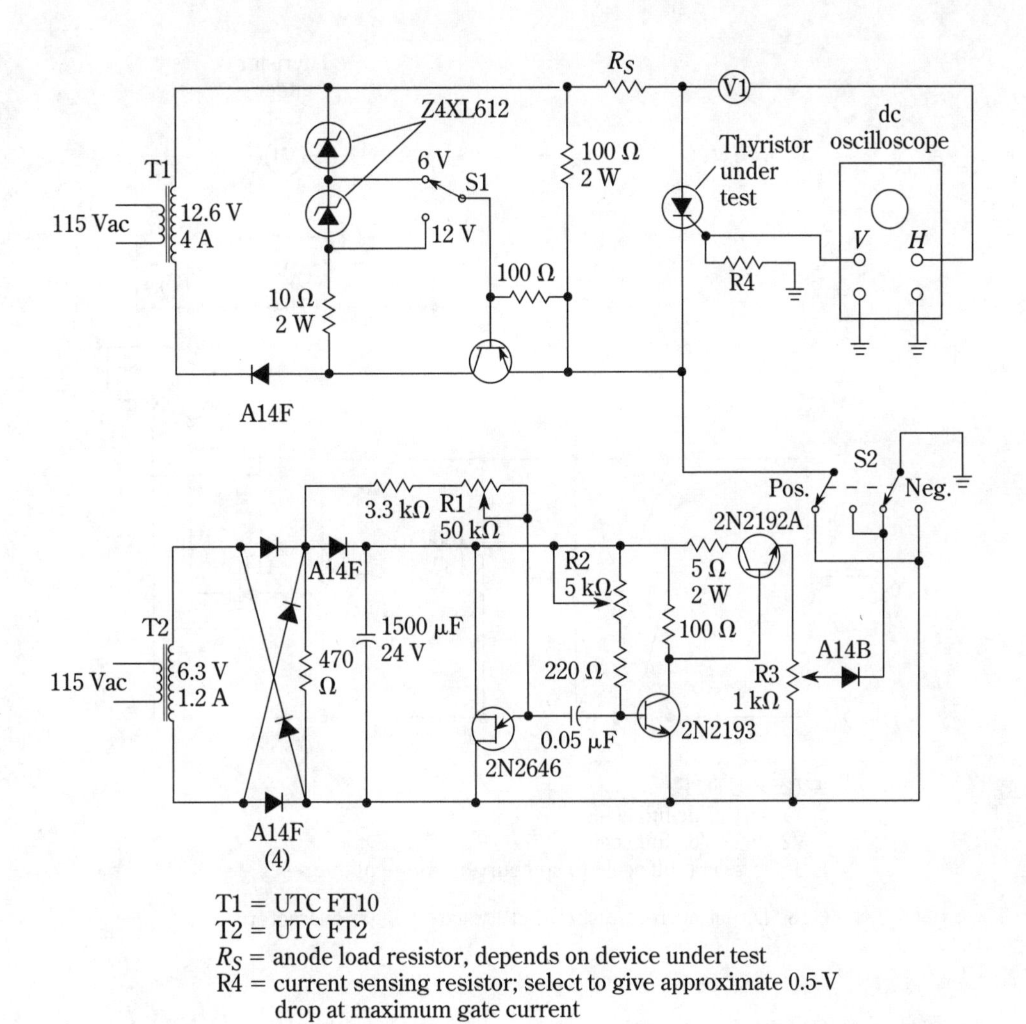

11-17 Test circuit for gate-trigger characteristics (operating currents above 2 A).

The circuit of Fig. 11-19 is designed specifically for testing SCRs with current ratings below 2 A. In this circuit, a variable half-wave voltage is applied to the gate (from a controlled impedance source), and the gate voltage-current characteristics are monitored on a scope. The triggering point is detected by the sudden change in gate impedance that occurs when the SCR switches on.

Figure 11-19B shows a typical scope presentation of gate voltage-current characteristics during test. The trace is shown dashed beyond the trigger point. In an actual scope display, the trace suddenly jumps at the triggering point because of the change in gate impedance. The portion of the trace beyond the triggering point becomes somewhat reduced in intensity.

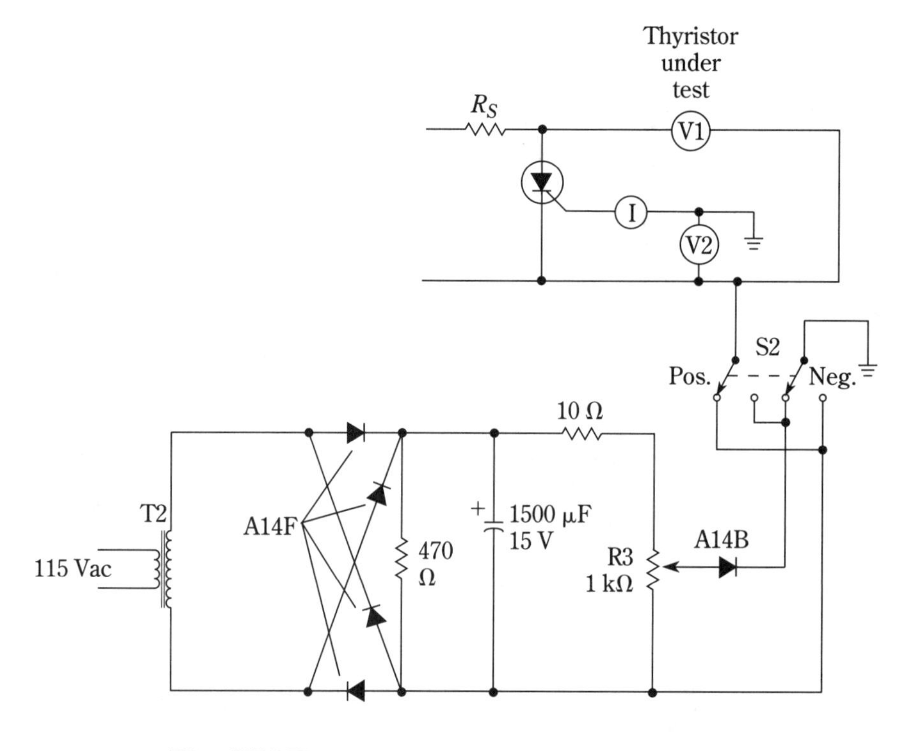

T2 = UTC F2
V1 = 10 Vdc full scale
V2 = 12 Vdc full scale
I = select full scale to suit current range of interest

11-18 Direct-current test circuit for gate-trigger characteristics.

The gate-trigger voltage is that value that just causes the device to switch on. As shown in Fig. 11-19B, the gate-trigger voltage is read just prior to the switching point. Unlike trigger voltage, the gate-trigger current value must be read at the point where current is maximum. Although this is not necessarily the actual triggering point, the trace must first pass through this maximum and the firing-circuit design must take this into account.

Many low-current SCRs show triggering with a negative value of gate current. Figure 11-19C shows a typical gate voltage-current trace for this type of SCR. Note that the gate-trigger voltage is always positive. On such SCRs, only the gate-trigger voltage is of interest.

11.6 Latching and holding current tests

This section describes test circuits and procedures to measure latching, and holding currents for control rectifiers and thyristors. Because latching and holding current values depend on gate conditions and anode-supply voltage, these parameters must

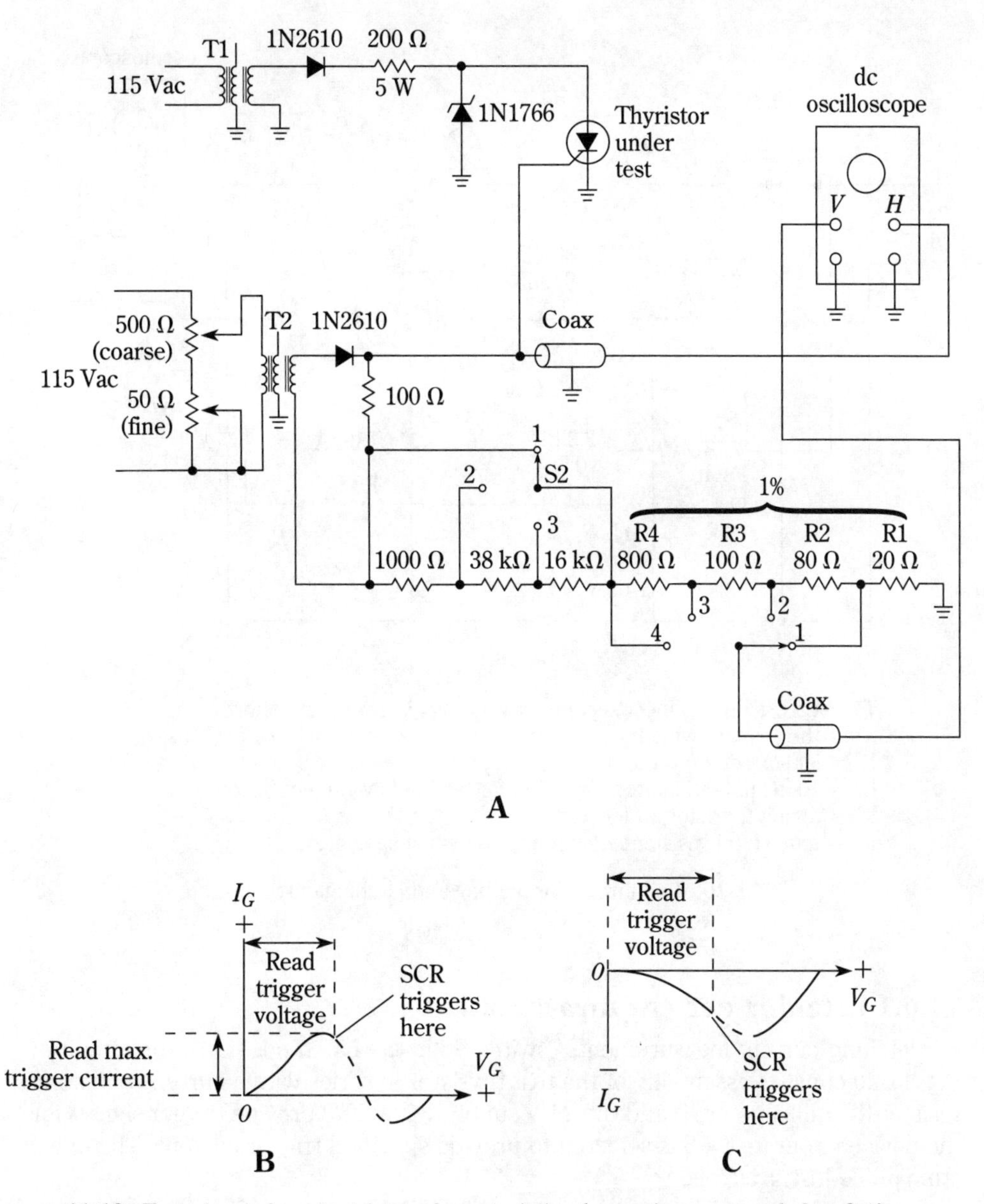

11-19 Test circuit for gate-trigger characteristics (operating currents below 2 A).

be specified as test conditions. A thyristor can show more than one value of on-voltage at a given forward-current level—especially if the maximum value of anode current never rises much above holding level. This can lead to several values of holding current for the same device. To properly perform a holding-current test, turn the device on initially with high-current pulse, and then reduce the current down to the holding level. The test circuit of Fig. 11-20 fulfills this requirement, and it can also be used for latching-current tests.

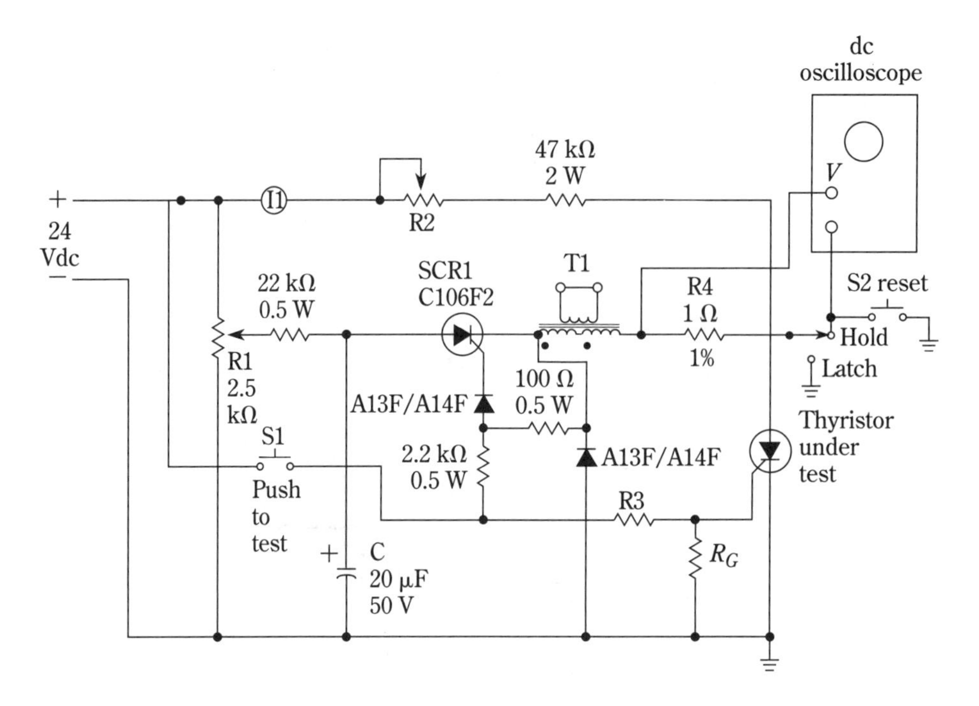

T1 = UTC FT10, connect secondary windings in series and short the primary winding
I1 = 500-mA dc movement
R2 = 10-kΩ potentiometer (or 50-kΩ for testing low-current SCRs)
R3 = to suit thyristor under test
R_G = gate shunt resistance (as required by test specification)

11-20 Test circuit for holding and latching current.

11.6.1 Latching-current measurement

For latching-current measurements (switch S3 in the Latch position), the circuit of Fig. 11-20 consists essentially of the test thyristor in series with a current-adjusting resistor R2, milliameter I1, and the 24-V supply. S1 and R3 provide trigger signals for the device under test. R3 is selected to provide specified trigger current. The operating procedure is as follows:

1. Set R2 to the maximum resistance value.
2. Gradually reduce the resistance of R2 while pressing and releasing S1. Each time S1 is pressed, I1 should deflect and then drop back to zero when S1 is released, as long as anode current flowing through the test thyristor is less than the latching current.
3. When latching finally occurs, I1 should deflect and remain deflected as S1 is released. The value of current indicated by I1 at the transition point is the latching current of the thyristor.

11.6.2 Holding-current measurement

For holding-current measurements, (S3 in Hold position), R2 is initially left at the setting determined during the latching-current test. Each time S1 is pressed to trigger the test thyristor, SCR1 is also triggered and passes an additional pulse of current through the test thyristor. The magnitude of this initial current pulse is determined by the setting of R1, and is specified for each thyristor type. The current value is monitored by R4 and displayed on the scope. Pulse width is fixed by C1 and T1. The operating procedure is as follows:

1. Set R2 to the value determined for latching current.
2. Press and release S1, while gradually increasing R_2 until I1 drops suddenly to zero. The reading on I1 just prior to this point is the holding current of the device under test.

11.7 Average forward-voltage test

The circuit of Fig. 11-21 can be used to test the forward voltage drop of a thyristor in the conducting state. This forward voltage drop is often listed as $V_{F(AVG)}$. In this

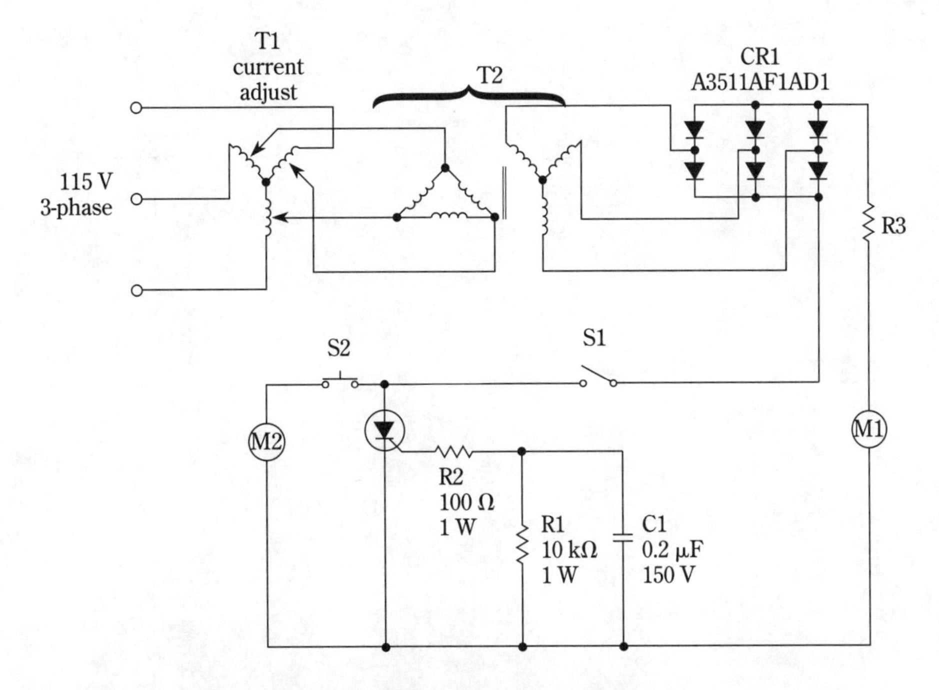

11-21 Test circuit for average forward-voltage measurement.

test, the thyristor is subjected to a direct current, as read by meter M1, and voltage drop measured by M2. To make the measurement, close S1 with T1 is a position approximately midway between the end positions. Adjust T1 to the level of current desired for test, as read on M1. Press S2 and read M2.

These readings (at M1 and M2) should be less than the maximum-rated current and voltage for the device under test. A test current somewhere near the continuous-duty current rating of the thyristor is recommended. Under these conditions, the device under test should be mounted on a heatsink during test. If a heatsink is not used, the readings should be completed within 2 or 3 seconds to prevent overheating of the device under test. The tests can be conducted at any ambient temperature within the operating range of the thyristor.

12
Audio and op-amp tests

This chapter is devoted entirely to test procedures for audio and op-amp circuits. The first sections describe audio and op-amp characteristics and test procedures from the practical standpoint. The information in these sections permits you to test all the important audio and op-amp characteristics using basic shop equipment. The sections also help you understand the basis for such tests. The remaining sections of the chapter describe how the same tests, and additional tests, are performed using more sophisticated equipment.

12.1 Audio circuit tests

This section covers the basic tests for audio circuits, starting with a review of typical audio test equipment. The section then goes on to describe test procedures that can be applied all types of audio equipment, using the actual-circuit test results, where practical.

12.2 Wow and flutter meters

Wow and flutter are turntable or tape-transport speed fluctuations that can cause a quivering or wavering effect in the sound during play. In tape systems, wow and flutter can occur both during record and playback. The longer fluctuations in sound (below about 3 to 6 Hz) are called *wow*. Shorter fluctuations (typically from 3 to 6 Hz up to 20 Hz) are called *flutter*.

Although wow and flutter problems are common to all turntables and tape decks, it is only when the problems go beyond a certain tolerance that they are objectionable. So, when testing for "excessive wow and flutter," the actual amount must be measured. The basic method for measuring wow and flutter is to measure the reproduced frequency of a precision tone or signal that is recorded on a test record or tape. Any deviation in frequency of the reproduced or playback tone, from the recorded tone, is an indication of wow and flutter.

In the case of a tape system, you can record your own precision signal instead of using a test tape. However, remember that both the recorded and playback signals are subject to wow and flutter. This might lead you to think that there is a serious problem when the actual fluctuation is within tolerance.

There are four primary standards for wow and flutter measurement: JIS (Japan), NAB (USA,), CCIR (France), and DIN (Germany). JIS, NAB, and CCIR use a prerecorded test tone of 3 kHz, whereas DIN uses 3.15 kHz.

A frequency counter (chapter 4) can be used to monitor the test tone during playback. However, a commercial wow and flutter meter is usually more convenient and accurate. Even though the counter might be quite accurate, the problem is one of resolution. As an example, the wow and flutter tolerance for a typical consumer-electronics turntable is 0.025 percent. This means that a playback frequency of 3000.75 Hz is acceptable, but 3000.76 is not acceptable.

In addition to the accuracy and resolution problems, there is a matter of weighting, or test conditions under which wow and flutter are measured. Each standard has a separate system of weighting. Likewise, each standard measures wow and flutter in different values. Peak values are used for CCIR and DIN. NAB uses mean values, while JIS uses rms values. Fortunately, all of the weighings and values are accounted for in a commercial wow and flutter meter.

12.2.1 Typical wow and flutter meter characteristics

The typical wow and flutter meter has several ranges (at least six) to provide good resolution from about 0.003 to 10 percent. Usually, the variations in playback (or reproduced frequency) are separated from the test tone and then are measured.

The frequency of the test tone is displayed on a frequency counter (at least four digits). The percentage of wow and flutter is indicated on an analog meter. Filters are used to separate both wow and flutter components (separately, if desired) from the test signal. A mode selector permits the user to select measurement of flutter only, wow only, or combined wow and flutter (both weighted and unweighted).

Most wow and flutter meters can accommodate a wide range of input signal levels, permitting measurement at nearly any desired point in the audio equipment. For example, one meter has two selectable sensitivities of 0.5 to 100 mV and 5 mV to 30 V. These sensitivity levels permit measurement of small signals directly from tape heads and phono cartridges, if desired. Some meters have a level monitor that turns on if the input signal level is adequate for wow and flutter measurements. No other measurements or adjustments are required. The level monitor is a time-saving feature for setting up tests. Other typical conveniences include built-in crystal-controlled oscillators that provide very stable sources of 3 and/or 3.15 kHz. These signals are used for recording wow and flutter test tapes.

Most wow and flutter meters have auxiliary jacks that provide outputs to other instruments, such as scopes or chart recorders, if desired. These jacks provide a dc voltage that is proportional to rotational speed error or deviation, and both ac and dc voltages proportional to the wow and flutter meter reading. The ac wow and flutter voltage is used with an auxiliary scope. The dc wow and flutter voltage is used with a chart recorder to make copies of the measurements. The dc rotational-speed error-deviation voltage can be used with either a chart recorder or a scope.

Rotational-speed error is not to be confused with wow and flutter. Speed error (if any) means that the record or tape is not being driven at the correct speed. For example, speed tolerance for a typical consumer turntable is 0.003 percent. This means that a rotational speed of 45.00135 rpm (for a 45-rpm record) is acceptable. Generally, speed error implies a constant value or one that changes slowly, whereas wow and flutter constantly fluctuate.

For added versatility, many wow and flutter meters have a built-in frequency counter that can be used (independent from wow and flutter measurements) for general measurement of audio frequencies. Typically, either a crystal-controlled or line-frequency (50 or 60 Hz) time base can be selected for the counter.

12.2.2 Basic wow and flutter measurement

This section does not cover the detailed operation of any particular meter. Such information is available in the instruction manual. However, for those totally unfamiliar with wow and flutter measurement, the following is a typical procedure. Figure 12-1A shows the basic test connections between the meter and turntable or cassette.

1. Select the type of measurement standard, JIS, NAB, CCIR, or DIN.
2. Use a prerecorded wow and flutter test tape or record on the equipment to be tested.
3. If you want to record a test tape, connect the oscillator output terminals to the input of the tape record, as shown in Fig. 12-1A. Then, set the 3- or 3.15-kHz switch to on (3 kHz for JIS, NAB, or CCIR; 3.15 kHz for DIN).
4. Remember that wow and flutter present during record are also present during playback and are added to the playback wow and flutter. Under these conditions, measured wow and flutter represents the total wow and flutter introduced for record and playback. One-half the measured value represents wow and flutter for playback only.
5. Connect the output of the equipment under test (cassette deck, or turntable) to the input terminal, as shown in Fig. 12-1A.
6. The level-monitor indicator (if any) should turn on if the signal is of adequate level and contains the 3- or 3.15-kHz test signal. If the level-monitor indicator does not turn on, change the sensitivity-switch setting. If the level monitor still does not turn on, check to make sure that there is some signal present and that the signal is at the correct test frequency. It is possible that you are not on the correct portion of the test record or tape, or that there is no signal at the point being measured.
7. Check that the frequency display shows the wow and flutter test frequency (3 or 3.15 kHz). Note that if the signal being monitored is at a high level; the level monitor might turn on at frequencies other than 3 or 3.15 kHz (on some meters).

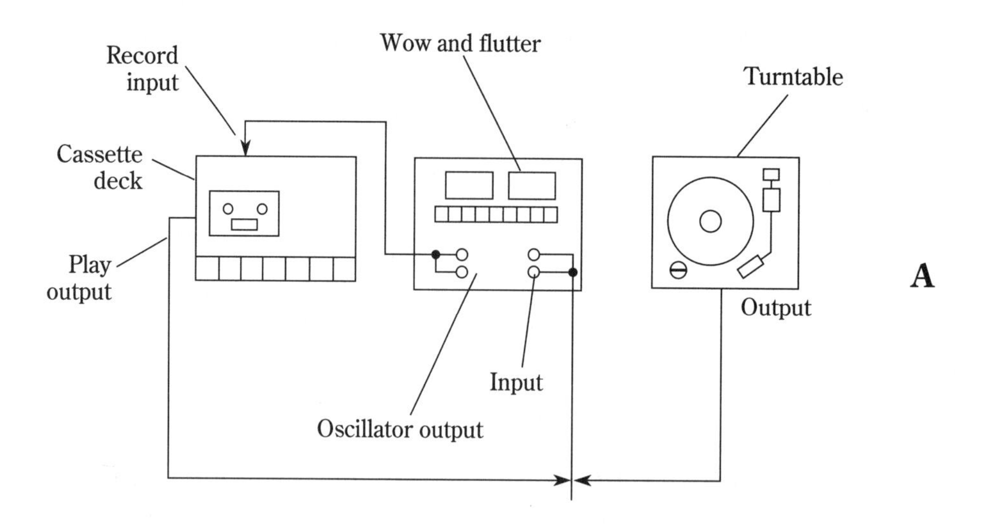

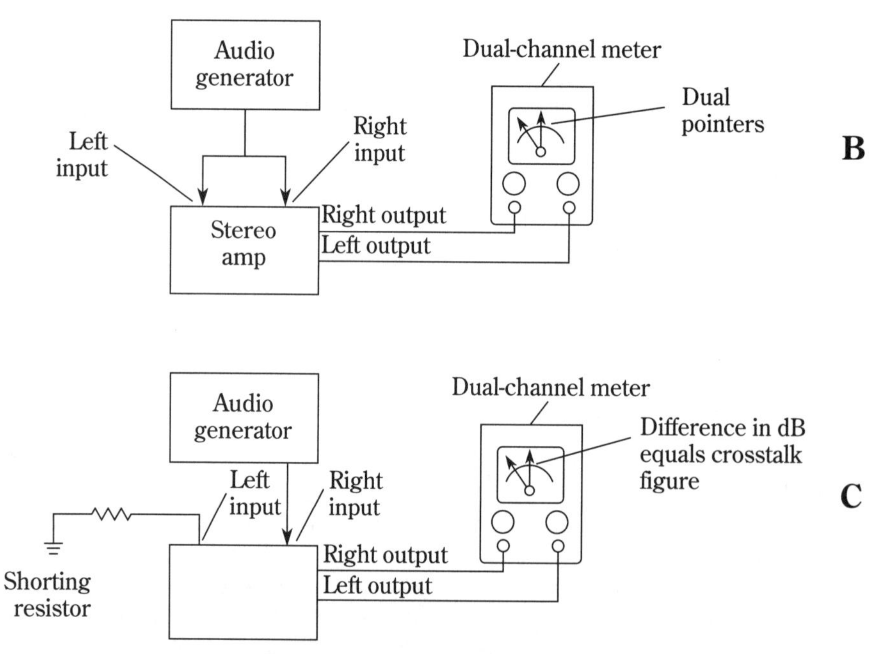

12-1 Basic test connections for wow flutter and stereo measurements.

8. Set the weight-function switch for the desired frequency component.

   Use the weighted switch for a combination of wow and flutter (with a typical frequency response of about 0.2 to 200 Hz).

   Use the wow switch for wow components only (0.5 to 6 Hz for NAB or JIS; 0.3 to 6 Hz for CCIR or DIN).

Use the flutter switch for flutter components only (with a typical frequency response of about 6 to 200 Hz).

Use unweighted for an unweighted combination of wow and flutter (with 0.5 to 200 Hz for NAB or JIS and 0.3 to 200 Hz for CCIR or DIN).

9. Set the range for the highest obtainable meter reading, without going off scale. Read the percentage of wow and flutter from the meter, using a scale that corresponds to the range.

12.3 Audio voltmeters

It is possible that you can get by with any ac meter (even a multimeter, analog or digital) for all audio work. However, for accurate measurements, I recommend a wideband ac meter, preferably a dual-channel mode. The dual-channel feature makes it possible to monitor both channels of a stereo circuit simultaneously. This is particularly important for stereo frequency-response and crosstalk measurements.

In addition to making routine voltage and resistance checks, the main functions of voltmeters in audio work are to measure frequency response and to trace audio signals from input to output. Many technicians prefer scopes for these procedures. The reasoning is that scopes also show distortion of the waveform during measurement or signal tracing. Other technicians prefer the simplicity of a meter, particularly in such procedures as voltage-gain and power-gain measurements.

12.3.1 Typical audio-voltmeter characteristics

The typical audio voltmeter has ranges (at least 10, possibly 12) that cover audio voltages from about 0.2 mV to 100 V. The usual wideband frequency permits measurements from 5 Hz to 1 MHz. High sensitivity permits measurements down to 30 μV (and lower in some meters).

Audio meters have two voltage scales: a dB scale and a dBm scale (0 dBm = 1 mW across 600 Ω). These scales are used for making relative measurements (gain, attenuation, etc.). Typical ranges are –90 to +40 for dB and –90 to +42 for dBm. Most audio meters have built-in low-distortion amplifiers that are used to drive the meter movements. This allows the meter to be used as a calibrated high-gain preamp. Typically, the output is calibrated at 1 Vrms for a full-scale reading.

Generally, audio meters use absolute mean-value (average) sensing, but are calibrated to read the rms value of a sine-wave voltage. The input voltage is capacitively coupled, which permits measurement of an ac signal superimposed on a dc voltage. A typical 10-MΩ input impedance on all ranges with low shunting capacitance, assures minimum circuit loading to the audio circuit under test. Dual-channel audio meters, which are capable of measuring two voltages simultaneously, have dual pointers (often red and black) for convenient direct comparison of two levels (such as is required during stereo balance measurements).

12.3.2 Basic audio measurements and signal tracing

Present-day audio meters are suitable for all forms of audio measurement and signal tracing. The following are some typical examples.

Stereo measurements A meter with two input channels applied to a dual-pointer scale is more useful for direct-comparison measurements than two separate voltmeters. An excellent example is the measurement of left- and right-channel characteristics of stereo circuits. To show the advantages of a dual-channel, wide-band meter, the typical stereo-crosstalk and stereo-response measurement (with a dual-channel meter) is covered here.

Stereo-response measurement With the equipment connected as shown in Fig. 12-1B (equal audio signals applied to the left- and right-channel inputs of the stereo circuit), set the controls to the desired range.

It might be necessary to set a ground-mode switch for minimum hum when making this test. For a stereo circuit that has no connection common to the left and right channels, this usually means setting the ground-mode switches to open. In most other cases (where there is some connection common to both channels), use the ground position of the ground-mode switch.

With the connections made and controls set, vary the frequency of the audio generator. Notice any difference in frequency response between the left and right channels, as indicated by unequal deflection of the two pointers. Even those not familiar with frequency-response measurements will see how much simpler a dual-pointer meter is to use (than two separate meters).

Stereo-crosstalk measurements With the equipment connected as shown in Fig. 12-1C (audio signal applied to one channel and the other channel shorted), set the controls to the desired range. Again, it might be necessary to set a ground-mode switch for minimum hum. Notice that the value of the shorting resistor should be equal to the input impedance of the amplifier.

With the connections made and controls set, notice the difference in dB between the signal level measured on the channel with the signal input. This is the crosstalk figure. Crosstalk is generally specified in dB (such as –40 dB).

For a complete crosstalk measurement, reverse the input connections to the stereo circuits and repeat the measurement. There should be no substantial difference in the meter readings with the input connections reversed. Make certain that the signal level, frequency, and all other input connections are identical when the connections are reversed.

12.4 Frequency response

The frequency response of an audio circuit can be measured with a generator and a meter or scope. When a meter is used, the generator is tuned to various frequencies, and the resultant output response is measured at each frequency. The results are then plotted in the form of a graph or response curve, such as shown in Fig. 12-2B. The procedure is essentially the same when a scope is used. The scope gives the added benefit of visual analysis of distortion.

The basic procedure (with emitter meter or scope) is to apply a constant-amplitude signal while monitoring the circuit output. The input signal is varied in frequency (but not in amplitude) across the entire operating range of the circuit. The voltage output at various frequencies across the range is plotted on a graph.

1. Connect the equipment (Fig. 12-2A).
2. Initially, set the generator frequency to the low end of the range. Then, set the generator amplitude to the desired input level.
3. In the absence of a realistic test input voltage, set the generator output to an arbitrary level. A simple method of finding a satisfactory input level is to monitor the circuit output (with a meter or scope) and increase the generator amplitude at the circuit center frequency (or at 1 kHz) until the circuit is overdriven. This point is indicated when further increases in generator output do not cause further increases in meter reading, (or the output waveform peaks begin to flatten on the scope display). Set the generator output just below this point. Then, return the meter or scope to monitor the generator voltage (at the circuit input) and measure the voltage. Keep the generator at this voltage throughout the test.
4. If the circuit is provided with any operating or adjustment controls (volume, loudness, gain, treble, balance, and so on), set the controls to some arbitrary point when making the initial frequency-response measurement. The response measurements can then be repeated at different control settings, if desired.

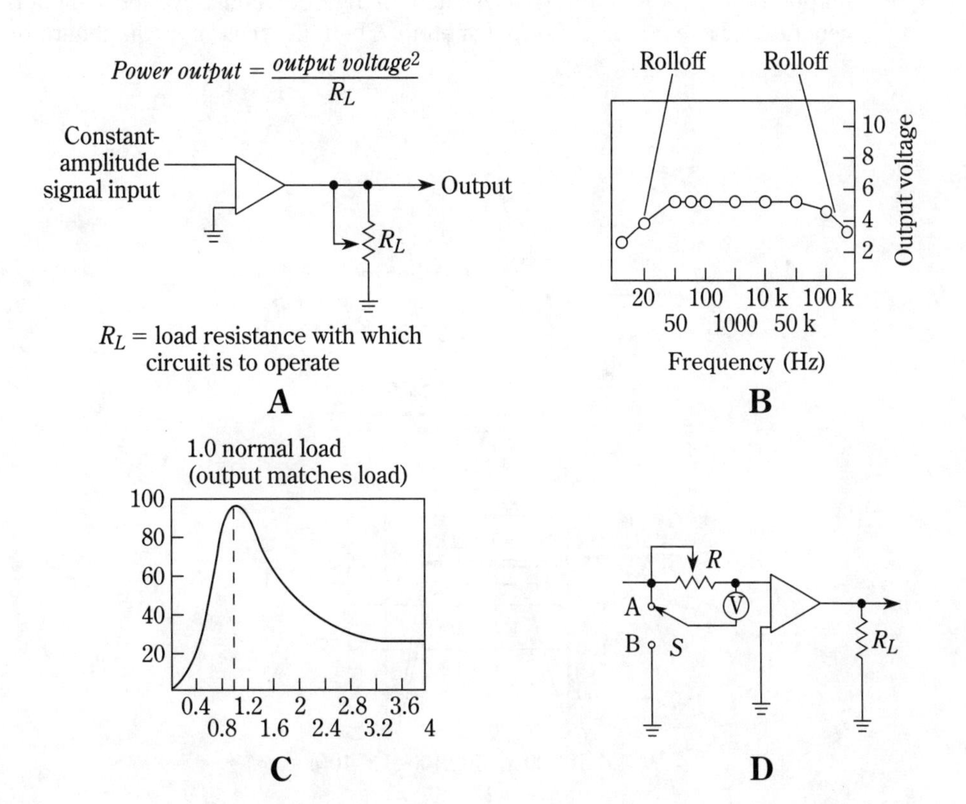

12-2 Basic audio-circuit test connections.

5. Record the circuit-output voltage on the graph. Without changing the generator amplitude, increase the generator frequency by some fixed amount, and record the new circuit-output voltage. The amount of frequency increase between each measurement is an arbitrary matter. Use an increase of 10 Hz, where rolloff occurs and 100 Hz at the middle frequencies.
6. Repeat the process; check and record the circuit-output voltage at each of the checkpoints. Figure 12-3A shows the frequency response for the amplifier of Fig. 12-3B.
7. After the initial frequency-response check, the effect of operating or adjustment controls should be checked. Volume, loudness, and gain controls should have the same effect all across the frequency range. Treble and bass controls might have some effect at all frequencies. However, a treble control should have the greatest effect at the high end, whereas bass controls should have the greatest effect at the low end.
8. Note that the generator output can vary with changes in frequency (a fact that is often overlooked in making frequency response tests during troubleshooting). Even precision lab generators can vary in output with changes in frequency, and thus can produce a considerable error. The generator output should be monitored after each change in frequency (most modern generators have a built-in output meter). Then, if necessary, the generator-

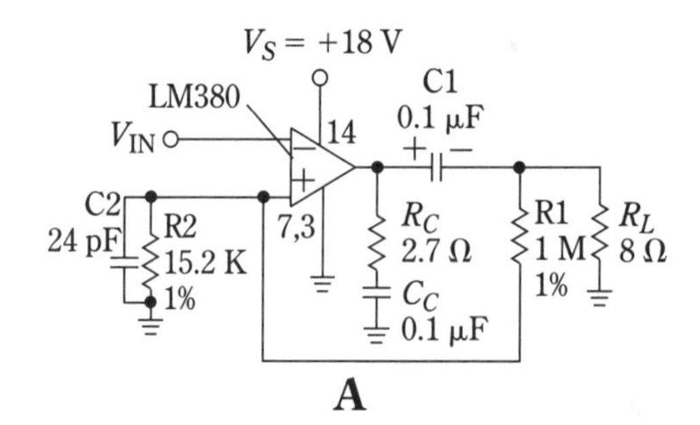

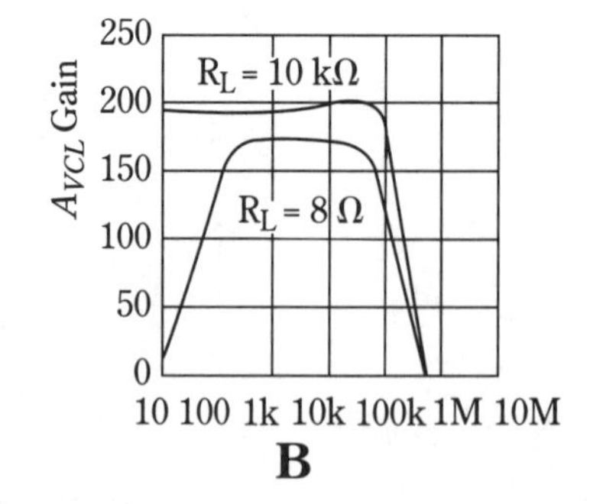

12-3 Typical frequency-response characteristics.

output amplitude can be reset to the correct value. It is more important for the generator amplitude to remain constant, rather than set at some specific value when making a frequency-response check.

12.5 Voltage gain

Voltage gain is measured in the same way as frequency response. The ratio of output voltage to input voltage (at any given frequency or across the entire frequency range) is the voltage main. Because the input voltage (generator output) is held constant for a frequency-response test, a voltage-gain curve should be identical to a frequency-response curve (such as shown in Fig. 12-3B).

12.6 Power output and power gain

The power output of an audio amplifier is found by noting the output voltage across the load resistance (Fig. 12-2A) at any frequency or across the entire frequency range. The power output is calculated, as shown in Fig. 12-2B. For example, if the output voltage is 10 V across a load resistance of 50 Ω, the power output is $10^2/50 = 2$ W.

Never use a wire-wound component (or any component that has reactance) for the load resistance. Reactance changes with frequency, and causes the load to change. Use a composition resistor or potentiometer for the load.

To find the power gain of an amplifier, first find both input and output power. Input power is found in the same way as output power, except that the input impedance must be known (the procedure for finding dynamic input impedance is described later in this chapter). With the input power known, the power gain is the ratio of output power to input power.

12.7 Input sensitivity

The datasheets of some amplifiers include an input-sensitivity specification. This means that there must be a given power output for a given input voltage (such as 100-W output with a 1-V input). The test connections shown in Fig. 12-2A are used. The specified input voltage is applied, and the actual power output measured.

12.8 Power bandwidth

The datasheets of some amplifiers include a power-bandwidth specification. This means that the amplifier must deliver an even power output across a even frequency range. For example, an amplifier might produce full-power output up to 20 kHz, even though the frequency response is flat up to 100 kHz. That is, voltage (without a load) remains constant up to 100 kHz, whereas power output (across a normal load) remains constant up to 20 kHz. Figure 12-4 shows a typical power-bandwidth curve or graph.

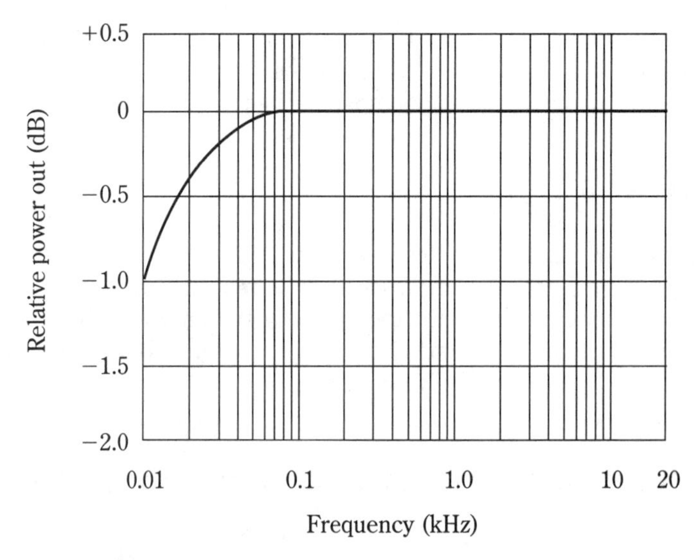

12-4 Typical power-bandwidth curve or graph.

12.9 Load sensitivity

All amplifiers are sensitive to changes in load. This is particularly true of power amplifiers (both IC and discrete component). An amplifier produces maximum power when the output impedance is the same as the load impedance. The test circuit for load-sensitivity measurement is the same as the circuit for frequency response (Fig. 12-2A), except that the load resistance is variable. Again, never use a wire-wound load resistance. The reactance can result in considerable error.

Measure the power output at various load-impedance values. That is, set the load resistance to various values (including a value equal to the true amplifier-output impedance) and notice the voltage and/or power gain at each setting. Repeat the test at various frequencies.

Figure 12-2C shows a generalized load-sensitivity response curve. Notice that if the load is twice the output impedance (as indicated by a 2:1 ratio, or a normalized load impedance of 2), the output power is reduced to about 50 percent.

12.10 Dynamic output impedance

The load-sensitivity test can be reversed to find the dynamic output impedance of an amplifier. The connections (Fig. 12-2A) and the procedures are the same, except that the load resistance is varied until maximum output power is found. Power is removed, the load resistance is disconnected from the circuit, and the resistance is measured with an ohmmeter. This resistance is equal to the dynamic output impedance of the amplifier (but only at the measurement frequency). The test can be repeated across the entire frequency range, if desired.

12.11 Dynamic input impedance

Use the circuit of Fig. 12-2D to find the dynamic input impedance of an amplifier. The test conditions are identical to those for frequency response, power output and so on. Move switch S between points A and B, while adjusting resistance R until the voltage reading is the same in both positions of S. Disconnect R and measure resistance R. This resistance is equal to the dynamic impedance of the amplifier input.

Accuracy of the impedance measurement depends on the accuracy with which resistance R is measured. Again, a noninductive (not wire-wound) resistor must be used for R. The impedance found by this method applies only to the frequency used during the test.

12.12 Sine-wave analysis

Amplifier distortion can be checked by sine-wave analysis using a scope. The primary concern in distortion analysis is deviation of the output from the input waveform. If there is no change (except in amplitude), there is no distortion. If there is a change in waveform, the nature of the change will often reveal the cause of distortion. Unfortunately, in the real world, analyzing sine waves to pinpoint amplifier problems is a difficult job that requires considerable experience. Unless distortion is severe, it might pass unnoticed. Square-wave analysis is far more practical.

12.13 Square-wave analysis

Distortion analysis is more effective with square waves because of the high odd-harmonic content in square waves (and because it is easier to see a deviation from a straight line with sharp corners than from a curving line). Square waves are introduced into the amplifier input, and the output is monitored with a scope, as shown in Fig. 12-5. The primary concern is deviation of the output waveform from the input waveform (which is also monitored on the scope). If the scope has a dual-trace feature, the input and output can be monitored simultaneously. If there is change in waveform, the nature will often reveal the cause of the distortion.

Notice that the drawings of Fig. 12-5 are generalized and that the same waveform can be produced by different causes. For example, poor LF (low frequency) response appears to be the same as HF (high frequency) emphasis. Figure 12-6 shows the waveforms that are produced by an actual circuit (on a dual-trace scope). Notice that the output (trace B) is inverted from the input, and there is no gain, or unity gain. This means that the output is following the input (trace A), but that the output is inverted (or –1 as it might be shown on the datasheet). Also notice that there is some slightly reduced high-frequency response in output trace B, but not the exaggerated response of Fig. 12-5.

The third, fifth, seventh, and ninth harmonics of a clean square wave are emphasized. If an amplifier passes a given frequency and produces a clean square-wave output, it is reasonable to assume that the frequency response is good up to at least 9 times the square-wave frequency.

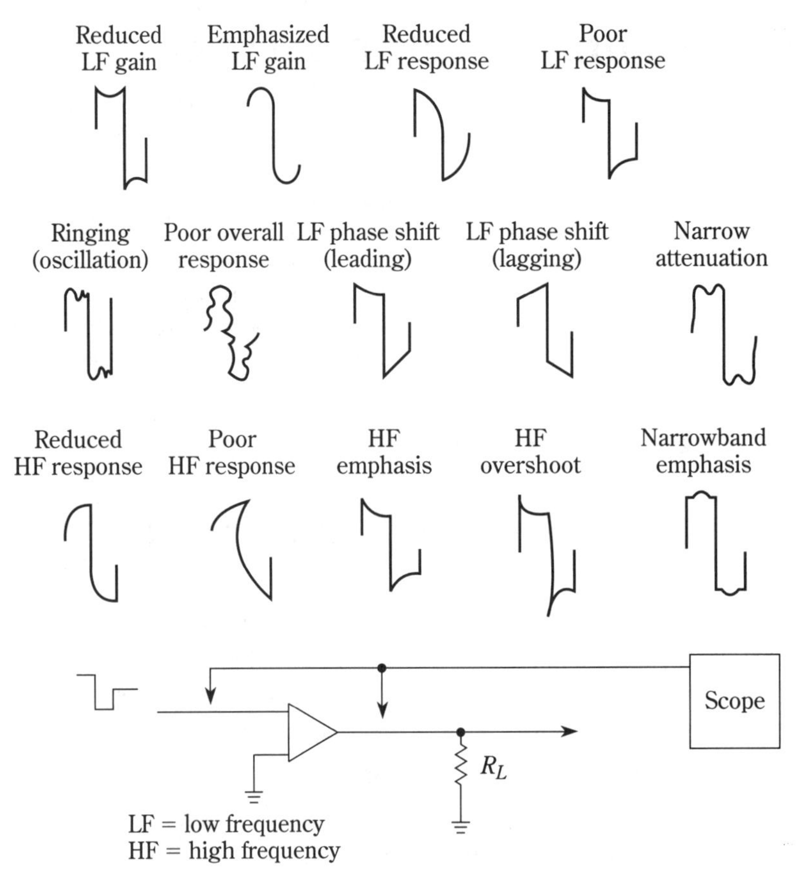

12-5 Basic square-wave distortion analysis.

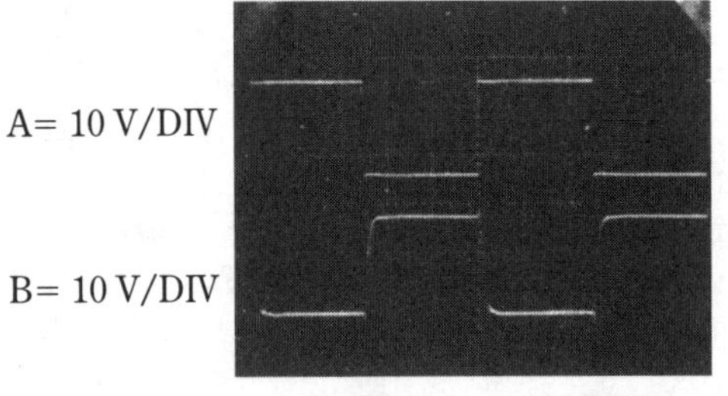

12-6 Audio-circuit response to square waves. Linear Technology Corporation

12.14 Harmonic distortion

With any amplifier, there is the possibility of odd or even harmonics being present with the fundamental. These harmonics combine with the fundamental and produce distortion, as is the case when any two signals are combined. The effects of second- and third-harmonic distortion are shown in Fig. 12-7.

Commercial harmonic-distortion meters operate on the fundamental-suppression principle. A sine wave is applied to the amplifier input, and the output is measured on the scope. The output is then applied through a filter that suppresses the fundamental frequency. Any output from the filter is then the result of harmonics.

The output is also displayed on the scope. Some commercial harmonic-distortion meters use a built-in meter instead of, or in addition to, an external scope. When the scope is used, the frequency of the filter-output signal is checked to determine harmonic content. For example, if the input is 1 kHz and the output (after filtering) is 3 kHz, third-harmonic distortion is indicated. Reduce the scope horizontal sweep down so that you can see one input cycle. If there are three cycles at the output for the same time period as one input cycle, this indicates third harmonic distortion.

The percentage of harmonic distortion is also determined by this method. For example, if the output is 100 mV without the filter, and 3 mV with the filter, a 3-percent harmonic distortion is indicated. Figure 12-8 shows the total harmonic distortion (or THD) for a typical audio circuit. Notice that THD varies with the power output of the circuit over a wide frequency range. Also, notice that THD depends on the load.

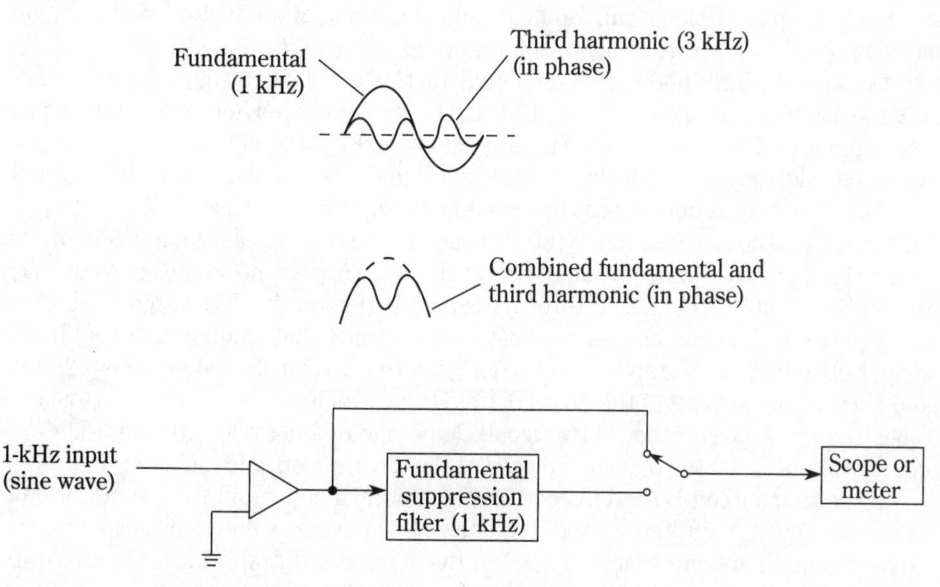

12-7 Basic harmonic-distortion analysis.

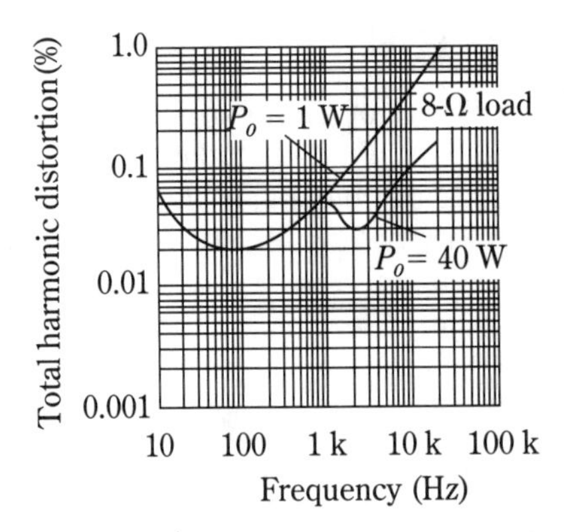

12-8 Typical total harmonic distortion (THD) curve.

On some commercial harmonic-distortion meters, the filter is tunable so that the amplifier can be tested over a wide range of fundamental frequencies. On other instruments, the filter is fixed in frequency, but they can be detuned slightly to produce a sharp null.

12.15 Intermodulation distortion

When two signals of different frequencies are mixed in an amplifier, it is possible that the lower-frequency signal will modulate the amplitude of the higher-frequency signal. This produces a form of distortion known as *intermodulation distortion.*

Commercial intermodulation-distortion meters consist of a signal generator and a high-pass filter, as shown in Fig. 12-9A. The generator portion of the meter produces a higher-frequency signal (usually about 7 kHz) that is modulated by a low-frequency signal (usually 60 Hz). The mixed signals are applied to the circuit input. The circuit output is connected through a high-pass filter to a scope. The high-pass filter removes the low-frequency (60 Hz) signal. The only signal that appears on the scope should be the 7-kHz signal. If any 60-Hz signal is present on the scope display, the 60-Hz signal is being paged through as modulation on the 7-kHz signal.

Figure 12-9B shows an intermodulation test circuit that can be fabricated in the shop. Notice that the high-pass filter is designed to pass signals that are about 200 Hz and above. The purpose of the 40- and 10-kΩ resistors is to set the 60-Hz signal at 4 times the amplitude of the 7-kHz signal. Some audio generators provide a line-frequency output, at 60 Hz, that can be used as the low-frequency modulation source.

If the shop circuit is used instead of a commercial meter, set the generator line-frequency (60 Hz) output to 2 V (if adjustable) or to some value that does not overdrive the amplifier being tested. Then, set the generator output (7 kHz) to 2 V (or to the same value as the 60-Hz output).

Calculate the percentage of intermodulation distortion using the equation shown in Fig. 12-9. For example, if the maximum output (shown on the scope) is 1 V and the minimum is 0.9 V, the percentage of intermodulation (listed as IM or IMD on most datasheets) is approximately:

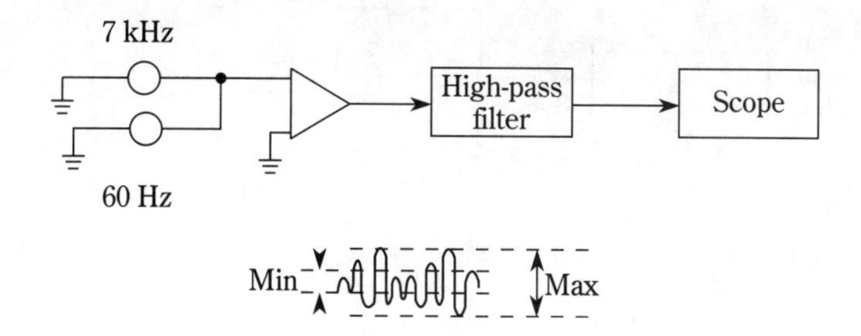

$$\% \textit{ intermodulation distortion} = 100 \times \frac{max - min}{max + min}$$

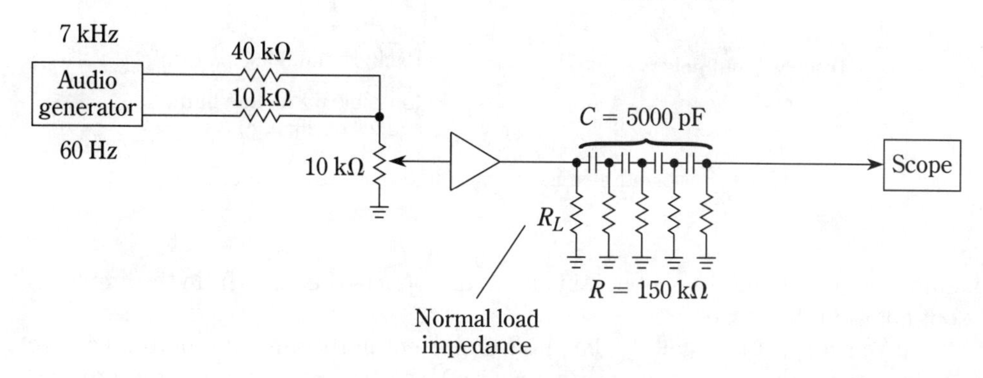

12-9 Basic intermodulation-distortion analysis.

$$\frac{1.0 - 0.9}{1.0 + 0.9} = 0.05 \times 100 = 5\%$$

Notice that a 5% IMD is quite high. Many home-entertainment audio amplifiers have IMDs of 0.09 or less.

12.16 Background noise

If the scope is sufficiently sensitive, it can be used to check and measure the background-noise level of an amplifier, as well as to check for the presence of hum, oscillation, and the like. The scope should be capable of a measurable deflection with an input below 1 mV (and considerably less if an IC amplifier is involved).

The basic procedure consists of measuring amplifier output with the volume or gains controls (if any) at maximum, but without an input signal. The scope is superior to a meter for noise-level measurements because the frequency and nature of the noise (or other signal) are displayed visually.

The basic connections for measuring the level of background noise are shown in Fig. 12-10. The scope gain is increased until there is a noise or "hash" indication. It is possible that a noise indication can be caused by pickup in the leads between the

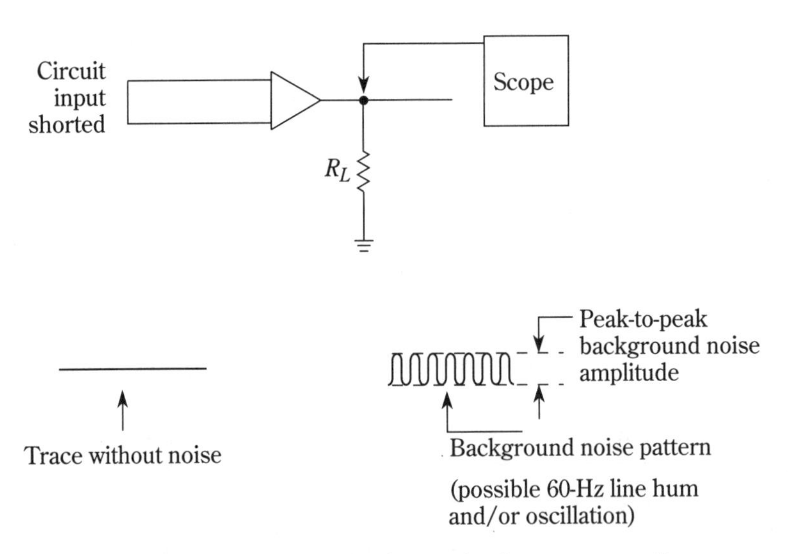

12-10 Basic audio-circuit background-noise test connections.

amplifier output and the scope. If in doubt, disconnect the leads from the amplifier, but not from the scope.

If you suspect that a 60-Hz line hum is present in the amplifier output (picked up from the power supply or any other source), set the scope sync control (or whatever other control is required to synchronize the scope trace to the line frequency) to line. If a stationary signal pattern appears, the signal is the result of line hum getting into the circuit.

If a signal appears that is not at the line frequency, the signal can be the result of oscillation in the amplifier or stray pickup. Short the amplifier input terminals. If the signal remains, suspect oscillation in the amplifier circuits.

With present-day IC amplifiers, the internal or background noise is considerably less than 1 mV, and it is impossible to measure, even with a sensitive scope. If it is necessary, use a circuit that amplifies the output of the IC under test before the output is applied to the scope. Figure 12-11 is such a circuit (and is used for noise tests of an OP-77 IC). The IC under test is connected for high gain, as is the following amplifier. This makes it possible to set a typical scope to the X1 position. It is also possible to monitor (and record) noise on a chart recorder (Fig. 12-12). Noise is usually measured over a 10-second interval, noting the peak-to-peak value.

12.17 Op-amp tests

The test procedures for the amplifiers described thus far can generally be applied to op-amp circuits. That is, the tests for frequency response, voltage gain, bandwidth, load sensitivity, input/output impedance, distortion, and background noise apply to op amps. However, there are some additional tests that apply specifically to op amps. These tests are summarized in the following paragraphs.

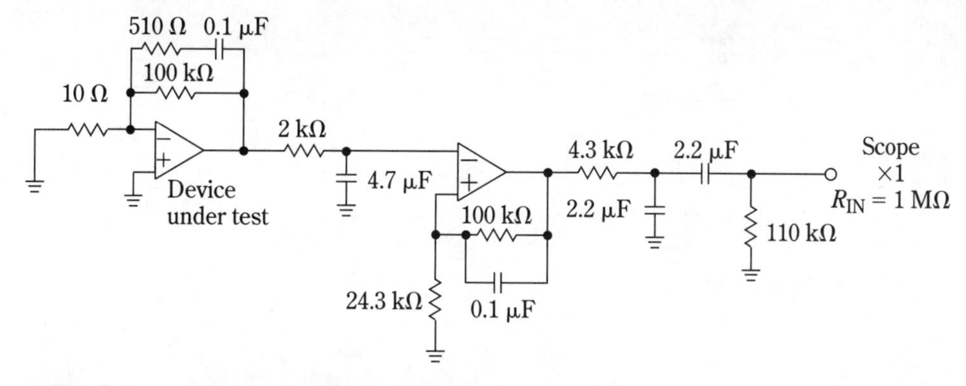

12-11 0.1- to 10-Hz noise test circuit.

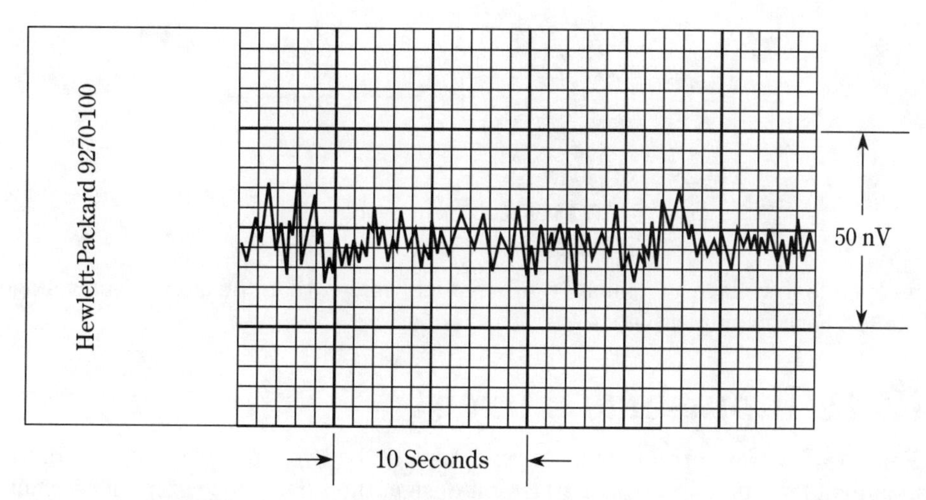

12-12 Typical amplifier noise display on a chart recorder.

12.18 Feedback measurement

Because op-amp circuits usually include feedback, it is sometimes necessary to measure feedback voltage at a given frequency with given operating conditions. The basic feedback measurement connections are shown in Fig. 12-13. Although it is possible to measure the feedback voltage, as shown in Fig. 12-13A, a more accurate measurement is made when the feedback lead is terminated in the normal operating impedance (Fig. 12-13B). If an input resistance is used in the circuit, and this resistance is considerably lower than the IC input impedance, use the resistance value. If in doubt, measure the input impedance of the IC, as described. Then, terminate the feedback lead in that value to measure open-loop feedback voltage. Remember that

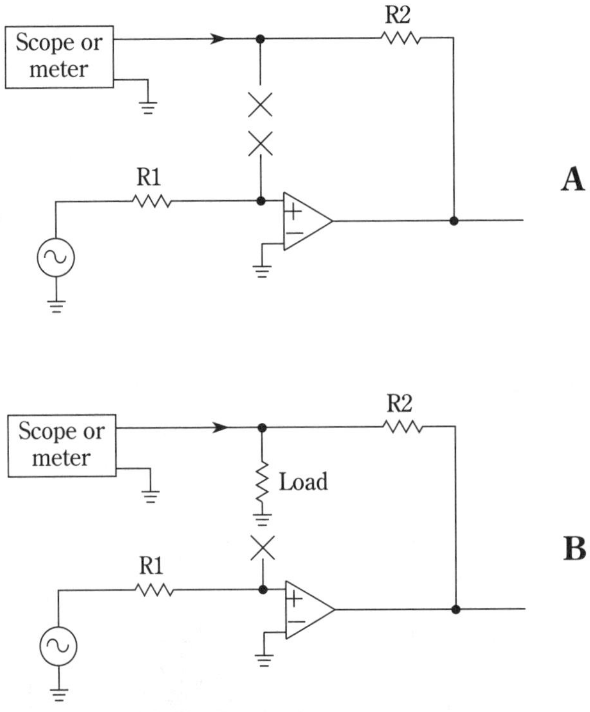

12-13 Feedback measurement.

the open-loop voltage gain must be substantially higher than the closed-loop voltage gain for most op-amp circuits to perform properly.

12.19 Input-bias current

Op-amp input-bias current is the average value of the two input-bias currents of the op-amp differential-input stage. In circuit design, the significance of input-bias current is that the resultant voltage drops across input resistors can restrict the input common-mode voltage range at higher impedance levels. The input-bias current produces a voltage drop across the input resistors. This voltage drop must be overcome by the input signal (which can be a problem if the input signal is low and the input resistors are large).

Input-bias current can be measured using the circuit of Fig. 12-14. Any resistance value for R1 and R2 can be used, provided that the value produces a measurable voltage drop, and that the resistance values are equal. A value of 1 kΩ with a tolerance of 1% or better, for both R1 and R2, is realistic for typical op amps.

If it is not practical to connect a meter in series with both inputs (as shown), measure the voltage drop across R1 and R2. Once the voltage drop is found, the input-bias current can be calculated. For example, if the voltage is 3 mV across 1 kΩ, the input bias current is 3 μA. Try switching R1 and R2 to see if any difference is the result of differences in resistor values.

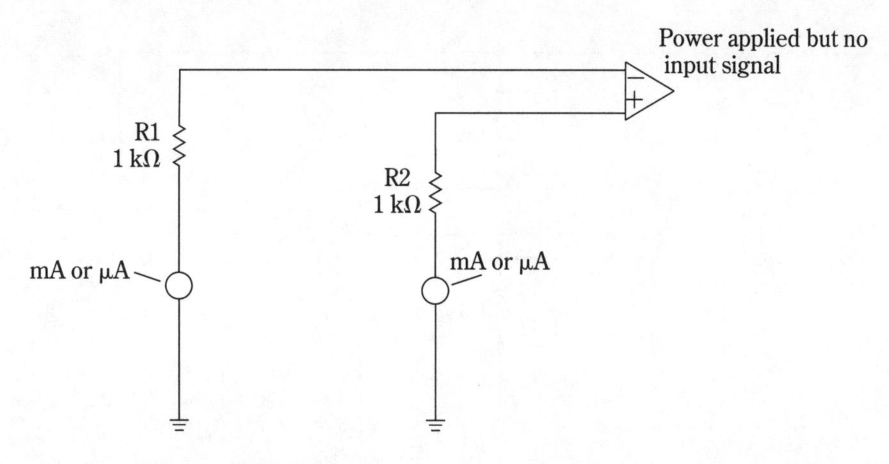

12-14 Input-bias current measurement.

In theory, the input-bias current should be the same for both inputs. In practice, the bias currents should be almost equal. Any great difference in input bias is the result of unbalance in the input differential amplifier of the IC, and can seriously affect circuit operation (and usually indicates a defective op-amp IC).

12.20 Input-offset voltage and current

Op-amp input-offset voltage is the voltage that must be applied at the input terminals to get zero output voltage, whereas input offset current is the difference in input-bias current at the op-amp input. Offset voltage and current are usually referred back to the input because the output values depend on feedback.

The effect of input-offset on op-amp circuits is that the input signal must overcome the offset before an output is produced. Likewise, with no input, there is a constant shift in the output level. For example, if an op amp has a 1-mV input-offset voltage and a 1-mV signal is applied, there is no output. If the signal is increased to 2 mV, the op amp produces only the peaks.

Input-offset voltage and current can be measured using the circuit of Fig. 12-15. As shown, the output is alternately measured with R3 shorted and with R3 in the circuit. The two output voltages are recorded as E1 (S1 closed, R3 shorted) and E2 (S1 open, R3 in the circuit).

With the two output voltages recorded, the input-offset voltage and input-offset current can be calculated using the equations of Fig. 12-15. For example, assume that $R_1 = 51\ \Omega$, $R_2 = 5.1\ \text{k}\Omega$, $R_3 = 1000\ \text{k}\Omega$, $E_1 = 83$ mV, and $E_2 = 363$ mV:

$$\textit{Input offset voltage} = \frac{83\ \text{mV}}{100} = 0.83\ \text{mV}$$

$$\textit{Input offset current} = \frac{280\ \text{mV}}{100\ \text{k}\Omega\ (1 + 100)} = 0.0277\ \mu\text{A}$$

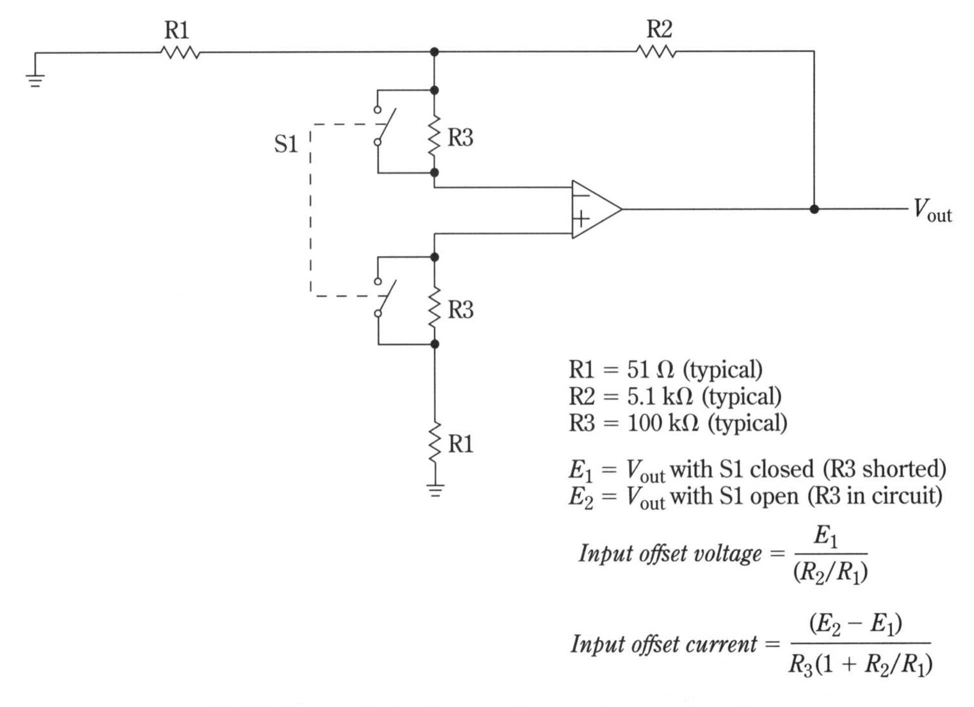

12-15 Input-offset voltage and current measurements.

12.21 Common-mode rejection

There are many definitions for *common-mode rejection, CMR* (also known as *CMRR, common-mode rejection ratio*). One definition is the ratio of differential gain (usually large) to common-mode gain (usually a fraction). That is, the amplifier might have a large gain of differential signals (different signals at each input terminal, or with one input terminal grounded and the opposite terminal with a signal), but little gain (or possibly a loss) of common-mode signals (same signal at both terminals). Another definition for CMR is the relationship of change in output voltage to change in input common-mode voltage producing the change, divided by the open-loop gain (amplifier gain without feedback).

No matter what definition is used, the first step to measure CMR is to find the open-loop (no feedback) voltage gain of the IC at the desired operating frequency (as described for amplifier voltage gain). Then, connect the IC in the common-mode circuit of Fig. 12-16. Increase common-mode voltage (at the same frequency used for open-loop gain test) V_{in} until a measurable output (V_{out}) is obtained. Be careful not to exceed the maximum specified input common-mode voltage swing. If no such value is shown, do not exceed the normal input voltage of the IC. Then, find CMR using the equation on Fig. 12-16.

To simplify calculation, increase the input voltage until the output is 1 mV. With an open-loop gain of 100, this provides an equivalent differential input signal of 0.00001 V. Then, measure the input voltage. Move the input-voltage decimal point over five places to find CMR.

V_{in}

V_{out} (1 mV)

$$\frac{V_{out}\text{ (1 mV)}}{\textit{open-loop gain}} = \textit{equivalent differential input signal}$$

$$\textit{Common-mode} = \frac{V_{in}}{\textit{equivalent differential input signal}}$$

12-16 Common-mode rejection measurements.

12.22 Slew rate

Op-amp slew rate is the maximum rate of change in output voltage, with respect to a time, that the op amp is capable of producing when maintaining linear characteristics (symmetrical output without clipping).

Slew rate is expressed in terms of difference in output voltage divided by difference in time, d_{V_O}/d_T. Usually, slew rate is listed in terms of volts per microsecond. For example, if the output voltage from an op amp is capable of changing 7 V in 1 μs, the slew rate is 7 (which might be listed as 7 V/μs). The major effect of slew rate in op-amp circuits (and in most audio amplifiers) is on power output. All other factors being equal, a lower slew rate results in lower power output.

A simple way to observe and measure op-amp slew rate is to measure the slope of the output waveform of a square-wave input signal (Fig. 12-17). The input square-

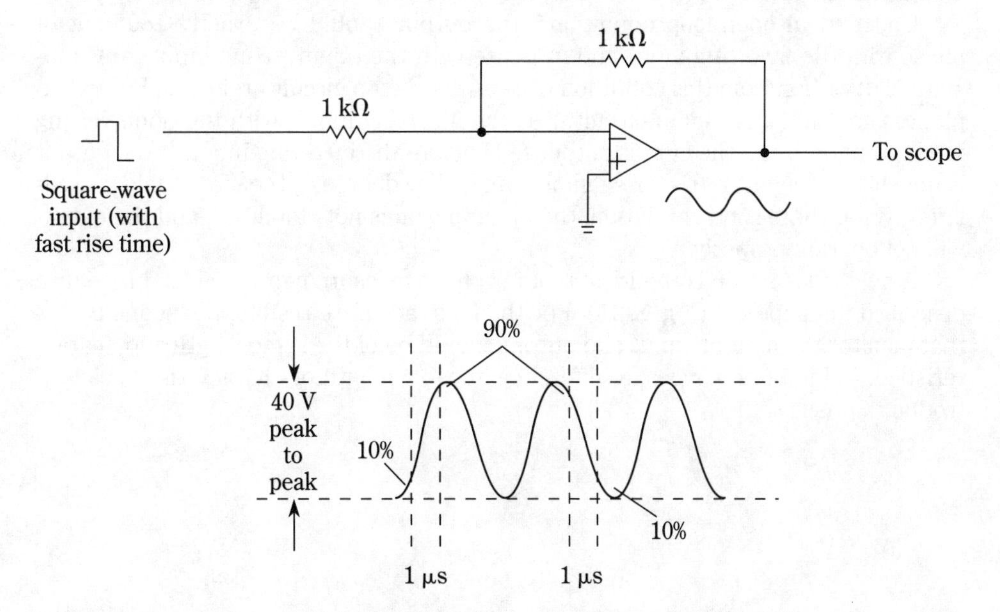

Example shows a slew rate of about 40 (40 V/μs) at unity gain

12-17 Slew-rate measurements.

wave must have a rise time that exceeds the slew-rate capability of the op amp. As a result, the output does not appear as a square wave, but as an integrated wave. In the example shown, the output voltage rises (and falls) about 40 V in 1 μs. Notice that slew rate is usually measured in the closed-loop condition. Also, the slew rate increases with higher gain.

12.23 Power-supply sensitivity

Op-amp power-supply sensitivity is the ratio of change in input-of-set voltage to the change in supply voltage that produces the change, with the remaining supply held constant. The term is expressed in millivolts or microvolts per volt (mV/V or μV/V, respectively), which represents the change of input-offset voltage (in mV or μV, respectively) to a change (in volts) of one power supply.

Power-supply sensitivity can be measured using the circuit of Fig. 12-15 (the same test circuit as for input-offset voltage). The procedure is the same as for measurement of input-offset voltage, except that one supply voltage is changed (in 1-V steps) while the other supply voltage is held constant. The amount of change in input-offset voltage for a 1-V change in one power supply is the power-supply sensitivity (or the input-offset voltage sensitivity, as it might be called).

12.24 Phase shift

The *phase shift* between input and output of an op-amp circuit is usually more critical than with the amplifiers described in the previous sections. This is because an op amp generally uses the principle of feeding basic output signals to the input.

Under ideal open-loop conditions, the output should be exactly 180° out of phase with the inverting input and in phase with the noninverting input. Any substantial deviation from this condition can cause op-amp circuit problems. For example, assume that an op-amp circuit uses the inverting input, with the noninverting input grounded, and the circuit output fed back to the inverting input. If the output is not shifted the full 180° (for example, only a few degrees), the circuit might oscillate. Even if there is no oscillation, the op-amp gain is not stabilized, and the circuit will not operate properly.

A dual-trace scope is the ideal tool for phase measurement. Use the procedure described in chapter 2 (Fig. 2-21). For the most accurate results, the scope cables that monitor the op-amp input and output should be of the same length and characteristics. At high frequencies, a difference in cable length or characteristics can introduce a phase shift.

13
Power-supply tests

This chapter is devoted entirely to test procedures for power-supply circuits and components. The first sections describe power-supply characteristics and test procedures from the practical standpoint. The information in these sections permits you to test all the important power-supply characteristics using basic shop equipment. The sections also help you understand the basis for such tests. The remaining sections of the chapter describe how the same tests, and additional tests, are performed using more sophisticated equipment.

13.1 Transformer tests

Most power supplies contain some form of transformer. The obvious test for transformers is to measure the windings for opens, shorts, and the proper resistance with an ohmmeter. In addition to these basic resistance checks, it is possible to test a transformer's polarity markings, regulation, impedance ratio, and center-tap balance with a voltmeter. The following paragraphs summarize these tests.

13.2 Measuring transformer phase relationships

When two supposedly identical transformers must be operated in parallel, and the transformers are not marked as to phase or polarity, the phase relationship of the transformers can be checked using a voltmeter and power source. The test circuit is shown in Fig. 13-1A. For power transformers, the source should be line voltage. Other transformers can be tested with a lower voltage dropped from a line source or (in extreme cases) from an audio generator.

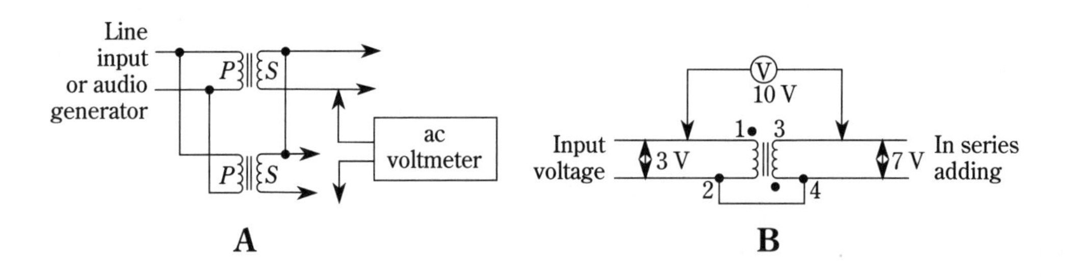

A **B**

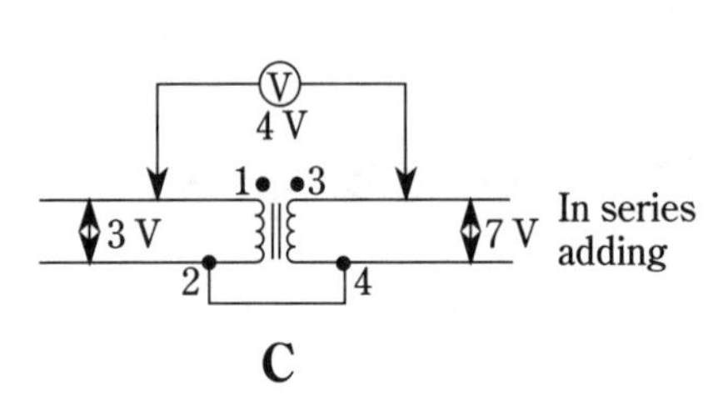

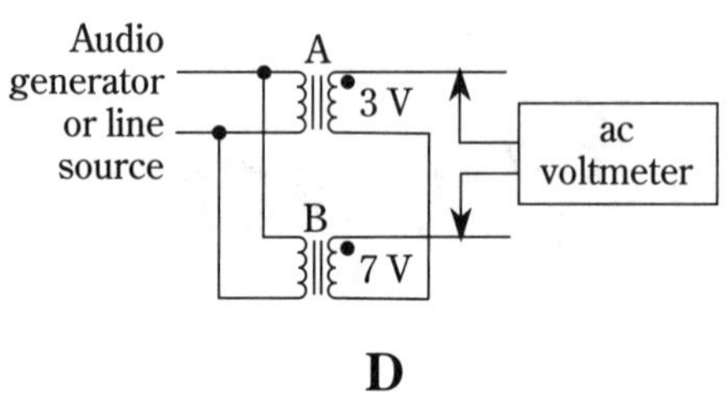

C **D**

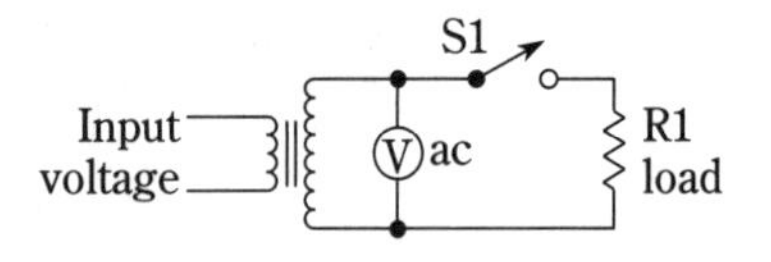

$$\% \ regulation = \frac{no \ load \ V - full\text{-}load \ V}{full\text{-}load \ V} \times 100$$

or

$$\% \ regulation = \frac{no \ load \ V - full\text{-}load \ V}{no\text{-}load \ V} \times 100$$

E

Input voltage V_1 V_2 V ac

Turns ratio $= V_1/V_2$ or V_2/V_1

Impedence ratio $= (V_1/V_2)^2$ or $(V_2/V_1)^2$

F

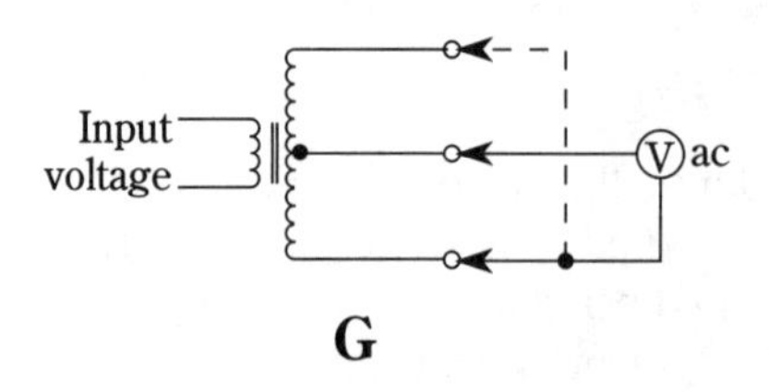

G

13-1 Transformer test circuits.

The transformers are connected in proper phase relationship if the meter reading is zero or very low. The transformers are out of phase if the secondary output voltage is double that of the normal secondary output. This condition can be corrected by reversing either the primary or secondary leads (but not both) of one transformer (but not both transformers).

If the meter indicates some secondary voltage, it is possible that one transformer has a greater output than the other. This condition results in considerable local current in the secondary winding and produces a power loss (if not actual damage to the transformer).

13.3 Checking transformer polarity markings

Many transformers are marked as to polarity or phase. These markings consist of dots, color-coded wires, or a similar system. Unfortunately, transformer polarity markings are not always standard. This can be very confusing during the experimental stage!

Generally, transformer polarities are indicated on schematics as dots next to the terminals. When standard markings are used, the dots mean that if electrons are flowing into the primary terminal with the dot, the electrons flow out of the secondary terminal with the dot. Therefore, the dots have the same polarity as far as the external circuits are concerned. No matter what system is used, the dots or other markings show relative phase because instantaneous polarities are changing across the transformer windings.

From a practical test standpoint, there are only two problems of concern: the relationship of primary to secondary, and the relationship of markings on one transformer to those on another.

The phase relationship of primary to secondary can be found using the test circuit of Fig. 13-1B and 13-1C. First, check the voltage across terminals 1 and 3, then across 1 and 4 (or 1 and 2). Assume that there is 3 V across the primary, with 7 V across the secondary. If the windings are as shown in Fig. 13-1B, the 3 V is added to the 7 V and appears as 10 V across terminals 1 and 3. If the windings are as shown in Fig. 13-1C, the voltages oppose each other, and appear as 4 V (7 – 3) across terminals 1 and 3.

The phase relationship of one transformer marking to another can be found using the test circuit of Fig. 13-1D. Assume that there is a 3-V output from the secondary of transformer A, and a 7-V output from transformer B. If the markings are consistent on both transformers, the two voltages oppose, and 4 V is indicated on the meter. If the markings are not consistent, the two voltages add, resulting in a 10-V reading.

13.4 Checking transformer regulation

All transformers have some regulating effect, even though not used in a power supply with a regulator. That is, the output voltage of the transformer tends to remain constant with changes in load. The preferred method of expressing regulation is:

$$\% \ regulation = \frac{(No\text{-}load\ voltage) - (Full\text{-}load\ voltage)}{Full\text{-}load\ voltage} \times 100$$

However, some manufacturers rate their transformers by:

$$\% \ regulation = \frac{(No\text{-}load\ voltage) - (Full\text{-}load\ voltage)}{No\text{-}load\ voltage} \times 100$$

Whichever factor is used, remember that all transformers do not provide good regulation, even though designed for power-supply use. Transformer regulation can be tested using the circuit of Fig. 13-1E. The value of R_1 (load) should be selected to draw maximum-rate current from the secondary. For example, if the transformer is to be used with a power supply designed to deliver 100 W, the secondary output voltage is 25 V; the value of R_1 should be such that 4 A flows, or $R_1 = 25/4 = 6.25\ \Omega$.

First, measure the secondary output voltage without a load, and then with a load. Use either equation to find the percentage of regulation.

13.5 Checking transformer impedance ratio

The impedance ratio of a transformer is the square of the winding ratio. For example, if the winding ratio of a transformer is 15:1, the impedance ratio is 225:1. Any impedance value placed across one winding is reflected onto the other winding by a value equal to the impedance ratio.

Impedance ratio is related to turns ratio (primary to secondary). However, turns-ratio information is not always available, so the ratio must be calculated using a test circuit, as shown in Fig. 13-1F. Measure both the primary and secondary voltage. Divide the larger voltage by the smaller, noting which is primary and which is secondary. For convenience, set either the primary or secondary to some exact voltage.

The turns ratio is equal to one voltage divided by the other. The impedance ratio is the square of the turns ratio.

For example, assume that the primary shows 115 V, with 23 V at the secondary. This indicates a 5:1 turns ratio and a 25:1 impedance ratio.

The reflected impedance ratio is determined by a combination of impedance ratio and the impedance connected at the primary or secondary. Assuming an impedance ratio of 225:1 and an 1800-Ω impedance placed on the primary, the secondary then has a reflected impedance of 8 Ω (1800/225 = 8). Similarly, if a 10-Ω impedance is placed on the secondary, the primary has a reflected impedance of 2250 Ω (225 × 10).

13.6 Checking transformer-winding balance

There is always some imbalance in center-taped transformers. That is, the turns ratio and impedance are not exactly the same on both sides of the center tap. A large imbalance can impair operation of any circuit, but especially a circuit, such as that used in switch-mode power supplies, dc-dc converters, etc.

It is possible to find a large imbalance by measuring the dc resistance on either side of the center tap. However, a small imbalance might not show up. It is more practical to measure the voltage on both sides of a center tap, as shown in Fig.

13-1G. If the voltages are equal, the transformer winding is balanced. If a large imbalance is indicated by a large voltage difference, the winding should be checked with an ohmmeter for shorted turns, poor design, and so on.

13.7 Basic power-supply tests

If a power supply delivers the full-rated output voltage into a full-rated load, the basic power supply function is met. It is also helpful to measure the regulating effect of a power supply, the power-supply internal resistance, and the amplitude of any ripple at the power-supply output. The following paragraphs summarize the procedures for such tests.

13.8 Power-supply output tests

Figure 13-2 shows the basic power-supply test circuit. This arrangement permits the power supply to be tested at no load, half load, and full load, depending on the position of S1. With S1 at position 1, no load is placed on the supply. At positions 2 and 3, there is half load and full load, respectively.

Using Ohms law, $R = E/I$, R1 and R2 are chosen on the basis of output voltage and load current (maximum or half load). For example, if the supply is designed for an output of 25 V at 500 mA (full load), the value of R_2 is 25/0.5 = 50 Ω. The value of R_1 is 25/0.25 = 100 Ω. Where more than one supply is to be tested, R1 and R2 should be variable, and adjusted to the correct value before testing (using an ohmmeter with the power removed).

The resistors must be noninductive (not wire wound) and must be capable of dissipating the rated power without overheating. For example, using the previous values for R1 and R2, the power dissipation of R1 is 25 × 0.5 or 12.5 W (use at least 15 W), and the dissipation for R2 is 25 × 0.25 or 6.25 W (use at least 10 W).

1. Connect the equipment, as shown in Fig. 13-2. Set R_1 and R_2 to the correct value.
2. Apply power. Set the input voltage to the correct value. Use the midrange value for the input voltage if a variable transformer or variac is available.

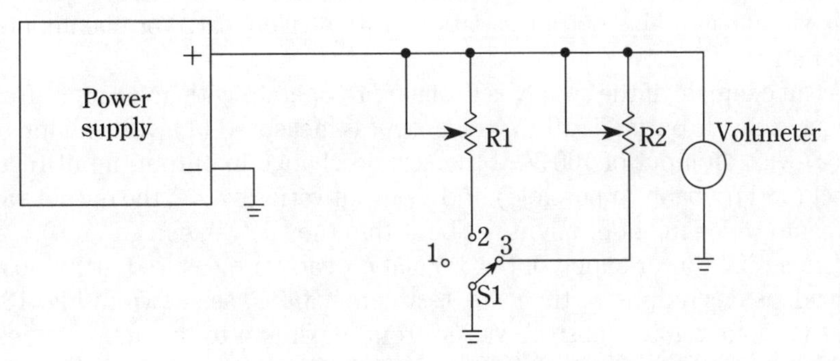

13-2 Basic power-supply test circuit.

3. Measure the output voltage at each position of S1.
4. Calculate the current at positions 2 and 3 of S1. For example, assume that R1 is 100 Ω, and the meter indicates 22 V at position 2 of S1. The actual load current is 22/100 = 220 mA. If the supply output is 25 V at position 1 (no load), and drops to 22 V at position 2, the supply is producing full output with a load (a half load in this case). This is an indication of poor regulation, possibly resulting from poor design (in the case of an experimental supply) or component failure.

13.9 Power-supply regulation tests

Power supplies can be checked for both output regulation and input regulation. Generally, output regulation is the most important in practical applications.

13.9.1 Output regulation

Power-supply regulation is usually expressed as a percentage, and is determined by the equation:

$$\%\ regulation = \frac{(No\text{-}load\ voltage) - (Full\text{-}load\ voltage)}{Full\text{-}load\ voltage} \times 1$$

A low percentage-of-regulation figure is desired because this indicates that the output voltage changes very little with load.

1. Connect the equipment, as shown in Fig. 13-1. Set R_2 to the correct value.
2. Apply power. Measure the output voltage at position 1 (no load) and position 3 (full load). Make certain that input voltage to the power supply is correct.
3. Using the equation, calculate the percentage of regulation. For example, if the no-load voltage is 26 V, and the full-load voltage is 25 V, the percentage of regulation is (26 – 25)/25, or 4 percent (very poor regulation). Notice that power-supply output regulation is usually poor (high percentage) when the internal resistance is high.

13.9.2 Input regulation

Power-supply input regulation is usually expressed as a percentage, and represents the maximum allowable output variation (with a given load) for maximum-rated input variation.

As an example, if the supply is designed to operate with an ac input from 110 to 120 V, and the dc output is 100 V, the output is measured (1) with an input of 120 V, and (2) with an input of 100 V. If there is no change in output, input regulation is perfect (and probably impossible). If the output varies by 1 V, the output variation is 1 percent (which might or might not be within the allowable maximum).

The actual power-supply input regulation can be measured at full load and/or half load, as desired, using the same test connections, as shown in Fig. 13-2. However, the input voltage must be varied from maximum to minimum values (with a variable transformer or variac) and monitored with an accurate voltmeter.

13.10 Power-supply internal-resistance tests

Power-supply internal resistance is determined by the equation:

$$\text{Internal resistance} = \frac{(\text{No-load voltage}) - (\text{Full-load voltage})}{\text{Current}} \times 100$$

A low internal resistance is most desirable because this indicates that the output voltage changes very little with load.

1. Connect the equipment, as shown in Fig. 13-2. Set R_2 to the correct value.
2. Apply power. Measure the actual output voltage at position 1 (no load) and at position 3 (full load).
3. Calculate the actual load current at position 3 (full load). For example, if R_2 is adjusted to 50 ohms, and the output voltage at position 3 is 25 V (as in the preceding example), the actual load current is 25/50 = 500 mA.

With no-load voltage, full-load voltage, and actual current established, find the internal resistance using the equation. For example, with no-load voltage at 26 V, a full-load voltage at 25 V, and a current of 500 mA (0.5 A), the internal resistance is (26 – 25)/0.5 = 2 Ω.

13.11 Power-supply ripple tests

Any power supply, no matter how well regulated or filtered, has some ripple. This ripple can be measured with a meter or scope. Usually, the factor of most concern is the ratio between ripple and full-output voltage. For example, if 3 V of ripple is measured together with a 100-V output, the ratio is 3 to 100 = 3%.

1. Connect the equipment, as shown in Fig. 13-2, if ripple is to be measured with a meter, or as shown in Fig. 13-3, if a scope is used to monitor ripple. The scope will show peak ripple, whereas most meters will show rms ripple (unless a peak-reading meter is used).
2. Set R_2 to the correct value. Set S1 to position 3 (full load). Ripple is usually measured under full-load power.
3. Apply power. Measure the dc output voltage at full load.
4. Set the meter to measure alternating current. Any voltage measured under these conditions is ac ripple.
5. Find the percentage of ripple, as a ratio between the two voltages (ac/dc).
6. One problem often overlooked in measuring ripple is that ripple voltage is not necessarily a pure sine wave. Most meters provide accurate ac voltage indications only for pure sine waves. A more satisfactory method of measuring ripple is with a scope (Fig. 13-3), where the peak value can be measured directly.
7. Adjust the scope controls to produce two or three stationary cycles of ripple on the screen. Notice that a full wave produces two ripple "humps" per cycle, whereas a half-wave supply produces one "hump" per cycle.

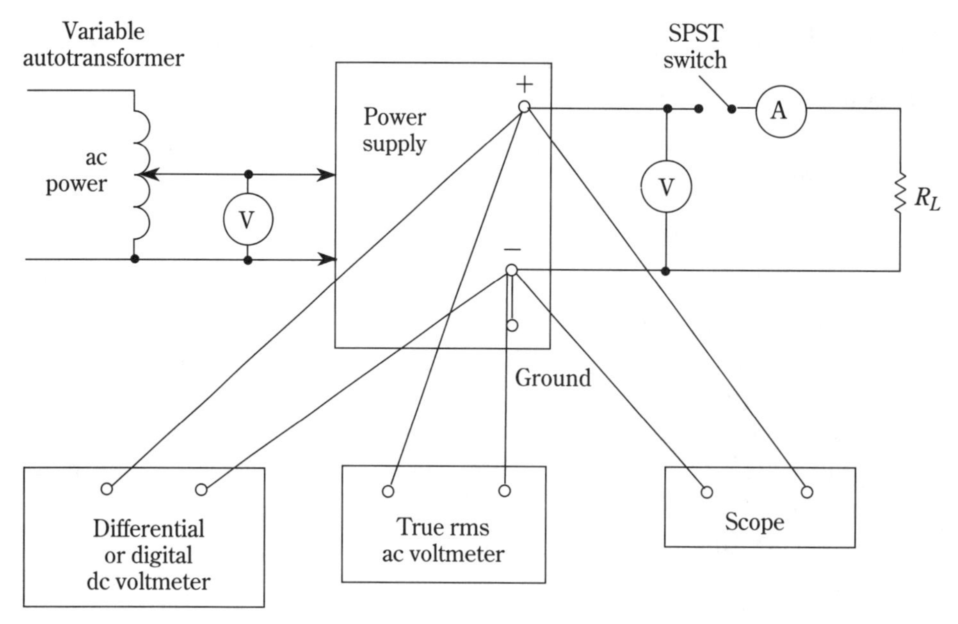

13-3 Test connections for measurement of source effect, load effect, PARD, drift, and temperature coefficient.

8. A study of the ripple waveform can sometimes show defects in an experimental circuit. Here are some examples. If the supply is unbalanced (one rectifier passes more current than the others), the ripple humps are unequal in amplitude. If there is noise or fluctuations in the supply (particularly in zener diodes), the ripple humps will vary in amplitude or shape. If the ripple varies in frequency, the ac source is varying, or (in switch-mode supplies) the switching frequency is varying. If a full-wave supply produces a half-wave output (one ripple hump per cycle), one rectifier is not passing current.

13.12 Advanced power-supply tests

The basic tests described in the previous sections are generally sufficient for most practical applications (usually for the experimenter and serious hobbyist). However, many other tests can be used to measure the performance of commercial and lab power supplies. The most common (and most important) of such tests are covered in the following paragraphs. Figure 13-3 shows test connections for the measurement of the five most important power-supply operating specifications: source effect, load effect, PARD, drift, and temperature coefficient.

Test equipment All of the tests described here can be performed with only four test instruments: a variable autotransformer, a differential or digital dc voltmeter, a true ac voltmeter, and a scope. Of course, on those supplies that are battery operated (typical switch-mode supply), the variable autotransformer can be omitted. When used, make certain that the autotransformer has an adequate current rating. If not, the input voltage applied to the supply might be severely distorted and rectifying/regulation circuits within the supply might operate improperly.

The dc voltmeter should have a resolution of 1 mV (or better) at voltages up to 1000 V. The scope should have a sensitivity of 100 μV/cm and a bandwidth of 10 MHz (although bandwidth is not critical for most supplies).

Proper connections For the most accurate measurements, the test connections should be permanent (not clip leads), and should be made on the exact point on the supply terminals. Clip-lead connections often produce measurement errors. Instead of measuring pure supply characteristics, you are measuring supply characteristics plus the resistance between output terminals and point of connection. Even using clip leads to connect the load to the supply terminals can produce a measurement error.

Ground-clip pickup Do not make any test measurements on a switch-mode supply with standard (alligator) ground-clip leads. Replace the alligator clip with a special soldered-in probe attenuator, which can be obtained from many probe manufacturers. The standard alligator ground clip lead can act as an "antenna," and can pick up magnetic and other radiated signals. Make the test described for ground loops if you suspect pickup by the scope probe.

Separate leads All measurement instruments must be connected directly by separate pairs of leads to the monitoring points, as shown in Fig. 13-3. This avoids the subtle mutual-coupling effects that can occur between measuring instruments (unless all are returned to the low-impedance terminals of the supply). Twisted pairs or shielded cable should be used to avoid pickup on the measuring leads.

Measure at the component Make all measurements (output voltage, ripple, etc.) at the component, not at a wire that is connected to a component. This is because wires are not shorts. For example, a switching regulator delivering square waves to an output capacitor can generate about 2-V per inch "spikes" in the lead inductance of the capacitor. The further you measure from the capacitor, the greater the spike voltage.

Ac voltmeter connections Connect the ac voltmeter as close as possible to the input ac terminals of the supply. The voltage indication is then a valid measurement of the supply input, without any error introduced by the drop present in the leads that connect the supply input to the ac line.

Scope probe compensation Always check that the scope probe is properly compensated when testing switch-mode supply/regulators. It is especially important for the ac attenuation (on a 10× probe, for example) to match the dc attenuation exactly. If not, low-frequency signals will be distorted and high-frequency signals will have the wrong amplitude. Remember that at typical switch-mode frequencies, the waveshape might look good because the probe appears purely capacitive, so the wrong amplitude might not be immediately obvious.

Load resistance Make certain that the load resistance is adequate for the supply and test requirements. Typically, the load resistance and load wattage should permit operation of the supply at maximum-rated output voltage and current.

Current limit If the supply has a current limit or adjustment control, set the control well above the maximum output current of the supply. In many supply circuits, the initial regulating action can cause a drop in output voltage, increased ripple, and other performance changes that could make a good supply appear bad.

Pickup and ground-loop effects Always check test-connection setups for possible pickup and/or ground-loop problems. As a simple test, turn off the supply and observe the scope for any unwanted signals (particularly at the line frequency) with the scope leads connected directly on the supply output terminals. Then, connect both scope leads to either terminal (+ or –), whichever is grounded to the chassis or to the common ground. If there is any noise in either test condition, with the supply off, you have possible pickup and/or ground-loop effects.

Figure 13-4 shows a typical ground-loop condition. A generator is driving a 5-V signal into 50 Ω on a power-supply regulator, which results in 100-mA current. The return path for this current divides between the ground from the generator (typically, the shield on a BNC cable) and the secondary ground "loop" that is created by the scope probe ground clip (shield), and the two "third-wire" connections on the generator and scope. In this case, assume that 20 mA flows in the parasitic ground loop. If the scope ground lead has a resistance of 0.2 Ω, the scope will show a 4-mV "bogus" signal. The problem gets much worse for higher currents and for fast-signal edges, where the inductance of the scope probe shield is important. The most practical solution is to use an isolation transformer for the scope.

Again, as a quick check, touch the scope probe tip to the probe ground clip, with the clip connected to the circuit or component ground. The scope should show a flatline. Any signal displayed on the scope is a ground-loop problem or a pickup problem.

Line regulator Do not use any form of line regulator when testing a supply and when using the supply (unless specifically recommended for a particular supply). This is especially true for switch-mode supplies and regulators. A line regulator can change the shape of the output waveform in a switching supply/regulator, thus offsetting any improvement produced by a constant line input to the supply.

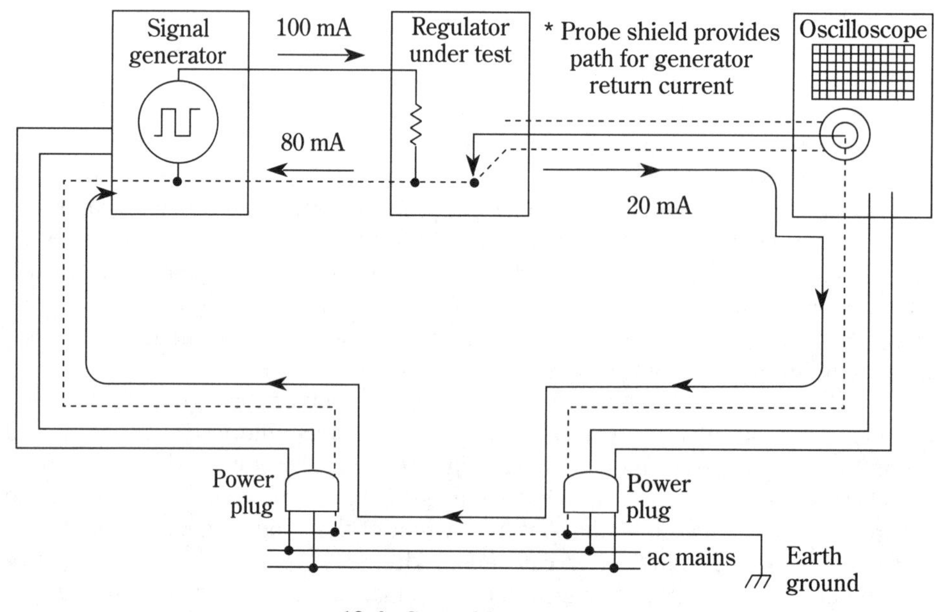

13-4 Ground-loop errors.

13.13 Source effect or line regulation

No matter what the test is called, the measurement is made by turning the variable autotransformer (Fig. 13-3) throughout the specified range from low-line to high-line, and noting the change in voltage at the supply output terminals. The test is performed with all other test conditions constant. The supply should stay within specifications for any rated output voltage, combined with any rated output current. The extreme source-effect test is with maximum output voltage and maximum output current.

13.14 Load effect or load regulation

No matter what the test is called, the measurement is made by closing and opening switch S1 (Fig. 13-3), and noting the resulting static change in output voltage. The test is performed with all other test conditions constant. The supply should stay within specifications for any rated output voltage, combined with any rated input line voltage. The extreme load-effect test is with maximum output voltage and maximum output current.

13.15 Noise and ripple (or PARD)

In many cases, *PARD (periodic and random deviation)* has replaced the terms noise and ripple, and represents deviation of the dc output voltage from the average value, over a specified bandwidth, with all other test conditions constant. For example, with Hewlett-Packard lab supplies, PARD is measured in rms and/or peak-to-peak values over a 20-Hz to 20-MHz bandwidth. Fluctuations below 20 Hz are considered to be drift. Peak-to-peak measurements are of particular importance for applications where noise spikes can be detrimental (such as in digital logic circuits). The rms measurement is not ideal for noise because output noise spikes of short duration can be present in ripple, but not appreciably increase the rms value. Always use twisted-pair leads (for single-ended scopes) or shielded two-wire leads (for differential scopes) when making PARD or noise/ripple tests.

13.16 Drift (stability)

Drift measurements are made by monitoring the supply output on a differential or digital voltmeter over a stated measurement interval (typically 8 hours, after a 30-minute warmup). In some cases, a strip chart is used to provide a permanent record. A thermometer is placed near the supply to verify that the ambient temperature remains constant during the period of measurement. The supply should be at a location that is immune from stray air currents (away from open doors or windows and from air-conditioning vents). If practical, place the supply in an oven and hold the temperature constant. A well-regulated supply will drift less during the 8-hour period than during the 30-minute warmup.

13.17 Temperature coefficient

Temperature-coefficient measurements are made by placing the supply in an oven and varying the temperature over a given range, following a 30-minute warm-up. The supply is allowed to stabilize at each measurement temperature. In the absence of other specifications, the temperature coefficient is the output-voltage change that results from a 5°C change in temperature. The measuring instrument should be placed outside the oven, and must have a long-term stability that is adequate to ensure that any voltmeter drift does not affect measurement accuracy.

14
Radio-frequency tests

This chapter is devoted entirely to test procedures for radio-frequency (RF) circuits and components (communications-equipment circuits are covered further in chapter 15). The first sections of this chapter describe RF characteristics and test procedures from the practical standpoint. The information in these sections permits you to test all the important RF characteristics using basic shop equipment. The sections also help you to understand the basis for such tests. The remaining sections of the chapter describe how the same tests, and additional tests, are performed using more sophisticated equipment.

For the experimenter or hobbyist, the tests described in this chapter should be made when the circuit is first completed in experimental form. If the test results are not as desired, the component values can be changed as necessary to get the desired results. Also, RF circuits should always be retested in final form (with all components soldered in place). This shows if there is any change in circuit characteristics because of the physical relocation of components.

Although this procedure might seem unnecessary, it is especially important at higher radio frequencies. Often, there is capacitance or inductance between components, from components to wiring, and between wires. These stray "components" can add to the reactance and impedance of circuit components. When the physical location of parts and wiring is changed, the stray reactances change and alter circuit performance.

14.1 Basic RF-voltage measurements

As covered in chapter 5, when voltages to be measured are at radio frequencies and are beyond the frequency capabilities of the meter circuits or scope amplifiers, an RF probe is required. Such probes rectify the RF signals into a dc output, which is almost equal to the peak RF voltage. The dc output of the probe is then applied to the meter or scope and is displayed as a voltage readout in the normal manner.

If a probe is available as an accessory for a particular meter, that probe should be used in favor of any experimental or homemade probe. The manufacturer's probe is matched to the meter in calibration, frequency compensation, and so on. If a matching probe is not available for a particular meter or scope, probes can be made as described in chapter 5 (Fig. 5-1).

14.2 Resonant circuits

Both RF amplifiers and RF oscillators use resonant circuits (or tank circuits) that consist of a capacitor and coil (inductance) connected in series or parallel, as shown in Fig. 14-1. Such resonant circuits are used to tune the RF amplifier/oscillator network. At the resonant frequency, the inductive and capacitive reactances are equal, and the circuit acts as a high impedance (in a parallel circuit) or a low impedance (in a series circuit). In either case, any combination of capacitance and inductance has some resonant frequency.

Either (or both) the capacitance or inductance can be variable to permit tuning of the resonant circuit over a given frequency range. When the inductance is variable, tuning is usually done by a metal (powdered-iron) slug or coke inside the coil. The slug is screwdriver-adjusted to change the inductance (and thus the inductive reactance), as required. Typical RF circuits include two resonant circuits in the form of a transformer (RF or IF transformer and the like). Again, either the capacitance or inductance can be variable.

14.2.1 Resonant frequency versus *Q* or selectivity

All resonant circuits have a resonant frequency and a *Q* factor. The circuit *Q* depends on the ratio of reactance to resistance. If a resonant circuit has pure reactance, the *Q* would be high (infinite). However, this is not practical because any coil has some resistance, as do the leads of a capacitor.

The resonant-circuit *Q* depends on the individual *Q* factors of inductance and capacitance used in the circuit. For example, if both the inductance and capacitance have a high *Q*, the circuit has a high *Q*, provided that a minimum of resistance is produced when the inductance and capacitance are connected to form a resonant circuit.

Figure 14-1 has equations that show the relationships among capacitance, inductance, reactance, and frequency, as these factors relate to resonant circuits. Note that there are two sets of equations. One set includes reactance (inductive and capacitive); the other omits reactance. The reason for the two sets of equations is that some design approaches require the reactance to be calculated for resonant networks.

From a practical standpoint, a resonant circuit with a high *Q* produces a sharp resonance curve (narrow bandwidth), whereas a low *Q* produces a broad resonance curve (wide bandwidth). For example, a high-*Q* resonant circuit provides good harmonic rejection and efficiency in comparison with a low-*Q* circuit, all other factors being equal.

The selectivity of a resonant circuit is related directly to *Q*. A very high *Q* (or high selectivity) is not always desired. Sometimes, it is necessary to add resistance to a resonant circuit to broaden the response (increase the bandwidth, decrease selectivity).

Typically, resonant-circuit *Q* is measured at the point on either side of the resonant frequency, where the signal amplitude is down 0.707 of the peak resonant

Resonance and impedance

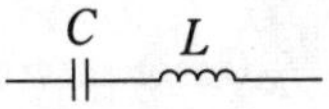

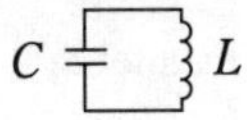

Series (zero impedance) Parallel (infinite impedance)

$$F\,(\text{MHz}) = \frac{0.159}{\sqrt{L\,(\mu\text{H}) \times C\,(\mu\text{F})}} \qquad L\,(\mu\text{H}) = \frac{2.54 \times 10^4}{F\,(\text{kHz})^2 \times C\,(\mu\text{F})}$$

A

$$C\,(\mu\text{F}) = \frac{2.54 \times 10^4}{F\,(\text{kHz})^2 \times L\,(\mu\text{F})}$$

Capacitive reactance

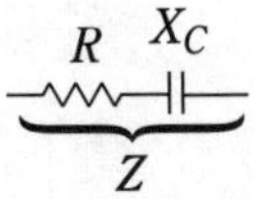

Series:

$$Z = \sqrt{R^2 + X_C{}^2} \qquad Q = \frac{X_C}{R} \qquad C = \frac{1}{6.28F\,X_C} \qquad F = \frac{1}{6.28C\,X_C}$$

Parallel:

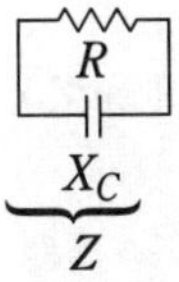

$$Z = \frac{RX_C}{\sqrt{R2 + X_C{}^2}} \qquad Q = \frac{R}{X_C} \qquad X_C = \frac{159}{F\,(\text{kHz}) \times C\,(\mu\text{F})}$$

B

Inductive reactance

Series:

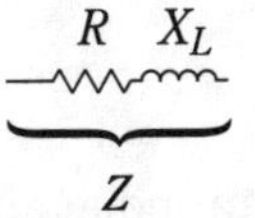

$$Z = \sqrt{R^2 + X_L{}^2} \qquad Q = \frac{X_L}{R} \qquad L = \frac{X_L}{6.28F} \qquad F = \frac{X_L}{6.28L}$$

C

Parallel:

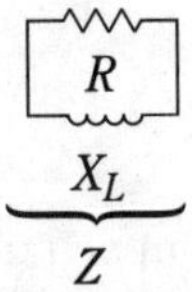

$$Z = \frac{RX_L}{\sqrt{R_2 + X_L{}^2}} \qquad Q = \frac{R}{X_L} \qquad X_L = 6.28 \times F(\text{MHz}) \times L\,(\mu\text{F})$$

14-1 Resonant-circuit equations.

value, as shown in Fig. 14-2A. Resonant-circuit Q measurements are covered in later paragraphs of this chapter.

Notice that Q must be increased for increases in resonant frequency if the same bandwidth is to be maintained. For example, if the resonant frequency is 10 MHz, with a bandwidth of 2 MHz, the required circuit Q is 5. If the resonant frequency is increased to 50 MHz, with the same 2-MHz bandwidth, the required Q is 25.

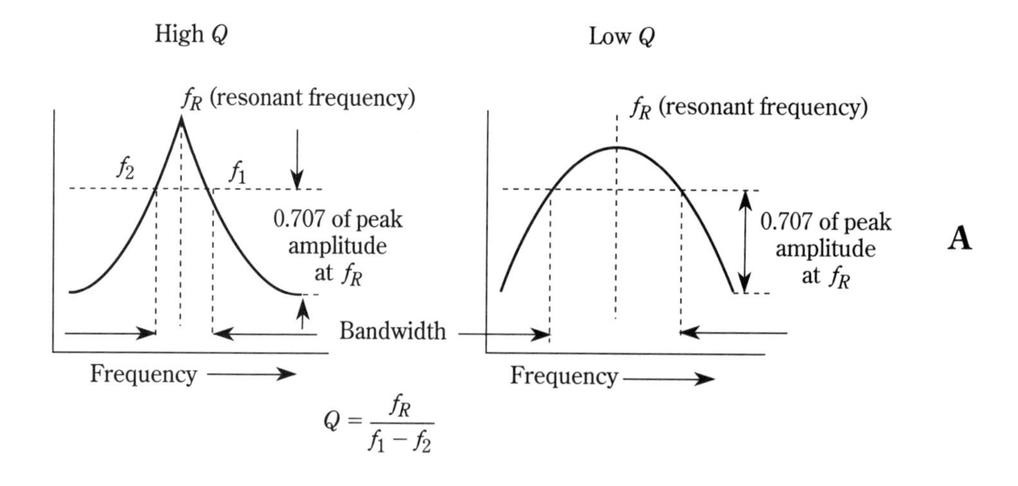

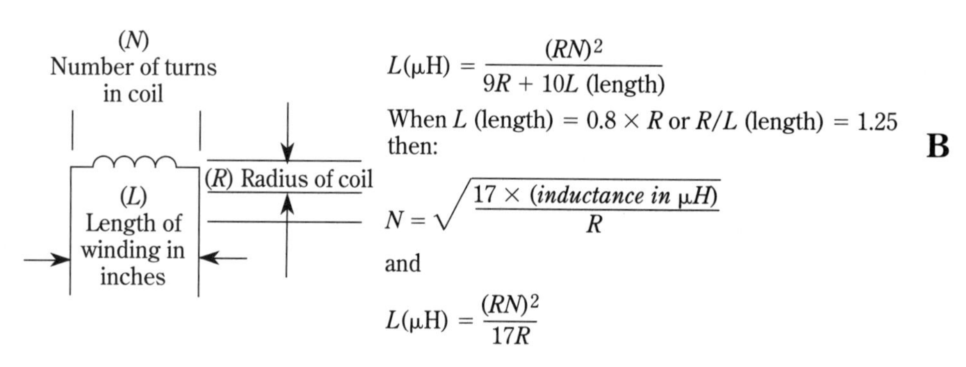

14-2 Q and inductance calculations for RF coils.

14.2.2 Calculating resonant values

Assume that you want a circuit that resonates at 400 kHz with an inductance of 10 μH. What value of capacitor is necessary? Using the equations in Fig. 14-1:

$$C = \frac{2.54 \times 10^4}{400^2 \times 10} = 0.0158 \ \mu F$$

Use the nearest standard value of 0.016 μF.

Assume that you want a circuit that resonates at 2.65 MHz with a capacitance of 360 pF. What value of inductance is necessary? Using the equations in Fig. 14-1:

$$L = \frac{2.54 \times 10^4}{2650^2 \times (360 \times 10^{-6})} = 10 \ \mu H$$

Assume that you must find the resonant frequency of a 0.002-μF capacitor and a 0.02-mH inductance. Using the equations in Fig. 14-1, first convert 0.02 mH to 20 μH, then:

$$f = \frac{0.159}{\sqrt{20 \ \mu H \times 0.002 \ \mu F}} = \frac{0.159}{\sqrt{0.04}} = \frac{0.159}{0.2} = 0.795 \text{ MHz or } 795 \text{ kHz}$$

Assume that an RF circuit must operate at 40 MHz with a bandwidth of 8 MHz. What circuit Q is required? Using the equations in Fig. 14-2A:

$$F_R = 40 \quad F_1 - F_2 = 8 \quad Q = \frac{40}{8} = 5$$

14.2.3 RF coils

Figure 14-2B shows the equations necessary to calculate the self-inductance of a single-layer, air-core coil (the most common type of coil used in RF circuits). Maximum inductance is obtained when the ratio of coil-radius to coil-length is 1.25 (when the length is 0.8 of the radius). RF coils wound for this ratio are the most efficient (maximum inductance for minimum physical size).

Assume that you must design an RF coil with 0.5-µH inductance on a 0.25-in. radius (air core, single layer). Using the equations in Fig. 14-1B, for maximum efficiency, the coil length must be 0.8*r*, or 0.2 in. Then:

$$N = \sqrt{\frac{17 \times 0.5}{0.25}} = \sqrt{34} = 5.8 \text{ turns}$$

For practical purposes, use six turns and spread the turns slightly. The additional part of a turn increases the inductance, but the spreading decreases inductance. After the coil is made, the inductance should be checked with an inductance bridge (chapter 6).

14.3 Basic resonant-frequency measurements

A meter can be used with an RF generator to find the resonant frequency of either series or parallel LC circuits. The following steps describe the measurement procedure:

1. Connect the equipment, as shown in Fig. 14-3. Use the connections in Fig. 14-3A for parallel-resonant LC circuits or the connections in Fig. 14-3B for series-resonant LC circuits.
2. Adjust the generator output until a convenient midscale indication is obtained on the meter. Use an unmodulated signal output from the generator.
3. Starting at a frequency well below the lowest possible frequency of the circuit under test, slowly increase the generator output frequency. If there is no way to judge the approximate resonant frequency, use the lowest generator frequency.
4. If the circuit being tested is parallel-resonant, watch the meter for a maximum, or peak, indication.
5. If the circuit being tested is series-resonant, watch the meter for a minimum, or dip, indication.
6. The resonant frequency of the circuit under test is the one at which there is a maximum (or parallel) or minimum (for series) indication on the meter.

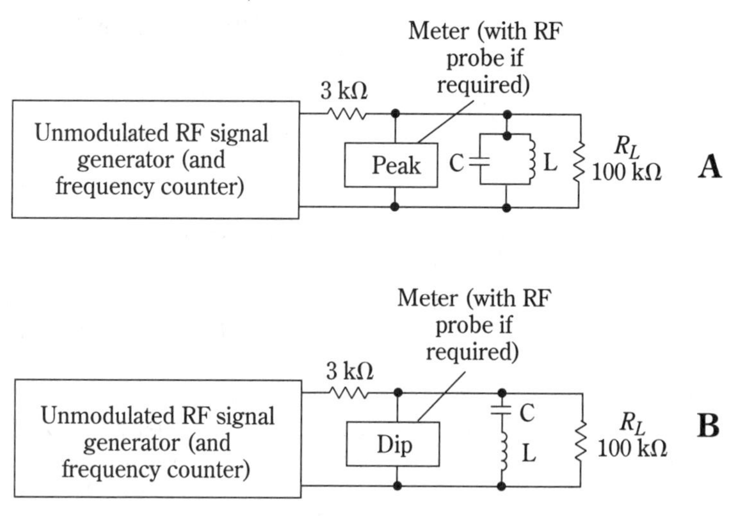

14-3 Basic RF voltage measurement.

7. There might be peak or dip indications at harmonics of the resonant frequency. Therefore, the test is most efficient when the approximate resonant frequency is known.
8. The value of load resistor R_L is not critical. The load is shunted across the LC circuit to flatten, or broaden, the resonant response (to lower the circuit Q), causing the voltage maximum or minimum to be approached more slowly. A suitable trial value for R_L is 100 kΩ. A lower value of R_L sharpens the resonant response, and a higher value flattens the curve.

14.4 Basic coil inductance measurements

A meter can be used with an RF generator and a fixed capacitor (of known value and accuracy) to find the inductance of a coil. The following steps describe the measurement procedure:

1. Connect the equipment as shown in Fig. 14-4. Use a capacitive value such as 10 µF, 100 pF, or some other even number to simplify the calculation.
2. Adjust the generator output until a convenient midscale indication is obtained on the meter. Use an unmodulated signal output from the generator.
3. Starting at a frequency well below the lowest possible resonant frequency of the inductance-capacitance combination under test, slowly increase the generator frequency. If there is no way to judge the approximate resonant frequency, use the lowest generator frequency.
4. Watch the meter for a maximum, or peak, indication. Note the frequency at which the peak indication occurs. This is the resonant frequency of the circuit.
5. Using this resonant frequency, and the known capacitance value, calculate the unknown inductance using the equation of Fig. 14-4.
6. The procedure can be reversed to find an unknown capacitance value, when a known inductance value is available.

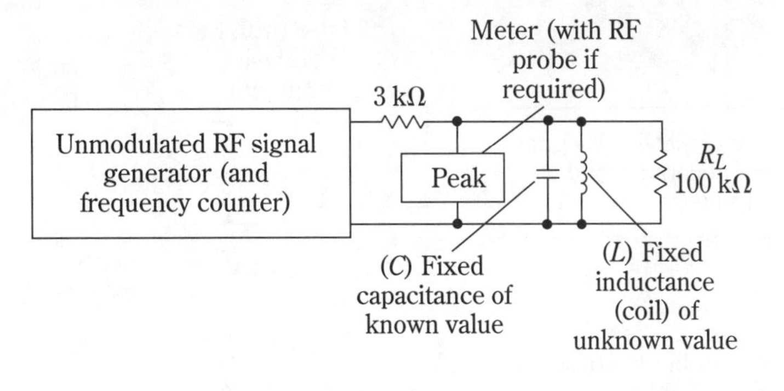

$$(C)\ (\text{in } \mu\text{F}) = \frac{2.54 \times 10^4}{f(\text{kHz})^2 \times L(\mu\text{H})} \qquad (L)\ (\text{in } \mu\text{H}) = \frac{2.54 \times 10^4}{f(\text{kHz})^2 \times C(\mu\text{F})}$$

14-4 Basic coil inductance measurements.

14.5 Basic coil self-resonance and distributed-capacitance measurements

No matter what design or winding method is used, there is some distributed capacitance in any coil. When the distributed capacitance combines with the coil inductance, a resonant circuit is formed. The resonant frequency is usually quite high in relation to the frequency at which the coil is used. However, because self-resonance might be at or near a harmonic of the frequency to be used, the self-resonant effect can limit the usefulness of the coil in LC circuits. Some coils, particularly RF choices, might have more than one self-resonant frequency.

A meter can be used with an RF generator to find both the self-resonant frequency and the distributed capacitance of a coil. The following steps describe the measurement procedure:

1. Connect the equipment, as shown in Fig. 14-5.
2. Adjust the generator output amplitude until a convenient midscale indication is obtained on the meter. Use an unmodulated signal output from the generator.
3. Tune the generator over the entire frequency range, starting at the lowest frequency. Watch the meter for either peak or dip indications. Either a peak or a dip indicates that the inductance is at a self-resonant point. The generator output frequency at that point is the self-resonant frequency. Make certain that peak or dip indications are not the result of change in generator output level. Even the best lab generators might not produce a flat (constant level) output over the entire frequency range.
4. Because there might be more than one self-resonant point, tune through the entire generator range. Try to cover a frequency range up to at least the third harmonic of the highest frequency involved in a resonant-circuit design.

Once the resonant frequency is found, calculate the distributed capacitance using the equation of Fig. 14-5. For example, assume that a coil with an inductance of

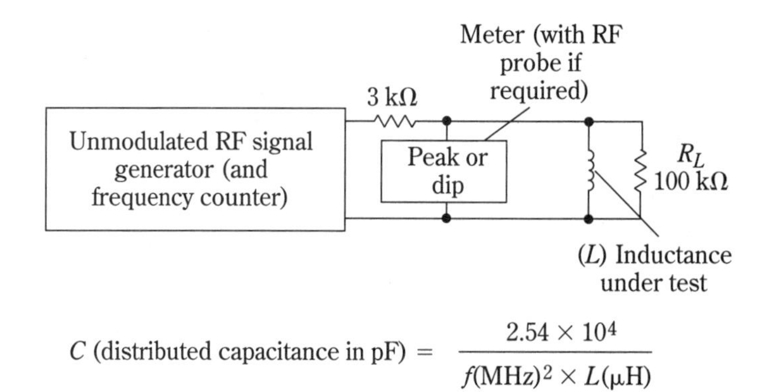

14-5 Basic coil self-resonance and distributed-capacitance measurements.

7 μH is found to be self-resonant at 50 MHz:C (distributed capacitance) (2.54×10^4 $50^2 \times 7$) = 1.45 pF.

14.6 Basic resonant-circuit *Q* measurements

The *Q* of a resonant circuit can be measured using a generator and a meter with an RF probe. A high-impedance digital meter generally provides the least loading effect on the circuit and thus provides the most accurate indication.

Figure 14-6A shows the test circuit in which the generator is connected directly to the input of a complete stage. Figure 14-6B shows the indirect method of connecting the signal generator to the input.

When the stage, or circuit, has sufficient gain to provide a good reading on the meter with a nominal output from the generator, the indirect method (with isolating resistor) is preferred. Any generator has some output impedance (typically 50 Ω). When this resistance is connected directly to the tuned circuit, the *Q* is lowered, and the response becomes broader (in some cases, the generator output seriously detunes the circuit).

Figure 14-6C shows the test circuit for a single component (such as an IF transformer). The value of the isolating resistance is not critical and is typically in the range of 100 kΩ. The procedure for determining *Q* using any of the circuits in Fig. 14-6 is as follows:

1. Connect the equipment, as shown in Fig. 14-6. Note that a load is shown in Fig. 14-6C. When a circuit is normally used with a load, the most realistic *Q* measurement is made with the circuit terminated in that load value. A fixed resistance can be used to simulate the load. The *Q* of a resonant circuit often depends on the load value.
2. Tune the generator to the circuit resonant frequency. Operate the generator to produce an unmodulated output.

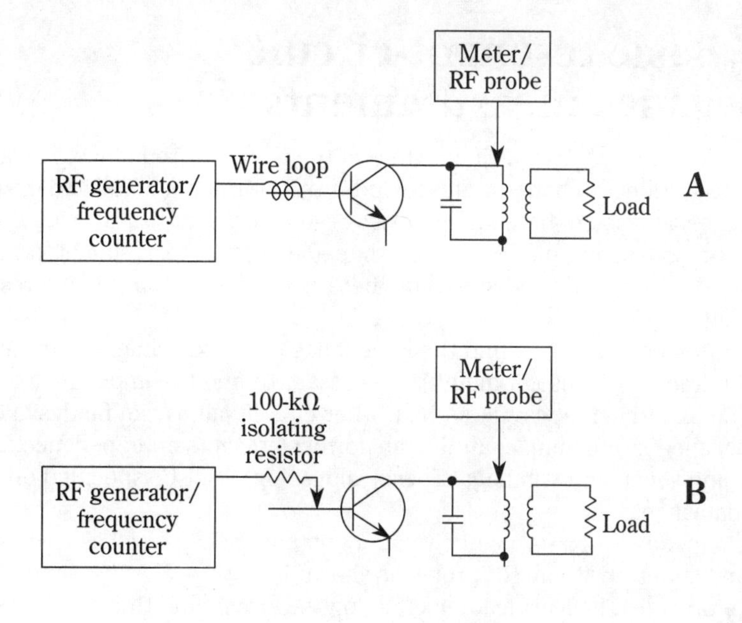

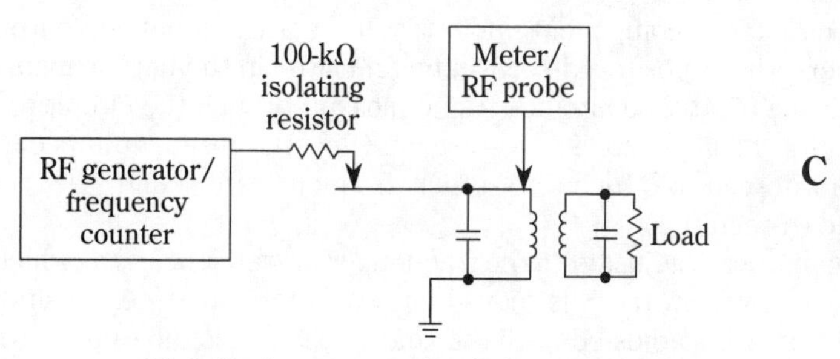

14-6 Basic resonant-circuit Q measurements.

3. Tune the generator frequency for maximum reading on the meter. Note the generator frequency.
4. Tune the generator below resonance until the meter reading is 0.707 times the maximum reading. Note the generator frequency. To make the calculation more convenient, adjust the generator output level so that the meter reading is some even value, such as 1 or 10 V, after the generator is tuned for maximum. This makes it easy to find the 0.707 mark.
5. Tune the generator above resonance until the meter reading is 0.707 times the maximum reading. Note the generator frequency.
6. Calculate the resonant-circuit Q using the equation in Fig. 14-2. For example, assume that the maximum meter indication occurs at 455 kHz (f_R), the below-resonance indication is 453 (f_2), and the above-resonance indication is 458 kHz (f_1). Then, $Q = 455/(458 - 455) = 91$.

14.7 Basic resonant-circuit impedance measurements

Any resonant circuit has some impedance at the resonant frequency. The impedance changes as frequency changes. This includes transformers (tuned and untuned), RF tank circuits, and so on. In theory, a series-resonant circuit has zero impedance, and a parallel-resonant circuit has infinite impedance, at the resonant frequency. In practical RF circuits, this is impossible because there is always some resistance in the circuit.

It is often convenient to find the impedance of an experimental resonant circuit at a given frequency. Also, it might be necessary to find the impedance of a component in an experimental circuit so that other circuit values can be designed around the impedance. For example, an IF transformer presents an impedance at both the primary and secondary windings. These values might not be specified on the transformer datasheet.

The impedance of a resonant circuit or component can be measured using a generator and a meter with an RF probe, as shown in Fig. 14-7. A high-impedance digital meter provides the least loading effect on the circuit and thus produces the most accurate indication.

If the circuit of a component under test has both an input and output (such as a transformer), the opposite side or winding must be in the normal load, as shown. If the impedance of a tuned circuit is to be measured, tune the circuit to peak or dip, then measure the impedance at resonance. Once the resonant impedance is found, the generator can be tuned to other frequencies to find the corresponding impedance (if required).

The generator is adjusted to the frequency (or frequencies) at which impedance is to be measured. Switch S is moved back and forth between positions A and B, while resistance *R* is adjusted until the voltage reading is the same in both positions of the switch. Resistor R is then disconnected from the circuit, and the dc resistance of R is measured with an ohmmeter. The dc resistance of R is then equal to the impedance at the circuit output.

Accuracy of the impedance measurement depends on the accuracy with which the dc resistance is measured. A noninductive resistance must be used. The impedance found by this method applies only to the frequency used during the test.

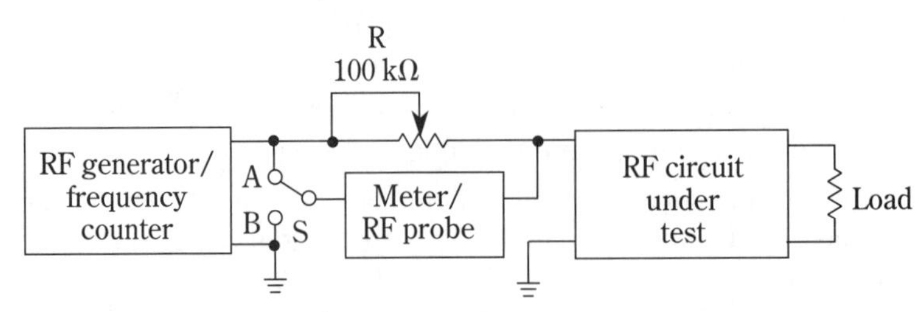

14-7 Basic resonant-circuit impedance measurements.

14.8 Basic transmitter RF-circuit testing

It is possible to test and adjust transmitter RF circuits using a meter and RF probe. If an RF probe is not available (or as an alternative), it is possible to use a circuit such as shown in Fig. 14-8A. This circuit is essentially a pickup coil (which is placed near the RF-circuit inductance) and a rectifier that converts the RF into a dc voltage or measurement on a meter. The basic procedures are:

1. Connect the equipment, as shown in Fig. 14-8B. If the circuit being measured is an RF amplifier, without an oscillator, a drive signal must be supplied by means of a generator. Use an unmodulated signal at the correct operating frequency.
2. In turn, connect the meter (through an RF probe or the special circuit in Fig. 14-8A) to each stage of the RF circuit. Start with the first stage (this is usually the oscillator if the circuit under test is a complete transmitter) and work toward the final (or output) stage.
3. A voltage indication should be obtained at each stage. Usually, the voltage indication increases each RF-amplifier stage as you proceed from oscillator to the final amplifier. However, some stages might be frequency multipliers and provide no voltage amplification.

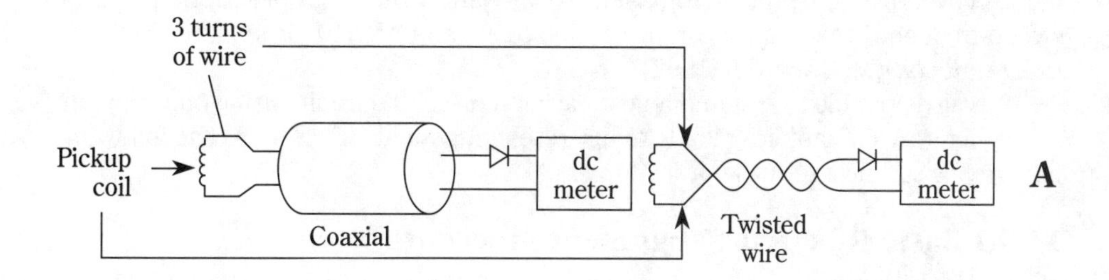

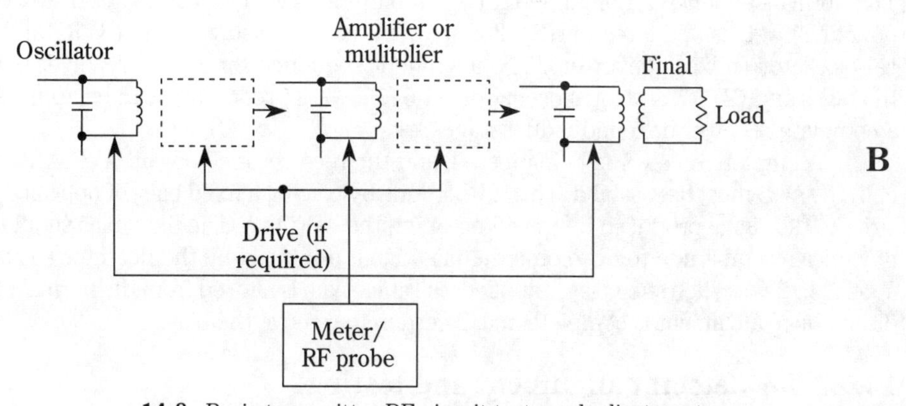

14-8 Basic transmitter RF-circuit tests and adjustments.

4. If a particular stage is to be tuned, adjust the tuning control for a maximum reading on the meter. If the stage is to be operated with a load (such as the final amplifier into an antenna), the load should be connected or a simulated load should be used. A fixed, noninductive resistance provides a good simulated load at frequencies up to about 250 MHz.
5. This tuning method or measurement technique does not guarantee that each stage is at the desired operating frequency. It is possible to get maximum readings on harmonics. However, it is conventional to design transmitter RF circuits so that the circuits cannot tune to both the desired operating frequency and a harmonic. Generally, RF-amplifier tank circuits tune on either side of the desired frequency, but not to a harmonic (unless the circuit is seriously detuned or the design calculations are hopelessly inaccurate).

14.9 Basic receiver RF-circuit testing and adjustment

The RF circuits of most present-day TV sets and tuners use some form of *frequency synthesis (FS)*. The same is true of many modern radio receivers, such as those used in communications and amateur radio. The procedures for testing and adjusting frequency-synthesis RF circuits are covered in chapter 3 (Figs. 3-2 and 3-3).

The RF circuits can also be tested and adjusted using the sweep-frequency techniques covered in chapter 3 (Fig 3-1). If you want a thorough description of how sweep-frequency techniques can be applied to TV and AM/FM circuits, read *Lenk's RF Handbook* (McGraw-Hill, 1992).

It is also possible to test and adjust basic receiver RF circuits using only a meter and generator. This applies mostly to discrete-component RF circuits. The following paragraphs summarize these procedures.

14.9.1 Basic RF-circuit alignment procedure

Both AM and FM receivers require alignment of the RF and IF stages. An FM receiver also requires alignment of the detector stage (discriminator or ratio detector). The normal sequence for alignment of a complete FM receiver (and a discrete-component TV set) is (1) detector, (2) IF amplifier and limiter stages, and (3) RF and local-oscillator (mixer/converter). The alignment sequence for an AM receiver is (1) IF stages and (2) RF and local oscillator. The following procedures can be applied to a complete receiver or to individual stages.

If a complete receiver is being tested, and the receiver includes an AVC-AGC circuit, the ACC must be disabled. This is best done by placing a fixed bias, of opposite polarity to the signal produced by the detector, on the AGC line. The fixed bias should be of sufficient amplitude to overcome the bias signal produced by the detector (usually about 1 or 2 V). When such bias is applied, the stage gain is altered from the normal condition. Once alignment is complete, make certain to remove the bias.

14.9.2 FM-detector alignment and testing

1. Connect the equipment, as shown in Fig. 14-9A (for a discriminator) or Fig. 14-9B (for a ratio detector).

2. Set the meter to measure dc voltage.
3. Adjust the generator frequency to the intermediate frequency (usually 10.7 MHz for a broadcast FM receiver). Use an unmodulated output from the generator.
4. Adjust the secondary winding (either capacitor or tuning slug) of the discriminator transformer for zero reading on the meter. Adjust the transformer slightly each way and make sure the meter moves smoothly above and below the exact zero mark. A meter at zero-center scale is most helpful when adjusting FM detectors of the type shown in Fig. 14-9.
5. Adjust the generator to some point below the intermediate frequency (to 10.625 MHz for an FM detector with a 10.7-MHz IF). Note the meter reading. If the meter reading goes downscale against the pin, reverse the meter polarity or test leads (the PF probe is not used for FM-detector alignment).
6. Adjust the generator to some point above the intermediate frequency, exactly equivalent to the amount set below the IF in step 5. For example, if the generator is set to 0.075 MHz below the IF (10.7 – 0.075 = 10.625), set the generator to 10.775 (10.7 + 0.075 = 10.775).
7. The meter should read approximately the same in both steps 5 and 6, except that the polarity is reversed. For example, if the meter reads 7 scale divisions below 0 for step 5 and 7 scale divisions above 0 for step 6, the detector is balanced. If a detector circuit under test cannot be balanced under these conditions, the fault is usually a serious mismatch of diodes or other components.
8. Return the generator output to the intermediate frequency (10.7 MHz).

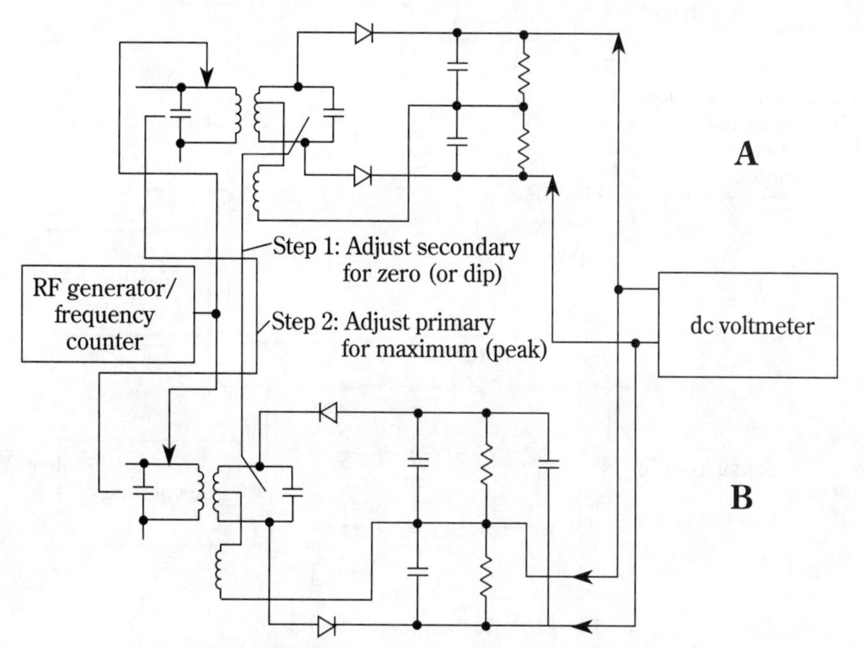

14-9 Basic FM-detector alignment and tests.

9. Adjust the primary winding of the detector transformer (either capacitor or tuning slug) for a maximum reading on the meter. This sets the primary winding at the correct resonant frequency of the IF amplifier.
10. Repeat steps 4 through 8 to make sure that adjustment of the transformer primary has not disturbed the secondary setting (the two settings usually interact).

14.9.3 AM and FM alignment and testing

The alignment procedures for the IF stages of an AM receiver are essentially the same as those for an FM receiver. However, the meter must be connected at different points in the corresponding detector, as shown in Fig. 14-10. In either case, the meter is set to measure direct current, and the RF probe is not used. In those cases where IF stages are being tested without a detector (such as during design), an RF probe is required. As shown in Fig. 14-10, the RF probe is connected to the secondary of the final IF output transformer.

1. Connect the equipment, as shown in Fig. 14-10.
2. Set the meter to measure direct current at the appropriate test point (with or without an RF probe as applicable).
3. Adjust the generator frequency to the receiver intermediate frequency (typically 10.7 MHz for FM and 455 kHz for AM). Use an unmodulated output from the signal generator.

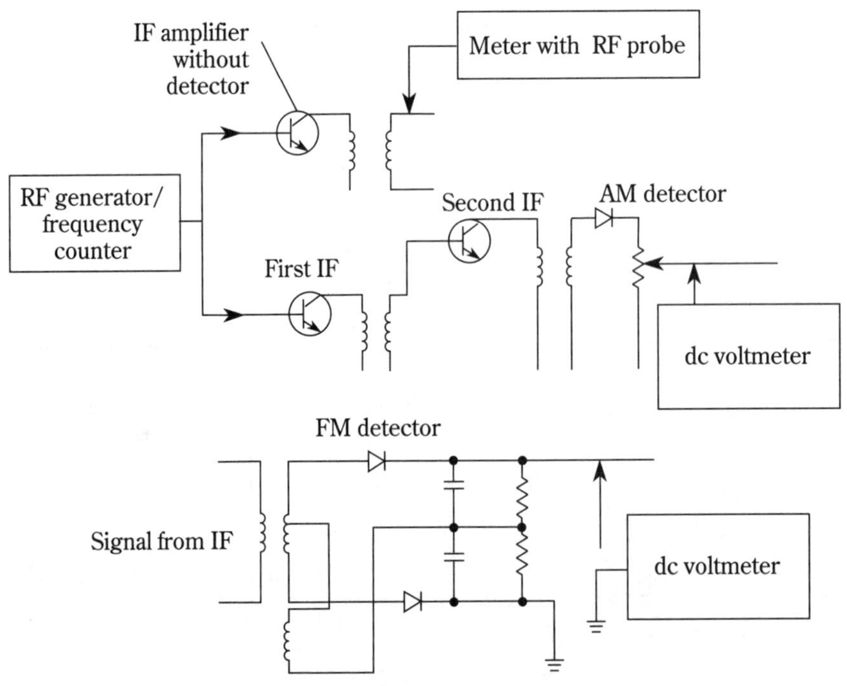

14-10 AM/FM IF alignment and tests.

4. Adjust the windings of the IF transformer (capacitor or tuning slug) in turn, starting with the last stage and working toward the first stage. Adjust each winding for maximum reading.
5. Repeat the procedure to make sure the tuning on one transformer has no effect on the remaining adjustments. Adjust each winding for maximum reading.

14.9.4 AM and FM RF alignment and testing

The alignment procedures for the RF stages (RF amplifier, local oscillator, mixer and converter) of an AM receiver are essentially the same as for an FM receiver. Again, it is a matter of connecting the meter to the appropriate test point. The same test points used for IF alignment can be used for aligning the RF stages, as shown in Fig. 14-11. However, if an individual RF stage is to be aligned, the meter must be connected to the secondary winding of the RF-stage output transformer, through an RF probe. The procedure is as follows:

1. Connect the equipment, as shown in Fig. 14-11.
2. Set the meter to measure direct current at the appropriate test point (with or without an RF probe, as applicable).
3. Adjust the generator frequency to some point near the high end of the receiver operating frequency (typically 107 MHz for an FM broadcast receiver and 1400 kHz for an AM broadcast receiver). Use an unmodulated output from the generator.
4. Adjust the RF-stage trimmer for reading on the meter.
5. Adjust the generator frequency to the low end of the receiver operating frequency (typically, 90 MHz for FM and 600 kHz for AM).

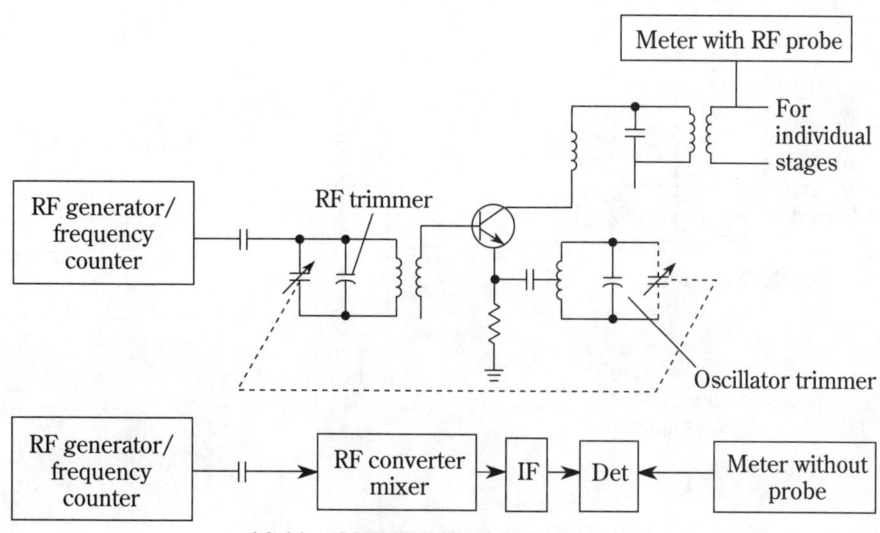

14-11 AM/FM RF alignment and tests.

6. Adjust the oscillator-stage trimmer for maximum reading on the meter.
7. Repeat the procedure to make sure that the resonant circuits "track" across the entire tuning range.

14.10 RF oscillator tests

The first step in testing an RF oscillator is to measure both the amplitude and frequency of the output signal. Many oscillators have a built-in test point. If yours does not, the signal can be monitored at the collector or emitter, as shown in Fig. 14-12A. Signal amplitude is monitored with a meter or scope using an RF probe. The simplest way to measure oscillator signal frequency is with a frequency counter.

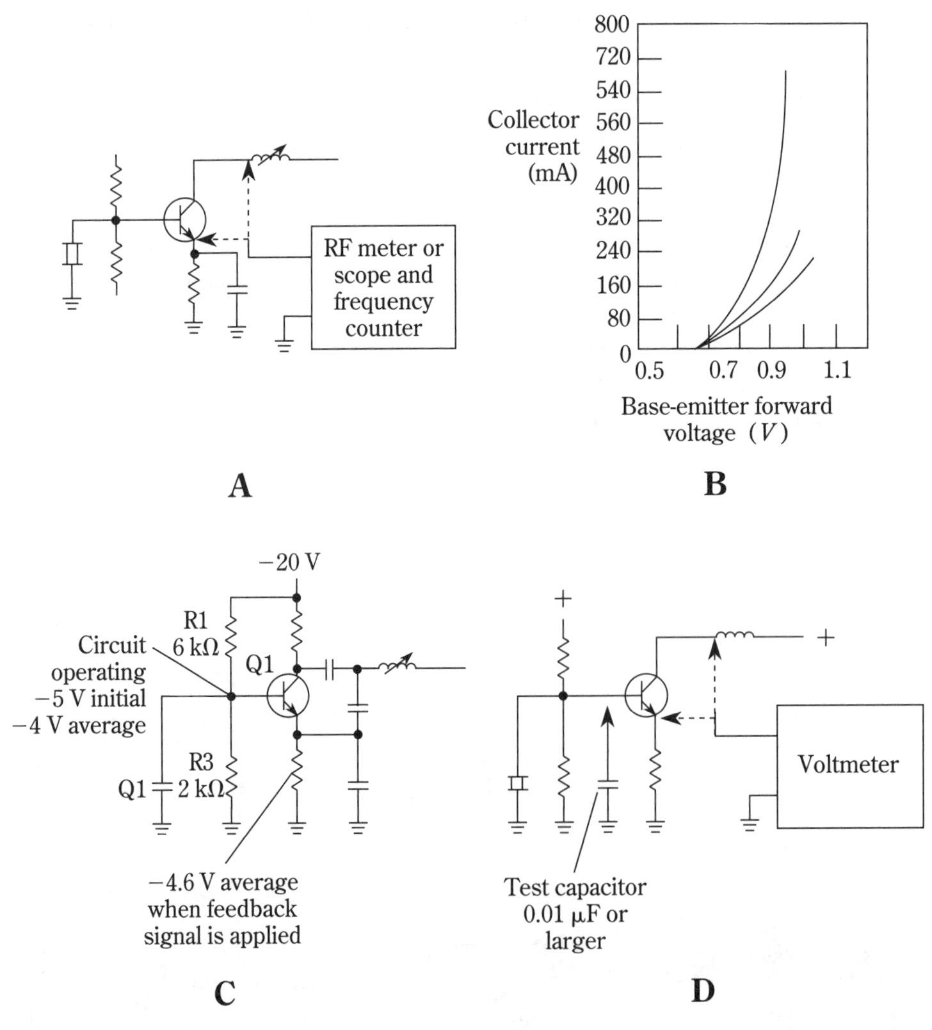

14-12 Oscillator testing and troubleshooting.

14.10.1 Oscillator frequency checks

When you check the oscillator signal, the frequency is (1) right on, (2) slightly off, or (3) way off. If the frequency is slightly off, it is possible to correct the problem with adjustment. Most oscillators are adjustable—even those with crystal control. Usually, the RF coil or transformer is slug-tuned. On many PLL ICs, there is an external adjustable capacitor for each crystal. The most precise adjustment is obtained by monitoring the oscillator signal with a frequency counter and adjusting the circuit for exact frequency. However it is also possible to adjust an oscillator using a meter or scope.

When the circuit is adjusted for maximum signal strength, the oscillator is at the crystal frequency. However, it is possible (but not likely) that the oscillator is being tuned for a harmonic (multiple or submultiple) of the crystal frequency. A frequency counter shows this, whereas a meter or scope does not.

If the oscillator frequency is way off, look for a defect, rather than improper adjustment. For example, a coil or transformer might have shorted turns, a transistor or capacitor might be leaking badly, or the wrong crystal might be installed in the right socket (this does happen).

14.10.2 Oscillator signal-amplitude checks

When you measure the oscillator signal, the amplitude is (1) right on, (2) slightly low, or (3) very low. If the amplitude is right on, leave the oscillator alone. If the amplitude is slightly low, it is possible to correct the problem with adjustment. Monitor the signal with a meter or scope, and adjust the oscillator for maximum signal amplitude. This also locks the oscillator on the correct frequency. If the adjustment does not correct the problem, look for leakage in the transistor or for a transistor with low gain.

If the amplitude is very low, look for defects (such as low power-supply voltages, badly leaking transistors and/or capacitors, and a shorted coil or transformer turns). Usually, when signal output is very low, there are other indications, such as abnormal voltage and resistance values.

14.10.3 Oscillator quick-test

It is possible to check if an oscillator is oscillating using a voltmeter and a high-value capacitor (typically 0.01 μF or larger). Measure either the collector or emitter voltage with the oscillator operating normally, then connect the capacitor from base to ground, as shown in Fig. 14-12D. This should stop oscillation, and the emitter or collector voltage should change. When the capacitor is removed, the voltage should return to normal.

If there is no change when the capacitor is connected, the oscillator is probably not oscillating. In some oscillators, you get better results by connecting the capacitor from collector to ground. Also, do not expect the voltage to change on an element without a load. For example, if the collector is connected directly to the power-supply voltage, or through a few turns of wire, as shown in Fig. 14-12D, this voltage does not change substantially, with or without oscillation.

15
Communications equipment tests

This chapter is devoted entirely to test procedures for communications equipment. The first sections of this chapter describe communications test equipment and procedures from the practical standpoint. The information in these sections permits you to test all of the important communications-equipment characteristics from the practical standpoint using basic shop equipment. The sections also help you to understand the basis for such tests. The remaining sections of the chapter describe how the same tests, and additional tests, are performed using more sophisticated equipment.

15.1 Communications-equipment test instruments

Most communications-equipment tests can be performed with the meters, scopes, generators, counters, and probes described in chapters 1 through 5. However, there are some specialized test instruments that greatly simplify communications test and service, such as dummy loads, RF wattmeters, field-strength meters, SWR meters, and dip adapters. Also, there are specialized versions of basic test equipment that have been specifically developed to simplify communications testing, such as spectrum analyzers and FM deviation meters. All such equipment is covered in the following sections.

15.2 RF dummy load for transmitters

Never adjust a communications transmitter (or any RF circuit that produces signals in excess of about 1 W) with a load connected to the output. This will almost certainly cause damage to the final RF circuit. For example, when a transmitter is connected to an antenna or load, power is transferred from the final RF stage of the transmitter to the antenna or load. Without an antenna or load, the final RF stage must dissipate the full power and will probably be damaged (even with heatsinks). Equally important, you should not make any major adjustments (except for a brief final tune-up) to a transmitter that is connected to a radiating antenna. You will probably cause interference.

These two problems can be overcome by a nonradiating load, commonly called a *dummy load*. There are a number of commercial dummy loads available. The RF wattmeters described in later paragraphs contain dummy loads. It is also possible to make dummy loads suitable for most communications testing.

There are two generally accepted dummy loads: the *fixed resistance* and the *lamp*. These loads are intended for routine tests, but they are not a substitute for the RF wattmeter (or any special test set designed for use with a particular type of communications equipment).

15.2.1 Fixed-resistor dummy load

The simplest dummy load is a *fixed resistor* capable of dissipating the full power output of the RF circuit. The resistor can be connected to the output of the RF circuit (say at the antenna connector of a transmitter) with a plug, as shown in Fig. 15-1A.

Most communications transmitters operate with a 50-ohm antenna and lead-in and this requires a 50-ohm resistor. The nearest standard resistor is 51 ohms. The 1-ohm difference is not critical. However, it is essential that the resistor be noninductive (composition or carbon), never wire wound. Wire-wound resistors have some inductance, which changes with frequency, causing the load impedance to change. Always use a resistor with a power rating higher than the anticipated maximum output power of the transmitter.

15.2.2 RF power output measurement with a dummy load

It is possible to get an approximate measurement of RF power output from a transmitter with a resistor dummy load and a suitable meter. Again, these procedures are not a substitute for power measurement with an accurate RF wattmeter.

The procedure is simple. Measure the voltage across the 50- or 50-ohm dummy-load resistor and find the power with the equation: $Power = Voltage^2/50$. For example, if the voltage measured is 14 V, the power output is: $14^2/51 = 3.84$ W.

Certain precautions must be observed. First, the meter must be capable of producing accurate voltage indications at the transmitter operating frequency. This usually requires a meter with an RF probe, preferably a probe calibrated with the meter. An AM or FM transmitter should be checked with an rms voltmeter, and with no modulation applied. An SSB transmitter must be checked with a peak-reading voltmeter and with modulation applied (because SSB produces no output without modulation). This usually involves connecting an audio generator to the microphone input of the SSB transmitter circuits. Always follow the service-literature recommendations for all RF power output measurements (frequency, channels, operating voltages, modulation, etc.).

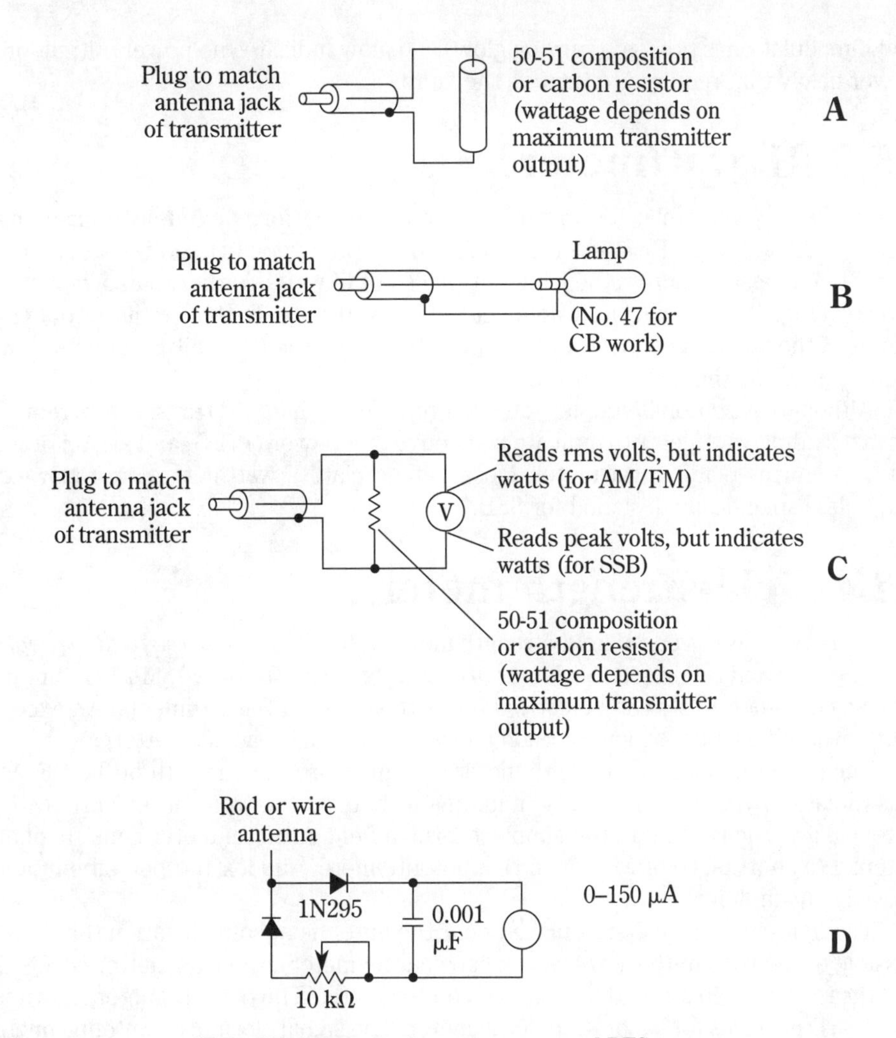

15-1 RF dummy loads, wattmeters, and RFS meters.

15.2.3 Lamp dummy load

Lamps have been the traditional dummy loads for communications equipment test and troubleshooting. For example, the #47 lamp (often found as a pilot lamp for older electronic instruments) provides the approximate impedance and power dissipation required as a dummy load for CB equipment. The connections are shown in Fig. 15-1B.

You cannot get an accurate measurement of RF power output when a lamp is used as the dummy load. However, the lamp provides an indication of the relative power and shows the presence of modulation. The intensity of the light produced by the lamp varies with modulation (more modulation produces a brighter glow), so you can tell at a glance if the transmitter is producing an RF carrier (steady glow) and if modulation is present (varying glow). No glow indicates no power output, or a power below that required to turn on the lamp.

15.3 RF wattmeter

A number of commercial RF wattmeters are available for communications-equipment test. The basic RF wattmeter consists of a dummy load (fixed resistor) and a meter that measures across the load, but reads out in watts (rather than in volts), as shown in Fig. 15-1C. You simply connect the RF wattmeter to the circuit output (for example, the antenna connector of a transmitter), key the transmitter, and read the power output on the wattmeter scale.

Although operation is simple, you must remember that SSB transmitters require a peak-reading wattmeter to indicate peak envelope power (PEP), and an AM or FM set uses an rms-reading wattmeter. Most commercial RF wattmeters are rms-reading, unless specifically designed for SSB.

15.4 Field-strength meter

There are two basic types of field-strength meters: the simple *relative field-strength (RFS) meter* and the *precision lab* or *broadcast-type instrument*. Most communications-equipment tests can be carried out with simple RFS instruments. An exception is where you must make precision measurements of broadcast patterns.

The purpose of a field-strength meter is to measure the strength of the RF signals radiated by an antenna. This simultaneously tests the transmitter output, the antenna, and the lead-in. In the simplest form, a field-strength meter consists of an antenna (a short piece of wire or rod), a potentiometer, diodes, a tuned circuit, and possibly an amplifier.

The field-strength meter is placed near the antenna at some location that is accessible to the transmitter (where you can see the meter), the transmitter is keyed, and the relative field strength is indicated on the meter. The potentiometer shown in Fig. 15-1D provides for calibration of the meter. The signal close to an antenna might be strong enough to drive the meter off scale. The potentiometer makes it possible to bring the reading back on scale. The potentiometer can then be reset when a field-strength reading is taken far from the antenna.

15.5 Standing-wave-ratio measurement

The *standing-wave ratio* of an antenna is actually a measure of match or mismatch for the antenna, transmission line (lead-in), and transmitter or RF output circuit. When the impedances of the antenna, line, and RF circuit are perfectly matched, all of the energy (RF signal) is transferred to or from the antenna, and there is no loss. If there is a mismatch (as is the case in any practical application), some of the energy, or RF signal, is reflected back into the line. This energy cancels part of the RF signal.

If the voltage is measured along the line, there are voltage maximums (where the reflected signals are in phase with the outgoing signals), and voltage minimums (where the reflected RF signal is out of phase, partially canceling the outgoing signal). The maximums and minimums are called standing waves. The ratio of the maximum to the minimum is the standing-wave ratio (SWR). The ratio can be related to either voltage or current. Because voltage is easier to measure, voltage is usually used (except in some RF lab work), resulting in the common term *voltage standing-wave ratio (VSWR)*. The theoretical calculations for VSWR are shown in Fig. 15-2A.

An SWR of 1 to 1, expressed as 1:1, means that there are no maximums (the voltage is constant at any point along the line) and that there is a perfect match for circuit, line, and antenna. As a practical matter, if the 1:1 ratio should occur on one frequency, the ratio will not occur at any other frequency because impedance changes with frequency. It is not likely that all three elements (circuits, antenna, line) will change impedance by exactly the same amount on all frequencies. Therefore, when checking SWR, always check on all frequencies or channels, where practical. As an alternative, check SWR at the high, low, and middle channels, or frequencies.

In the case of microwave RF signals being measured in the lab, a meter is physically moved along the line to measure maximum and minimum voltages. This is not practical at most communications-equipment frequencies because of the physical length of the wave. In communications equipment, it is far more practical to measure forward or outgoing voltage and reflected voltage and then calculate the reflection

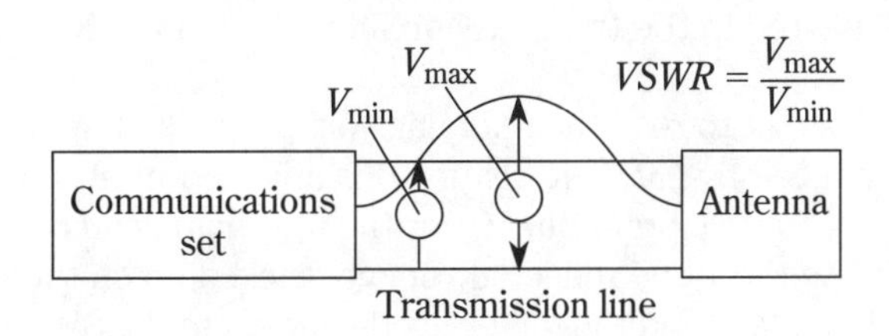

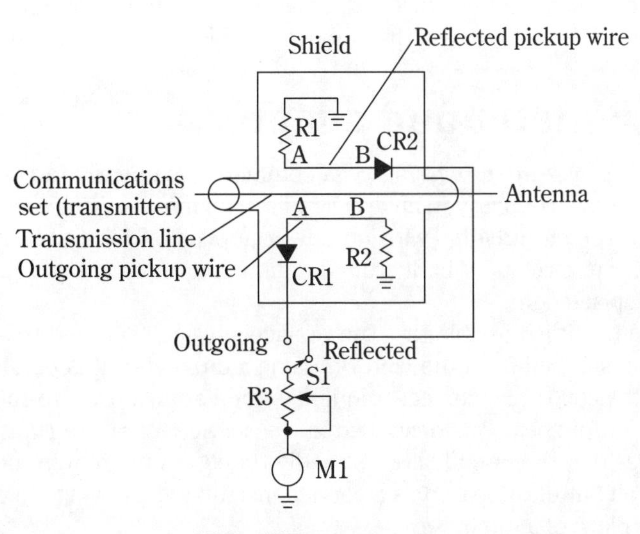

15-2 SWR measurements.

coefficient (reflected voltage/outgoing voltage). The relation of reflection coefficient to SWR is as follows: *Reflection coefficient = Reflected voltage/Forward voltage.* For example, using a 10-V forward and a 2-V reflected voltage, the reflection coefficient is 0.2(2/10).

Reflection coefficient is converted to SWR by dividing (1 + *Reflection coefficient*) by (2 – *Reflection coefficient*). For example, using the 0.2 reflection coefficient, the SWR is (1 + 0/2)/(1 – 0.2) = 1.2/0.8 = 1.5 SWR. This might be expressed as 1:1.5, 1.5:1, or simply as 1.5, depending on the meter scale. In practical terms, an SWR of 1.5 is poor because it means that at least 20 percent of the power is being reflected (and not radiated by the antenna in the case of a transmitter).

In commercial SWR meters used for communications tests, it is not necessary to calculate either reflection coefficient or SWR. This is done automatically by the SWR meter. (The meter is actually reading the reflection coefficient, but the scale indicates SWR. If you have a reflection coefficient of 0.2, the SWR readout is 1.5.)

Some communications sets, such as amateur radio and CB, have built-in SWR meters and circuits. The SWR function is often combined with other measurement functions (field strength, power output, etc.). The basic SWR circuit is shown in Fig. 15-2B. You could build such a meter, but it is generally not practical to do so. The directional coupler (the area in the shielded portion) is difficult to construct. A poorly designed coupler inserted in the transmission line results in power loss and inaccurate readings.

In use, switch S1 is set to read the outgoing voltage (picked up on the wire terminated by R2, and rectified by CR1). Resistor R3 is adjusted until the meter needle is aligned with some "set" or "calibrate" line (near the right-hand end of the meter scale). Switch S1 is then set to reach the reflected voltage (picked up on the wire terminated by R1, and rectified by CR2), and the meter needle moves to the SWR position.

A typical SWR meter does not read beyond 1:3 because a reading above a 1:3 indication is a poor match. Typical scale divisions are 1, 1.5, 2, and 3. These scale indications mean 1:1, 1:1.5, 1:2, and 1:3, respectively. The scale indications between 1 and 1.5 are the most useful because a good antenna system (antenna and lead-in) typically shows 1:1.1, or 1:1.2. Anything above 1:1.5 is on the borderline.

15.6 DIP meter and adapter

The *dip meter* (or *grid-dip meter*) was a common tool in early radio-communications service work, particularly in amateur radio. The most useful function is in presetting "cold" resonant circuits (with no power applied). This makes it possible to adjust the resonant circuits of badly tuned equipment, or where new coils must be installed as a replacement.

As an example, it is possible that the replacement coil is tuned to an undesired frequency when shipped from the factory. Using a dip meter, it is possible to install the coil, tune the coil to the correct frequency, then apply power to the circuit and adjust the circuit for "peak," as described in the service literature (most service literature assumes that the circuits are not badly tuned or only require peaking). The dip meter has all but disappeared as a service instrument. However, a *dip adapter* can be used in place of a dip meter.

15.6.1 Dip meter

Figure 15-3 shows the basic dip adapter. The circuit is basically an RF generator with an external pickup coil, a frequency counter, a diode, and a meter. When the coil is held near the RF circuit to be tested, and the generator is tuned to the resonant frequency of the RF circuit, part of the RF energy is absorbed by the RF circuit, and the meter reading "dips." The procedure can be reversed, where the generator is set to the desired frequency, and the circuit is tuned to produce a dip on the meter.

Resistor R1 should match the impedance of the generator (typically 50 ohms). Coil L1 consists of a few turns of insulated wire. The accuracy of the circuit depends on counter accuracy.

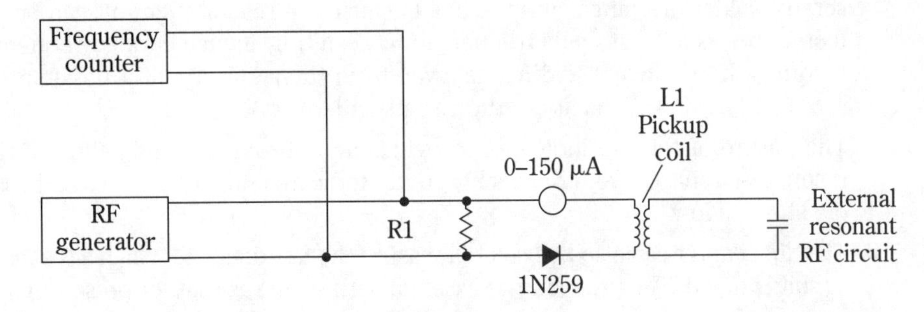

15-3 Dip adapter circuit.

15.6.2 Setting resonant frequency with a dip adapter

The following applies to both series and parallel resonant RF circuits.

1. Couple the dip adapter to the resonant circuit using L1. Usually, the best coupling has a few turns of L1 passed over the resonant-circuit coil. Make certain that no power is applied to the RF circuit.
2. Set the generator to the desired resonant frequency, as indicated on the counter. Adjust the generator output-amplitude for a convenient reading on the adapter meter.
3. Tune the resonant circuit for a maximum dip on the adapter meter (using the adjustable capacitors or the slugs within the resonant-circuit coil).
4. Most resonant circuits are designed so as not to tune across both the fundamental frequency and any harmonics. However, it is possible that the RF circuit will tune to a harmonic and produce a dip. To check this condition, tune the circuit for maximum dip and set the generator to the first harmonic (twice the desired resonant frequency). Notice the amount of dip at both frequencies. The harmonic should produce substantially smaller dips than the fundamental resonant frequency.
5. For maximum accuracy, check the dip frequency from both high and low sides of the resonant-circuit tuning. A significant difference in frequency readout from either side indicates overcoupling between the dip-adapter and

the resonant circuit. Move L1 away from the resonant circuit until the dip is just visible. This amount of coupling should provide maximum accuracy. If you have difficulty in finding a dip, overcouple the adapter until a dip is found, then loosen the coupling and make a final check of frequency. Generally, the dip is more pronounced when approached from the direction that causes the meter reading to rise.

6. If you have doubt as to whether the adapter is measuring the resonant frequency of the RF circuit, or some nearby circuit, ground the RF circuit under the test. If there is no change in the dip indication, the resonance of another circuit is being measured.
7. The area surrounding the circuit being measured should be free of wiring scraps, solder drippings, and the like because the resonant circuit can be affected (especially at high frequencies), resulting in inaccurate frequency readings. Keep fingers and hands away from the adapter coil if possible (to avoid adding body capacity to the circuit under test).
8. The nature of the dip indication provides an approximate indication of the resonant-circuit Q. Generally, a sharp dip indicates high Q, whereas a broad dip shows a low Q.
9. The dip adapter can also be used to measure the frequency to which a resonant circuit is tuned. The procedure is essentially the same as that for presetting the resonant frequency (steps 1 through 8), except that the generator is tuned for maximum dip with the circuit still cold (no power applied). The resonant frequency to which the circuit is tuned can then be read from the counter. When making this test, watch for harmonics (which also produce dip indications).

15.7 Spectrum analyzers and FM-deviation meters

Figure 15-4A shows the basic *spectrum-analyzer circuit.* Spectrum analyzers are most often used in communications work where FM is involved, although they can also be useful in AM and SSB equipment.

A spectrum analyzer is essentially a narrowband receiver, electrically tuned over a given frequency range, combined with a scope or display tube. The oscillator is swept over a given range of frequencies by a sweep generator. Because the IF amplifier passband remains fixed, the input circuit and mixer are swept over a corresponding range of frequencies.

For example, if the intermediate frequency is 10 kHz, and the oscillator sweeps from 100 to 200 kHz, the input is capable of receiving signals in the range 110 to 210 kHz. The IF amplifier output is applied to the vertical-deflection plates of the display tube. The horizontal plates receive a signal from the same sweep-generator circuit used to control the oscillator frequency. As a result, the length of the horizontal sweep represents the total sweep-frequency spectrum (if the sweep is from 110 to 210 kHz, the left-hand end of the trace represents 110 kHz, and the right-hand end represents 210 kHz). Any point along the horizontal trace represents a corresponding frequency (with the midpoint representing 160 kHz, in this example).

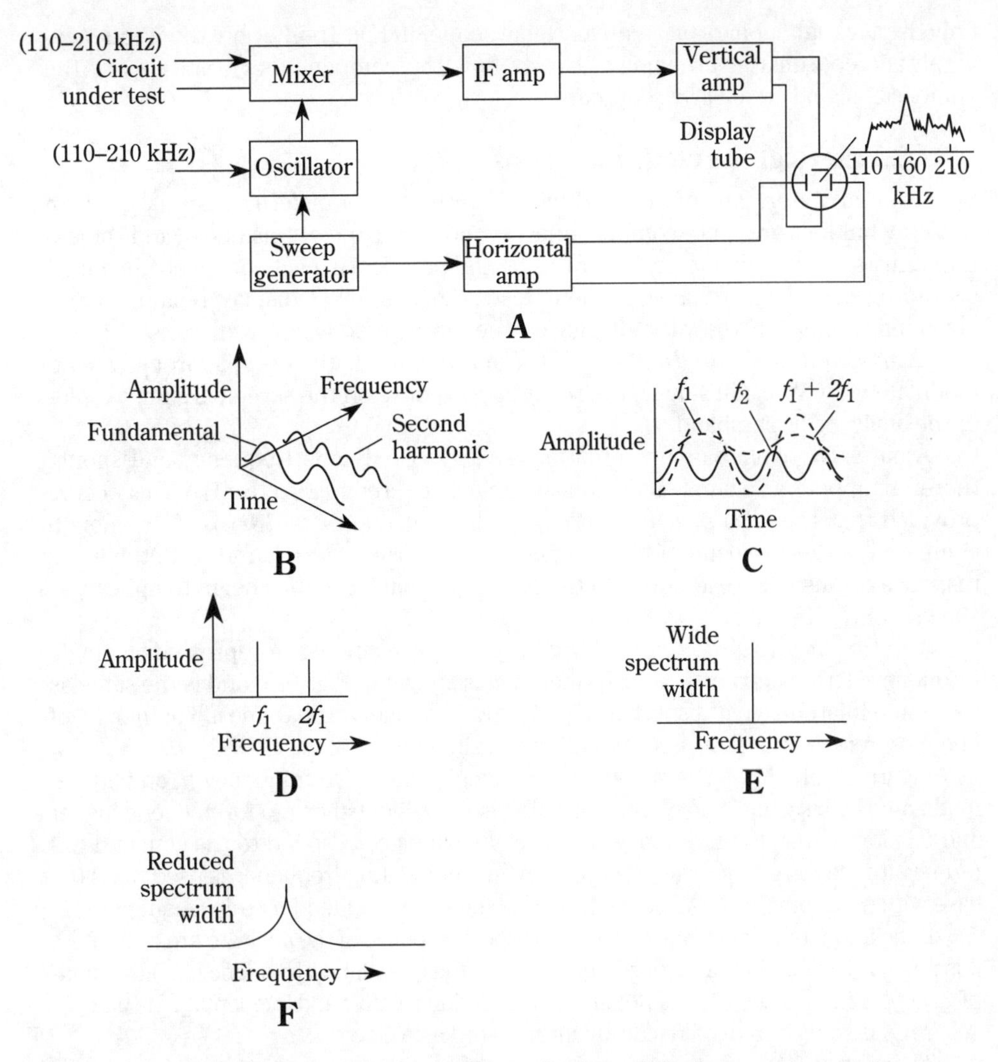

15-4 Spectrum analyzer circuits and displays.

15.7.1 Time amplitude versus frequency amplitude

A conventional scope (chapter 2) produces a time-amplitude display. For example, pulse rise time and width are read directly on the horizontal axis. A spectrum analyzer produces a frequency-amplitude display, where signals (unmodulated, AM, FM, or pulse) are broken down into individual components for display on the horizontal axis. Figure 15-4B through 15-4D show the relationship of time-amplitude and frequency-amplitude displays.

In Fig. 15-4B, both the time-amplitude and frequency-amplitude coordinates are shown together. The example given is that showing the addition of a fundamental frequency and the second harmonic. In Fig. 15-4C, only the time-amplitude coordinates are shown. The solid line (which is the composite of fundamental $f1$ and $2f1$ is the only display that appears on a conventional scope. In Fig. 15-4E, only the frequency-amplitude coordinates are shown. Notice how the components (f_1 and $2f_1$) of the composite signal are clearly seen here.

15.7.2 Practical spectrum analysis

Spectrum analyzers are often used with Fourier and transform analysis. Both of these techniques are quite complex and beyond the scope of this book (and the author). Instead, this section concentrates on the practical aspects of spectrum analysis during communications-equipment tests. That is, what display results from a given input signal and how the display can be interpreted is covered.

Unmodulated signal displays If the analyzer oscillator sweeps through an unmodulated or CW signal slowly, the resulting response on the screen is simply a plot of the analyzer IF passband.

A pure CW signal has, by definition, energy at only one frequency and should therefore appear as a single spike on the analyzer screen (Fig. 15-4E). This occurs provided that the total *sweep width* (so-called *spectrum width*) is wide enough compared to the passband of the analyzer. As spectrum width is reduced, the spike response begins to spread out until the passband characteristics begin to appear, as shown in Fig. 15-4F.

Amplitude-modulated signal displays A pure sine wave represents a single frequency. The spectrum of a pure sine wave is shown in Fig. 15-5 and is the same as the unmodulated signal display in Fig. 15-4E (a single vertical line). The height of line f_o represents the power contained in the single frequency.

Figure 15-5B shows the spectrum for a single sine-wave frequency f_0, amplitude-modulated by a second sine wave, f_1. In this case, two sidebands are formed, one higher and one lower than the frequency f_0. These sidebands correspond to the sum and difference frequencies, as shown. If more than one modulating frequency is used (as is the case with more practical AM signals), two sidebands are added for each frequency.

If the frequency, spectrum width, and vertical response of the analyzer are calibrated, it is possible to find (1) the carrier frequency, (2) the modulation frequency, (3) the modulation percentage, and (4) the nonlinear modulation (if any) and incidental FM (if any).

An AM spectrum display can be interpreted as follows:

The carrier frequency is determined by the position of the center vertical line f_0 on the horizontal axis. For example, if the total spectrum is from 100 to 200 kHz, and f_0 is in the center, as shown in Fig. 15-5B, the carrier frequency is 150 kHz.

The modulation frequency is determined by the position of the sideband line $f_0 - f_1$, or $f_0 + f_1$, on the horizontal axis. for example, if sideband $f_0 - f_1$ is at 140 kHz, and f_0 is at 140 kHz (as shown), the modulating frequency is 10 kHz. Under these conditions, the upper sideband ($f_0 - f_1$) should be 160 kHz. The distance between the carrier line (f_0) and either sideband is sometimes known as the *frequency dispersion* and is equal to the modulation frequency.

The modulation percentage is determined by the ratio of the sideband amplitude to the carrier amplitude. The amplitude of either sideband, with respect to the carrier amplitude, is one-half of the percentage of modulation. For example, if the carrier amplitude is 100 mV and either sideband is 50 mV, this indicates 100 percent modulation. If the carrier amplitude is 100 mV and either sideband is 33 mV, this indicates 66 percent modulation.

A

B

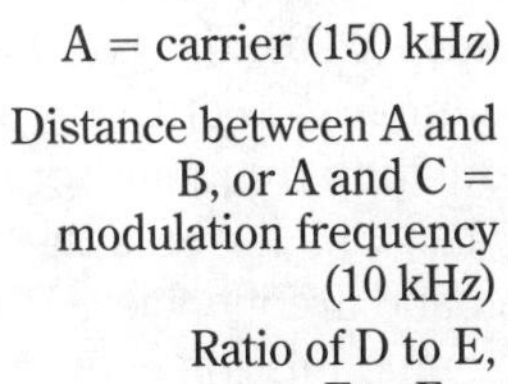

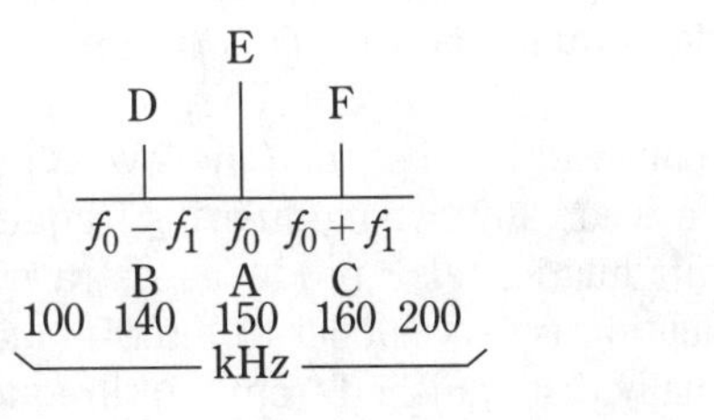

C

Carrier (center) frequency

Amplitude 1.0

f_0

Unmodulated

D

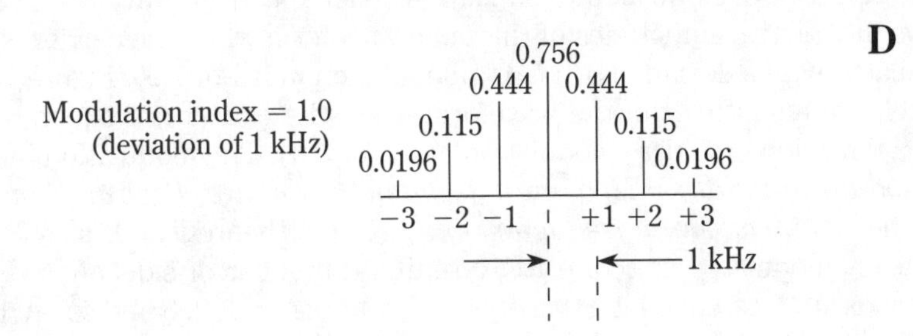

E

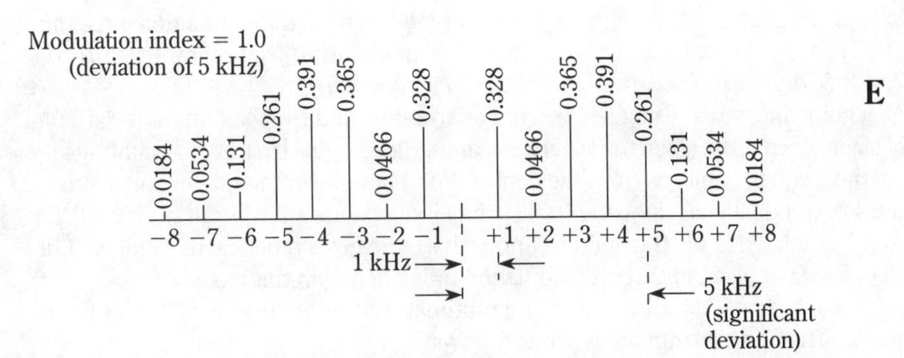

15-5 Spectrum analyzer AM/FM displays.

Nonlinear modulation is indicated when the sidebands are of unequal amplitude, or are not equally spaced on both sides of the carrier frequency. Unequal amplitude indicates nonlinear modulation that results from a form of undesired frequency modulation combined with amplitude modulation.

Incidental FM is indicated by a shift in the vertical signals along the horizontal axis. For example, any horizontal "jitter" of the signals indicates rapid frequency modulation of the carrier.

The rules for interpreting AM spectrum-analyzer displays are summarized in Fig. 15-5B. In practical tests, carrier signals are often amplitude-modulated at many frequencies simultaneously. This results in many sidebands (two for each modulating frequency) on the display. To resolve this complex spectrum, make sure that the analyzer bandwidth is less than the lowest modulation frequency or less than the difference between any two modulating frequencies, whichever is the smaller.

Overmodulation also produces extra sideband frequencies. The spectrum for overmodulation is very similar to multifrequency modulation. However, overmodulation is usually distinguished from multifrequency modulation by: (1) the spacing between overmodulated sidebands is equal, and multifrequency sidebands are arbitrarily spaced (unless the modulating frequencies are harmonically related); and (2) the amplitude of the overmodulated sidebands decreases progressively out from the carrier, but the amplitude of the multifrequency-modulated signals is determined by the modulation percentage of each frequency and it can be arbitrary.

Frequency-modulated displays The mathematical expression for an FM waveform is long and complex, involving a special mathematical operator known as the *Bessel function*. However, the spectrum representation of the FM waveform is relatively simple.

Figure 15-5C shows an unmodulated-carrier spectrum waveform. Figure 15-5D shows the relative amplitudes of the same waveform when the carrier is frequency-modulated with a deviation of 1 kHz (modulation index of 1.0). Figure 15-5E shows the relative amplitudes of the waveform when the carrier is frequency-modulated with a deviation of 5 kHz (modulation index of 5.0). The modulation index is given by: *Modulation index = Maximum frequency deviation/Modulation frequency*.

The term *maximum frequency deviation* is theoretical. If a SW signal (f_c) is frequency-modulated at a rate RF, an infinite number of sidebands result. These sidebands are located at intervals of $f_c \pm N_f$, where $N = 1, 2, 3$ and so on. However, in practical tests, only the sidebands containing significant power are considered. For a quick approximation of the bandwidth occupied by the significant sidebands, multiply the sum of the carrier deviation and the modulating frequency by 2: *Bandwidth* = 2 (*Carrier deviation + modulating frequency*).

As a guideline, when using a spectrum analyzer to find the maximum deviation of an FM signal, locate the sideband where the amplitude begins to drop and continues to drop as the frequency moves from the center. For example, in Fig. 15-5E, sidebands 1, 2, 3 and 4 rise and fall, but sideband 5 falls, and all sidebands after 5 continue to fall. Because each sideband is 1 kHz from the center, this indicates a practical or significant deviation of 5 kHz. It also indicates a modulation index of 5.0, in this case.

As in the case of AM, the center and modulation frequencies for FM can be determined with the spectrum analyzer as follows:

The FM carrier frequency is determined by the position of the center vertical line on the horizontal axis (the centerline is not always the highest amplitude, as shown in Fig. 15-5E).

The FM modulating frequency is determined by the position of the sidebands in relation to the centerline or the distance between sidebands (frequency dispersion).

15.7.3 FM-deviation meter

Unless a service shop specializes in FM communications or broadcast work, an *FM-deviation meter* can be used as a substitute for a spectrum analyzer. The operating controls and procedures for an FM-deviation meter are far less complex than those of a spectrum analyzer. Typically, FM-deviation meter controls include a meter scale marked in terms of frequency, a tuning control, and a zero control.

In use, the FM-deviation meter is connected to monitor the output of the communications equipment (typically, the transmitter output). The transmitter is first keyed without modulation so that the meter can be tuned to the exact carrier frequency. Then the transmitter is frequency-modulated with a tone (typically in the range of 1 to 5 kHz), and the exact amount of frequency modulation is indicated on the FM-deviation meter.

15.8 Miscellaneous communications test equipment

In addition to checking that communications equipment produces the correct signal on all frequencies or channels, it is helpful to know that the equipment is not producing any other signals. For example, the final RF amplifier in a transmitter might break into oscillation (if not properly neutralized) and produce signals at undetermined frequencies. These signals might not show up on the channel being used, but could interfere with other communications or with TV (TV interference is a very common problem in CB communications). All such extra signals (generally referred to as *spurious signals* in FCC regulations and service literature) are illegal and certainly undesirable.

The ideal instrument for detecting undesired signals from an RF circuit is the spectrum analyzer. Unfortunately, spectrum analyzers are expensive and generally restricted to use in labs or broadcast work. You can do essentially the same job with a good communication-type receiver. The receiver should have a BFO and an S-meter (chapter 4). In addition to using the receiver for signal checks, you can monitor transmissions of equipment being serviced (and check WWV signals).

One of the most frequent types of interferences caused by communications equipment is on TV channels (especially CB). A TV set in the shop quickly indicates if equipment being serviced is causing interference (and can settle disputes concerning TV interference problems).

Before you become overconfident using a TV set, remember the following. Most TV interference enters through the IF stages, and all TV IFs do not operate on the same frequency. Some use the range 22 to 28 MHz, whereas other sets use the range 41 to 47 MHz (and some very old TV sets use other IF ranges). So, it is possible for a communications set to produce interference on one TV and not on another, with both TV sets located in the same room, and tuned to the same channel.

15.9 Modulation checks with a scope

There are many variations of the basic technique for modulation measurement in communications work. The following is a summary of the most commonly used techniques.

15.9.1 Direct measurement with high-frequency scopes

If the vertical-channel response of the scope is capable of handling the frequency without distortion, the communications-equipment (transmitter) signal can be applied directly through the scope vertical amplifier. The basic test connections are shown in Fig. 15-6. The procedure is as follows:

1. Connect the equipment, as shown in Fig. 15-6.
2. Key the transmitter and adjust the scope controls to produce displays, as shown. You can either speak into the microphone (for a rough check of modulation), or you can introduce an audio signal (typically 400 or 1000 Hz) at the microphone input (for a precise check of modulation). Notice that Fig. 15-6 provides simulations of typical scope displays during modulation tests.
3. Measure the vertical dimensions shown as A and B in Fig. 15-6. Calculate the percentage of modulation using the equation. For example, if the crest amplitude (A) is 63 (63 screen divisions) and the trough amplitude (B) is 27, the percentage of modulation is:

$$\frac{63 - 27}{63 + 27} \times 100 = 40\%$$

Make certain to use the same scale for both crest (A) and trough (B) measurements. An RF wattmeter can be substituted for a dummy load.

15.9.2 Linear-detector measurement

If you must use a scope that cannot pass the transmitter frequency, use a linear detector, as shown in Fig. 15-7. The scope must have a dc input. The procedure is as follows:

1. Connect the equipment, as shown in Fig. 15-7.
2. With the transmitter not keyed, adjust the scope position control to place the trace on a reference line near the bottom on the screen, as shown (carrier off).
3. Key the transmitter but do not apply modulation. Adjust the scope gain control to place the top of the trace at the center of the screen (carrier on). It might be necessary to switch the transmitter off and on several times to adjust the trace properly because the position and gain controls of most scopes interact.

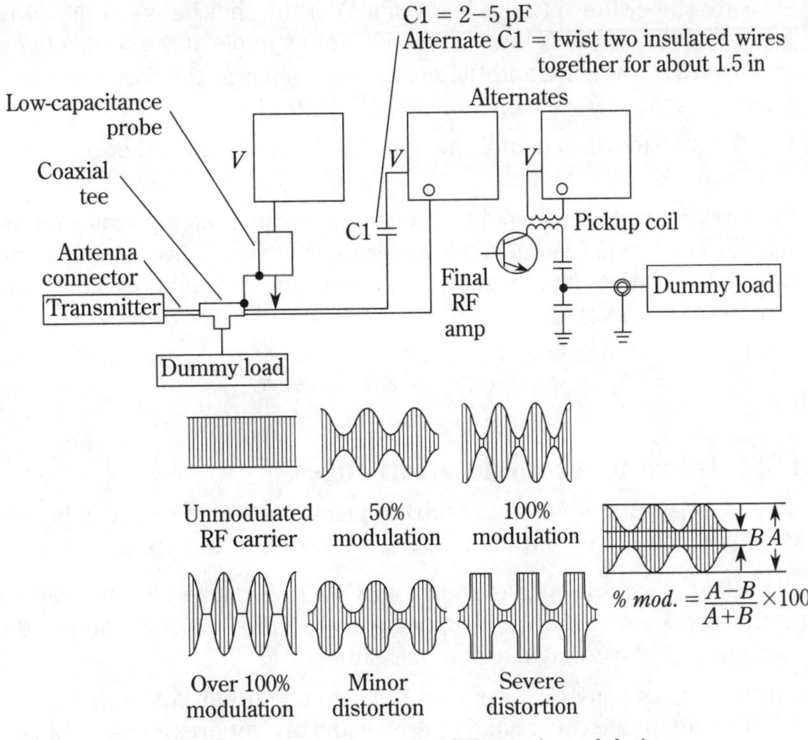

15-6 Direct measurement of RF-circuit modulation.

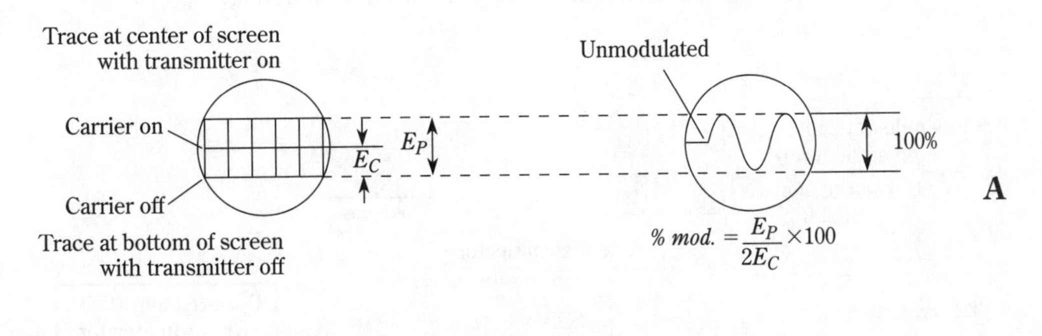

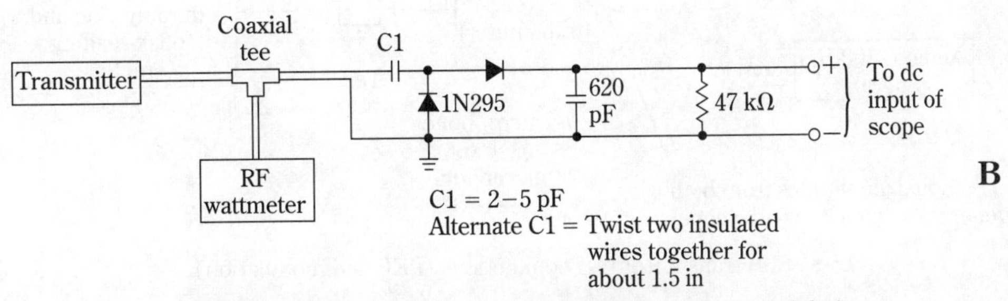

15-7 Linear-detector measurement of RF-circuit modulation.

4. Measure the distance (in scale divisions) of the shift between the carrier-on (step 3) and carrier-off (step 2) traces. For example, if the screen has a total of 10 vertical divisions, and the no-carrier trace is at the bottom or zero line, there is a shift of 5 scale divisions to the center line.
5. Key the transmitter and apply modulation. Do not touch either the position or gain controls of the scope.
6. Find the percentage of modulation using the equation. For example, assume that the carrier-on (center) and carrier-off (bottom) is five divisions, and that the modulation produces a peak-to-peak envelope of eight divisions. The percentage of modulation is:

$$\frac{8}{2 \times 5} \times 100 = 40\%$$

15.9.3 SSB transmitter modulation check

Figure 15-8 shows a circuit for test of SSB transmitters. The circuit can be used for both modulation and power output, if desired. The procedures are as follows:

1. Connect the equipment, as shown in Fig. 15-8. Make certain that the RF wattmeter is capable of reading peak-power (PEP). If not, use the dummy load and a peak-reading meter to measure power.
2. Operate the transmitter to produce maximum output power. Remember, in SSB operation, the carrier and one sideband are suppressed, and all power is carried on the other sideband. Also, there is no output when the transmitter is unmodulated.
3. Apply two simultaneous, equal-amplitude audio signals for modulation, such as 500 and 2400 Hz. Other frequencies can be used for modulation, provided

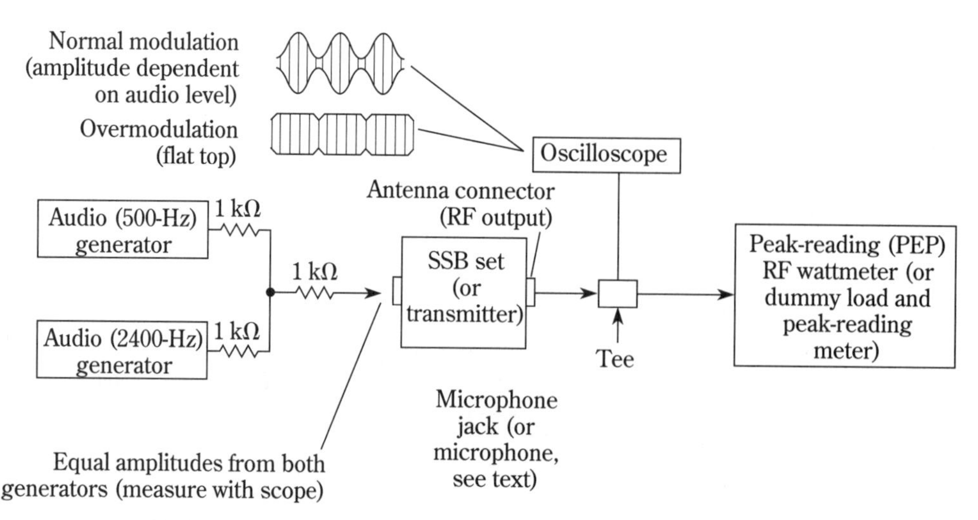

15-8 SSB transmitter test connections (PEP and modulation).

that the two signals do not have a direct harmonic relationship. If a precise modulation check is not required, you can use a microphone instead of audio signals. If the transmitter has an adjustable microphone gain, set the gain to midposition.

4. Key the transmitter and adjust the scope for a stable display of the modulation envelope.
5. Check the modulation-envelope patterns against the patterns of Fig. 15-8, or (preferably) against patterns shown in the SSB transmitter service data. Notice that the typical SSB modulation envelope resembles the 100-percent AM-modulation envelope, except that the amplitude of the entire SSB waveform varies with the strength of the audio signal. Thus, the percentage of modulation calculation that applies to AM cannot be applied to SSB.
6. Increase the amplitude of both audio-modulation signals, making certain to maintain both signals at equal amplitudes. When peak SSB power output is reached, the modulation envelope will "flat-top," as shown. That is, the instantaneous peaks of the SSB reach saturation—even with less than peak audio modulation applied. This overmodulated condition results in distortion.
7. With the transmitter at maximum, but before overmodulation or flat-topping occurs, read the power on the peak-reading RF wattmeter (or dummy load and meter). This is the transmitter PEP.
8. If the transmitter is capable of both upper and lower sideband (USB and LSB) operation, check both sidebands of at least one channel, or all channels, if you desire. The PEP reading should be the same for all channels, on both USB and LSB.

15.10 Antenna and transmission-line checks with limited test means

In general, antennas and transmission lines (lead-ins or cables) used with communications equipment are best tested with built-in power/SWR meters, commercial SWR meters, field-strength meters, etc. However, it is possible to make a number of significant tests using basic meters (voltmeters, ohmmeter, and ammeter). The following paragraphs describe how these procedures can be performed when commercial antenna-test devices are not available.

15.10.1 Antenna length

Most antennas are cut to a length that is related to the wavelength of the signal being transmitted or received. Generally, antennas are cut to one-half wavelength (or one-quarter wavelength) of the center operating frequency.

As a practical matter, the electrical length of an antenna is always greater than the physical length, because of capacitance and end effects. Therefore, two sets of calculations are required: one for electrical length and one of physical length. The calculations for antenna length and resonant frequency are shown in Fig. 15-9.

← Half-wave antenna →

Basic hertz type (Typical for TV antennas)

Electrical length:

$$\text{Meters} = \frac{150}{\text{Frequency (MHz)}}$$

$$\text{Feet} = \frac{492}{\text{Frequency (MHz)}}$$

$$\text{Inches} = \frac{5906}{\text{Frequency (MHz)}}$$

Physical length (approximate) = *electrical length* × *K-factor*
K-factor = 0.96 for frequencies below 3 MHz
= 0.95 for frequencies between 3 and 30 MHz
= 0.94 for frequencies above 30 MHz

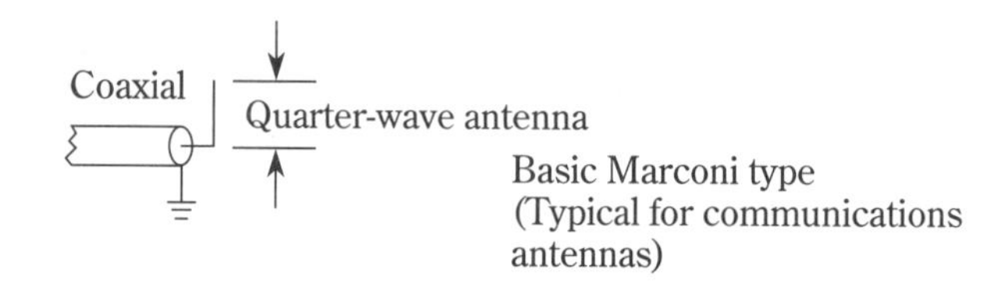

Electrical length:

$$\text{Meters} = \frac{75}{\text{Frequency (MHz)}}$$

$$\text{Feet} = \frac{246}{\text{Frequency (MHz)}}$$

$$\text{Inches} = \frac{2953}{\text{Frequency (MHz)}}$$

15-9 Calculations for antenna length and resonant frequency.

15.10.2 Antenna resonance, impedance and radiated-power measurements

All antennas have a resonant frequency and an impedance. The *resonant frequency* is that frequency where maximum power is transferred from the transmitter to the antenna (and from antenna to receiver, although this is difficult to measure). Antenna impedance is found using Ohm's law $Z = E/I$, with voltage and current being measured at the antenna feed point. Figure 15-10A shows these relationships. *Radiation resistance* is a more meaningful term than *antenna impedance.* When the dc resistance of an antenna is disregarded, the antenna impedance can be considered as the radiation resistance. Radiated power can then be determined using the basic Ohm's law equation $P = I^2R$.

On those antennas that use coax transmission lines (the usual case for communications equipment), the antenna and transmission-line impedance must be matched. In this case, the impedance match between transmission line and antenna is of greater importance than actual impedance value (both antenna and line must be 50 ohms, 72 ohms, etc). The condition of match (or mismatch) can best be tested by SWR measurement (covered in this chapter).

However, the following procedure can be use to find resonance, impedance, and radiated power of any antenna system. The values found to apply to the complete antenna system (antenna and transmission line), are as seen from the measurement end (which is generally where the transmission line connects to the communications equipment).

1. Connect the equipment, as shown in Fig. 15-10B. Disconnect the transmission line from the antenna terminal of the set, then reconnect the line to the generator, with the series ammeter in place, as shown. The transmitter can be used in place of the generator. However, the transmitter usually operates on fixed frequencies, thus limiting the measurements to those frequencies (the transmitter also produces too much power and causes interference).
2. Adjust the generator to the supposed center frequency at which the antenna is used (or for any desired operating frequency to which the antenna can be tuned).
3. Tune the generator for maximum indication on the ammeter. This is the resonant frequency of the antenna system. Ideally, this should be the center of the transmitter frequency range. For example, if the transmitter covers channels in the range from 10 to 20 MHz, the antenna system should be resonant at 15 MHz.
4. If the antenna can be tuned, set the generator at the transmitter center frequency, and then tune the antenna for maximum indication on the ammeter. Record the indicated current. If the antenna cannot be tuned, adjust the generator for maximum on the ammeter, and record thc current. If the antenna cannot be tuned and you use a fixed-frequency transmitter, simply record the indicated current.
5. If a frequency counter is available, verify that the generator (or transmitter) and antenna are tuned to the correct frequency (after the antenna and/or generator are tuned for maximum indication on the ammeter).
6. Disconnect the transmission line from the ammeter and generator. Connect the generator to the dummy load through the ammeter, as shown. If the antenna is tuned in step 4, adjust inductor L or capacitor C for maximum indication on the ammeter. If the antenna is tuned with a variable inductance, use capacitor C to tune the dummy load, and vice versa.
7. Adjust the dummy-load resistance (R) until the indicated current on the ammeter is the same as the antenna current recorded in step 4.
8. Remove power from the test circuit. Measure the dc resistance of R with an ohmmeter. This resistance is equal to the antenna-system impedance (or radiation resistance) at the measurement frequency.
9. Calculate the actual power delivered to the antenna (or radiated power) using: $I^2 \times R$ (or Z), where I is indicated current (amperes) and R is radiation resistance (or Z is antenna-system impedance), found in step 8.
10. An alternative method must be used when the operating frequency is beyond the range of the available ammeter, when no ammeter is available, or when the ammeter presents an excessive load.

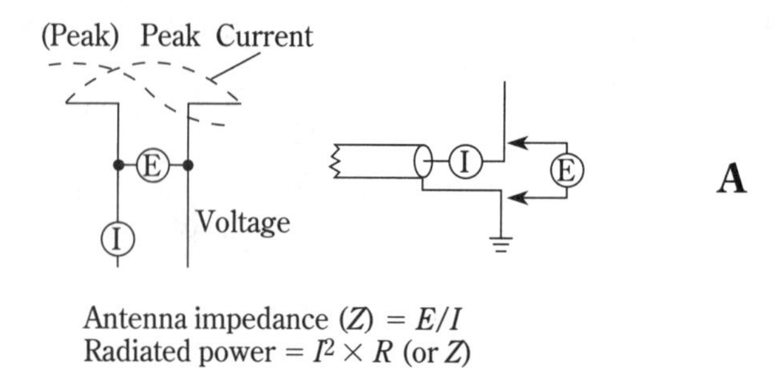

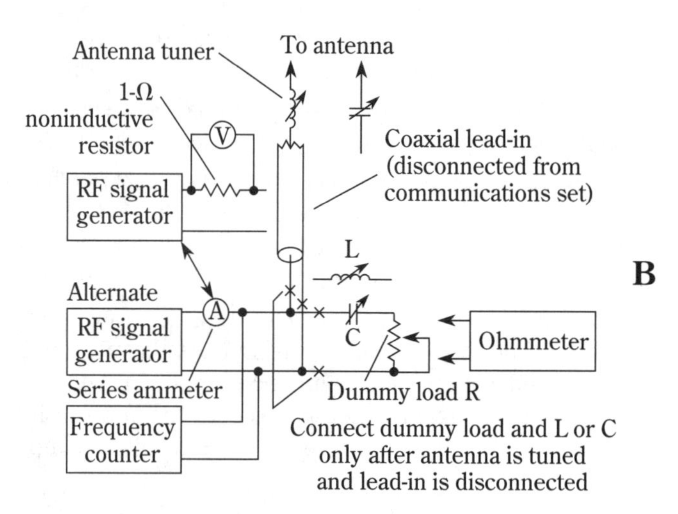

15-10 Antenna measurements.

A precision 1-ohm noninductive resistor and voltmeter can be used in place of the ammeter, as shown in Fig. 15-10B. With a 1-ohm resistor, the indicated voltage is equal to the current passing through the resistor (and antenna system). Except for the meter and resistor connections, the procedure is identical to that of using the ammeter.

Index

D

E

F

About the author

John D. Lenk has been a self-employed consulting technical writer specializing in practical troubleshooting guides for more than 40 years. A long-time writer of international best-sellers in the electronics field, he is the author of 77 books on electronics, which together have sold more than one million copies in nine languages. Mr. Lenk's guides regularly become classics in their fields and his most recent books include the *Lenk's Video Handbook, Lenk's Audio Handbook, Lenk's Laser Handbook, Lenk's RF Handbook, Lenk's Digital Handbook,* and the *McGraw-Hill Circuits Encyclopedia.*